Joint IMEKO TC1-TC7-TC13-TC18 Symposium 2019

Journal of Physics: Conference Series Volume 1379

St. Petersburg, Russia
2 - 5 July 2019

Editors:

Kseniia Sapozhnikova
Roald Taymanov

ISBN: 978-1-7138-1135-0
ISSN: 1742-6588

Printed from e-media with permission by:

Curran Associates, Inc.
57 Morehouse Lane
Red Hook, NY 12571

Some format issues inherent in the e-media version may also appear in this print version.

Institute of Physics
Dirac House, Temple Back
Bristol BS1 6BE UK

Phone: 44 1 17 929 7481
Fax: 44 1 17 920 0979

techtracking@iop.org

Additional copies of this publication are available from:

Curran Associates, Inc.
57 Morehouse Lane
Red Hook, NY 12571 USA
Phone: 845-758-0400
Fax: 845-758-2633
Email: curran@proceedings.com
Web: www.proceedings.com

TABLE OF CONTENTS

KEYNOTE PAPERS

ORAL PAPERS

POSTER PAPERS

Journal of Physics: Conference Series **1379** (2019) 011001 doi:10.1088/1742-6596/1379/1/011001

Joint IMEKO TC1-TC7-TC13-TC18 Symposium 2019
July 2-5, 2019
St. Petersburg, Russia

PREFACE

The Symposium was organized by the D.I. Mendeleyev Institute for Metrology (VNIIM) with the assistance of the Russian Academy of Metrology in the persons of Anton Pronin, Acting Director of VNIIM and Vladimir Okrepilov, President of the Russian Academy of Metrology, who co-chaired the National Organizing Committee.

The organizers of the Symposium took into account that the future sets interdisciplinary challenges entailing a growing need for a better mutual understanding and interaction of professionals in various disciplines.

For the first time, members of four IMEKO Technical Committees were invited to participate in the Symposium:

TC1 Education and Training in Measurement and Instrumentation
TC7 Measurement Science
TC13 Measurements in Biology and Medicine
TC18 Measurement of Human Functions

Symposium objectives:

- To confirm that measurement science responds to the emerging needs of society.
- To address the new challenges of the Industry 4.0 through cooperation of scientists under various TCs.

Symposium motto: "The future glimmers long before it comes to be".

In addition to the plenary meeting consisting of four sessions there were held at the Symposium:

- 13 thematic sessions and meetings with oral presentations on the subjects:
  - "Measurement science",
  - "Measurements in the digital era",
  - "Measurements in education and training",
  - "Measurement of human functions",
  - "Measurements of various quantities and their features",
- a poster session,
- a round table "Terminology issues related to expanding boundaries of measurements",
- a Meeting of TC1, where professors of Russian and foreign universities discussed educational issues related to measurements.

The Symposium was attended about 100 people from 23 countries:

Australia, Austria, Belgium, Brazil, Bulgaria, China, France, Germany, Croatia, Israel, Italy, Japan, Malaysia, North Macedonia, Poland, Portugal, Russia, Slovak Republic, Spain, Sweden, Ukraine, United Kingdom, and USA.

We managed to get together scientists working in measurement science, professors engaged in teaching undergraduate and graduate students, and specialists in measurements of multidimensional quantities characterizing humans. Such joint participation of various specialists was very beneficial to all of them and their interaction will undoubtedly have a synergistic effect.

The Symposium made it possible to focus on the most relevant areas of measurement, among which we can note:

- measurements (including those that use artificial intelligence and big data) of multidimensional quantities characterizing a person's properties, such as his cognitive and psychological characteristics;
- measurements of multidimensional quantities that determine properties of social and economic systems, as well as the quality of services, including those related to education;
- methods to improve the objectivity of expert assessments to a level ensuring the compatibility and comparability with an uncertainty acceptable for practical purposes;
- methods to improve the reliability of measurement results with a smaller cost of the metrological maintenance due to the automation of the metrological service and, if necessary, to the automated application of corrections.

The Collected Articles of the Symposium include 78 papers peer-reviewed by at least two members of the International Program Committee.

Journal of Physics: Conference Series **1379** (2019) 011001 doi:10.1088/1742-6596/1379/1/011001

The Symposium was held on the premises located in the central part of St. Petersburg - one of the most beautiful cities in the world. In honor of the Symposium, the Co-chairperson of the National Organizing Committee A. Pronin and the Chairperson of the International Program Committee E. Benoit fired the traditional midday shot from the old cannon installed on the territory of the Peter-and-Paul Fortress.

Participants of the Symposium took part in a sightseeing tour of St. Petersburg and in technical visits to leading technical universities of St. Petersburg arranged by the Organizers.

According to many participants, the organizational decisions made, have largely contributed to the friendly and creative atmosphere of the Symposium and its success.

In conclusion, we believe it is our duty to thank the members of the IMEKO General Council for their support in choosing the venue for the Symposium, and the members of the National Organizing Committee, International Program Committee, Advisory Board, and all the participants for their active and fruitful work.

Anton Pronin, Co-Chairperson of the Symposium

Kseniia Sapozhnikova, Roald Taymanov, members of the National Organizing and International Program Committees, members of the Editorial Board

Journal of Physics: Conference Series **1379** (2019) 011002 doi:10.1088/1742-6596/1379/1/011002

Peer review statement

All papers published in this volume of *Journal of Physics: Conference Series* have been peer reviewed through processes administered by the proceedings Editors. Reviews were conducted by expert referees to the professional and scientific standards expected of a proceedings journal published by IOP Publishing.

Journal of Physics: Conference Series **1379** (2019) 012001 doi:10.1088/1742-6596/1379/1/012001

Measurement science in methodological instruction of Ph.D. students of technoscience

Roman Z Morawski

Warsaw University of Technology, Faculty of Electronics and Information Technology, Institute of Radioelectronics and Multimedia Technology, Nowowiejska 15/19, 00-665 Warsaw, Poland

E-mail: r.morawski@ire.pw.edu.pl

Abstract. Technoscience is a result of integration of technology with empirical sciences, including not only physical sciences, but also life sciences, social sciences and cognitive sciences. The development of technoscience is unthinkable without measurement and mathematical modelling, being two interrelated operations used for acquisition and representation of knowledge. Measurement fundamentals must, therefore, belong to the core of methodological instruction of Ph.D. students of technoscience. This keynote speech will provide an outline of a graduate course on methodological aspects of technoscientific research, whose programme includes conceptual analysis of the dialectical relationship between measurement and mathematical modelling, of the role of measurement in scientific explanation, of the role of measurement in the context of discovery and in the context of justification, as well as of the role of measurement in mitigating uncertainty of scientific knowledge.

1. Introduction

Many today's research practitioners seem to share the famous opinion: "*Philosophy of science is about as useful to scientists as ornithology is to birds*", once expressed by the 1965 Nobel Prize winner, Richard P. Feynman [1]. They are usually not aware that on another occasion, in his famous paper "Cargo cult science", the latter complained about researchers who imitate research "rituals", *i.e.* carry out research activities in an apparently correct way, but without understanding their essence, and – consequently – without any chance for obtaining meaningful results and using them in a productive manner [2]. Does philosophy of science really matter for research practitioners? The following opinion on the significance and educational value of the history and philosophy of science – expressed in 1944 by another great physicist Albert Einstein – cannot be ignored: "... *many people today – and even professional scientists – seem to me like somebody who has seen thousands of trees but has never seen a forest. Knowledge of the historic and philosophical background gives that kind of independence from prejudices* [...] *from which most scientists are suffering. This independence created by philosophical insight is – in my opinion – the mark of distinction between a mere artisan or specialist and a real seeker after truth.*"[3] Paradoxically, those "artisans" are today quite systematically trained in their professional skills, while the candidates for the profession of the "seeker after truth", *i.e.* Ph.D. students, are not: the regular academic courses on philosophy of science and research methodology are rarely offered to the students of science and technology. Moreover, the vast majority of those courses are given by philosophers of science who, unsurprisingly, are inclined to ignore practical issues of research, as being trivial, and to focus on abstract controversies, as being intellectually more attractive. Since research methodology is actually an interdiscipline combining philosophy with science and technology, it would be very beneficial for research practitioners to follow the courses inspired by all those three sources of ideas. There is a need for a less sophisticated and more practical approach which does not focus on the overgrown terminology of philosophy of science – *e.g.* on classification of the panoply of stances promoted by various schools of thinking – but offers guidance in the everyday activities aimed at generation and evaluation of knowledge. On the other hand, it should be, however, avoided another extremity: the course on the philosophy of science and research methodology should not be replaced with a quick-reference guide for Ph.D. students in a desperate need to make their theses satisfy minimum formal requirements being in force in their domain of study…

2. Science, technology and technoscience

Science is usually understood as a systematic enterprise aimed at generating and organising knowledge in the form of testable explanations and predictions about the natural world which comprises all the components of the physical universe (atoms, ecosystems, people, societies, galaxies, *etc.*), as well as the natural forces at work on those components. According to the Anglo-Saxon tradition, humanities do not count as sciences. So, the studies of various artefacts of human culture and various cultures themselves – philosophy and the studies of languages, literature, religion, arts, history, law, *etc.* – remain beyond science. The corresponding terms in some other European languages – *e.g. Wissenschaft* in German or *nauka* in Polish – have a broader meaning since they encompass all the academic disciplines, including humanities.

The sciences whose explicit goal is to generate knowledge and support understanding, regardless of potential applications, are called pure sciences or basic sciences. The sciences whose explicit goal is to solve practical problems, *e.g.* to develop new technologies, are called applied sciences. The boundary between pure and applied sciences is rather blurred: the research findings in pure sciences often have useful applications, and intentionally applied research may contribute to better understanding of the natural world. Applied sciences play today an important role as a bridge between pure sciences and technology usually understood as the fabrication, modification and usage of tools, machines, techniques or systems – aimed at solving practical problems or improving existing solutions to those problems. Science and technology frequently contribute to one another: scientific advances lead to the development of new technologies, and new technologies broaden the experimental potential of science, enabling advancement of research. This is a substantiation behind introduction of the concept of *technoscience* addressing the integration of science and technology, which has been progressing from the beginning of the 20th century (for more details see the recent book by the American philosopher and historian of science David F. Channel [4]).

3. Philosophy of science and research methodology

Philosophy of science is the study of assumptions, foundations and implications of science. Philosophy and science are often perceived as quite separate enterprises, seemingly having little to do with one another: philosophy is viewed as a search for answers to abstract questions, concerning such issues as meaning of life, while science is understood as a study of more down-to-earth matters. It turns out, however, that philosophy is omni-present in science. There are not only philosophical problems concerning foundational concepts and tools used by scientists but also philosophical issues which are directly relevant to, and directly affect, the day-to-day work of scientists.

Almost until the end of the 20th century, the development of philosophy of science had been predominantly shaped by the advancement of physics (including astrophysics), especially by the research practices specific of that domain. Since the evolution of other scientific disciplines – in spite of the 19th-century prophecies and 20th-century endeavours – had not always followed the paradigms of physics, it has turned out at the end of the 20th-century that certain methodological guidelines of the philosophers of science did not match the reality of most dynamic fields, such as information sciences, life sciences and cognitive sciences. The philosophical discourse was, in particular, still focused on laws and theories while for those sciences semantic, mathematical and computational models were more important. Today, an urgent need for accommodation of methodological follow-ups of this situation is generally recognised.

In English, two related although different meanings are associated with the term methodology: on the one hand, it is understood as the system of methods and principles used in a particular discipline of science or art; on the other – as the branch of philosophy concerned with the study of methods and procedures used in science (or in a philosophical system) [5]. In some languages, there is no such ambiguity because two different terms are associated with those two meanings, e.g. in Polish metodyka means the system of methods and principles, while metodologia is the name for the branch of philosophy, focused on the scientific method; a similar distinction exists in German [6]. To avoid ambiguity, the philosophical study of the scientific method is called in English philosophy of science and the system of methods and principles used in a particular discipline of science (or group of disciplines, such as empirical sciences) is referred to as research methods and research methodology. To avoid the necessity to enter into unnecessary distinctions, the combined term philosophy of science and research methodology will be used hereinafter.

4. Cornerstones of teaching philosophy of science and research methodology

The philosophy of science and research methodology is about so-called *scientific method*, *i.e.* the most general paradigm of doing scientific research, followed by Western civilisation systematically since the 17th century, and in some elements – since the times of Ancient Greece. Its naïve image, presented usually in the secondary schools, may be outlined as follows:

- An objective (*i.e.* impartial and unbiased) researcher makes observations and *measurements* concerning a selected object of study.
- On the basis of the results of observations and *measurements*, the researcher works out a theory which is able to explain those results and to enable prediction of the future states and behaviours of the object of study.
- On acquiring additional results of observations and *measurements*, which are not coherent with the already accepted theory, the researcher corrects this theory (possibly, in an iterative process).

An academic course, devoted to the philosophy of science and research methodology, must provide much more complete and precise description of the scientific method, as well as its critical analysis. This requires introduction of a number of highly abstract concepts (and related terms) which play an important role in research practice, in communication of research results and in scientific criticism. At first glance, this may seem a bit burdensome for the students trained on purely technoscientific literature, but they quickly discover the beneficial impact of this difficult experience when approaching more practical problems, especially those which require profound logical argumentation. As the British neuroscientist Rodrigo Q. Quiroga explained in his 2017 book [7], we not only think with concepts, but we also use them for remembering our external and internal experience. This enables us to both compress memorised data and make thinking more efficient. The famous statement "*By giving the wrong name to an object, we are contributing to the misfortune of this world*" [8] is pointing the essence of this problem.

The logical structure of the scientific method is the basis for structuring the contents of an academic course devoted to the philosophy of science and research methodology. The most typical and – in fact – indispensable components are the following issues: mathematical modelling and measurement, scientific explanation, generation of hypothesis (the so-called context of discovery), their transformation into scientific knowledge (the so-called context of justification) and uncertainty of scientific knowledge.

5. Measurement in teaching philosophy of science and research methodology

It follows from the rough outline of scientific method, provided in Section 4, that mature science is unthinkable without measurement. However, despite widespread recognition of the importance of measurement for research practice and technoscientific progress, the covering of metrological issues in the handbooks of the philosophy of science and research methodology is rather limited. The statistical data provided in Table 1 seem to well corroborate this observation. They indicate that the philosophers of science are inclined to think about measurement as a source of uncertainty-free information…

Table 1. The numbers of occurrence of three key terms of measurement science in three recent handbooks of philosophy of science.

Key term	[9] 275 pages	[10] 425 pages	[11] 200 pages
measurement	33	240	2
measurement error	0	0	0
measurement uncertainty	0	0	0

The famous mathematician and polymath John von Neumann exaggerated, most probably, when saying: "The sciences do not try to explain, they hardly even try to interpret, they mainly make models"[12], and having in mind mathematical models. In this way, however, he made us aware of the crucial role of mathematical models in science. We realise the importance of mathematical modelling even more when getting aware that not only identification of mathematical models requires results of measurement but also any measurement method or technique requires several models to be identified, viz. the model of a system under measurement, the model of the measurand conversion into treatable signals and the model of the measurand reconstruction. A deeper analysis of the mutual interdependence of measurement and mathematical modelling is showing that the epistemic status of those two operations of vital significance for science is comparable and that evaluation of measurement uncertainty is necessary but not sufficient for evaluation of the uncertainty of scientific knowledge (cf. Chapters 5 and 6 of [13] for more details).

Measurement and mathematical modelling play a key role in scientific explanation. According to the Western intellectual tradition, to explain a phenomenon, an event or a process means: to remove perplexity, or to change the unknown to the known, or to identify causes of events or phenomena. According to the Greek philosopher of science Stathis Psillos, the explanation of a fact (explanandum) is achieved by stating some causal or nomological links between it and the facts that are called upon to do the explaining (explanans). There are two broad approaches to explanation. According to the first, explanations are argumentative: to explain an event is to construct such a line of argumentation that the explanandum follows – logically or with high probability – from certain universal or statistical laws and initial conditions. Most typical species of this genus are deductive-nomological schemes of scientific explanation. According to the second approach, an explanation needs not refer to any scientific law to be complete; it is enough that it specifies the relevant causal mechanisms [14]. Although not directly mentioned, measurement and mathematical modelling are parts of both approaches since both scientific laws and causal mechanisms have the form of mathematical models, and initial (or boundary) conditions must be defined on the basis of measurement results (cf. Chapter 7 of [13] for more details).

A fundamental distinction, emphasising a qualitative difference between two stages of scientific work, viz.: generation of hypotheses (in the context of discovery) and their transformation into theories or other pieces of scientific knowledge (in the context of justification), is usually attributed to the German philosopher of science Hans Reichenbach [15]. According to some interpretations, the context of discovery, unlike the context of justification, should be a subject of study belonging to psychology and sociology rather than to philosophy of science because the act of conceiving a new idea is not necessarily a rational process, and therefore cannot be regulated [16]. Alternative interpretations, however, stress the importance of the presence of logical components in the process of discovery, which are of interest for philosophers of science. Thus, the context of discovery is an interdisciplinary topic being a subject of reflection whose conclusions may have (and actually have) significant impact on technoscientific research. Although creative mind is the most important "tool" for generation of scientific hypothesis, results of measurements may be the most important among the sources of its inspiration (cf. Chapter 8 of [13] for more details).

The transformation of hypothesis into scientific knowledge in the context of justification requires extensive use of measurement methods and tools. Since the times of Plato, knowledge is understood as "justified true belief". So, its practical definition depends on what is considered to be true, and what is considered to be justified. On the one hand, we refer to at least five co-existing definitions of truth (the correspondence, coherence, consensus, evidence and pragmatic definitions of truth), on the other – we use various patterns of justification, such as: verification, validation, falsification, substantiation, corroboration and proof. Each of them refers to the results of measurement and basic methods of logical reasoning: deduction, induction and abduction (cf. Chapter 9 of [13] for more details).

Regardless of the applied pattern of justification, any piece of information included in the body of scientific knowledge remains uncertain. Unlike scholars of the Ancient Greece, Middle Ages and early modern era, we do not expect certainty of scientific knowledge but rather that it is accompanied by some indicators of its uncertainty. Such indicators are well known in measurement science, and therefore they are primary tools for evaluation of the quantitative uncertainty of scientific knowledge. By extension they are applicable for characterisation of the quantitative uncertainty of mathematical modelling. Unfortunately, they are of little usefulness if qualitative evaluation of the uncertainty associated with mathematical modelling and logical inference is concerned (cf. Chapter 10 of [13] for more details).

6. Concluding remarks

Measurement is an indispensable tool of any mature branch of technoscience, and the philosophy of measurement is an integral part of the philosophy of science. Those facts are not properly reflected in the courses on philosophy of science and research methodology, addressed to the entrants in the research profession (such as Ph.D. students) and to the research practitioners in technoscientific disciplines. On the one hand, the number of such courses across leading academic institutions of the world is relatively limited, on the other – they are given by philosophers who are inclined to underestimate the importance of measurement for productive implementation of the scientific method. While replacing philosophers with experienced research practitioners would be counterproductive, a hybrid solution consisting in the inclusion of the latter in the teaching process could be a partial remedy for the difficulties of the current situation.

There is also an important ethical motivation behind attempts to reintroduce philosophy of science and research methodology in the graduate curricula: ignorance in this respect may be (and most probably is) a cause of growing incidence of scientific misconduct. The first imperative of any professional ethics is "to honestly do one's job". Unlike in the case of a ditch-digger job, the explanation of what does it mean to honestly do a research job requires a semester-long academic course rather than five-sentence instruction. The young researchers, who have not been exposed to an appropriate methodological training, are at risk to err in real-world research situations whose complexity is growing from day to day. They are also prone to run into scepticism when seeing on the every-day basis how much research practice diverges form the naïve definition of the scientific method – its only definition that they have learned at school.

References and endnotes

[1] https://en.wikiquote.org/wiki/Talk:Richard_ Feynman
[2] Feynman R P 1974 *Eng. Sci.* **37**(7) pp 10–13
[3] https://aeon.co/essays/the-string-theory-wars-show-us-how-science-needs-philosophy
[4] Channell D F 2017 *A History of Technoscience: Erasing the Boundaries Between Science and Technology* (London – New York: Taylor & Francis)
[5] *Random House Dictionary* 2017; *Collins English Dictionary – Complete & Unabridged* 2012 (Digital Edition)
[6] *die Methodik* = "die Gesamtheit der Techniken der wissenschaftlichen Vorgehensweisen"; *die Methodologie* = "die Theorie bzw. Lehre von den wissenschaftlichen Methoden", *cf. http://www.univie.ac.at/ksa/elearning/cp/ksamethoden/ksamethoden-30.html*

Journal of Physics: Conference Series **1379** (2019) 012001 doi:10.1088/1742-6596/1379/1/012001

[7] Quiroga R Q 2017 *The Forgetting Machine: Memory, Perception, and the Jennifer Aniston Neuron* (Dallas: BenBella Books)
[8] This is the author's translation of the sentence "*Mal nommer un objet c'est ajouter au malheur de ce monde*" which appears in: Camus A 1944 *Poésie* **44**(17) p 22
[9] Christian A, Hommen D, Retzlaff N and Schurz G 2018 *Philosophy of Science: Between the Natural Sciences, the Social Sciences, and the Humanities* (Cham: Springer)
[10] DeWitt R 2018 *Worldviews: An Introduction to the History and Philosophy of Science* (Oxford: Wiley & Sons)
[11] Dicken D 2018 *Getting Science Wrong: Why the Philosophy of Science Matters* (London – New York: Bloomsbury Pub)
[12] von Neumann J 1955 *The Unity of Knowledge* (New York: Doubleday) pp 157–164
[13] Morawski R Z 2019 *Technoscientific Research: Methodological and Ethical Aspects* (Berlin – Boston: de Gruyter)
[14] Psillos S 2007 *Philosophy of Science A–Z* (Edinburgh: Edinburgh University Press) p 85
[15] Glymour C and Eberhardt F 2016 in *The Stanford Encyclopedia of Philosophy* Winter 2016 Edition, https://plato.stanford.edu/archives/win2016/entries/reichenbach/ [accessed on 2019-02-02].
[16] Psillos S 2007 *Philosophy of Science A–Z* (Edinburgh: Edinburgh University Press) p 49.

Journal of Physics: Conference Series **1379** (2019) 012002 doi:10.1088/1742-6596/1379/1/012002

Probabilistic measurement modelling may overcome the opposition between the Bayesian and the frequentistic views

Giovanni Battista Rossi and Francesco Crenna

Università degli Studi di Genova –DIME/MEC, Laboratorio di Misure, Genova, Italy

E-mail: gb.rossi@dimec.unige.it.

Abstract. The probabilistic modelling of measurement systems is discussed, in regards to uncertainty evaluation and different models representing the same real system from different standpoints are compared. We suggest that proper probabilistic modelling directly yields evaluation of measurement uncertainty, without making any commitment to specific philosophical schools, such as the frequentistic and the Bayesian ones. We suggest that this model-based approach may have advantages in education and in the development of recommended practices in measurement.

1. Introduction

The Guide to the Expression of Uncertainty in Measurement was published in 1993 [3][1], in order to find an agreeable common way of evaluating and expressing uncertainty in measurement, and the requirement of accounting for both random variations and systematic effects that contribute to such uncertainty.

The GUM had an enormous impact in the field of measurement. On the theoretical side, it stimulated studies on the foundation and best methods for evaluating measurement uncertainty. Such studies were often concerned to the adoption of special visions of probability and of probabilistic inference such as the frequentistic and the Bayesian [4, 6]. Other studies were instead more concerned in addressing measurement modelling in general terms [9, 10].

On the application side, although the practice of actually evaluating uncertainty had a significant diffusion, difficulties were experienced, mainly due to a lack of practical guidelines for its application in the many scientific, technical and even social areas in which measurement is required.

After so many years from its original publication, a revision is now foreseen [12-16]. Yet current proposals seem to be closely related to the debate of the visions mentioned above. We suggest instead that the focus should rather be put to *addressing measurement modelling*, which would allow to overcome the traditional opposition between the Bayesians and the frequentists, and lead to a more user-oriented approach.

Therefore, in this communication, we consider the case of a measuring system which although being very simple, is general enough for our proposed, and we show how it can be properly modelled in different ways. Then we discuss such models, in respect also to the two mentioned competing visions. We suggest that *the choice of a proper model is what really matters,* since this naturally leads

[1] References are listed in chronological order, to give a sense of the historical development of the subject.

to the solution by the application of the rules of probability calculus, without any additional philosophical commitment.

2. Different probabilistic models for the same measuring system

Consider the case of a linear measuring system for some quantity x, for example a force (N), and producing an (instrument) indication y, expressed for example as a voltage reading (mV), and having a sensitivity, k (mV/N).

Consider just two sources of uncertainty: a noise, v, at the output of the voltmeter, and an uncertainty on the sensitivity, δk, typically due to the calibration uncertainty for the device under consideration.

In order to obtain the final result, i.e., the measurement value, $\hat{x}$, the (rough) instrument indication has to be scaled by dividing it by k. Let us call *observation* the (physical) transformation operated by the measuring system, and *restitution* the (numerical) transformation that maps the indication into the measurement value. An illustration of this system is presented in figure 1, where $c = k^{-1}$.

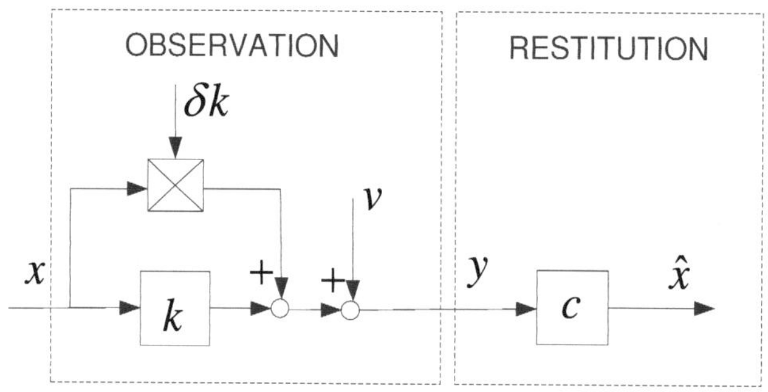

Figure 1. The measuring system: model $\mathcal{M}_1$.

In spite of its simplicity, this model allows us to discuss some features of measurement uncertainty. In fact v and δk, although both constituting uncertainty source, are quite different entities. Indeed, v represents random noise and this variable is the typical component of the "classical" theory of errors and of the frequentistic approach, where probabilities are interpreted as expressing relative frequencies in the occurrence of events. In other words, probability as a description of "randomness" is perfectly suited here [5].

On the other hand, δk is assumed to express our "imperfect" knowledge of the value of the sensitivity, which is instead assumed to keep a constant value during the measurement process. Dealing with δk is typical of a Bayesian attitude, where probability is interpreted as expressing the state of knowledge of the experimenter, rather than the state of thing of the (real) system under consideration [2].

Considering the extreme cases, we could argue that if we had only uncertainty sources similar to v, the classical, error-theory, frequentist approach would be perfectly suited. On the opposite side, if all the uncertainty sources would be similar to δk, we may admit that a Bayesian, subjectivists, approach would be more appropriate. But the point is that in measurement *both* kinds of uncertainty sources typically coexist!

Let us then now discuss how this can be managed in three approaches:

- frequentistic,
- Bayesian, or
- model-based (the one we support).

Prior of entering into this discussion, let us first consider the model in figure 1, let us call it $\mathcal{M}_1$. Introducing the measurement error, e,

$$e = \hat{x} - x, \tag{1}$$

Journal of Physics: Conference Series **1379** (2019) 012002 doi:10.1088/1742-6596/1379/1/012002

we obtain the sensitivity equation:

$$e = x\frac{\delta k}{k} + cv\,, \tag{2}$$

which yields, for standard uncertainty:

$$u = \sigma_e = \sqrt{c^2\sigma_v^2 + \frac{x^2\sigma_{\delta k}^2}{k^2}} \cong \sqrt{c^2\sigma_v^2 + \frac{\hat{x}^2\sigma_{\delta k}^2}{k^2}}\,, \tag{3}$$

since the (unknown) value of the measurand, x, can surely be approximated by the measurement value, $\hat{x}$.

Yet there are alternative possibilities for modelling the same "situation" described in figure 1. One possibility is to leave v in the observation phase, since it is may be considered as an "objective" source of "randomness", and to move δk to the restitution phase, since it is related to the state of knowledge of the measurer, who have an imperfect knowledge of the measuring system. Note that in figure 1, k is the "nominal" value of the sensitivity, i.e., the value we know for it, whilst the actual value is $k+\delta k = k'$. Therefore, uncertainty arises, because we use k instead of k' in the restitution phase. Therefore, after considering that

$$c = \frac{1}{k} = \frac{1}{k'-\delta k} = k'^{-1}\frac{1}{1-\dfrac{\delta k}{k'}} \cong k'^{-1}\left(1+\frac{\delta k}{k'}\right),$$

we obtain the model presented in figure 2. Let us call this model $\mathcal{M}_2$.

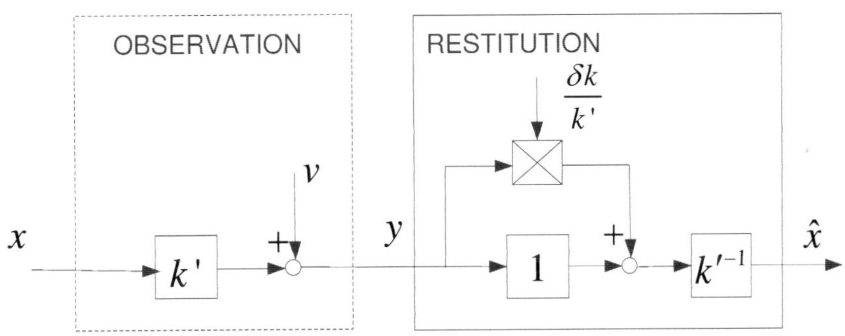

Figure 2. The measuring system: model $\mathcal{M}_2$.

Here the sensitivity equation reads:

$$e = x\frac{\delta k}{k'} + \frac{1}{k'}v\,, \tag{4}$$

and standard uncertainty can be evaluated by

$$u = \sigma_e = \sqrt{k'^{-2}\sigma_v^2 + \frac{x^2\sigma_{\delta k}^2}{k'^2}} \cong \sqrt{c^2\sigma_v^2 + \frac{\hat{x}^2\sigma_{\delta k}^2}{k^2}}\,, \tag{5}$$

since k' is, strictly speaking, unknown, and it can be replaced by k. Therefore, *model $\mathcal{M}_2$ yields the same evaluation of standard uncertainty as model $\mathcal{M}_1$.*

Lastly, let us consider a third model that expresses uncertainty as expressing the incomplete knowledge of the measurer, in respect of the measurement process. In this perspective, uncertainty sources are more appropriately referred to the restitution phase, as shown in figure 3, where model $\mathcal{M}_1$ is presented.

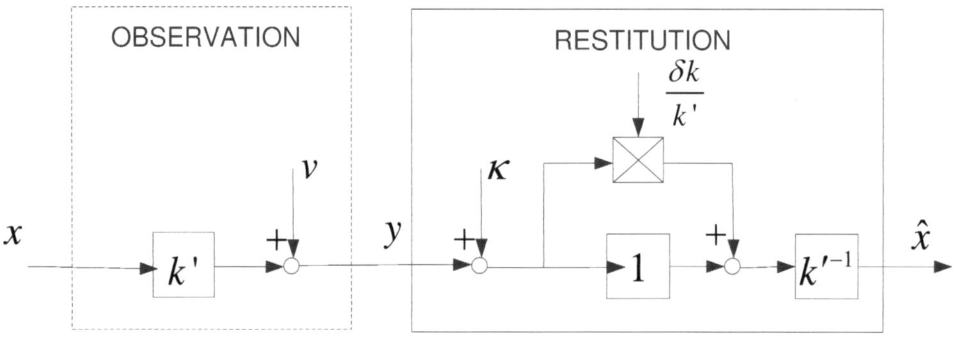

Figure 3. The measuring system: model $\mathcal{M}_3$.

Here a "correction" $\kappa = -v$ is virtually applied in order to compensate the effect of the noise v. But if such noise is assumed to have a null expected value, no compensation is actually apply, and the variable κ is only considered in the uncertainty budget. That is to say: we are uncertain because we are unable to actually apply the compensation κ, as we would wish.

Here standard uncertainty is understood as the standard deviation of the measurement value, $\hat{x}$, for each given value of the indication y. We obtain:

$$u = \sigma_{\hat{x}/y} = \sqrt{k'^{-2}\sigma_\kappa^2 + \frac{x^2\sigma_{\delta k}^2}{k'^2}} \cong \sqrt{c^2\sigma_v^2 + \frac{\hat{x}^2\sigma_{\delta k}^2}{k^2}}\,. \tag{6}$$

Again evaluated standard uncertainty is the same. Let us now discuss these three models.

3. Discussion of the models

Model $\mathcal{M}_1$ is typical of the classic or frequentistic approach, since it ascribes all the uncertainty sources to the observation phase. Yet its application rises a problem: how can a fixed systematic effect such δk be treated in probabilistic terms, if probability has to be related to some observable relative frequency in the occurrence of some event? In fact, the classical theory of (measurement) errors, assumed the absence of non-negligible systematic effects. Indeed, this is the main objection raised by the Bayesians to the frequentistic approach, in particular in the case of measurement.

Yet this objection can be circumvented, by referring to a kind of *though experiment*.

In fact, we can consider the instrument we are using as randomly sampled by the set of instrument of the same kind, independently calibrated! If we repeated the observation not with the same instrument, but randomly sorting each time an instrument from the class, the variable δk would also produce an observable variability, that corresponds to the standard deviation $\sigma_{\delta k}$. Therefore, even in the case we use a single instrument, we can properly associate to it such dispersion, that it "inherit" from the class it belongs to.

On the opposite side, model $\mathcal{M}_3$ is clearly related to a Bayesian approach.

This model also works, since it also yields the same result as the others, yet it also encounters some difficulties. For example, it treats the random effect by introducing a "virtual" correction that it is not actually applied: this is not such a straightforward argument, in our opinion. In particular, it is not less difficult to accept that the though experiment we have mentioned above, in relation to the frequentistic approach. Furthermore, standard uncertainty is now defined as a conditional quantity, but the conditioning term is never reported, as far as we now, in the final measurement result.

Lastly, model $\mathcal{M}_2$ is somewhat intermediate between the two, although probably closer to $\mathcal{M}_1$, for example in the way standard uncertainty is defined. It has the advantage of outline the different way in which the two uncertainty sources affect the measurement process (one being a systematic effect, the other a source of random variations), but at the cost of a greater complexity.

In any case we can note that once a model has been adopted, it is possibly to evaluate the uncertainty by simply applying to the model the – universally accepted – rules of probability calculus. So where is the benefit of committing ourselves to one of the other philosophical approaches to probability? What this add to modelling? In the examples above we have seen that it just adds difficulties in their interpretation.

Our proposal is thus *to approach measurement uncertainty evaluation by simply assuming a probabilistic model and then to drive the implied consequences*.

If we agree on this, what model is to be preferred?

Let us that the problem is not dramatic, since at least in case we have considered all the models yield the same result. Yet there may be other reason for preferring one solution. In this regard, educational aspects also play a role.

4. Educational aspects

In University courses on measurement and instrumentation, modelling is essential: it is through models that students understand the functioning principles of devices that constitute measuring chains, and learn how to design them, for performing required tasks and to evaluate their performance [1, 7]. Therefore, when alternatives are possible, as it happens here, models should be chosen also accounting for their ability to effectively communicate the information.

In we compare, from this standpoint, the three models we have presented above, it is apparent how model $\mathcal{M}_3$ is by far the most difficult to understand.

Models $\mathcal{M}_1$ and $\mathcal{M}_2$ are both simpler and more straightforward than the previous one, and are quite similar to each other. The definition of the error, of the sensitivity function and of standard uncertainty as the standard deviation of the error, are clear and intuitive. Comparing them, we may note that $\mathcal{M}_2$ is perhaps more apt to highlight the difference between systematic effects and random variations, yet this is obtained at the expense of a higher complexity. For this reason, we use, in our courses, preferably model $\mathcal{M}_1$.

Concerning best practice guidelines, in our opinion what works better in education is, in our opinion likely to be more appropriate. For this reason we have recently made a proposal for revising the GUM in order to make it more user oriented, in which we have proposed model $\mathcal{M}_1$ as the most appropriate [16].

5. Conclusions

We have shown how the same measurement system can be correctly modelled in different ways, all of which allow us to obtain the same evaluation of standard uncertainty.

We suggest thus that uncertainty evaluation can be presented, in education and in recommended practice, with a focus on probabilistic modelling, without involving particular visions of probability such as the frequentist or the Bayesian, that may be confusing and do not help to solve the problem. The choice between different, although substantially equivalent models, should be made considering aspect of simplicity and effectiveness of communication.

Acknowledgements

This work was partially funded by the University of Genova through an Academic Research Program (FRA 2017).

References

[1] Doebelin E 1983 *Measurement Systems* (Singapore: Mc Graw Hill)

[2] Press S J 1987 *Bayesian Statistics* (New York: Wiley)
[3] BIPM, IEC, IFCC, ISO, IUPAC, IUPAP, OIML 1993 *Guide to the Expression of Uncertainty in Measurement* Corrected and reprinted 1995 (Geneva, Switzerland: ISO)ISBN 92-67-10188-9
[4] Weise K and Wöger W 1993 A Bayesian theory of measurement uncertainty *Meas. Sci. Technol.* **4** pp 1-11
[5] Hacking I 2001 *An Introduction to Probability and Inductive Logic* (Cambridge, UK: Cambridge University Press)
[6] Lira I 2002 *Evaluating the Measurement Uncertainty* (Bristol: IOP)
[7] Bentley J P 2005 *Principles of Measurement Systems* (Harlow: Pearson-Prentice Hall)
[8] Morris A S and Langari R 2011 *Measurement and Instrumentation* (Academic Press)
[9] Frigerio A, Giordani A and Mari L 2010 Outline of a general model of measurement *Synthese* **7** pp 123-149
[10] Rossi G B 2012 Toward an interdisciplinary probabilistic theory of measurement *IEEE T. Instrum. Meas.* **61** pp 2097-106
[11] Rossi G B 2014 *Measurement and Probability* (Dordrecht: Springer)
[12] Bich W, Cox M and Michotte C 2016 Towards a new GUM – An update *Metrologia* **53** pp 149-59
[13] Lira I 2016 The GUM revision: The Bayesian view toward the expression of measurement uncertainty *Eur. J. Phys.* **37** 025803
[14] White D R 2016 In pursuit of a fit-for-purpose uncertainty guide *Metrologia* **53** pp 107-124
[15] Possolo A and Pintar A L 2017 Plurality of Type A evaluations of uncertainty *Metrologia* **54** pp 617-632
[16] Rossi G B, Crenna F and Palazzo A 2018 A proposal for a new approach to the GUM *Proc. IEEE Int. Instrumentation and Measurement Technology Conf. (I2MTC 2018)*, Houston, Texas, USA, May 14-17 - 2018, pp 1569-73

Journal of Physics: Conference Series **1379** (2019) 012003 doi:10.1088/1742-6596/1379/1/012003

High-end sensors: challenges for the future

Sergey Y. Yurish[1,2]

[1] International Frequency Sensor Association (IFSA), c/Esteve Terradas, 1, RDIT-K2M Castelldefels, Barcelona, Spain
[2] F2D, Ltd., Dublin, Ireland

E-mail: SYurish@sensorsportal.com

Abstract. The paper describes the coming challenges in high-end sensors and transducers design. These challenges are coming from the recent technological limitations in microelectronics. The proposed approach to overcome such limitations is based on programmable parameter-to-frequency converters as a digital sensor's core and advanced method of measurement in order to move from a traditional analog-to-digital conversion to alternative frequency (period, duty-cycle, time interval)-to-digital conversion. Working in the frequency-time signal domain simplifies design, and obviates some technical and technological problems, due to the properties of frequency as informative parameter of sensors and transducers. The major benefits offered by such approach are high reliability, high metrological performance, wide functionality, cost effectiveness and scalability. The perspectives, which will bring such approach to various sensors and transducers for physical and chemical, electrical and non-electrical quantities are illustrated by examples of multivariable sensors, sensors for smart health care and wearable sensors.

1. Introduction

Further technical progress in microelectronics, instrumentation, robotics, automation control and systems requires the design and development of various high-end sensors and sensor systems on its basis for different physical and chemical, electrical and non-electrical quantities. In addition, high-end sensors will play significant role in the coming IoT 2.0 and Industry 4.0 trends.

High-end sensors are advanced detecting devices, which have high metrological performances and technical (operational) characteristics. Recently, the functionality extension, namely, self-testing, self-validation and self-adaptation are also observed in some high-end sensors. This trend along with the increase of metrological performances will prevail in the next few years in the field of high-end sensors and sensor systems on its basis.

Speaking about the high-end sensors and transducers, the basic metrological performances are the following: specified measuring range; measuring time, and various components of error of measurement. Namely metrological performances directly affect the cost of high-end sensors, transducers and sensor systems. So, for example, increasing the accuracy of a sensor in two times, can cause an increase in cost in 10 times. In order to make high-sensors that would meet modern requirements and at the same time have a reasonable price, new technologies that are cutting costs while increasing effectiveness are necessary.

The best metrological performance of high-end sensors can be achieved by using the frequency as the output informative parameter of the sensor, and the subsequent use of a frequency-to-digital converter (FDC) when creating digital sensors based on its basis. Such sensors have undeniable advantages over voltage, current and digital output sensors based on analog-to-digital converters (ADC) due to the well-known frequency properties as an information parameter of the sensor signals: high noise immunity; high power signal; wide dynamic range; high accuracy of reference; simple interfacing, integration and coding, and multiparametricity.

There are many frequency, period, duty-cycle, pulse width modulated, time interval and pulse number output sensors on the modern sensor market [1].

2. State-of-the-art and technology challenges

The first frequency output transducers (string distant thermometer and tensometer) and time-to-digital converter (TDC) – the ADC for the narrow time intervals, have developed and patented in the Former USSR in 1930, 1931 and 1941 respectively [2-4].

In 1961 professor P. V. Novitskiy wrote: *"... In the future we can expect, that a class of frequency output sensors will get such development, that the number of now known frequency output sensors will exceed the number of now known amplitude sensors..."* [5]. It was also predicted in a technology forecast in 1971: *"Many basic measurements will be related to or measured by time/frequency techniques. This situation will occur because of the increasing facility and accuracy provided through time frequency measurements... Frequency and computing counters can be expected to become smaller, lighter and cheaper, provide direct frequency readout and will be more universally used... Progress will continue in the field of time and frequency during the next five years and will contribute materially to the state of the art in electronics"* [6].

Despite the fact that this forecast was not fully justified till now, it is really the situation when *"the future glimmers long before it comes to be"*. Today, the next big challenge has come from microelectronics. So, in 2010 professor Staphan Henzler did write: *"Below the 100 nm technology node the design of analog and mixed-signal circuits becomes perceptibly more difficult. This is particularly true for low supply voltage near 1 V or below"* [7]. The result is not only an increased design effort, long development time, high risk, cost and the need for very high volumes [8] but also a growing power consumption [7].

There are many reasons that analog doesn't scale as readily [9]. The area shrinks considerably less than predicted by the digital scaling factor. Obviously, both effects are contradictory to the original goal of scaling. However, digital circuits become faster, smaller, and less power hungry [7]. The fast switching transitions reduce the susceptibility to noise, e.g. flicker noise in the transistors. There are also a few drawbacks, such as the generation of power supply noise or the lack of power supply rejection. Because of these drawbacks the analog circuits do not become much smaller in area in smaller technologies.

In the saturation region, there is a much larger change in current in 40 nm process in comparison with the 130 nm process. This change leads to the fact that the transistor has much less control over the current. The design of critical analog circuits that don't scale easily, for example, amplifiers, analog front ends and ADCs, is therefore much more difficult, as the realizable gain per stage is significantly reduced. To compensate for that reduction, more sophisticated circuits are required, leading to lager area and higher consumption [9].

Another factor is the reduced supply voltage and dynamic range. As the supply voltage is reduced with the migration to lower process nodes, the available signal range shrinks. This lower signal range requires proportionately lower noise levels to maintain the same dynamic range. In mixed-signal circuits, such as ADCs using switches capacitors, the reduction in the noise level can be achieved with larger capacitors. To compensate for a 2X reduction in signal range, the capacitors must be increased 4X – making it quite difficult to scale down the area. In addition, the larger loading capacitances require larger currents for charging and discharging, invariably leading to higher consumption. In addition, the designs will become more complicated in order to yield good performance at those low voltages [9].

Their power consumption and speed performance improve significantly due to reduced parasitic capacitance. On the other hand, this results in major challenges for the design of interface electronics:

matching and noise becomes a serious issue, while the dynamic range is reduced due to the supply voltage reduction [10].

The scaling for the analog/mixed ICs can be achieved using a number of different design techniques, not using design CAD tools but rather different architectures. At 28 nm, second- and third-order effects have become serious concerns. As a result, existing design and verification techniques are starting to fall short. Smarter solutions are needed to manage issues arising from tighter spacing [11].

However, digital circuits become faster, smaller, and less power hungry and are scaled very well with scaling CMOS technologies. So, another trend is the transition of traditional analog functions to the quasi-digital domain, for example, to use a frequency (time)-to-digital conversion instead of analog-to-digital conversion, and implement as many system components as possible in the digital or quasi-digital domain. It lets get over technological limitation at scaling.

Frequency/time domain signal processing and so TDC and especially FDC converters are well suitable for high volume microelectronics in order to circumvent analog impairments in nanometer-scale CMOS technologies. In principle TDC only consist of sampled delay-lines. The challenges arise when continuous measurement, long time intervals, high resolution, linearity, low power consumption and robustness come into the play. The architectures addressing one or more of these issues rapidly become more complex [7]. The TDC can exploit some advantages of digital circuits but still behaves like a mixed-signal circuit. Hence, only the FDC concepts are promising that have the ability to exploit fully the advantage of digital circuits [12]. The FDC will be a trend-setting technology for ultimately scaled CMOS technologies in the nearest future, and can be used for many applications such as smart digital sensors, IoT and Industry 4.0.

3. Advanced FDC for digital high-end sensors, transducers and sensor systems

Nevertheless the frequency measurement technique is well established, converting frequency to a digital world for real time systems is not an easy task. For a long time this task was complicated by the absence of any standard peripheral integrated circuit chip for this application. From the other side to meet all the above requirements for smart high-end sensors, modern FDC must be based on advanced frequency-to-digital conversion methods. Only the advanced method can guarantee the low and constant quantization error in all specified frequency range; wide dynamic range (130 dB and more), and non-redundant conversion time. Such measuring technologies (the patented methods and ICs of FDC) were introduced by the F2D, Ltd. (Ireland) – the IFSA Group company, and realized in their series of integrated circuits, suitable for various physical and chemical smart digital high-end sensors and sensor systems.

The integrated Frequency-to-Digital Converters of USTI Series of IC (Figure 1) have the wide frequency range from 0.05 Hz to 9 MHz without prescalling and up to 144 MHz with prescalling, selectable relative error from ±1 to ±0.0005 %. The ICs can convert to digital 26 different frequency-time parameters of signal including rotational speed, resistance, capacitance and resistive bridge values. The conversion time is non-redundant and lets to design various high-end sensors, transducers and sensor systems with intelligent features, such as self-adaptation to select appropriate accuracy, conversion time or power consumption. The ICs have I2C, SPI and RS232 popular interfaces.

The FDC and frequency output sensing elements, as well as voltage output sensing element, voltage-to-frequency converter (VFC) and FDC can be integrated into a single chip or System-on-Chip (SoC). The proposed design approach lets eliminate the mentioned above technological limitations because of the FDC is pure digital component, which can be easily realized in a standard CMOS technology below 100 nm.

Figure 1. USTI Series of ICs in 28-lead PDIP, 32-lead TQFP and 5×5 mm 32-pad MLF packages.

4. Perspectives and applications

Nevertheless the described sensor technologies open a lot of various sensor related markets, let focus on following:

- Sensors for smart health care;
- Multivariable sensors for physical and chemical quantities;

4.1. Sensors for smart health care

There are big expectations for the potential of Smart Health Technologies to support healthcare. This type of technology is still in its infancy, but Smart Health Technologies are expected to be commonplace in the future. However, further research and development is needed, e.g. to ensure the accuracy of data [13].

According to the recent market report published by Persistence Market Research, titled 'IoT Sensors in Healthcare Market: Global Industry Trend Analysis 2013 to 2017 and Forecast 2018–2026', the global IoT sensors in healthcare market was valued at US$ 2,208.9 Mn in 2017, and is expected to register a CAGR of 12.2% from 2018 to 2026 [14]. Improvements in patient engagement, increased accuracy in data analysis, enhanced disease management and treatment results, and reduction in treatment costs are major factors driving the growth of the global IoT sensors in healthcare market. However, privacy concerns and adherence to standard regulations, and varying communication protocols in different IoT sensors in healthcare equipment, hinder the growth of the global IoT sensors in healthcare market. Some of the recent trends for the IoT sensors in healthcare market are focus on implementation of artificial intelligence and IoT technologies, and the increasing demand for wearable devices in the IoT sensors in healthcare sector [14].

The common sensors used in smart healthcare are the following: temperature sensors, ECG, blood pressure, blood glucose, EMG, heart rate, SpO_2, gyroscope, motion sensors, and accelerometers [15]. Things-oriented architectures need to be adaptive based on the application, real time monitoring, on-time delivery, higher sensitivity, maintain higher efficiency at lower power dissipation, and embark on intelligent processing [15]. Combining sensors, actuators, microcontrollers, processors, along with cloud computing, the IoT helps in getting accurate results and makes healthcare attainable to everyone.

Wearables, especially in the form of smart watches or bands, have been revolutionizing the market. Significant amongst the healthcare products are smart watches. Smart watches are becoming more ubiquitous. The projected annualized rate is expected to reach 70 million units at a growth of 18% annualized rate by 2021 [15]. Recommended sensors are those that measure the vital signs - pulse, respiratory rate, and body temperature - as these are the essential signs for determination of critical health. Further sensors that could be implemented are blood pressure and blood oxygen sensors, as these parameters are often taken alongside the three vital signs. Special-purpose sensors such as blood-glucose, fall detection, and joint angle sensors could also be implemented for systems targeting a specific condition [16].

The perspectives to usage of such digital sensors based on the FDC lets to achieve a lower cost, which is one of the main requirements for any IoT sensor. Also, the use of so-called self-adaptive sensor systems based on FDC with an accuracy/power consumption trade-off can reduce the power consumption in several times [12].

4.2. Multivariable sensors for physical and chemical quantities

Multivariable sensors sometimes are called 'multifunctional sensors' or 'multiparametrical sensors' [12, 17]. Multivariable sensing capability enhances operations by measuring several physical or chemical quantities by a single device. There are many multivariable sensors on the market, for example, temperature, absolute and differential pressure multivariable transmitter (Rosemount 3051S) from Emerson; multivariable mass flow meter (PanaFlow MV82) with temperature, velocity and pressure measurements (Baker Hughes a GE company); pressure and temperature transducers with two square waves outputs from Parascientific (USA); multivariable pressure and temperature transducers QHB, SPB, QMB from Quartzdyne with two frequency outputs, and others. GE has also proposed chemical and bio multivariable resonant sensors for environmental, industrial, and security applications [18].

However, the most perspective multivariable sensors are sensors with a frequency output because of one, single sensor's output (pin) can contain information about two sensing variables: one of which is proportional to the frequency, and second – to the duty-cycle of rectangular pulse output signal. For example, in the optical sensor developed in Delft University of Technology the pulse frequency proportional to the optical intensity (luminance) and the duty cycle to the colour (chrominance) [19].

Another example of quasi-digital, multifunctional, multiparametric sensor is described in [20]. This piezoresistive pressure sensor has one output with frequency proportional to pressure and duty-cycle proportional to temperature.

The multivariable sensor for humidity, pressure and temperature is described in [21]. The system has three frequency outputs with a single informative parameter for each of the outputs.

The perspective multivariable sensors can be easy designed based on several frequency output sensing elements and digital multiplexer. In comparison with the analog multiplexed the last one is cheaper and does not introduce any addition error.

5. Conclusions

Today the most advanced high-end sensors, transducers and sensor systems for various physical and chemical, electrical and non-electrical quantities can be built on the novel frequency-to-digital converters, which are based on advanced methods for frequency-to-digital conversion.

Working in the frequency-time signal domain simplifies design, and obviates some technical and technological problems, due to the properties of frequency as informative parameter of sensors and transducers. The major benefits offered by such approach are high reliability, high metrological performance, wide functionality, cost effectiveness and scalability. So, time has come to manufacture high-end sensors.

References

[1] Sensors Web Portal: http://www.sensorsportal.com/HTML/Sensor.htm
[2] Davydenkov N and Yakutovich M 1930 String distant thermometer Patent No. 61727 USSR
[3] Golovachov D, Davydenkov N and Yakutovich M 1931 String distant tensometer Patent No. 21525 USSR
[4] Filipov V N and Negnevitskiy S B 1941 ADC for the narrow time intervals Patent No. 68785 USSR
[5] Novitskiy P V 1961 Frequency output sensors design problem for all electrical and non-electrical quantities *Measurement Technology* **4** 1621 (in Russian)
[6] Stone R R, Phillips D H and Berg W B 1971 Technology forecasts: frequency and time *Proc. the IEEE Conf. on Technology Forecasts and Assess* New York USA 5455
[7] Henzler S 2010 Time-to-digital converters *Springer*
[8] Backram H-E 2010 Scared of mixed-signal and RF ASICS ? Think twice ! *EE Times Europe* 50
[9] Azeredo-Leme C and Nandra N 2011 Analog circuits benefit from scaling trends *Chip Design,*

Spring 1823

[10] Denneels H, Piette F, Smedt V De, Dehaene W, Gielen G and Novel A 2011 PLL-based frequency conversion mechanism for sensor interfaces *Sensor. Actuat.- A: Phys.* **172**(1) 220227

[11] Joe Davis 2011 The Challenge of Analog, mixed-signal and custom physical implementation at 28 nm, *Tech Design Forum* **8**(3) pp 40–43

[12] Yurish S Y 2011 *Digital Sensors and Sensor Systems: Practical Design* (Barcelona: IFSA Publishing)

[13] Smart health technology: the next step in healthcare technology https://path2025.dk/

[14] Global market study on IoT sensors in healthcare: North America to remain largest regional market through 2026 (Persistence market research), June 2018 https://www.persistencemarketresearch.com/market-research/iot-sensors-healthcare-market.asp

[15] Sundaravadivel P, Kougianos E, Mohanty S P and Ganapathiraju M 2018 Everything you wanted to know about smart healthcare 2018 *IEEE Consumer Electronics Magazine* **7** January 2018 pp 18–28

[16] Baker S B, Xiang W and Atkinson I 2017 Internet of things for smart healthcare: technologies, challenges, and opportunities *IEEE Access* **5** pp 26521–44

[17] Kirianaki N V, Yurish S Y, Shpak N O and Deynega V P 2002 *Data Acquisition and Signal Processing for Smart Sensors* (Chichester: John Wiley & Sons)

[18] Potyrailo R A, Boudries H and Naik R R 2012 Multivariable MHz and GHz Wireless Chem/Bio Sensors for Environmental, Industrial, and Security Applications *Proc. the 14th Int. Meeting on Chemical Sensors (IMCS' 2012)* **1** pp 399-402

[19] De Graaf G, Riedijk F and Wolffenbuttel R F 1997 Colour Sensor System with a Frequency Output and an ISS or 12C Bus Interface *Sensor. Actuat.- A: Phys.* **61** pp 441-445.

[20] Crescini D *et al.* 2003 Design and Development of a Piezoresistive Pressure Sensor on Micromachined Silicon for High-temperature Applications and of a Signal-conditioning Electronic Circuit *Microsyst. Technol.* **9** pp 431–435

[21] Lang T T 1991 *Computerized Instrumentation* (Chichester: John Wiley & Sons)

The complex concept of quantity in the past and (possibly) the future of the International Vocabulary of Metrology

L Mari[1][1], A Chunovkina[2] and C Ehrlich[3]

[1] Università Cattaneo – LIUC, Italy
[2] Mendeleyev Institute for Metrology (VNIIM), Russian Federation
[3] National Institute of Standards and Technology (NIST), United States of America

E-mail: lmari@liuc.it

Abstract. The *International Vocabulary of Metrology* (VIM) is a foundational document of measurement science, with the ambition of providing a system of "basic and general concepts and associated terms" of metrology. Such a system has evolved with the evolution of measurement science, and this is particularly manifest in the case of the cluster of concepts around 'quantity'. The present paper presents some aspects of this development, by first remarking on the importance of terminology for measurement science and then introducing the basic problem *what is a quantity?* as a way to provide a well-grounded explanation to the usual claims that two quantities can be equal, or two values of quantities can be equal, or a quantity of an object can be equal to a value of a quantity. The analysis develops along three questions: (1) what is the relation between general properties and individual properties? (2) what is the relation between properties and quantities? (3) what is the relation between quantities of objects and values of quantities?

1. Why terminology is important for our subject

Our knowledge of the world around us is based on two kinds of entities: *objects* (such as electrons, rods, galaxies, persons and companies, industrial processes and software systems) and *properties* (physical quantities, such as mass, temperature, etc, but also non-physical properties, such as beauty, wealth, competence, and wellbeing – "object" and "property" are not standard terms: by "object" we generalize here what in [1] is called "phenomenon, body, or substance" (e.g., in def. 1.1); more or less equivalent to "property" are "attribute", "observable", "variable", etc.; we use the term "entity" for "anything perceivable or conceivable", as in [2], def. 3.1.1, where the term "object" is used instead; for a terminological analysis of this foundational topic, see [3]).

Objects and properties are so embedded in our knowledge systems that this basic epistemology becomes an ontology: we behave as if what is knowable *is made of* objects that have properties. Accordingly, a key knowledge construct is *property of an object*. Indeed, properties of objects

- can be causally relevant: it is the mass of the rod that stretches the spring of the dynamometer,

[1] The authors are members of the Joint Committee on Guides in Metrology (JCGM) Working Group 2 (VIM). The opinion expressed in this paper does not necessarily represent the view of this Working Group, nor of the authors' home organizations.

- establish the behavior of objects in their comparison with other objects: the comparison of two rods by length and by mass generally produces different outcomes, and
- define the identity of objects: the rod can maintain its identity even if some of its properties change, provided that some other of its properties remain the same.

Some properties of some objects are empirical: we model them on the basis of the available knowledge, but the underlying hypothesis is that they exist independently of our models. For example, we accept that the sun emitted radiant power also before life appeared on the earth, and therefore also before the concept 'power' was formulated. As a consequence, such properties are not conceptual entities: we construct concepts to understand them, but they are not concepts. And of course they are not linguistic entities: we construct terms to designate them, but they are not terms. Indeed, the *semiotic triangle* applies both to objects and properties (in the diagram of figure 1).

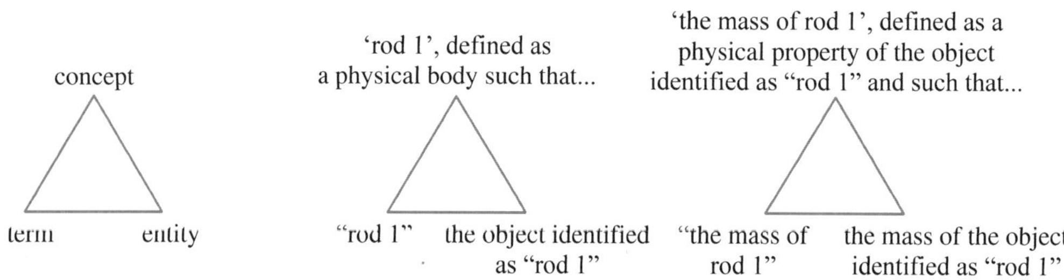

Figure 1. The semiotic triangle, and two relevant examples.

This is an effective representation of the basic function of language as a tool of intersubjective communication. We understand each other when we use the same term ("rod", "mass") to refer to the same object (a given rod, the property mass). While sometimes this can be stipulated and learned by direct reference ("from now on "rod ABC123" is the name of this rod", uttered while pointing to a given object, and thus providing a contextual meaning of "this rod"), in most cases such a strategy is impractical or just unfeasible. Concepts are then the critical connectors between terms and objects: we refer to the same object when uttering "rod" if we share the same concept of what a rod is. Terms are introduced and agreed for the common understanding of concepts and the common reference to objects. This highlights the importance of terminology, in scientific and technical contexts and not only: the definitions contained in vocabularies are linguistic expressions aimed to characterize concepts. Hence a good definition of, say, 'measurement' (the concept, with single quotes) is expected to provide a criterion to identify measurements (the objects), as designated as "measurement" (the term, with double quotes). In the case of property-related concepts this is a particularly delicate result to achieve, due to their elusive nature. On the other hand, properties, and specifically quantities, have a pivotal role in measurement – what is measured are properties of objects – and not surprisingly therefore the *International Vocabulary of Metrology* (VIM) since its first edition (VIM1, [4]; we refer to the subsequent editions as VIM2 [5] and VIM3 [1]) devoted great attention to them, as witnessed by the number of property-related definitions included in the Vocabulary.

2. Introducing the problem

The cluster of concepts around 'property' that need to be clarified and defined can be introduced in a simple and operative way with an example as in the following diagram.

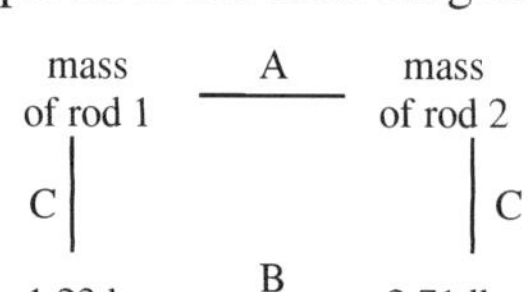

Figure 2. The relations about properties.

There are three kinds of relations here, and for each of them we use the "=" sign if appropriate:

A. mass of rod 1 = mass of rod 2

B. 1.23 kg = 2.71 lbs

C. mass of rod 1 = 1.23 kg

The key question here is: what is the meaning of those "=" signs in the three cases?

3. First key point: general properties and individual properties

Relations A, B, and C convey some information because they are related to *meaningful* comparisons. While the mass of rod 1 and the mass of rod 2 can be compared, and the outcome may be, in particular, that they are discovered to be equal, or at least indistinguishable, nothing similar could be said of the mass of rod 1 and the length of rod 2: they are neither equal nor different, because they cannot be compared. This condition of comparability justifies introducing the distinction between *general properties*, such as mass, and *individual properties*, such as any given mass (these terms are not standard; in the specific context of quantities, the three editions of the VIM define the concept 'quantity' as encompassing both entities such as mass and the mass of a given object, and their distinction is only noted: in the VIM1 in terms of "quantities in a general sense" and "specific quantities" (def. 1.01 note 1), in the VIM2 in terms of "quantities in a general sense" and "particular quantities" (def. 1.1 note 1), and in the VIM3 in terms of "kinds of quantities" (def. 1.2 note 3, even though the definition seems to suggest something else) and "individual quantities" (def. 1.1 note 1)).

Any individual property is the instance of a general property, and two individual properties are comparable if they are instances of the same general property or, equivalently then, if they are properties of the same kind. Hence the mass of rod 1 and the mass of rod 2 are comparable because they are individual properties that are both instances of the general property mass. A problem of foundational ontology is about how to interpret the statement of equality of two individual properties, e.g., relation A above, and two positions can be envisaged. In the example, does it mean that

- *Position 1*: there is one individual mass that rod 1 and rod 2 share? or that,
- *Position 2*: each rod has its own individual mass, and the two masses are indistinguishable?

While this problem cannot be solved by experimental means, we will see that *these positions lead to distinct ways to conceive measurement results and the way we report them.*

The VIM does not provide separate definitions for the two concepts, 'general quantity' and 'individual quantity', and uses the same term "quantity" for both, but the context helps solving this polysemy. For example, a measurand, i.e., a "quantity intended to be measured" (VIM3, def. 2.3), is an individual quantity (one can measure the mass of a given rod, not mass as such), whereas the distinction between base quantities and derived quantities (VIM3, def. 1.4 and 1.5) applies to general quantities (a general quantity of the SI is mass, not the mass of a given rod). Future editions of the VIM could provide better explanations, but introducing distinct terms and separate definitions does not seem to be a priority.

4. Second key point: properties and quantities

We have already mentioned that quantities are specific properties (so that 'quantity' is defined as "property such that..."). The VIM has assumed so far that only quantities are measurable, i.e., that non-quantitative properties are not measurable ("Measurement does not apply to nominal properties.", VIM3, def. 2.1 note 1). Hence identifying what characterizes a quantity as a specific property is important. According to the VIM1 and the VIM2, quantities are properties (they use the equivalent term "attribute") which "may be distinguished qualitatively and determined quantitatively" (VIM1, def. 1.01 and VIM2, def. 1.1). This is unfortunately not so useful, given that defining 'quantity' as something "determinable quantitatively" is basically tautological. Hence the VIM3 changed the definition of 'quantity': "property of a phenomenon, body, or substance, where the property has a magnitude that can be expressed as a number and a reference" (def. 1.1). Let us analyze it.

First, the expression "phenomenon, body, or substance" refers to what we are generically calling here an object. Second, the condition "that can be expressed as a number and a reference" seems to be a clarification more than an actual specification, given that everything can be expressed as a number and a reference if the expression is not somehow constrained, and therefore can be dropped as inessential (the VIM3 considers ordinal properties as quantities, so that in the case of ratio quantities the reference is a measurement unit (def. 1.19 note 1)). With these simplifications, we obtain that *a quantity is a property of an object, where the property has a magnitude*. The VIM does not define what a magnitude is, but a plausible hypothesis is that "magnitude" can be intended here as "amount", so that for example mass is a quantity because it is a property that objects have in amounts (as contrasted, e.g., to shape, that is a property that objects do not have in amounts). Accordingly, the phrase "the magnitude of the mass of this rod" means "the amount of mass of this rod". On the other hand, we commonly refer to the mass of objects, not to the magnitude of the mass of objects: why? The distinction between *Position 1* and *Position 2* provides an explanation.

According to *Position 1*, "mass of rod 1 = mass of rod 2" means that there is one individual mass that rod 1 and rod 2 share. Hence for two objects having the same (individual) mass and having the same amount of mass are identical conditions: a magnitude of mass is just an individual mass.

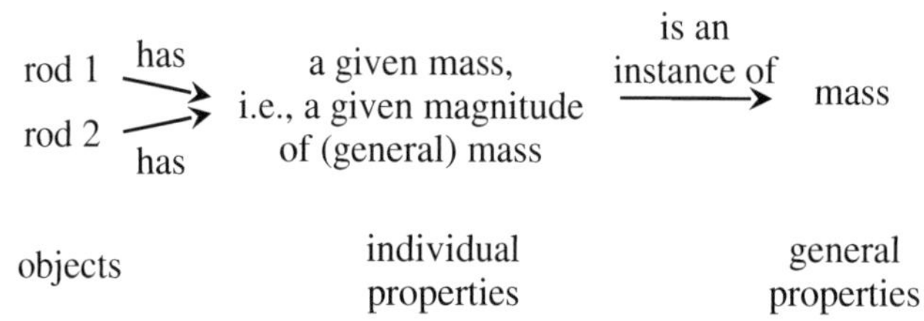

Figure 3. The interpretation of relation mass of rod 1 = mass of rod 2 according to *Position 1*.

This explains why measurement results are customarily reported by referring to quantities of objects, such as the (individual) mass of rod 1, instead of to magnitudes of general quantities as considered of objects, such as the magnitude of (general) mass as considered of rod 1.

According to *Position 2*, "mass of rod 1 = mass of rod 2" means that each rod has its own individual mass and the two masses are indistinguishable. Since two objects having the same (individual) mass is not possible, they may only have the same amount of mass: a magnitude of mass is what indistinguishable masses have in common.

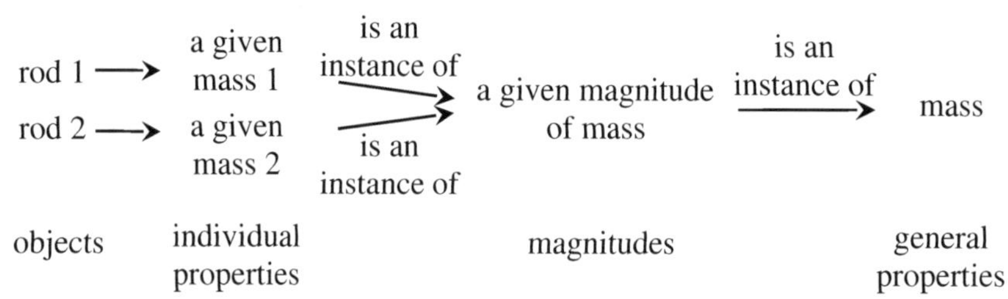

Figure 4. The interpretation of relation mass of rod 1 = mass of rod 2 according to *Position 2*.

In this sense, the expression "mass of rod 1 = mass of rod 2" cannot be literally true, and must be intended as an elliptical reference to the condition that the two masses have the same magnitude, i.e., magnitude of mass of rod 1 = magnitude of mass of rod 2.

Under this interpretation, the definition of 'quantity' given by the VIM3 is then compatible with both *Position 1* and *Position 2*. Future editions of the VIM could provide even better explanations, making this concept of magnitude/amount more explicit by characterizing it in terms of invariance conditions among comparable individual properties. For example, ratio quantities could be defined as general properties whose instances can be compared by ratio (where then double amount corresponds to ratio = 2, and so on). Furthermore, while the VIM3 definition of 'nominal property' (def. 1.30) is phrased in negative terms, "property of a phenomenon, body, or substance, where the property has no magnitude", such a strategy would lead to characterize nominal properties as general properties whose instances can be compared by equality: a simple, operational option.

A further comparative analysis of *Position 1* and *Position 2* requires taking also values of quantities into account.

5. Third key point: quantities of objects and values of quantities

We have provided some explanation about relation A in the diagram of figure 2, but relations B (1.23 kg = 2.71 lbs) and C (mass of rod 1 = 1.23 kg) still require our analysis: again, what is the meaning of those "=" signs? Both relations involve values of quantities, and therefore let us consider how the concept 'value of a quantity' – the VIM3 has "quantity value" as the preferred term, but the difference is only lexical – is defined in the three editions of the VIM (we leave to other contexts the more general analysis about values of properties):

VIM1: "expression of a quantity in terms of a number and an appropriate unit of measurement" (def. 1.17);

VIM2: "magnitude of a particular quantity generally expressed as a unit of measurement multiplied by a number" (def. 1.18);

VIM3: "number and reference together expressing magnitude of a quantity" (def. 1.19).

In their marked differences, each of them grasps an aspect of what values are.

The VIM1 emphasizes the function of values to convey information, hence considering them "expressions". On the other hand, the concept 'expression' is generic, and seems to be related to linguistic entities, so that in relation B, "1.23 kg" and "2.71 lbs" are plainly different expressions (say, only the former contains the character "3"). Indeed, what the relation states to be equal are not the expressions but what they express, i.e., the entities that are individual quantities. Analogously, in relation C, an empirical entity, such as the mass of rod 1, cannot be equated to an expression, so that what is stated to be equal is an individual quantity and what an expression such as "1.23 kg" expresses. With its operative flavor, the VIM1 does not maintain the distinction between expressions and what expressions express, and values are what expressions such as "1.23 kg" and "2.71 lbs" express. Not surprisingly, then, the VIM2 takes a different track, by emphasizing the ontological status of values: according to the analysis in the previous section, *Position 2* is endorsed here and values are considered magnitudes/amounts of individual ("particular") quantities, even though the characterization of such magnitudes is provided in terms of an expression like in the VIM1. Finally, the VIM3 definition does not actually specify what a value is ('value and reference' is not a single concept), and rather specifies what values *are made of*: numbers and references, and therefore numbers and units in the case of ratio quantities.

Even this short analysis shows that future editions of the VIM could bring significant improvements to our understanding of what values *are*. Let us then better explore this subject.

6. What are values of quantities?

Instead of dealing with the general case of values of properties, we focus here on values of ratio quantities, like 1.23 kg. As the VIM3 explicitly states (def. 1.19 note 1), such a value is the product of a number and a unit: *Position 1* and *Position 2* provide related but distinct interpretations of this, both based on the acknowledgment that 1.23 kg is an entity identified by the mathematical technique of somehow selecting a mass unit (the task of those who define what is the kilogram) and then identifying all other related entities as multiples or submultiples of the selected unit.

According to *Position 1*, a mass unit is unproblematically an individual mass, so that 1.23 kg is the individual mass identified as 1.23 times the kilogram, where the identification, which is in principle mathematical (in fact for assessing that, e.g., 1.23 kg < 2.34 kg no empirical processes are required), may be also realized by constructing an object – thus a measurement standard – that materializes a mass that is 1.23 times the kilogram. This provides a simple explanation of the meaning of the relations between individual quantities identified as quantities of objects and individual quantities identified as multiples or submultiples of units. Being entities of the same nature, they can be actually compared, so that for example mass of rod 1 < 1.23 kg means exactly what it is written: it is the comparison of two masses, such that the mass of rod 1 is less than the mass that is 1.23 times the kilogram. Analogously, relation C in the diagram of figure 2, mass of rod 1 = 1.23 kg, means that there is one individual mass, that has been identified as both the mass of rod 1 and 1.23 times the kilogram (as then in figure 3). In summary, according to *Position 1* values *are individual quantities* identified as multiples or submultiples of units, which *are themselves individual quantities*.

Position 2 introduces the further entity, magnitude, that is what indistinguishable individual quantities have in common, so that 1.23 kg is a magnitude of mass, identified as 1.23 times the kilogram, which now has to be considered a magnitude in turn. Exactly as for relation A between masses of objects, also in this case relation C needs to be interpreted as elliptical: it is written "mass of rod 1 = 1.23 kg" but it must be meant as magnitude of mass of rod 1 = 1.23 kg (as in figure 4). In summary, according to *Position 2* values *are magnitudes* identified as multiples or submultiples of units, which *are themselves magnitudes*.

Interestingly, this analysis leads us also to interpret relation B, i.e., the equality of values having different units, like 1.23 kg = 2.71 lbs: it is the claim that there is one individual mass (in *Position 1*) or a magnitude of mass (in *Position 2*) that is identified by reference to different units. Back to the semiotic triangle, in both cases the equality of 1.23 kg and 2.71 lbs states that there is one entity (an individual mass, in *Position 1*; a magnitude of mass, in *Position 2*) identified by means of two different concepts, the concept of being 1.23 times the kilogram and the concept of being 2.71 times the pound.

7. Conclusion

Quantities have a pivotal role in measurement, and the three editions published so far of the *International Vocabulary of Metrology* witness the evolution of the understanding around the three basic questions: (1) what is the relation between general properties and individual properties? (2) what is the relation between properties and quantities? (3) what is the relation between quantities of objects and values of quantities? At least two general positions can be consistently maintained about them, and along these paths the future editions of the VIM will plausibly move forward in providing an always better foundation of measurement science.

Acknowledgments

The Authors are grateful to David Flater and Alessandro Giordani for their useful comments to the draft of this paper.

References

[1] JCGM 200:2012 *International Vocabulary of Metrology—Basic and General Concepts and Associated Terms (VIM)* 3rd ed (2008 version with minor corrections) (Sèvres: Joint Committee for Guides in Metrology) (www.bipm.org/en/publications/guides/vim.html)

[2] ISO 1087-1:2000 *Terminology Work – Vocabulary – Part 1: Theory and Application* (Geneva: International Organization for Standardization)

[3] ISO 704:2009 *Terminology Work – Principles and Methods* (Geneva: International Organization for Standardization)

[4] ISO 1984 *International Vocabulary of Basic and General Terms in Metrology* (VIM), published in the name of BIPM, IEC, ISO and OIML 1st ed (Geneva: International Organization for Standardization)

[5] ISO 1993 *Guide 99:1993 International Vocabulary of Basic and General Terms in Metrology* (VIM) published in the name of BIPM, IEC, IFCC, ISO, IUPAC, IUPAP and OIML 2nd ed (Geneva: International Organization for Standardization)

Journal of Physics: Conference Series **1379** (2019) 012005 doi:10.1088/1742-6596/1379/1/012005

Novel multidimensional biomarkers for non-invasive medical diagnostics

M S Vonsky[1,2,3], **A L Runov**[1,2,3] **and O M Moiseeva**[1]

[1] Almazov Medical Research Centre, St.-Petersburg, 197431, Russia
[2] D.I. Mendeleev Institute for Metrology, St.-Petersburg, 190005, Russia
[3] Institute of Cytology RAS, St.-Petersburg, 194064, Russia

E-mail: m.vonsky@gmail.com

Abstract. Value of clinical laboratory diagnostics results are often underestimated in medical practice due to the fact that result of laboratory analysis in most cases represents a single biomarker of the organism homeostasis complex pathological alteration. New systemic approaches emerging on the basis of Omic's knowledge and techniques allow development of multidimensional biomarkers, reflecting multiple pathobiological axes which characterize defined pathological condition. Combination of liquid biopsy technology with multidimensional biomarkers allows to obtain clinically valuable personalized information on the patient by non-invasive approach even for hard-to-reach organs and tissues.

1. Introduction

Biological markers, or biomarkers, are defined as "a characteristic that is objectively measured and evaluated as an indicator of normal biological processes, pathogenic processes, or responses to an exposure or therapeutic intervention" [1]. Even biomarkers cannot substitute clinical endpoints completely, their importance increases in medical laboratory diagnostics and medical research.

Laboratory diagnostics historically plays a secondary role in clinical practice. In most cases results of laboratory studies are taken into consideration, but they are not decisive in the diagnosis demanding supplementary instrumental examinations. This position of the clinical laboratory originated from the fact that each individual qualitative or quantitative result of laboratory analysis represents a single biomarker of the complex homeostasis of the organism alteration. A result, provided by laboratory, as a rule, does not carry, or carries indirect diagnostically significant information about the pathology, the causes of its development or localization. For example, elevated activity level of alanine transaminase (ALT) for a long time was concerned as markers of hepatitis or muscular disease. Nowadays, ALT is considered only as a biomarker of a violation of cells integrity - elevated ALT activity may be caused by different pathophysiological events. Therefore, a laboratory report on elevated ALT activity provides evidently insufficient information to explain the clinical implication underlying the pathology. Another cardiac biochemical marker – troponin I (cTn I) – previously was considered as the preferred biomarker for the diagnosis of myocardial infarction. By fact, cTn I is a contractile protein, a component of cardiac myocytes filaments. Release of cTn I into circulation may occur due to ischemia, necrosis or damage of cardiomyocyte and may be related to underlying cardiac or non-coranary and extracardiac disease, such as severe renal dysfunction. But in order to verify the acute coronary syndrome diagnosis, combination of increased cTn I level, echocardiography and electrocardiography data are necessary.

The exceptions are biomarkers that carry the necessary and sufficient information for medical decision making. These include, for example, biomarkers of infectious diseases - nucleic acids and antigens of pathogenic microorganisms and viruses, identification of which prove the presence of the pathogen and allows to evaluate the microbial or viral load, to monitor the results of therapy in dynamics. Improving the information load of biomarkers could simplify the diagnosis algorithms and can be implemented through a systematic approach that uses a combination of biomarkers to target the noncommunicable disease.

The implementation of the Human Genome project has led to the explosive development of Omic's, systemic approaches that consider biological processes in the complex, taking into account their direct and indirect influence on each other. Thus, genomics considers the genome not as a sum of genes, but as their organized system, proteomics — a complex of cell proteins in their interaction, and transcriptomics reflects the current pattern of gene expression due to the processes occurring in the cell. Further on, metabolomics deals with a comprehensive evaluation of metabolites, substrates and by-products of enzymatic reactions which reflect all functional activities ongoing in biological systems. Taken together, all this approaches allow to discuss patholomics, as the systemic study covering intertwining of genes, proteins and metabolites underlying, promoting and reflecting the pathology development Omic's approach together with bioinformatics allows development of multiplex, or multidimensional biomarkers, reflecting different pathobiological axes which characterize defined pathological condition.

2. Transcriptomic biomarkers

Development of pathological processes is necessarily associated with changes in the transcriptome of the cells of the affected tissue, manifested through a decrease or increase of the expression level of a defined set of genes. The expression profiles of a diagnostic panel of genes determined in this way represent perspective biomarkers reflecting defined pathophysiological processes. Microarray analysis (high density nucleic acids hybridization on microchip) is often performed to study transcription profile, but from analytical point of view this data can be considered as qualitative and preliminary, allowing to identify but not validate candidate biomarkers. Quantitative RNA measurements nowadays can be assessed after the reverse transcription by means of established DNA analysis methods – real-time and digital quantitative polymerase chain reaction (RT-qPCR and RT-dPCR) as well as RNA sequencing or next-generation sequencing (NGS).

2.1. RNA quantitative measurements

Comparability of RNA copy number ratio measurements for single and multiple gene targets were studied previously in the Consultative Committee for Amount of Substance (CCQM) international comparison studies CCQM P103 "Quantification of RNA transcript" and P103.1 "Measurement of multiple RNA transcripts" performed under the auspices of the CCQM Nucleic Acids Working Group [2]. Measurements of three human endogenous gene targets were performed in the background of human cell line RNA. Good agreement was observed between reported results obtained by means of RT-qPCR, RT-dPCR and NGS in this model study.

Capabilities to perform RNA copy number ratio measurements itself are not sufficient for transcriptomic biomarker development, validation and their application in medical purposes. Identification of internal control - reference genes, characterized by stable level of expression, unaffected by pathology, is a cornerstone of RNA quantitative measurements in biological material. These genes (usually in a combination of 2, up to 5) are used for normalization of RT-qPCR results, allowing to perform copy number ratio, or relative measurements of target genes expression with required accuracy. Usually a small subset of so called "housekeeping" candidate reference genes is tested in medical studies – glyceraldehydes 3-phosphate dehydrogenase (*GAPDH*), hypoxanthine phosphorybosyl transferase (*HPRT*), β-actin, tubulin, and ribosomal RNA genes are typical examples of frequently used reference genes [3]. Stability of candidate gene expression usually evaluate applying software applications GeNorm [4], BestKeeper [5] or NormFinder [6] to process RT-qPCR experimental data.

Typical problems linked with the reference genes selection, include a limited number of candidate reference genes tested and requirements for comparable expression level of target and reference genes. To ensure comparability of RT-qPCR results, obtained in different laboratories, it is accepted to follow the guideline for minimum information for publication of quantitative real-time PCR experiments (MIQE) [7].

2.2. RNA biomarkers of cardiovascular disease
Cardiovascular disease is a major public health problem globally and causes in many countries twice more death then cancer, promoting development of new diagnostic and therapeutic approaches. A number of protein-based circulating biomarkers are well characterized and used in routine clinical practice for diagnosis and monitoring of cardiovascular disease. In case of known target it is possible to substitute analyte (protein or peptide) by corresponding gene transcript. This approach allow to measure target gene expression in tissues of interest directly, verifying localization of pathological process. For example, soluble ST2 protein – receptor of interleukin 1 – can be measured in peripheral blood and used in risk stratification of both acute and chronic heart failure. In the early study of Weinberg et al. it was shown, that expression of *IL1RL1* (ST2 receptor coding gene) is increased in cardiac myocytes in case of stretch [8]. Even this study was performed by semiquantitative end-point PCR analysis, it result in a novel RNA biomarker of heart failure development and made one of the first steps on the way from single, “stand alone” to multidimensional biomarkers.

2.3. Multidimensional RNA biomarkers
Development of real multidimensional transcriptomic biomarkers begins on the basis of microarray technology since 1999 - investigators at MIT identified a signature of 50 genes out of 6817 surveyed that could distinguish between acute myeloid leukemia and acute lymphoblastic leukemia [9]. Since then, in the area of oncology transcriptomic biomarkers – unique gene expression profiles – were applied for classification of hematologic malignancies, solid tumors, prediction of survival, clinical outcome, recurrence risk, and response to therapy. In the area of cardiology – another important group of noncommunicable disease, comparison of global gene expression profiles in endomyocardial biopsy (EMB) specimens from clinically and histopathologically characterized patients demonstrated ability of microarrays to facilitate the differential diagnosis of cardiomyopathies [10, 11]. It was shown, that about 45% of genes are differentially expressed between patients groups. In some cases transcriptomic biomarkers allow to detect pathology better then histological approaches regarded as “golden standards” in laboratory diagnostic [11].

3. Multidimensional biomarkers for myocarditis diagnosis
Myocarditis is commonly defined as inflammation of the myocardium followed by cardiomyocyte necrosis. Distribution of the myocarditis in human population reaches 4 %, however, real prevalence of the disease suggested to be underestimated as it is often asimptomatic or has unclear clinical manifestations. Clinical diagnostic of myocarditis based on histolopathological analysis of EMB, so called Dallas criteria, proposed in 1986 by the European Society of Cardiology. It was suggested in 2006, that due to EMB sampling errors, variation in expert interpretation, variance with other markers of viral infection and immune activation in the heart, and variance with treatment outcomes the Dallas criteria are no longer adequate [12].

Microarray analysis on EMB samples from patients with histologically proven myocarditis (n=16) and dilated cardiomyopathy (n=32) reveal 9878 differentially expressed genes [13]. A multidimensional biomarker consists of 62 genes, which distinguished myocarditis with 100 % sensitivity and 100% specificity, was identified. Results of microarray analysis were validated by of RT-qPCR; some candidate genes were not confirmed and the final panel included 13 genes involved in inflammation and cardiac remodeling pathways. No data was presented in the paper on reference genes used or RT-qPCR normalization.

In our study a set of 6 candidate reference genes was assessed for normalization of expression in cardiac tissues from patients with myocarditis, a combination of *GAPDH* and *HPRT1* was selected [14]. EMB samples (n=23, M:15, F:8) from myocardial patients were included into the study together with control samples – tissue fragments obtained during orthotopic heart transplantation (n=6, M:4, F:2). Target panel included 61 candidate genes. It was shown, that myocarditis induced statistically significant ($p<0,035$) modulation of gene expression in cardiac tissue of male and female patients (10 and 12 genes, correspondingly). Unexpectedly, we found significant gender-dependent difference in gene expression profiles: only 6 genes were co-modulated for both genders (*FCER1G*, *IL2*, *NOTCH2*, *SEC24A*, *SIGLEC1*, *TMOD3*). Even in this co-modulated subset expression of *FCER1G* was up-regulated 7-fold higher in males, expression *TMOD3* was down-regulated 3-fold lower in females. Expression of *ADCY7* was modulated for both genders, up-regulated for males and down-regulated for females. Three genes were differentially expressed only in male patients (*GLIPR*, *ITGB2*, *NFkB*), five – only in female (*ACOX1*, *CD1D*, *CD14*, *CD8B1*, *HRH3*). A correlation was found between expression modulation and disease progress for 4 genes, so this subset can be used for therapeutic monitoring purpose.

Unfortunately, this perspective approach includes invasive endomyocardial biopsy step, but it can be used as a supplementary study to improve histopathological examinations of EMB samples.

4. Non-invasive multidimensional biomarkers

Combination of liquid biopsy technology with transcriptional multidimensional biomarkers allows to realize non-invasive differential diagnostic of hard-to-reach organs and tissues pathologies, including pathologies of cardiovascular system and oncopathologies. Potential correlation of simultaneous gene expression in the cardiac tissue and peripheral blood mononuclear cells (PBMC) is widely discussed: blood considered as a surrogate tissue, a liquid organ, being in an intimate proximity to the heart. Up to 80% of known genes are expressed in PBMCs; blood cells, being in constant contact with numerous exo- end endogenous factors, have to respond to them via differential gene expression.

4.1. Cardiac pathology biomarkers

Doxorubicin (DOX) is a commonly used antibiotic for malignancies treatment. It may cause unpredictable doze-dependent cardiotoxicity begins with the first doze: patients, receiving chemotherapy, are at increased risk of developing cardiac dysfunction. Differential 235 genes expression was revealed in PBMC of breast cancer patients treated with single dose of DOX by RT-qPCR method with single 18S mRNA reference gene [15]. Out of these 235 genes, Todorova et al. identified 63 genes, differentially expressed in patients with abnormal heart left ventricle ejection fraction as a marker of cardiotoxicity. Candidate non-invasive multidimensional biomarker was developed as a result of the study, proving applicability of PBMC gene expression profiling for diagnosis of cardiac pathologies.

In the ongoing in our laboratory study we are developing candidate transcriptomic biomarkers of myocarditis examining gene expression in PBMC of patients and healthy donors. No correlation up to the moment was found between gene expression in EMB samples and PBMCs. Gender-dependent target gene expression difference is not as significant in PBMC, as in cardiac tissue. More then 20 differentially expressed genes ($p<0,03$) were identified at the moment by RT-qPCR, which allows us to hope for a successful completion of the project.

5. Conclusion

Multidimensional biomarkers are one of the key tools of personalized medicine. Considering liquid biopsy as non-invasive source of biomaterial, we can develop RNA biomarkers using coding and non-coding nucleic acids. Application of Omic's approach together with bioinformatics allows to develop not only diagnostic biomarkers, but to monitor disease and treatment progress taking into account the individual properties of patophysiological processes.

Ethics Statement

This study was carried out in accordance with the ethical guidelines of the 1975 Declaration of Helsinki and was approved by the ethics committee of the Almazov Medical Center. All subjects signed an IRB approved informed consent where they were informed for the use of their EMB, blood samples and medical records for research purposes.

References

[1] FDA-NIH Biomarker Working Group 2016 *BEST (Biomarkers, EndpointS, and other Tools) Resource* (Silver Spring (MD): Food and Drug Administration (US); Bethesda (MD): National Institutes of Health (US))
[2] Devonshire A *et al.* 2016 *Biomol. Detect. Quantif.* **8** 15
[3] Suzuki T, Higgins P and Crawford D 2000 *BioTechniques* **29** 332
[4] Thellin O, Zorzi W, Lakaye B, De Borman B, Coumans B, Hennen G, Grisar T, Igout A and Heinen E 1999 *J. Biotechnol.* **75** 291
[5] Vandesompele J, De P, Pattyn F, Poppe B, Van R, De P and Speleman F 2002 *Genome Biol.* **3** RESEARCH0034
[6] Pfaffl M, Tichopad A, Prgomet C and Neuvians T 2004 *Biotechnol. Lett.* **26** 509
[7] Andersen C, Jensen J and Orntoft T 2004 *Cancer Res.* **64** 5245
[8] Bustin S *et al* 2009 *Clin. Chem.* **55** 611
[9] Weinberg E, Shimpo M, De Keulenaer G, MacGillivray C, Tominaga S, Solomon S, Rouleau J and Lee R 2002 *Circulation* **106** 2961
[10] Ginsburg G, Seo D and Frazier C 2006 *JACC* **48** 1618
[11] Barth A *et al.* 2006 *J. Am. Coll. Cardiol.* **48** 1610
[12] Ruppert V, Meyer T, Pankuweit S, Möller E, Funck C, Grimm W, Maisch B and German Heart Failure Network 2008 *J. Thorac. Cardiovasc. Surg.* **136**(2) 360
[13] Baughman K 2006 *Circulation* **113** 593
[14] Heidecker B, Kittleson M, Kasper E, Wittstein I, Champion H, Russell S, Hruban R, Rodriguez E, Baughman K and Hare J 2011 *Circulation* **123** 1174
[15] Runov A, Kurchakova E, Khaschevskaya D, Moiseeva O and Vonsky M 2015 *Cell Tissue Biol.* **9** 330.
[16] Todorova V, Makhoul I, Siegel E, Wei J, Stone A, Carter W, Beggs M, Owen A and Klimberg°V 2016 *PLoS One* **11** e0160224

Exemplifying natural science measurement in the social sciences with Rasch measurement theory

David Andrich

Graduate School of Education, The University of Western Australia, Crawley WA 6009, Australia

E-mail: david.andrich@uwa.edu.au

Abstract. This paper presents a social science example of measurement in which the integer count from a convenient origin is shown to be in equal units. This feature is identical to direct measurement in the natural sciences. The example rests on applying the probabilistic Rasch measurement model to two assessments of the same persons which (i) realises measurement as a real valued comparison between persons invariant with respect to instruments; (ii) characterises random variation of replicated measurements by the Gaussian distribution; (iii) shows that the variance of the replications is the inverse of the size of the unit.

1. Introduction

Although elementary measurement can be understood by school children, it is central to the remarkable advancement of the natural sciences [1]. In principle, measurement is the mapping of an amount of some property of an object onto a line partitioned into equal intervals, called *units*. The count of units from a convenient origin is the measurement of the object.

2. The abstraction of measurement

The mapping of a property onto a continuous, abstract line joining two points with no magnitude is a deep abstraction. No objects, or their properties, immediately fit such a description. Even a *drawn line* examined through a microscope shows gaps. Natural objects rarely approximate a line. The elementary measurement of mass with a beam balance requires the control of gravity with three dimensional objects which can be of all shapes and sizes. The measurement of temperature is more complex, both in theory and instrument design. The measurement of the concentration of sugar in juice, where the reading is corrected for temperature, is even more complex, as is the measurement of various electro-magnetic properties. A common feature is reading measurements directly from instruments. Direct reading, termed *direct* measurement, hides the theory and design which manifest the property measured and controls those properties that might disturb this manifestation.

Despite being ubiquitous, natural scientists seem able to take the common characteristics of measurement for granted, for example, those common in measuring the different properties of mass and temperature of an object – they do not agonise as to what measurement really is. This contrasts with social scientists, both in the successes they have, and in the agonising they do, with measurement.

3. Assuming measurement: Carl Friedrich Gauss (1777 – 1855)

One characteristic of measurement that continues to concern natural scientists is uncertainty, the evidence that either literal or inferred replications of measurements do not give the same measurement [2]. To summarise replications required an intellectual leap to hypothesise that the variation was *random*, and what were perceived as errors should cancel rather than propagate. Establishing a probability distribution *…to describe the distribution of the errors arising in repeated measurement of a fixed quantity by the same procedure under constant conditions…*, [3 p1] occupied the best mathematicians of the time and culminated

> in the quadratic exponential law of Gauss, $f_Z(z) = (h/\sqrt{\pi})\exp(-h^2 z^2)$,…which became almost universally regarded in the nineteenth century as "the law of error". [3 p1]

In the above quote, $z = x - \mu$ is a real number, μ is the mean of the distribution and $f_Z(z)$ is continuous even though x is an integer count in the unit of the instrument. In updated notation, $h = 1/\sigma$ where σ^2 is the variance of the distribution. This distribution then provided a criterion that the variation in the measurements was no more than random, with no systematic factor disturbing the instrument. An observation relevant for this paper is that

> Gauss commented that (1) (*the distribution*) cannot represent a law of error in full rigor because it assigns probabilities greater than zero to errors outside the range of possible errors, which in practice always has finite limits; that such a feature is unavoidable because one can never assign limits of error with absolute rigor; but this shortcoming is of no importance in the case of (1), because it "decreases so rapidly, when hz has acquired a considerable magnitude, that it can safely be considered as vanishing." [4 p2]

This paper takes it that *when all of the known or suspected components of error have been evaluated* [2], and the variation in replicated measurements is no more than random, then the Gaussian distribution provides a measure of *uncertainty* rather than of error. The result of the paper accommodates that the range of an instrument is finite, that measurements are discrete integer, unit counts, and that within the range $x = 0,1,2,...,m$ of measurements the probability of a measurement may *vanish*. In anticipation of this characterisation, consider the discrete Gaussian

$$P_{nix} = y_x / (\sum_{x=0}^{m} y_{nix}) = \{\exp[-(x-\mu_n)^2 / 2\sigma_i^2]\} / \gamma_{ni} \qquad (1)$$

where $x = 0,1,2,...,m$ are possible measurements, $y_{nix} = \{\exp[-(x-\mu_n)^2 / 2\sigma_i^2]\} / \sqrt{2\pi}\sigma$ is the ordinate of the continuous Gaussian, σ_i^2 is its variance and μ_n is the mean of the measurements of object n. The normalisation $\gamma_{ni} = \sum_{x=0}^{m} y_{nix}$ eliminates $1/\sqrt{2\pi}\sigma$ but retains σ_i^2 in Eq. (1).

4. Problematizing measurement: Georg Rasch (1901 – 1980)

In developing the Gaussian distribution, mathematicians took for granted that they had measurements. In contrast, and arising from his work in the social sciences, the Danish mathematician Rasch problematized measurement itself. In contrast to a representational characterisation which describes measurement as a process in which the operations on the measurements match the possible operations on the properties of the objects, for example amalgamating masses and adding their measurements [5], Rasch abstracted the principle of invariant comparisons: *that within a specified frame of reference, the comparison in terms of a real valued number should be invariant with respect to which instruments are used, and vice versa* [6]. Andrich [7] shows that this principle has consistent results with other approaches to understanding measurements and that it not only explains deterministic measurement, but in its probabilistic formulation, inherently subsumes distributions of replicated measurements.

If an instrument i has the measurement range $x = 0,1,2,...,m$, then the Rasch distribution is

$$P_{nix} = \{\exp[(x(m-x)/2)\Delta_i + x(\beta_n - \delta_i)]\} / \gamma_{ni} \qquad (2)$$

where P_{nix} is the probability of measurement x, β_n is the measure of object n, $\Delta_i = \tau_{i(x+1)} - \tau_{ix} > 0$ where τ_{ix}; $x = 0,1,2...,m$ are the m thresholds of instrument i which partition the continuum into $m+1$ equal units and where measurement x is the count of the number of thresholds exceeded , $\delta_i = (\sum_{x=1}^{m} \tau_{ix}) / m$ is the mean of the thresholds, and γ_{ni} is the normalising constant ensuring equation (1) is a probability distribution [8]. If $\beta_n = \delta_i$, that is β_n is at the centre of the instrument, then $E[X_{ni}] = m/2$, the distributions in equations (1) and (2) are identical and $\Delta_i = 1/\sigma_i^2$ [9]. Thus the Rasch distribution specialises to the discrete Gaussian and its unit is the inverse of the variance of the complementary continuous Gaussian. Andrich [9] shows an example in which the probability of extreme measurements 0 and m *vanish* and in which the Rasch and Gaussian distributions remain identical. This paper applies this observation to a real example and shows that in the range in which the Gaussian and Rasch distributions are identical, the integer count can be read directly as a measurement in the unit of the instrument just as it is in the natural sciences.

5. An example of measurement as in the natural sciences

The example involves the assessments of two, three-hour Mathematics examinations of 1093 Year 12 students in Western Australia who competed, together with performances in other disciplines, for entry into universities. In this paper they are referred to simply as Tests A and B. The possible scores of 0 to 100 are integers with a substantial number of scores having zero frequency. To study for Test B, it is necessary to study for Test A, but not the other way around. As a consequence, it is expected that Test B is more difficult and discriminates more, than Test A, that is, its unit is smaller. The two examinations together can be seen as two replications of the assessment of a higher order variable of mathematics knowledge. The syllabuses, the class of persons eligible to take the tests, and the construction and administration of the tests, form part of Rasch's *specified frame of reference*.

Table 1 shows estimates $(\hat{\delta}_i, \hat{\Delta}_i)$ for each test. The conditional maximum likelihood (ML) estimates using the algorithm described in Andrich and Luo [10], are independent of the distribution of the proficiencies β. It is evident that $\hat{\delta}_A + \hat{\delta}_B = 0$, which is the only identifying constraint required in the estimation. As expected, Test B is more difficult, and has a smaller unit, than Test A.

Table 1. Estimates of the test parameters.

Test	δ	Δ	$\sigma^2 = 1/\Delta$	$\sigma = 1/\sqrt{\Delta}$
A	-0.245	0.033	29.989	5.476
B	0.245	0.026	38.844	6.233

Figure 1 shows the plot of the scores against the estimates of the person proficiencies β for each possible integer score on each test. The proficiencies are also ML estimates, with the estimates of the test parameters taken as given. The plot also shows, for each test, the observed means highlighted by dots (●) for 10 class intervals. This graphical test of fit shows the observed and theoretical means are virtually identical. The value $m/2 = 50$, the centre score of each test, and the range of scores $20 \le x \le 80$, are also highlighted. The former is highlighted because it is the point where the Rasch distribution is algebraically identical to the Gaussian, and the latter is highlighted because it is the common range of scores which appear linear with the person estimates. Outside that range, which accounts for the finiteness of the scores, the relationship is curvilinear. However, there is a further justification for highlighting the range $20 \le x \le 80$.

Figure 2 shows the Rasch, discrete and continuous Gaussian distributions for two locations for Test A, $\beta = -1.246$, $E[x] = 20$ and $\beta = 0.775$, $E[x] = 80$. Two observations are relevant. First, in the left

Journal of Physics: Conference Series **1379** (2019) 012006 doi:10.1088/1742-6596/1379/1/012006

distribution, $E[x]=\mu=20$, the probability of the extreme response, $P_{nix}=0,\ x=0$, *vanishes*, and in the right distribution, $E[x]=\mu=80$, the extreme response $P_{nix}=0,\ x=100$ vanishes. Second, the three distributions overlap, with the two probabilities of the discrete distributions having values of the ordinates of the continuous Gaussian. All distributions in the range $20\leq x\leq 80$ have these features.

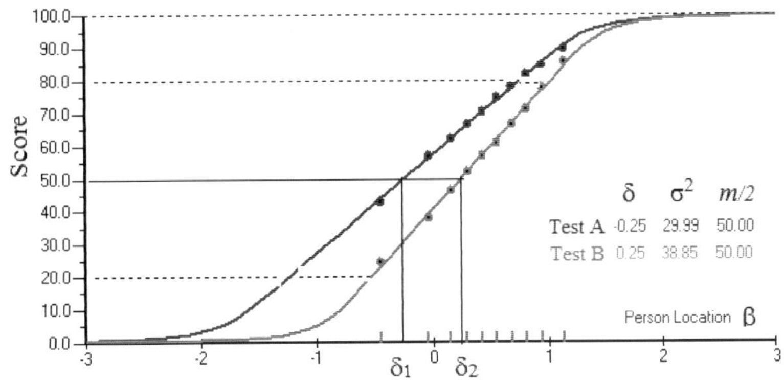

Figure 1. Plot of the observed score against the person estimate β.

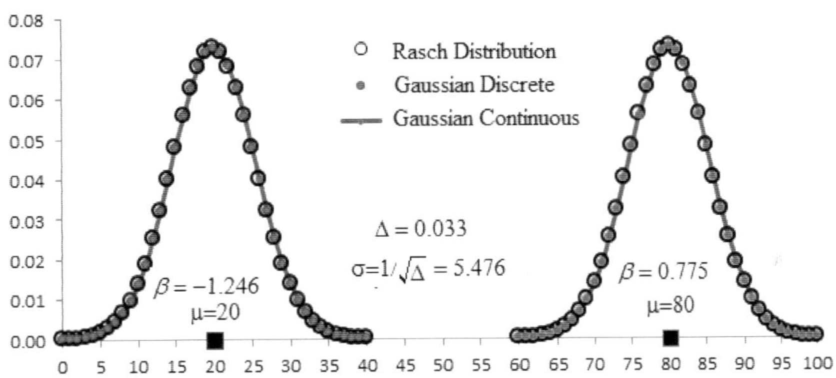

Figure 2. Distributions of Test A in which $P_{nix}=0$(vanishes), $x=0; x=100$ respectively.

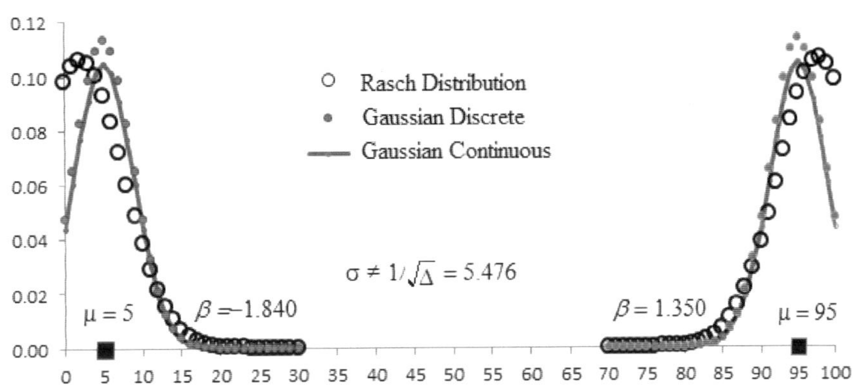

Figure 3. Distributions of Test A in which $P_{nix}\neq 0,\ x=0,\ 100$ respectively.

In contrast, figure 3 shows the distributions for two locations more extreme for Test A: $\beta = -1.840$, $E[x] = \mu = 5$ and $\beta = 1.350$, $E[x] = \mu = 95$. Neither of the features which hold in figure 2, hold in figure 3: the probability of an extreme response does *not vanish* and the three distributions diverge. Equivalent distributions for Test B are not shown.

6. Measurement within the range where the Gaussian holds

The paper now focusses on the range of scores $20 \le x \le 80$ for both tests within which the discrete Gaussian and Rasch distributions are identical and overlap with the continuous Gaussian, the range in which Gauss's assumption, that the extreme probabilities *vanish,* holds. The left panel of figure 4 shows the relationship between the score on each test and the estimated values of the person locations β. Both are perfectly linear. This may have been anticipated from figure 1. In addition, the slope of the respective lines is exactly their unit Δ. Thus for each successive integer score increment, the increment on the estimated variable β is exactly the unit distance Δ between the thresholds in the Rasch distribution of equation (2). The right panel of figure 4 shows the linear regression in the prediction of the score on Test A from an integer score on Test B. The slope of this regression line is the ratio of the units of the tests, Δ_B / Δ_A. Thus the reading of the integer score on the test in the range $20 \le x \le 80$, which is the count of the number thresholds in equal units exceeded, is directly analogous to reading degrees Centigrade on a thermometer, and then the conversion of scores between these two tests is directly analogous to converting degrees Centigrade to degrees Fahrenheit.

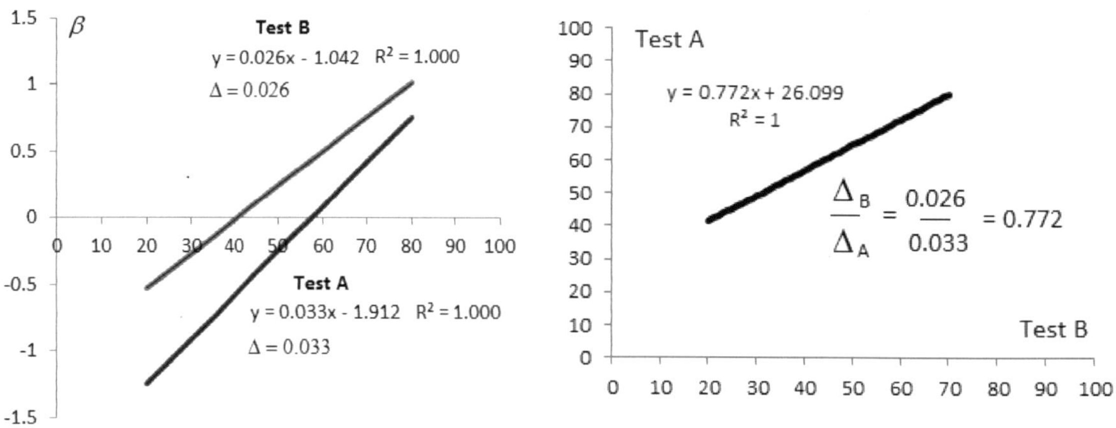

Figure 4. Conversion of each test's measurements to β (left graph); conversion of Test B integer measurement to Test A (right graph).

The Rasch model permits estimating person locations beyond the linear limits shown above, but that topic is not dealt with in this paper, which focusses on an example showing an equivalence between Rasch and natural science measurement.

7. Comment on the abstraction and realities of measurement

These concluding comments consider further the abstraction of measurement, and in particular not being wedded to the literal examples of concatenation considered in representational measurement. This abstraction is consistent with the Bureau of International Weights and Measures redefining the base units, such as the Kilogram, in terms of universal constants rather than an object. However, the requirement of invariant measurements with respect to instruments and the Gaussian as the criterion of uncertainty which is no more than random, remain.

First, Δ_A, Δ_B are estimated *average* unit distances between successive thresholds, not literally the distance between each pair of thresholds which is the same in their respective instruments. However, this simply applies the principle of probability distributions to hypothesised randomness, which can be tested using the Rasch model.

Second, the integer scores from 0 to 100 for each test were constructed from multiple items with different maximum scores for each item and the sum of the scores was presented to the nearest integer. Again the principle of randomness in cancelling what may be perceived as errors is implicit.

Third, there is no literal replication of measurements of the same person with the same instrument. In the development of the Gaussian, the number of examples with replicated measurements was small, no larger it seems than 27 [11]. In the above example, there are two measurements from two instruments. From the estimated relative locations and units of each instrument, the Rasch model is the probability distribution of inferred replications for any possible measurement of a person.

Fourth, it is relevant to emphasise the relationship between the measure of uncertainty and the Rasch and Gaussian distributions: the variance is the inverse of the size of the unit, showing that the smaller the unit, which implies greater precision, the greater the variance in the smaller unit.

As indicated in Section 1, measurement of physical properties requires substantial controls and accounting of confounding variables, which in turn is informed by understanding relevant scientific theories. Kuhn [1, p 189] stresses that *The road from scientific law to scientific measurement can rarely be travelled in the reverse direction* (italics in the original).

In the example of the two measurements of mathematics proficiency, there were also substantial controls which had an eye to something like measurement. The curriculum and syllabuses, which the teachers taught the students who studied for the high stakes examinations, were tightly specified, the exam writers were experienced and wrote questions consistent with the syllabus, the administration of the examination was closely controlled, and the markers were experienced and trained. In this example they also seemed able to produce instruments which measure in a constant unit across a substantial range. This construction and manifestation of a variable is consistent with the kinds of controls on both objects and instruments of measurements in the physical sciences. The difference between those and the above example seems to be primarily in the breadth of the frame of the reference to which the measurement can be applied.

References

[1] Kuhn T S 1961 The function of measurement in modern physical science *Isis* **52** 161-90
[2] JCGM 2008 *JCGM 100:2008 Evaluation of Measurement Data - Guide to the Expression of Uncertainty in Measurement* (GUM 1995 with minor corrections) (Joint Committee for Guides in Metrology)
[3] Eisenhart C 1983a Law of error I: development of the concept *Encyclopedia of Statistical Sciences* vol 4 ed S Kotz and NL Johnson (Toronto: Wiley) pp 530–47
[4] Eisenhart C 1983b Law of error II: Development of the concept *Encyclopedia of Statistical Sciences* vol 4 ed S Kotz and NL Johnson (Toronto: Wiley) pp 547–62
[5] Krantz D H, Luce R D, Suppes P and Tversky A 1971 *Foundations of Measurement* vol 1 (New York: Academic Press)
[6] Rasch G 1961 On general laws and the meaning of measurement in psychology *Proc. 4th Berkeley Symp. on Mathematical Statistics and Probability* vol 4 ed J Neyman (Berkeley CA: University of California) pp 321-34
[7] Andrich D 2018 Advances in social measurement: A Rasch measurement theory *Perceived Health and Adaptation in Chronic Disease: Stakes and Future Challenge* ed F Guillemin *et al.* (London: Routledge) chapter 7 pp 66–91
[8] Andrich D 1982 A two parameter extension of the Rasch model for ratings *Psychometrika* **47**(1) 105–13
[9] Andrich D 2018 On an identity between the Gaussian and Rasch measurement error distributions: making the role of the instrument explicit *J. Phys.: Conf. Ser.* (accepted for publication)
[10] Andrich D and Luo G 2003 Conditional estimation in the Rasch model for ordered response categories using principal components *J. Appl. Meas.* **4** 205–21

Journal of Physics: Conference Series **1379** (2019) 012006 doi:10.1088/1742-6596/1379/1/012006

[11] Stigler S M 1986 *The History of Statistics: The Measurement of Uncertainty before 1900* (Cambridge, Massachusetts: The Belknap Press of Harvard University Press)

Journal of Physics: Conference Series **1379** (2019) 012007 doi:10.1088/1742-6596/1379/1/012007

Is context the hidden spanner in the works of educational measurement: exploring the impact of context on mode of learning preferences

Elena Fitkov-Norris[1] **and Ara Yeghiazarian**

Kingston Business School, Kingston University, Kingston Hill, Kingston upon Thames KT2 7LB, UK

E-mail: e.fitkov-norris@kingston.ac.uk

Abstract. This paper examines the VARK learning preferences evaluation tool for evidence of context effect as a result of the questionnaire design using a common scenario stimulus to elicit learning preferences. Confirmatory Factor Analysis (CFA) is used to test the overall model fit of a multidimensional hierarchical model which included a context scenario effect and a non-hierarchical model which does not explicitly model a context effect. The results show that half of the scenarios in the VARK learning preferences questionnaire have significant context effect, which in turn introduce response bias. The presence of context response bias in some of the questions offers a possible explanation to the lack of consistent empirical evidence for the existence of preference styles flagged by educators, which contradicts findings from neuroscience and psychology of a convergence effect between learning modality preferences and learning approaches. The findings have significant implications for designers of learning preference tools in particular and educational measurement tools in general as they highlight the risks of the inadvertent introduction of possible response bias when implementing scenario-based evaluation tools.

1. Introduction

Measurement in educational research is continuing to thrive. For example, standardised testing in core subjects such as reading and mathematical attainment is now well established and applied throughout the entire length of educational career. In the meantime, the pervasive use of educational technology is making more data readily available, potentially shedding light on best educational practices through the use of meta analytic studies. However, the increased sample sizes and large data sets have failed to produce the empirical evidence to support some well-established and long held beliefs in what are best practices in education. For example, meta analytic studies into learning styles have failed to find empirical support to the hypothesis that adjusting the mode of presentation materials to the preferred mode of learning preference, also known as convergence theory, will ultimately lead to better educational attainment [1]. Despite the lack of empirical evidence, this learning preferences myth is thriving [2] and articles validating tools for assessing the strength of learning preference for visual, auditory, read/write and kinesthetic (VARK) inputs continue to be published [3] possibly due to the intuitive appeal of these concepts.

[1] To whom any correspondence should be addressed.

However, in direct contradiction to the empirical meta analytic findings, evidence has emerged from neuroscientific and psychological direct observational studies into information processing. For example, functional Magnetic Resonance Imaging (fMRI) scans have shown that the convergence theory holds true for people with strong visual preference, who have to convert words into pictures when presented with text input in order to understand the input. Similarly, the convergence theory holds for people with strong read/write text preference, who have to convert pictures into text to help with comprehension [4]. Further support to the existence of a convergence effect in preference for visual or text based (read/write) information intake is provided by eye-tracking studies in psychology, showing that visualizers (students with a preference for pictures and diagrams) and verbalisers (students with preference for text) generally examine the areas on the screen where information is presented in line with their preference and for instance visualizers tend to focus for longer on information rich areas that hold visual cues (such as diagrams) than verbalisers [5]. Conversely, verbalisers tend to focus on areas that hold textual cues (such a text). The presence of this contradictory evidence suggests that there could exist a confounding variable(s) or process(es) that is unaccounted for in the design of the studies.

One possible explanation for the apparent contradiction between empirical evidence gathered in the classroom and functional and processing empirical evidence gathered by psychologists and neuroscientists in controlled lab conditions is the impact of context. The confounding effect of context has been reported in a number of meta analytic studies in a number of diverse areas, such as energy efficiency [6], healthcare [7], ecology [8] and advertising [9], leading to apparently contradictory effects of interventions and affecting the significance of hypothesized relationships.

This article tests whether context could play a significant confounding role in the measurement of learning preferences using questionnaires and thus potentially leading to contradictory findings to the ones made by direct observation, this article examines the responses given to the VARK learning preferences questionnaire for consistency across 16 different scenario contexts with a view of providing some insights into the impact of context on the strength of self-reported mode of learning preference. Confirmatory Factor Analysis (CFA) is used to compare the fit of a model that takes into account the context effect and a model that do not explicitly take into account the context effect.

The article is organised as follows: the next section gives a formal definition of context and an overview of the literature on the empirical evidence for the impact of context and other design factors that affect the reliability self-reported measuring instruments in the area of educational research. Section 3 discusses the methodology used to evaluate the context effect for the VARK preference measurement tool. Section 4 reports the results from the analysis and discusses the implications from the findings in the context of educational measurement. The final section summarises the research findings and makes recommendations for further research.

2. Context in educational measurement

2.1. Definition of context

Awareness of the role of context in education is growing and educators have recommended approaches such as Problem Based Learning (PBL) in a range of contexts as a better way to encourage acquisition and retention of transferable knowledge [10, 11]. However, despite this growing awareness, there is no universally accepted definition of context in educational research and researchers utilise a variety of terms such as context effect in decision sciences when evaluating different alternatives [12] or context specificity to refer to the impact of context on educational attainment [11, 13] or personality tests [14]. For example, context specificity has been demonstrated across a range of teaching related situations such as perceived teacher enthusiasm [13], learning motivation [15], doctoral supervision [16]. More recent works such as [16, 17] has highlighted the issues arising from decontextualization and suggests that applying the lens of decontextualization to educational research leads to poor understanding of the complexities of the educational interactions. In fact, empirical evidence is emerging showing that context is best viewed as a complex multidimensional construct, rather than a single factor [13, 18]. To reflect the multidimensional nature of context this this paper will follow the definition by Parker et. al.

[19, page 46]. who define context as something that "... can be derived from anything that is significant in a given moment and potentially including the environment, an item within that environment, a user, or even an observer." This is an all-encompassing definition that can cover a range of potential situations and scenarios, as well as reflect various research and measurement tool designs.

2.2. The VARK learning preference measurement tool

The VARK learning modalities questionnaire measures the degree to which learners prefer to absorb information in visual (V), auditory (A), Read/Write (R) and Kinaesthetic (K) form in different educational settings [20]. Visual form includes preference for diagrams, charts to explain ideas as opposed to text which is the preferred method for people with high Read/Write preference or spoken words which are the preference for people with high Auditory preference. People with high kinaesthetic preference prefer to use experience, examples or trial and error. The VARK questionnaire (http://vark-learn.com/the-vark-questionnaire/) consists of 16 items and each item has four possible answer, each indicating a particular preference. Users are encouraged tick all items that apply in each case. Two questions taken from an earlier version (2014) of the questionnaire have been shown below:

Q1: A group of tourists wants to learn about the parks or wildlife reserves in your area. You would:
a. show them maps and internet pictures. (answer represents V)
b. talk about, or arrange a talk for them about parks or wildlife reserves. (answer represents A)
c. give them a book or pamphlets about the parks or wildlife reserves. (answer represents R)
d. take them to a park or wildlife reserve and walk with them. (answer represents K)

Q2: You have a problem with your heart. You would prefer that the doctor:
a. used a plastic model to show what was wrong. (answer represents K)
b. showed you a diagram of what was wrong. (answer represents V)
c. gave you something to read to explain what was wrong. (answer represents R)
d. described what was wrong. (answer represents A)

Following on from the definitions of context provided in section 2.1, each question is positioned in a particular context (encountering a group of tourists (Q1) or having a problem with one's heart (Q2)) that could potentially bias the responses of participants. For, example, in the context of encountering a group of tourists, an introvert may choose to act differently to an extrovert, despite their actual learning modality preference. In the context of concern about one's heart, a respondent may wish to utilise all four options given, in order to maximise their understanding. Context is also present in the range of response options. For example, a male or a female respondent could have different risk propensity to walking with strangers, which could affect their response to item Q1(d). In fact, attempts to validate the questionnaire have shown that fitting a model that takes into account the scenarios leads to a better overall model fit [21], although the relative contribution of each scenario was not examined or reported.

Despite the potential to introduce bias, contextualization can have a positive impact on measurement and, for example, it was found to reduce the error variances in personality measurement instruments [14]. Research has also shown that the common stimuli (context) design across different dimensions can mitigate the negative impact of local item dependence although the single stimuli (context) design across a single dimension can introduce a significant bias [22]. Thus, untangling the impact of context on the validity of measurement tools is not trivial as the questionnaire design can also play a significant role in biasing responses.

3. Methodology

This research utilizes Fleming's VARK questionnaire [23] to indicate the mix and strength of students' preferred learning modes[2].

[2] This research used the 2014 version of the VARK Questionnaire. The Questionnaire has had a number of options and questions replaced as well as the wording of some options and question-stems changed and is now in version 8.0. Those changes will have affected some of the results shown.

The sample consists of responses by 17,413 participants who complete the VARK questionnaire on the VARK website between June and December 2014.

The majority of respondents were students (89 %) and female (65 %). 52 % of respondents were aged between 19 34 and the majority were classified as having a single preference mode (37 %), followed by respondents who utilized all 4 preferences (35 %). A smaller proportion of respondents utilized either 2 (15 %) or 3 modes (13 %) respectively.

The data were analysed using R version 3.4.4 (2018-03-15) -- "Someone to Lean On" [24] using the "mirt" package [25]. A single layer multidimensional confirmatory factor model which hypothesized the existence of the four main factors (Visual, Auditory, Read/Write and Kinesthetic) but did not take into account the existence of scenarios, was fitted to the data first. This was followed by the fitting of a hierarchical multidimensional model which considered the existence of the four main factors (Visual, Auditory, Read/Write and Kinesthetic) and 16 secondary scenario factors that introduced different contexts to each of the main dimensions. The strength of context effect for each scenario was evaluated using each secondary factor's eigen value. Figure 1 is a graphical representation of the examined two-tier hierarchical model with four primary latent VARK dimensions and 16 secondary or specific latent scenario dimensions, together with the 64-indicator variables (16 per each dimension). The non-hierarchical model consists of just four primary latent VARK dimensions, each with 16 indicator variables.

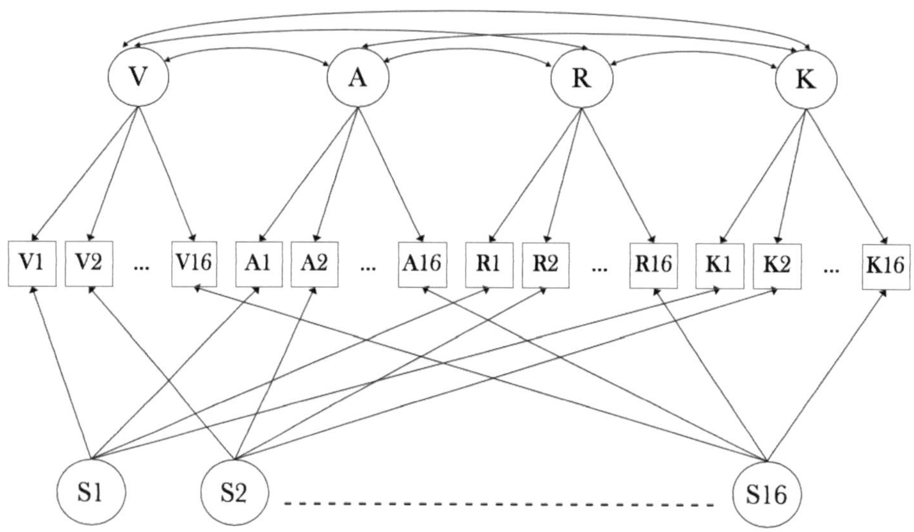

Figure 1. Two-tier hierarchical model with secondary context effects.

Following from [26] the conditional probability of correct endorsement response for a dichotomously scored item y_{ij} in a hierarchical model with p primary dimensions and s specific or secondary dimensions is given by:

$$P_j(y_{ij} = 1|\boldsymbol{\eta}_i, \xi_{is}, \boldsymbol{\theta}) = (1 + \exp\{-[\alpha_i(\boldsymbol{\theta}) + [\boldsymbol{\beta}_i(\boldsymbol{\theta})]'\boldsymbol{\eta}_i + \beta_{is}(\boldsymbol{\theta})\xi_{is}]\})^{-1} \quad (1)$$

where α_i is the intercept, $\boldsymbol{\beta}_i$ is the $p \times 1$ vector of item slopes on the primary factor vector $\boldsymbol{\eta}_i$, β_{is} is the item slope on specific factor ξ_{is} and $\boldsymbol{\theta}$ is a vector of all estimable and/or structural parameters in the two-tier model. The conditional probability for the incorrect/non-endorsement response is $P_j(y_{ij} = 0|\boldsymbol{\eta}_i, \xi_{is}, \boldsymbol{\theta}) = 1 - P_j(y_{ij} = 1|\boldsymbol{\eta}_i, \xi_{is}, \boldsymbol{\theta})$.

4. Results and discussion

The results from the two models are summarized in table 1. The two-tier hierarchical model with four primary dimensions for Visual, Auditory, Read/Write and Kinesthetics preferences and 16 secondary context dimensions representing a different context fits the data better than the non-hierarchical multidimensional model that does not consider context effects. In particular, the two-tier hierarchical model has lower Root Mean Square Error of Approximation (RMSEA) with a 95% CI [.04159, .04217] meets the standard requirement of being below 0.7 [27]. Furthermore, the two-tier hierarchical model has a higher Tucker Lewis Index (TLI) which at .87366 is closer to the desired value of 1. In addition, the two-tier model also has a higher Comparative Fit Index (CFI) measure and significantly lower Bayesian Information Criterion (BIC) value. Both models have eigenvalues greater than 1 for the four main factors of Visual, Auditory, Read/Write and Kinaesthetic preferences, suggesting that the 64 items measure a 4-dimensional structure. The magnitude of the factor eigenvalues for the secondary factors in the hierarchical model is greater than 1 for eight out of the 16 scenarios, suggesting a significant context effect. Further textual analysis of the scenarios with significant context effect may provide valuable insights into the type of context that is prone to introduce bias and so inform good design practices, although it falls beyond the scope of this work.

Table 1. Model results comparison.

Model Fit Measures	Non-Hierarchical Model	Two-tier hierarchical model with context effects
RMSEA	0.05345	0.04188
RMSEA LCL	0.05316	0.04159
RMSEA UCL	0.05373	0.04217
TLI	0.77963	0.86467
CFI	0.78728	0.87366
AIC	1337249	1304551
BIC	1338290	1306089 [a]

[a] Significant at 1% level

The results show that the context effect is significant and present in half of the scenarios used in the VARK questionnaire. This suggests that the strength of the context effect on the expression of learning preferences is not ubiquitous but specific to each situation. The result is surprising on the one hand, given that prior research has indicated that utilising a single stimulus design across dimensions reduces inter-item dependence significantly [22]. Thus, by adopting a single stimulus design and measuring the strength of all four preferences (Visual, Auditory, Read/Write and Kinaesthetic) within each scenario context, should have mitigated the impact of context. However, on the other hand this finding is in line with empirical research in the field of education which has identified differential context effect functioning in a range of measurements such as teacher enthusiasm [13], student motivation in online distance learning courses [15], dissertation supervisory practices [16] and memory recall tasks [11].

The presence of context response bias in some of the questions also offers a possible explanation to the lack of consistent empirical evidence for the existence of convergence effect between learning modality preferences and learning approaches flagged by educators [1]. Given that the context is not ubiquitous, its presence is challenging to identify and therefore its impact in meta analytical studies my overpower the learning modalities or learning preferences convergence effect. In turn this will provide some justification for contradictory findings from neuroscience [4] and psychology [5] on the presence of such convergence effect in the case of people with strong learning modality preference. This finding has overarching implications for the future design decisions of preference measurement instruments such as the VARK questionnaire, suggesting that utilization of single stimulus design across multiple dimensions may be necessary, but not sufficient requirement to avoid the impact of context bias on user responses. The results emphasise the inherent complexity of building reliable measuring tools in the presence of significant context effects and highlight the need for further investigation into the impact of context in educational research.

5. Conclusions

This paper examined the VARK learning preferences evaluation tool for evidence of context effect as a result of the questionnaire design using a common scenario stimulus to elicit learning preferences. The results suggest that half of the scenarios have significant evidence of context effect, which in turn introduce response bias. The presence of context effect has implications for the future design decisions of preference measurement instruments such as the VARK questionnaire, suggesting that utilization of single stimulus design across multiple dimensions may be not be sufficient to avoid the impact of context bias on user responses. These findings have significant implications for designers of learning preference tools in particular and educational measurement tools in general as they highlight the risks of introducing a significant response bias when implementing scenario-based evaluation tools.

References

[1] Husmann P, O'Loughlin V 2018 *Anat. Sci. Educ.* **12** 6–19
[2] Newton P 2015 *Front. Psychol.* **6** 491–5
[3] Stowe K, Clinebell S 2016 *J. Int. Bus.* **26** 258–72
[4] Kraemer D, Hamilton R, Messing S, DeSantis J, Thompson-Schill S 2014 *Front. Hum. Neurosci.* **8** 1–9
[5] Koć-Januchta M, Höffler T, Thoma G, Prechtl H, Leutner D 2017 *Comput. Hum. Behav.* C **68** 170–9
[6] Maidment C, Jones C, Webb T, Hathway E, Gilbertson J 2014 *Energy Policy* C **65** 583–93
[7] Kaplan H, Brady P, Dritz M, Hooper D, Linam W, Froehle C, Margolis, P 2010 *Milbank Q.* **88** 500–59
[8] Meli P, Rey Benayas J, Balvanera P, Martínez Ramos M 2014 *PLoS ONE* **9** 93507–10
[9] Kwon E, King K, Nyilasy G, Reid L 2019 *J. Advert. Res.* 59(1) 99–128
[10] Alkhasawneh I, Mrayyan M, Docherty C, Alashram S, Yousef H 2008 *Nurs. Educ. Today* **28** 572–9
[11] Smith S, Handy J 2014 *J. Exp. Psychol. Learn. Mem. Cogn.* **40** 1582–93
[12] Lee C-F, Chuang S-C, Chiu C-K, Lan K-H. 2017 *Curr. Psychol.* **36** 1–18
[13] Kunter M, Frenzel A, Nagy G, Baumert J, Pekrun R 2011 *Contemp. Educ. Psychol.* **36** 289–301
[14] Robie C, Schmit M, Ryan A, Zickar M. 2000 *Organ. Res. Methods* **3** 348–65
[15] Hartnett M, George A, Dron J 2011 *IRRODL* **12(6)** 20–38
[16] Bastalich W 2015 *Stud. High. Educ.* **42** 1145–57
[17] Forissier T, Bourdeau J, Mazabraud Y, Nkambou R 2013 *Context* **LNAI8175** 330–5
[18] Koens F, Mann K, Custers E, Cate Ten O 2005 *Med. Educ.* **39** 1243–9
[19] Parker J, Hollister D, Gonzalez A, Brézillon P, Parker S 2013 *Context***LNAI8175** 45–58
[20] Fleming N, Mills C. Not another inventory, rather a catalyst for reflection. 1992
[21] Leite W, Svinicki M, Shi Y 2010 *Educ. Psychol Meas.* **70** 323–39
[22] Wang W-C, Cheng Y-Y, Wilson M 2016 *Educ. Psychol. Meas.* **65** 5–27
[23] Fleming N. VARK: a Guide to Learning Styles. The VARK Questionnaire. 2001
[24] R Core Team. 2018 Available from: https://www.R-project.org/
[25] Chalmers R 2012 *J. Stat. Soft.* **48** 1–29
[26] Bonifay W 2014 *Handbook of Item Response Theory Modeling* Routlege
[27] Hair J Jr, Black W, Babin B, Anderson R 2009 *Multivariate Data Analysis* Pearson

Journal of Physics: Conference Series **1379** (2019) 012008 doi:10.1088/1742-6596/1379/1/012008

Musing on extreme quantity values in physics and the problem of removing infinity

Franco Pavese

Independent Scientist, 10139 Torino, Italy

frpavese@gmail.com

Abstract. Many physical quantities display range values apparently extending to infinity (unbounded on one or on both sides). In this respect, unit systems and measurement conventions do not place any constraint to their validity for a maximum (or minimum) value. In general, this happens because such extreme values are far from being reached on the earth or yet are reached in experimental settings. Nevertheless, the issue of extreme values (not in the usual mathematical analysis meaning here) is not irrelevant, since the same units are used also in countless fields of physics, chemistry or technology where extreme values do occur—namely, in the description of the universe in one frame, and in pico/nano-scale or particle physics in another. The issue, of direct interest also of measurement science and specifically of metrology, is discussed here illustrating, as an example, our currently accepted concept of temperature, i.e., the kelvin temperature scale based on Lord Kelvin's second definition, which encompasses the full range between bounds (0, +∞). In general, the occurrence of infinite values in physical equations, such as singularities in the description of black holes, is a painstaking problem that causes many theories to break down and/or being incapable of describing extreme events. Different methods, such as re-normalization (scaling) or logistic/geometrical, have been used in the assessment of physical observables in order to avoid the undesirable infinity.

1. Introduction

Physicists have already raised the point of infinity long since. For example, almost one century ago the physicist Bridgman discussed the issue for time in the frame of measurement science and, specifically, of metrology ("What is the meaning in saying that an electron when colliding with a certain atom is brought to rest in 10^{-18} s?" in 1927) [1] and length ("What is the possible meaning of the statement that the diameter of an electron is 10^{-13} cm?" in 1955), [2] then opting for operational definitions.

The same issue also attracted the philosophers of science. For example, in a recent book, Chang [3] asked "Is a [scale] definition valid for a quantity's full range?" and introduced the concept of "metrological extension", then proposing a "compatibility requirement" for measurement standards in different ranges, most often satisfied by "patching up disconnected standards". However, in principle, only a theory-based definition—i.e. a model-based definition with operational method(s) ("realisations") available—might satisfy the necessary conditions.

In the following, the issue is discussed illustrating, as an example, our currently accepted concept of temperature, i.e., the kelvin temperature scale based on Lord Kelvin's second definition, which encompasses the full range between bounds (0, +∞). In the following, some modern viewpoints in the frames where the present definition is not applicable are introduced.

In general, in physical equations the occurrence of infinite values, such as, e.g., singularity in the description of black holes, is a painstaking problem that causes many theories to break down and/or being incapable of describing extreme events. Different methods, such as re-normalization (scaling), have been used in the assessment of physical observables in order to avoid the undesirable infinity.

2. An example: the concept of temperature at extreme values

Let us take temperature as a good example of the above cases, for both its asymptotes for extreme low and extreme high values of the scale (present experimental limits: $\approx 10^{-9}$ K and $\approx 3 \cdot 10^{9}$ K). The question applies to the present kelvin scale, based on Lord Kelvin's second (II) definition (indicated in the following with T^{II}), based, as the first one, to (mechanical) work, historically linked to the use of a gaseous media and heat for producing work, i.e., "power motion". [4] The paper is specifically focusing on the lower extreme only, $-\infty$ (as in his first [I] definition, T^{I} —see Table 1), $\rightarrow 0$ in the kelvin scale T^{II}.

Table 1. Temperature values in Kelvin's first definition T^{I} and second definition T^{II} [a].

T^{II}/K	T^{I}/L	T^{II}/K	T^{I}/L
$\rightarrow +\infty$ [b]	$\rightarrow +\infty$ [b]	100	–322
1 000 000	2 630	10	–1 060
100 000	1 892	1	–1 798
10 000	1 154	0.1	–2 536
1 000	416	0.01	–3 274
373.15	100	0.001	–4 012
273.15	0	$\rightarrow 0$ [b]	$\rightarrow -\infty$ [b]

[a] T^{I} in unit L, T^{II} in unit K.
[b] But see text about approaching the asymptotes.

For the issue of the upper extreme, $+\infty$, it is enough to recall here that, quite far from experimental science, contemporary models of physical cosmology postulate an "absolute hot", i.e. that the highest possible temperature is the Planck temperature, 1.416 808(33) 10^{32} K (energy of the Planck mass for Boltzmann constant $k_B = 1$). [5] Above about 10^{32} K, particle energies become so large that gravitational forces between them would become as strong as other fundamental forces according to current theories. A quantum theory of gravity would be required [6] ("The point at which our physical theories run into most serious difficulties is that where matter reaches a temperature of approximately 10^{32} degrees, also known as Planck's temperature. The extreme density of radiation emitted at this temperature creates a disproportionately intense field of gravity. To go even farther back, a quantum theory of gravity would be necessary, but such a theory has yet to be written." [7] At those high energies, consequences may arise for some electromagnetic units of the revised SI, [8, 9] due to an increasing variability with energy of the hyperfine constant α value [10]).

Let us start now a bit of history, then this paper will tackle some of the current and modern views. In order to keep acceptable the length of this paper, the reader is directed for specific contents to the relevant references.

One can ask, for example, if mechanical work does really fit all needs of thermometry and the corresponding metrology. Actually, since the Caloric concept was defeated, the modern way-out is to use rather Energy, [11, 12] inclusive of Mechanical Work and Heat. [13, 14, 15] Energy is a "subtle concept", [16] possibly too subtle and pervasive, and so not so easy to define. For replacing "living force", Young proposed the term "energy" since 1807, basically meaning "potential work". Then Lord Kelvin introduced it formally only in 1851 (later, in 1865, Clausius introduced the term 'entropy'). Feynman popular definition is: "Energy is that-which-is-conserved". For Coopersmith, [16] "the energy of a system is the capacity of the system to do work"—a definition that requires the concept of 'force', not anymore popular in some branches of physics ("the concept of force is conspicuously absent from our [physical] most advanced formulations of the basic laws", e.g., see [17]); also, "energy has extensive (entropy) and intensive (temperature) attributes" [16] — but temperature is not an attribute of energy in the same way that in a reservoir level is not an attribute of an amount of substance. The recent revision of the definition of the International System of Units (SI) [18] endorses an energy-based temperature, where T^{II} is linear in energy, as indicated before: $\Delta T = 1$ K for $\Delta Q = k_B$, with $[k_B] = [\text{J K}^{-1}]$, the unit of a quantity called heat capacity.

2.1. Modern views on the classical domain

A well-known picture of statistical mechanics, here reported in figure 1, indicates the basic three statistical models on which the T^{II} kelvin scale is presently based at the lowest temperatures, where the statistics splits into different possibilities; note that there classical Boltzmann statistics are just limiting values of either the Fermi and Bose branches at low occupancy of available quantum states. This fact was commented since many years. For example, Simon, [19] while appreciating the Lord Kelvin First Definition (in more recent literature [20–24] one can find other examples of illustrations, and possible redeeming, of the Lord Kelvin first definition), noted specific basic problems related to the Second Definition in approaching the "absolute zero": (i) the modern justification of the choice of T^{II} is the kinetic theory and statistical mechanics. However, for $T^{II} \rightarrow 0$ one reaches a point where the statistical hypotheses are not anymore respected (and before that fluctuations occur); (ii) the above theories are normally dealing with material's lattice, while for $T^{II} \rightarrow 0$ one needs to distinguish between specific sub-systems. Actually, going toward $T^{II} \rightarrow 0$, energy is going toward zero logarithmically—and the level $E =: 0$ is unreachable except possibly for the whole universe energy balance, according to some recent theories (e.g., [25]).

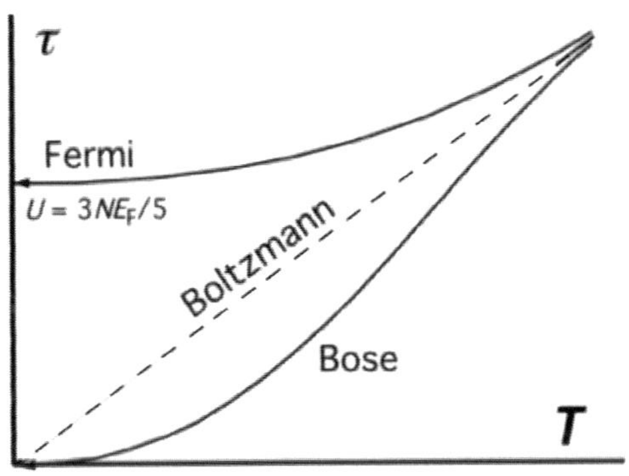

Figure 1. Statistical models for gas toward 0 K: for Bose gas $U \propto T^{5/2}$. The arrows indicate that the $T^{II} = 0$ cannot be reached. (modified from [26])

2.2. Two contemporary views on possible future temperature concepts

In nano-thermodynamics and the quantum frame, the recent extension of experimental work and technologies to very small dimensions (nano-technologies) and to very low temperatures (nano-temperatures), prompted new problems and the need of rethinking the very concept of temperature in the lowest range. When the concept of temperature does not apply to the whole system, the concept of "local temperature" is introduced, and one should study the "minimal length scales for the existence of local temperature" (e.g., see [27–29]). In these studies it is claimed, for example, that "This length scale is found to be constant for temperatures above the Debye temperature and proportional to T^{-3} below" so that "high temperatures can exist quite locally, while low temperatures exist on larger scales only" and, e.g., "in quasi one-dimensional systems, like carbon-nanotubes, room temperatures (300 K) exist on length scales of 1 mm, while very low temperatures (10 K) can only exist on scales larger than 1 mm". [27]

More generally concerning nano-thermodynamics, [30] and "small systems", [31] the issue was initially prompted by studies on thermodynamic "fluctuations", experimentally observed only since

1992, before the nanotechnology field started to develop. Several theories have been developed, [29, 32] like the non-extensive statistical mechanics, Hill's theory, [30] and tensorial approach. These studies also involve an effort to reconcile the quantum with the classical thermodynamics, [33] with controversial positions, similarity between quantum mechanics and thermodynamics. (e.g., see [34, 35]) It is found that if the Clausius equality is imposed on the Shannon entropy and the analogue of the quantity of heat, then the value of the Shannon entropy comes to formally coincide with that of the von Neumann entropy of the canonical density matrix, and pure-state quantum mechanics apparently transmutes into quantum thermodynamics. The corresponding quantum Carnot cycle of a simple two-state model of a particle confined in a one-dimensional infinite potential well was studied. However, some authors (e.g. [35]) contended that the statement is incorrect. In particular, they claim to have proved that the state at the beginning of the cycle is mixed due to the process of measuring energy. The imposition of the Clausius equality allows the connection between quantum mechanics and thermodynamics, thus resulting in quantum thermodynamics. An asserted experimental evidence of connection of the classical to quantum world is also available. [36]

Another approach (especially [37-39, 40]), also bringing to the analysis of negative temperatures, involves the concept of temperature under another perspective, with the introduction of a Gibbs thermodynamic temperature T_G, alternative to the Boltzmann one, T or T_G, an issue still controversial too (e.g., see [41–43]). For classical systems with many degrees of freedom, the difference in the value of the temperature based on entropies S_B and S_G is considered negligible. Yet, concerning entropy, some findings indicate that Gibbs entropy satisfies the three fundamental laws, while Boltzmann does not. [39] However, other authors [44] argue that Gibbs' entropy fails to satisfy a basic requirement of thermodynamics, that when two bodies are in thermal equilibrium they should be at the same temperature, while Boltzmann's one does.

The above discussion involves, as a consequence the acceptance, or not (in T^{II}), of negative temperatures—e.g., see [45, 46]. A recent paper [47], introducing generalised entropy, intended to be inclusive of Gibbs, Boltzmann and Shannon definitions, supports the latter position.

For sure one can already state that no "ultimate" solution exists. Any new Kuhn's "revolution" can provide, in one year or in 10^X years, new knowledge that, while extending the range where the concept of temperature can be managed, could also innovate, at least partially, in the ranges where today we think to be confident that no innovation is needed. This evidence already exists. After all, the previous concepts of time (subjective—of the observer), space (e.g., force (e.g., [18]), vacuum (e.g., [48]) — and related ones— and even 'ether' (e.g., [49]), 'universe' (e.g., [25] and space: "the space of astronomy is not a physical space of meter sticks, but is a space of light waves" [1, 50]),) have already been revisited.

3. Infinity in general physics

In their theories and models dealing with formulas that describe finite, measurable quantities, physicists overlook the occurrence of unwished infinite values. Indeed, with the exception of various forms of conformal infinity [51, 52], mathematical infinity (indeterminate infinite results in which, e.g., solutions of the gravitational field equations cannot be continued, [53] prevents scientific issues to provide practical formulas that correspond to, or at least approximate, the real observables). For example, in the case of bodies with infinite gravitational mass and/or energy, equations become intractable and useless, since their results would be always the same, regardless of objects' position, mass and movement. In some cases, infinite results mean that a theory is approaching the point where it fails. Therefore, although infinity can be used in physics, scientists require for practical purposes the final result to be physically meaningful: e.g., in quantum field theory, infinities are treated through procedures such as renormalization [54].

4. Infinity as a straight line in a geometrical approach

As stated above, unqualified infinity cannot be any of the physical observables that one either can assess or measure: when one sets out to investigate the infinity, one must leap beyond simple physical concepts and use mathematics. A way to use mathematical and geometrical features to undertake physical infinity is illustrated in [55]. If one wants to assess finite physical measurements leading to infinity, one needs at first to consider finite mathematical figures and topological manifolds, together with their features and relations. Next, one must apply these relations in a projective way. Thirdly, one must thereafter, in a still more highly transformed way, apply the relations of these infinite figures to the general concept of mathematical infinity, which is altogether independent even of all figures and manifolds.

Let us start, [55, 56] with the picture of mathematical infinite, which will be represented by a straight line. One can maintain that, if there were an infinite line, it would be a straight one, or, for example, an infinite triangle, circle or sphere. Since the latter three figures display infinite sides, as will be shown, they can also be described in terms of infinite lines. First of all, an infinite line would be a straight one. The circle's diameter is a straight line, and its circumference is a curved line greater than the diameter. If the curved line becomes less curved in proportion to the increased circle's circumference, then the maximum circle's circumference, which cannot be greater, is minimally curved and therefore maximally straight (Figure 2. *A:* for a positive-curvature manifold; *B:* for a positive-negative curvature manifold. [57]). Indeed, in the figure, the arcs of the larger circle are less curved than the smaller ones. Therefore, the straight line will be the arc of the maximum circle, which cannot be greater. An infinite line is necessarily the straightest; and to it no curvature is opposed. In the same way, every manifold with positive curvature, such as, for example, a triangle, or a circumference, or a sphere, can be described in terms of an infinite line standing for a maximum triangle, or a maximum circle, or a maximum sphere. In fact, an infinite line is whatever is present in the curvature of a finite line: a line finite in length can be longer and straighter; therefore the maximum line is the longest and straightest. If a finite line can describe figures, and if an infinite line is all-the-things-with-respect-to-which a finite line is in infinity, then it follows that an infinite line stands also for a triangle, a circle, and a sphere.

How is it possible that an infinite line is a side of a triangle? Since any two sides of any triangle cannot, if conjoined, be shorter than the third, this means that, in the case of a triangle whose one side is infinite, the other two sides are not shorter, i.e., they are both infinite. Further, since there cannot be more than one infinite thing, an infinite triangle cannot be composed of a plurality of lines, even though it is the greatest and simplest triangle. And because it is a triangle—something which it cannot be without three lines—it will be necessary that the one infinite line be three lines, and that the three lines be one most simple line. And similarly, regarding the angles: for there will be only one infinite angle; and this angle is three angles, and the three angles are one angle. Nor will this maximum triangle be composed of sides and angles; rather, both the infinite line and angle are one and the same thing, so that the line is the angle, because the triangle is the line. The larger the one angle is, the smaller are the other two. Now, any one angle can be increased almost but not completely up to the size of two right angles. Nevertheless, let us make the hypothesis that it is increased completely up to the size of two right angles, while the triangle remains nonetheless a triangle. In that case, it will be obvious that the triangle has one angle that comprises the three angles and that the three angles are one. In the same manner, one can state that a triangle is a line and an infinite line is a maximum triangle. For any two sides of a quantitative triangle are, if conjoined, as much longer than the third side as the angle which they form is smaller than two right angles. Hence, the larger the angle is, the less the lines and the smaller its surface. Therefore, if, by hypothesis an angle could be two right angles, the whole triangle would be resolved into a simple line. Hereby it is evident that an infinite line is a maximum triangle. Next, by applying the same reasoning and the proper rotations, it is feasible to show that an infinite triangle is also an infinite circle and an infinite sphere.

In sum, an infinite line has been shown to be all that which is in the possibility of every finite line and manifold: a triangle is educed from a line, and an infinite line from an infinite triangle. Hence, an important speculative consideration can be inferred: infinity is correlated with finite manifolds.

Because infinite curvature is infinite straightness, this means that an infinite manifold can be described in opposite terms: it is not a thing and is not any other thing; it is not here and is not there; it is unqualifiedly free from all things and is beyond all things; is above the negation of all things. By a physicist's standpoint, this explains why physical theories leading to infinite values are awfully problematic and difficult to cope with.

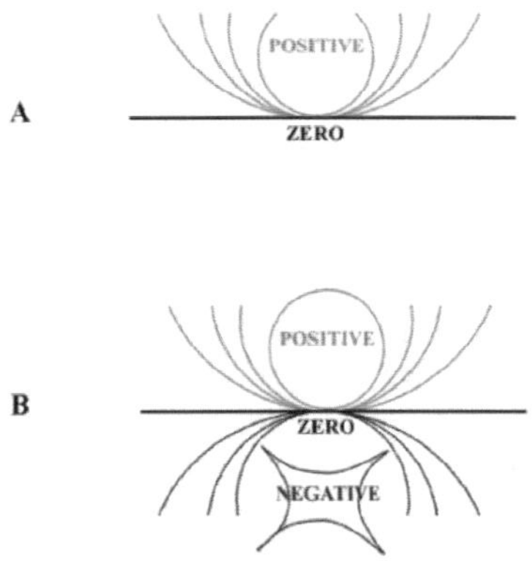

Figure 2. Manifolds toward infinity.
A: given a physical system described by progressively increasing curves on a positive-curvature manifold, the occurrence of infinity (straight line) can be removed by taking into account progressively decreasing curves on a negative-curvature manifold.
B: by placing physical observables or equations on a toroidal manifold, one achieves a correspondence between positive and negative curvatures, thus erasing the unwanted occurrence of infinity (from [57]).

It is assumed by physicists that, due to pragmatic issues, no measurable quantity or event has infinite values. Indeed, any physical theory needs to provide operational tools that correspond to, or at least approximate, reality. This also reflects on the corresponding models and methods of measurement science, in particular of metrology.

To solve the problem of the occurrence of infinity in physical equations and quantities novel conceptual frameworks are needed.

References

[1] Bridgman P W 1927 *The Logic of Modern Physics* (New York: Macmillan)

[2] Bridgman P W 1955 *Reflections of a Physicist* (New York: Philosophical Library)

[3] Chang H 2004 *Inventing Temperature: Measurement and Scientific Progress* (Oxford: Oxford University Press)

[4] Pavese F and Molinar Min Beciet G F 2012 *Modern Gas-Based Temperature and Pressure Measurement* (New York: Springer) 2nd ed (1st ed 1990)

[5] Mohr P J, Newell D B and Taylor B N 2014 2014 CODATA recommended values of the fundamental constants *Rev. Mod. Phys.* **88**(2) (arXiv:at-phys-1507.07956)

[6] Atick J J and Witten E 1988 The Hagedorn transition and the number of degrees of freedom of string theory *Nucl. Phys.* B **310** 291

[7] Reeves H 1991 *The Hour of Our Delight: Cosmic Evolution, Order and Complexity* (Oxford: W H Freeman Company)

[8] BIPM/CCU 2019 *9th SI Brochure* https://www.bipm.org/en/publications/si-brochure/

[9] Pavese F and Charky A 2016 Some important features of the proposed new definition of the International System of Units (SI) that the users should know about before 2018 *Int. J. Metrol. Qual. Eng.* **7** 403

[10] Lipovka A A 2016 Variation of the fine structure constant *arXiv:1608.04593v6 2016*

[11] Cleveland C J (Ed) 2009 *Concise Enciclopaedia of History of Energy* (New York: Elsevier) 283

[12] Truesdell C 1979 Absolute temperatures as a consequence of Carnot's general axiom *Arch. Hist. Exact Sci.* **20** 357

[13] Carnot N L S 1824 Réflexions sur la puissance motrice du feu et sur les machines propres à développer cette puissance: 1986 *Reflections on the Motive Power of Fire*. Translated & Ed Fox R (Manchester: Manchester University Press)

[14] Thomson W (Lord Kelvin) 1848 On an Absolute Thermometric Scale Founded on Carnot's Theory of the Motive Power of Heat, and Calculated from Regnault's Observations *P.*

Camb. Philos. Soc. **1** 66; also 1848 in: *Philos. Mag.*, 3rd Series, **33** 313

[15] Joule J P and Thomson W 1882 On the thermal effects of fluids in motion Part 2 in Thomson W, *Mathematical and Physical Papers* **1** (Cambridge: Cambridge University Press) 357 Originally published in: 1854 *Philos. T. R. Soc. Lond.* **144** 321

[16] Coopersmith J 2015 *Energy, The Subtle Concept* (Oxford: Oxford University Press)

[17] Wilczek F 2006 *Fantastic Realities: 49 Mind Journeys and a Trip to Stockholm* (Singapore: World Scientific)

[18] CGPM, Revision of the SI, Resolution 1, *26th Meeting, 16 November 2018,* http://www.bipm.org/ en/measurement-units/new-si/

[19] Simon F E 1955 The concept of temperature near absolute zero Ch 2 *Temperature: Its Measurement and Control* Wolfe HC ed **2** AIP (New York: Reinhold)

[20] Donald M B 1946 Thermodynamics and the logarithmic temperature scale *Nature* **157** 624

[21] Groves W D and Lielmezs J 1965 The linear temperature scale *Nature* **205** 489

[22] Pellicer J, Hernandez M-J and Dolz M A 1997 New definition of the logarithmic temperature scale based on the triple point of water *Rev. Mex. Fis.* **43** 837

[23] Müller I 2007 *A History of Thermodynamics: The Doctrine of Energy and Entropy* (Berlin: Springer) doi:10.1007/978-3-540-46227-9

[24] *Thermal Engineering Web Handbook* (partially in Russian) 2007 Ochkova V F Ed MEI Project (http://www.rffi.ru) http://twt.mpei.ac.ru/TTHB/2/KiSyShe/eng/Chapter3/3-5

[25] Berman M S 2009 The zero-energy universe *arXiv:gr-qc/0605063v3*

[26] Strehlow P and Seidel J 2007 Definition der Temperatur und ihre Grenzen *Temperatur, PTB Mitteilingen* **117** 16

[27] Hartmann H, Mahler G and Hess O 2004 Fundamentals of nano-thermodynamics. On the minimal length scales, where temperature exists *arXiv:0408033*

[28] Hartmann M and Mahler G 2005 Measurable consequences of the local breakdown of the concept of temperature *arXiv:0410526v2*

[29] Carlos L D and Palacio Parada F (Eds) 2016 Thermometry at the nanoscale: techniques and elected applications *Roy. Soc. Ch. Nanoscience & Nanotechnology* **38** (London: Royal Society of Chemistry Publ)

[30] Hill T L 2001 A different approach to nano thermodynamics *Nano Lett.* **1** 273

[31] Hill T L 1964 *Thermodynamics of Small Systems Parts 1 and 2* (New York: WA Benjamin)

[32] Reza Vakili-Nezhaad G 2005 Nanothermodynamics Ch E6-152-05-00 Nanoscience and Nanothechnologies (UNESCO-EOLSS Joint Committee ed) *Encyclopedia of Life Support Systems (EOLSS)* (Oxford: Eolss Publishers) http://www.eolss.net

[33] Caruso F and Tsallis C 2006 Nonadditive entropy reconciles the area law in quantum systems with classical thermodynamics *Phys. Rev.* E **78** 021102

[34] Abe S and Okuyama S 2011 Similarity between quantum mechanics and thermodynamics: entropy, temperature, and Carnot cycle *Phys. Rev.* E **83** 021121

[35] Gonzalez-Dıaz L A and Dıaz-Solorzano S 2015 Comment on 'Similarity between quantum mechanics and thermodynamics: Entropy, temperature, and Carnot cycle' *Phys. Rev.* E **91** 056101

[36] Langen T, Geiger R, Kuhnert M, Rauer B and Schmiedmayer J 2013 Local emergence of thermal correlations in an isolated quantum many-body system *Nat. Phys.* 2739 doi:10.1038/nphys2739

[37] Tsallis C 1988 Possible generalisation of Boltzmann-Gibbs statistics *J. Stat. Phys.* **52** 479

[38] Tsallis C 2009 *Introduction to Nonextensive Statistical Mechanics. Approaching a Complex World* Ch 2 (New York: Springer)

[39] Hilbert S, Hänggi P and Dunkel J 2014 Thermodynamic laws in isolated systems *Phys. Rev.* E **90** 062116

[40] Hänggi P, Hilbert S and Dunkel J 2015 Meaning of temperature in different thermostatistical ensemble *arXiv:1507.05713*

[41] Campisi M 2014 Construction of microcanonical entropy on thermodynamic pillars *arXiv: 1411.2425*; Notes on heat engines and negative temperatures *arXiv:1606.05244*

[42] Frenkel D and Warren P B G 2015 Gibbs, Boltzmann, and negative temperatures *Am. J. Phys.* **83** 163

[43] Miranda E N 2015 Boltzmann or Gibbs Entropy? Thermostatistics of Two Models with Few Particles *J. Modern Phys.* **6** 1051

[44] Dunkel J and Hilbert S 2014 Consistent thermostatistics forbids negative absolute temperatures *Nat. Phys.* **10** 67

[45] Braun S, Ronzheimer J P, Schreiber M, Hodgman S S, Rom T, Bloch I and Schneider U 2013 Negative absolute temperature for motional degrees of freedom *Science* **339** 52

[46] Ehrlich P 1982 Negative, infinite, and hotter than infinite temperatures *Synthese* **50** 233

[47] Treumann R A and Baumjohann W 2014 Beyond Gibbs-Boltzmann-Shannon: general entropies —the Gibbs-Lorentzian example *Frontiers in Physics (AIP Conf. Proc.)* **2** doi:10.3389/fphy.2014.00049 Reprinted in *Philosophical Problems of Modern Physics* ed by H S Plendl (Boston: Reidel Publishing Co. 1982)

[48] Haisch B, Rueda A and Dobyns Y 2001 Inertial mass and the quantum vacuum fields *Ann. Phys.* (Leipzig) **10** 393

[49] Granek G 2001 Einstein's Ether: Why did Einstein come back to the ether? *Apeiron* **8** 19; also, A. Einstein 1922 *Ether and the Theory of Relativity* (in German) Leiden lecture (London: Methuen & Co) http://www-history.mcs.st-andrews.ac.uk/Extras/Einstein_ether. html (in English)

[50] Shaf J 2019 The Meaning of Motions and Their Effects in Einstein's Empty Space and in the Scenario of the Higgs Quantum Fluid Space, *Journal of Modern Physics* **10** 256

[51] Frauendiener J 2000 Conformal infinity *Living Rev. Relativ.* http://www.springer. com/us/livingreviews/articles/volume3/2000-4frauendiener

[52] Sommers P 1978 The geometry of the gravitational field at space-like infinity *J. Math. Phys.* **19** 549

[53] Bergmann P G 1989 Quantum gravity at spatial infinity *Gen. Relat. Gravit.* **21** 271

[54] van't Hooft G 1971 Renormalizable Lagrangians for massive Yang-Mills fields *Nucl. Phys.* B **35** 167

[55] de Cusa N 1997 *De docta ignorantia* English translation in: H. Lawrence (ed.), *Nicholas of Cusa: Selected Spiritual Writings*, Western Spirituality Classics (New York: Paulist Press)

[56] Tozzi A and Peters J F 2017 What does it mean "the same"? *Prog. Biophys. Mol. Bio.* https://doi.org/10.1016/j.pbiomolbio.2017.10.005

[57] Tozzi A 2018 Curvature changes in biological dynamics *Preprints* doi:10.20944/preprints 201810.0299.v1

Aerodynamic loads acting on the M5/ONERA/IAE aeronautical standard model

M L C C Reis[1,2], M S Souza[1,2], M C C Araújo[2], R R Santos[1]

[1]Instituto de Aeronáutica e Espaço, Pr. Mal. Eduardo Gomes, 50, CEP 12228904, Brasil
[2]Instituto Tecnológico da Aeronáutica, Pr. Mal. Eduardo Gomes, 50, CEP 12228900, Brasil

E-mail: marialuisamlccr@fab.mil.br

Abstract. A test campaign to estimate the aerodynamic loads acting on the aeronautical standard model M5/ONERA/IAE is under progress. The model is being tested at the TA-2 subsonic wind tunnel using an external balance. The values of force and moment coefficients and associated uncertainties are evaluated by measuring aerodynamic loads read by the balance sensors and the parameters of the wind tunnel airflow. The campaign is part of an Interlaboratory comparison of results involving other wind tunnel facilities around the world. This paper presents load coefficient curves versus sideslip angle. At TA-2, the uncertainty associated with the pitching coefficient presents the highest value. The dominant contribution comes from the hysteresis.

1. Introduction

During real flight, the airflow exerts pressure and shear stress on the surface of an aeronautical vehicle. Wind tunnel facilities simulate real conditions encountered by the vehicle by driving a flow to a fluid which can either be air or another gas such as helium or nitrogen.

Wind tunnels are classified according to the speed range of the flow: they are subsonic, near sonic, transonic, supersonic and hypersonic. They are built in different sizes and configurations according to the test needs. The circuit can be open or closed, the test section can accommodate a full size, a scale model or part of a model [1].

As in any laboratory, the wind tunnel experimentalists must conduct activities to guarantee the metrological reliability of the test results. These activities include calibration of the measurement chain as well as the calibration of the airflow. The former is related to standard equipment and sensors and for the latter an aeronautical standard model can be used.

This paper is concerned with the second type, *i. e.*, the calibration of the airflow of the subsonic wind tunnel number 2, TA-2. For this purpose, the aeronautical model M5, designed by ONERA, the French Aerospace Laboratory, was used as the standard model. The M5 is a wing/body/tail model and belongs to the *M*-series family. In the original version, it receives a sting strut as the interface to an internal aerodynamic balance. The TA-2 team modified the design in order to allow different

---

1 To whom any correspondence should be addressed.

configurations for the use of external balances. Besides the original sting, the M5/ONERA/IAE model can also be fixed to the test section in single, double or three-strut interfaces (Figure 1).

The wind tunnel TA-2 is a subsonic, atmospheric, closed circuit wind tunnel. It was built in the 1960's and has contributed to the development of aerodynamic projects for the Brazilian aeronautical sector and the civil engineering industry. The facility is located at the Institute of Aeronautics and Space, Brazil.

The maximum airflow speed for an empty test section is around 140 m/s, *i. e.*, the maximum Mach number is approximately 0.40. The test section is 2.1 m high, 3.0 m wide and 3.2 m long. The airflow is driven by an 8.4 m diameter 7-blade-fan connected to a motor drive. The engine power is 1.2 Megawatts with 400 revolutions per minute. The strut configurations provide an angle of attack α up to ±30° and a turntable permits a yaw angle β of ±45°.

An intensive campaign is being carried out at the TA-2 to evaluate the aerodynamic loads acting on the M5/ONERA/IAE in the subsonic regime. The estimated parameters are force and moment coefficients, C_F and C_m, respectively. The input quantity related to the airflow is the free stream dynamic pressure, q. A six-component external balance composed of six strain-gauges is employed to measure the 3 forces and the 3 moments acting on the model under test (Figure 2). At TA-2, the drag, side and lift forces are denominated F_1, F_2 and F_3, while the rolling, pitching and yawing moments are denominated F_4, F_5 and F_6, respectively. A reference length, l, and a reference area, A, taken from the geometry of the model are also necessary to evaluate the 3 force coefficients C_F and the 3 moment coefficients C_m.

Once it has been tested at the TA-2 wind tunnel, the model will be transferred to other facilities in order to compare test results, in the scope of an Interlaboratory Program. The data repeatability, accuracy and uncertainty will be analyzed and the airflow quality of each wind tunnel will be checked [2].

Figure 1. The standard model M5 installed at the TA-2 test section in a two-strut configuration.

Figure 2. The TA-2 six-component external balance.

2. Methods

For the present study, the M5 model was tested at a nominal airflow speed of 200 km/h. The angle of attack α was maintained at 0° and the sideslip angle β varied from -5° to +5°. The test was comprised of one cycle of increasing β beginning at -5° and going in 1° steps to +5° and then returning in similar fashion to -5°. With this procedure, it is possible to analyze the hysteresis of the wind tunnel instrumentation thereby allowing a type A uncertainty evaluation.

A previous calibration of the external balance supplied the uncertainty values associated with the measured aerodynamic forces and moments [3]; this is a type B uncertainty evaluation. No wind tunnel correction and no tare results were applied to the experimental data, i. e., the information presented in this paper is the raw data. The results are related to the origin of the external balance and to the wind reference axes.

Journal of Physics: Conference Series **1379** (2019) 012009 doi:10.1088/1742-6596/1379/1/012009

The characteristics of the model are:

– Reference area, A, is the geometric area of the wing = 0.5260 m^2;

– Reference length, l, for the pitching moment is the mean aerodynamic cord of the wing = 0.295 m; and

– Reference length, l, for the rolling and yawing moments is the span of the wing = 1.964 m.

3. The Mathematical Modelling

The mathematical modelling of the aerodynamic forces and moments is the polynomial expressed by equation (1). The modelling considers second order interactions of the strain-gauge readings, R. The output quantity F_i represent drag, side or lift if force is being evaluated. For the evaluation of the aerodynamic moment, F_i represents pitching, rolling or yawing. β is the sideslip angle.

$$\begin{aligned}
F_i &= a_{i,1}R_1 + a_{i,2}R_2 + a_{i,3}R_3 + a_{i,4}R_4 + a_{i,5}R_5 + a_{i,6}R_6 + \\
&a_{i,7}R_1R_1 + a_{i,8}R_1R_2 + a_{i,9}R_1R_3 + a_{i,10}R_1R_4 + a_{i,11}R_1R_5 + a_{i,12}R_1R_6 + a_{i,13}R_1\sin\beta + a_{i,14}R_1\cos\beta + \\
&a_{i,15}R_2R_2 + a_{i,16}R_2R_3 + a_{i,17}R_2R_4 + a_{i,18}R_2R_5 + a_{i,19}R_2R_6 + a_{i,20}R_2\sin\beta + a_{i,21}R_2\cos\beta + \\
&a_{i,22}R_3R_3 + a_{i,23}R_3R_4 + a_{i,24}R_3R_5 + a_{i,25}R_3R_6 + a_{i,26}R_3\sin\beta + a_{i,27}R_3\cos\beta + \\
&a_{i,28}R_4R_4 + a_{i,29}R_4R_5 + a_{i,30}R_4R_6 + a_{i,31}R_4\sin\beta + a_{i,32}R_4\cos\beta + \\
&a_{i,33}R_5R_5 + a_{i,34}R_5R_6 + a_{i,35}R_5\sin\beta + a_{i,36}R_5\cos\beta + \\
&a_{i,37}R_6R_6 + a_{i,38}R_6\sin\beta + a_{i,39}R_6\cos\beta
\end{aligned} \tag{1}$$

The force and moment coefficients, C_F and C_m, are respectively expressed by [4]:

$$C_F = \frac{F}{qA} \tag{2}$$

and

$$C_m = \frac{m}{qAl} \tag{3}$$

The dynamic pressure q is the free stream dynamic pressure of the airflow measured by a Pitot tube located at the entrance of the TA-2 test section. The area A and the length l are reference values related to the geometry of the model.

The law of propagation of uncertainty is applied to equation (1) to estimate the associated uncertainties of aerodynamic forces and moments F_i, u_{F_i} [5]. Similarly, the law is applied to equations (2) and (3) for the aerodynamic coefficient uncertainties. As an example, for force coefficients C_F, the positive square root of expression (4) is the combined uncertainty u_{CF}:

$$u_{C_F}^2 = \left(\frac{1}{qA}\right)^2 u_F^2 + \left(\frac{-F}{q^2A}\right)^2 u_q^2 + \left(\frac{-F}{qA^2}\right)^2 u_A^2 \tag{4}$$

Least squares method with matrix formalism is applied to reduce both balance calibration and test data [6, 7].

4. Results and discussions

Codes in MatLab® and Excel® worksheets were elaborated to present curves C_Fxβ and C_mxβ for the flow regimes covered by the tests. In this study we present results which correspond to a nominal velocity of 200 km/h.

Table 1 shows the coefficient values for the drag and lift forces, C_{F1} and C_{F3}, respectively, and the pitching moment coefficient, C_{F5}. The associated uncertainties, u_{CF1}, u_{CF3} and u_{CF5}, are also presented in Table 1; they are the combined uncertainties resulting from the type A and type B evaluation. The dominant contribution comes from the type A evaluation, due to hysteresis. This can be seen in figures 3 and 4. Figure 3 shows the $C_{F1}\text{x}\beta$ curve for a one-cycle test which is comprised of the increasing and decreasing values of β. Error bars represent the uncertainties related to the calibration of the external balance. In figure 4, data correspond to the average value of both ascending and descending runs; in this figure, the contribution of the hysteresis to the uncertainty associated with the drag coefficient is also considered, combined with that related to the balance calibration. The curves for the drag, lift and pitching coefficients are shown together in figure 5.

Table 1. Load coefficients and associated uncertainties estimated at the 200 km/h airflow regime.

β (°)	C_{F1}	u_{CF1}	u_{CF1}/C_{F1} (%)	C_{F3}	u_{CF3}	u_{CF3}/C_{F3} (%)	C_{F5}	u_{CF5}	u_{CF5}/C_{F5} (%)
-5	0.03442	0.00016	0.5	0.2505	0.0020	0.8	0.0378	0.0021	5.6
-4	0.03301	0.00015	0.5	0.2521	0.0020	0.8	0.0400	0.0013	3.3
-3	0.03201	0.00028	0.9	0.248	0.004	1.6	0.043	0.003	7.0
-2	0.03127	0.00016	0.5	0.2467	0.0024	1.0	0.0439	0.0023	5.2
-1	0.03068	0.00022	0.7	0.2461	0.0027	1.1	0.0451	0.0015	3.3
0	0.03056	0.00017	0.6	0.2449	0.0026	1.1	0.0451	0.0005	1.1
1	0.03088	0.00016	0.5	0.2459	0.0018	0.7	0.0442	0.0021	4.8
2	0.03132	0.00017	0.5	0.2503	0.0022	0.9	0.0399	0.0006	1.5
3	0.03206	0.00016	0.5	0.2534	0.0029	1.1	0.0355	0.0019	5.4
4	0.03308	0.00019	0.6	0.2559	0.0013	0.5	0.0308	0.0010	3.2
5	0.03434	0.00012	0.3	0.2595	0.0009	0.3	0.02670	0.00014	0.5

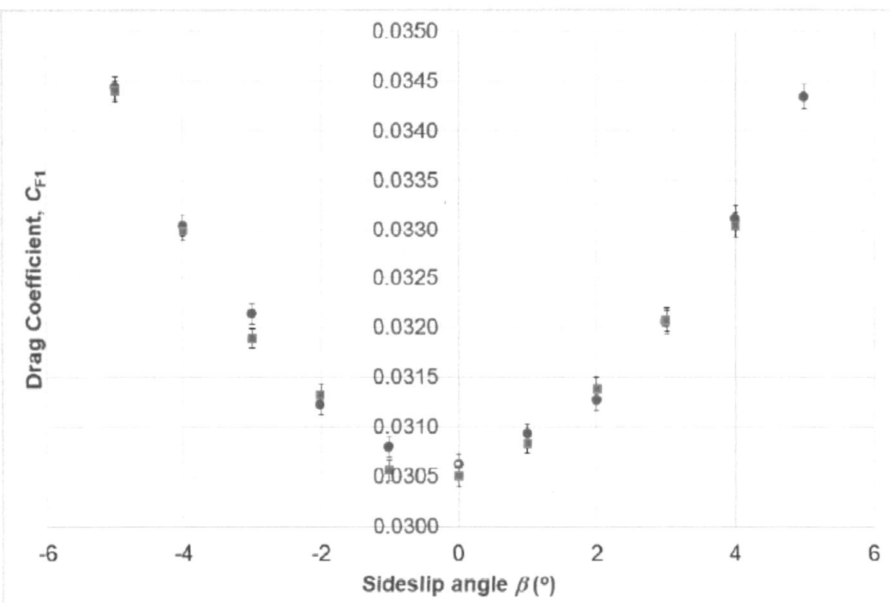

Figure 3. The drag force coefficient variation with sideslip angle, $C_{F1}\text{x}\beta$, for the ascending and descending runs.

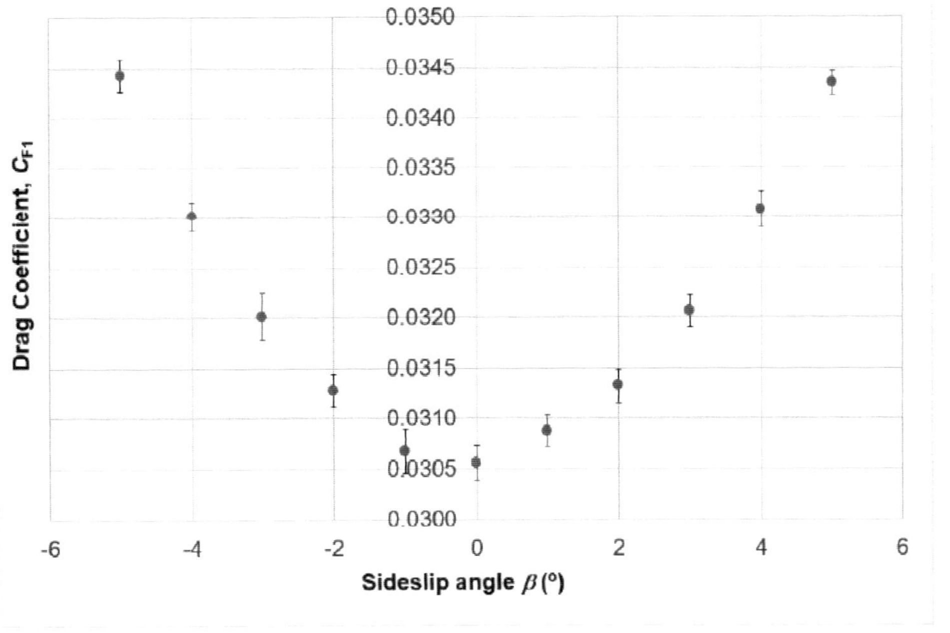

Figure 4. The drag force coefficient variation with sideslip angle, C_{F1}xβ.

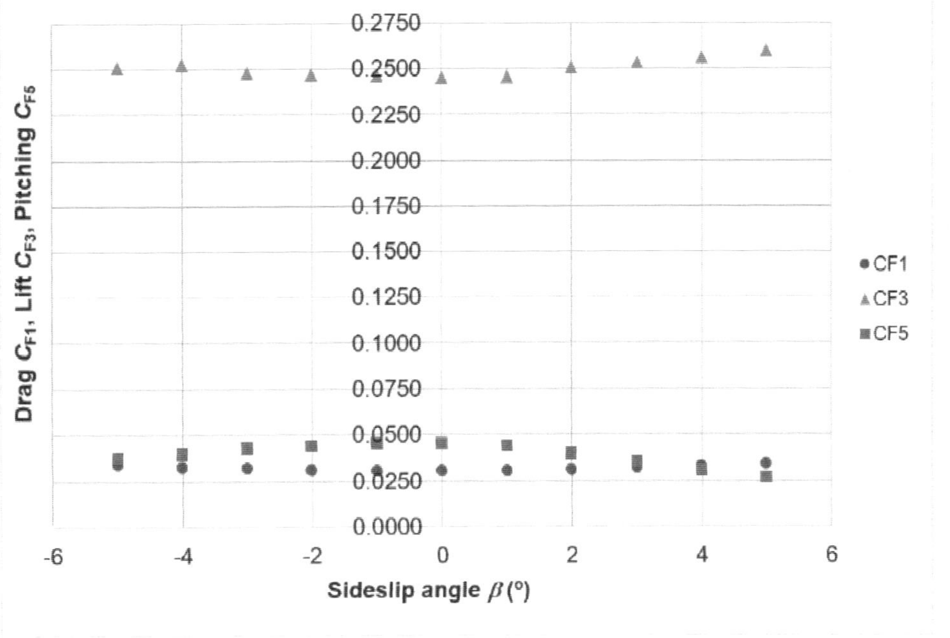

Figure 5. The drag force, side force and pitching moment coefficient variations with sideslip angle.

5. Conclusions

The results of the wind tunnel campaign employing the M5/ONERA/IAE aeronautical standard model were presented. Curves of drag, lift and pitching coefficients versus the sideslip angle were supplied. Data reduction revealed that the uncertainty component related to the hysteresis is the dominant contribution. The pitching moment coefficient presented the highest associated uncertainty. The test campaign has just started and the tests with varying the angle of attack α are under progress.

References

[1] Barlow J B, Rae Jr W H and Pope A 1999 *Low-Speed Wind Tunnel Testing* (John Wiley & Sons Inc.) p 713

[2] BIPM/JCGM 200:2012 *International Vocabulary of Metrology – Basic and General Concepts and Associated Terms (VIM)* (Joint Committee for Guides in Metrology, BIPM, Metrologia, IEC, IFCC, ILAC, ISO, IUPAC, IUPAP and OIML) p 91

[3] Reis M L C C, Castro R M and Mello O A F 2013 Calibration uncertainty estimation of a strain-gage external balance *Measurement* **46** pp 24–33

[4] Anderson Jr J D 2001 *Fundamentals of Aerodynamics* (New York: Mc Graw Hill) p 892

[5] BIPM/JCGM 100:2008 *Evaluation of Measurement Data – Guide to the Expression of Uncertainty in Measurement – GUM 1995 with Mminor Corrections* (Joint Committee for Guides in Metrology, BIPM, Metrologia, IEC, IFCC, ILAC, ISO, IUPAC, IUPAP and OIML) p 120

[6] Press W H, Teukolsky B P T and Vetterling W T 1990 *Numerical Recipes* (Cambridge University Press) chapter 14 pp 547–565

[7] Bevington P R and Robison D K 2003 *Data Reduction and Error Analysis for Physical Sciences* (New York: Mc Graw Hill Higher Education) p 320

Journal of Physics: Conference Series **1379** (2019) 012010 doi:10.1088/1742-6596/1379/1/012010

Multicriteria approach to design of strain gauge force transducers

S I Gavrilenkov[1], S S Gavriushin[1,2] and V A Godzikovsky[3]

[1]Department of Computer Systems of Manufacturing Automation, Bauman Moscow State Technical University, Moscow, Russian Federation
[2]Institute for Machine Science named after A.A. Blagonravov of the Russian Academy of Sciences, Moscow, Russian Federation
[3]JSC "Weight-measuring Company "Tenso-M", Moscow region, Russian Federation

E-mail: gavrilenkov_si@student.bmstu.ru, gss@bmstu.ru

Abstract. Force transducers based on strain gauges are ubiquitous in force measuring equipment. Manufacturers of force transducers have to strive for stronger, cheaper and more accurate force transducers in order to compete successfully on the global market. Transducers have to be designed as fast as possible and require as little experimental work as possible. To that end, a reliable design tool is required. In this paper, we propose a new approach to design of force transducers based on strain gauges. The approach is based on the Parameter Space Investigation method and methods of multicriteria optimization. The transducer's parameters used as design criteria (output signal at rated load, nonlinearity, etc.) are evaluated using Finite Element Analysis (FEA). The FEA results are used to find the optimal locations of strain gauges. This method was implemented in the software system for multicriteria design of force transducers based on strain gauges. The system was used to design a novel force transducer and can be used for developing new strain gauge-based force transducers and optimizing the existing ones.

1. Introduction

Strain gauge force transducers are widely used in force-measuring equipment. The advantages of using strain gauge force transducers include high measurement accuracy, small dimensions. These transducers have good resistance to electromagnetic disturbances and work well with voltage-measuring equipment which is crucial when using transducers in modern measuring systems. Despite the existing wide range of designs [1] the task of further improving force transducers' design to make them more competitive on the market is still relevant. To that end, the designers will need a new tool of force transducer design that shortens the development cycle and minimizes the amount of experimental work. The rest of the paper is organized as follows. Firstly, we introduce the working principle and design requirements for strain gauge force transducers. Then we will introduce the multicriteria approach to design of strain gauge force transducers; the approach is implemented in a special software system. After that, we demonstrate the work of the proposed approach with an example.

2. Working principle and design requirements for strain gauge force transducers

A strain gauge force transducer is a transducer that converts applied force into electrical signal. It is comprised of an elastic element which is being loaded via a thrust piece. The applied load deforms the elastic element. Mechanical strain in certain areas of the elastic element is measured using foil strain gauges. The measured strain depends on the place where the strain gauges are bonded to the elastic element. Strain gauges are connected to form a circuit knows as the Wheatstone bridge circuit. Strain gauges change their resistance because of strain; resistance change creates voltage at the output terminals of the Wheatstone bridge. The voltage is proportional to the applied load [1].

Designing strain gauge force transducers is a multicriteria problem. The input parameters include characteristic geometric dimensions of the force sensor's elastic element, mechanical properties of the elastic element's material, configuration of the Wheatstone bridge circuit (type of strain gauges, location of strain gauges, etc.). These parameters are called design parameters. The output parameters, also called objectives or design criteria, are the parameters determining the measurement error (nonlinearity, hysteresis, creep) and other parameters such as the overloading capacity of a transducer, the magnitude of the output signal, etc. These criteria often contradict each other. For instance, the increase of the output signal proportional to the magnitude of the measured strain leads to the increase of nonlinearity and decrease of the overloading capacity.

The design process comes down to choosing the topology of the transducer's elastic element, defining the dimensions of the elastic element, and determining the locations where the strain gauges will be installed so that the output signal is of desired magnitude, the measurement errors and weight are minimized, the overloading capacity of the transducer is greater than required by the designer.

3. Design method

We propose using the approach based on the Parameter Space Investigation Method [2, 3] and methods of multicriteria optimization [4]. The design process workflow (see figure 1) can be subdivided in two stages.

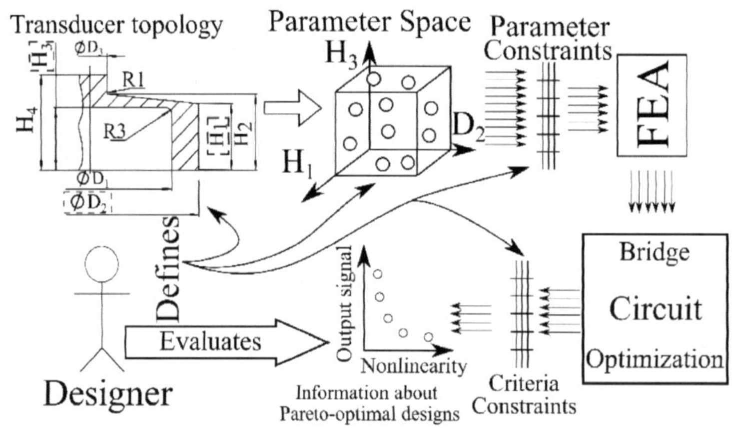

Figure 1. Design method workflow.

At the first stage, the designer chooses the parametric topology of the transducer's elastic element, for example, membrane or ring torsion elastic element. Then the designer chooses the transducer's parameters that are going to vary in the design process. The designer also specifies the limits within which the parameter may vary. These limits define the design parameter space – an *N*-dimensional hypercube for the case of *N* design parameters. The parameter space is investigated using LP_τ sequence [5]. Each element of the sequence corresponds to a particular geometry of the transducer's elastic element within the chosen topology. Each design is simulated using FEA in order to obtain information about the elastic element's stress-strain state. The analysis is taking into account geometric nonlinearity [6, 7] and nonlinearity arising from the contact interaction between force transducer and its loading pads [8].

After that, the optimal configuration of the Wheatstone circuit is determined for each elastic element design. This problem comes down to determining such locations and orientations of the strain gauges so that the measured linear strain is maximised while the resultant nonlinearity of the sensor, including nonlinearity of the Wheatstone bridge, is minimized. The problem is solved using the following algorithm. Firstly, a set of possible locations where the strain gauges can be installed is generated. Then the measured strain is calculated for each strain gauge location in the given strain field. To that end, a special mathematical model of strain gauge is used. The model takes into account the dimensions of the strain gauge and its orientation in the strain field. The model is similar to the one used in [9]. It is based on nodal strain data from FEA. After that, possible locations are split into two sets: gauges experiencing tension and gauges experiencing compression. The algorithm iterates through different combinations of tension and compression gauge locations forming full Wheatstone circuits and evaluates the output signal at the rated load and the nonlinearity of the output signal. Finally, the algorithm chooses the optimal configuration having the least absolute nonlinearity and the output signal greater than specified by designer. After that, the values of the remaining criteria (overloading capacity, mass, etc) are calculated using FEA results.

At the second stage, the designer imposes constraints on the design criteria values. The constraints look like "the output signal should greater than 1 mV/V" or "the maximum nonlinearity should not exceed 0.05 % of full scale". A set of feasible designs that satisfy all the constraints is constructed. After that, a set of Pareto-optimal designs is constructed based on the set of feasible designs. If the set of feasible designs is empty, the criteria constraints can be weakened; in this case the sets of feasible and Pareto-optimal designs are recalculated and re-evaluated. The parameter space can be investigated anew after changing its boundaries. A local search can be conducted in the neighborhood of a certain Pareto-optimal design.

The proposed design method was implemented in a software system for designing strain gauge force transducers. Some of the user interface forms of the developed software system are shown in figure 2.

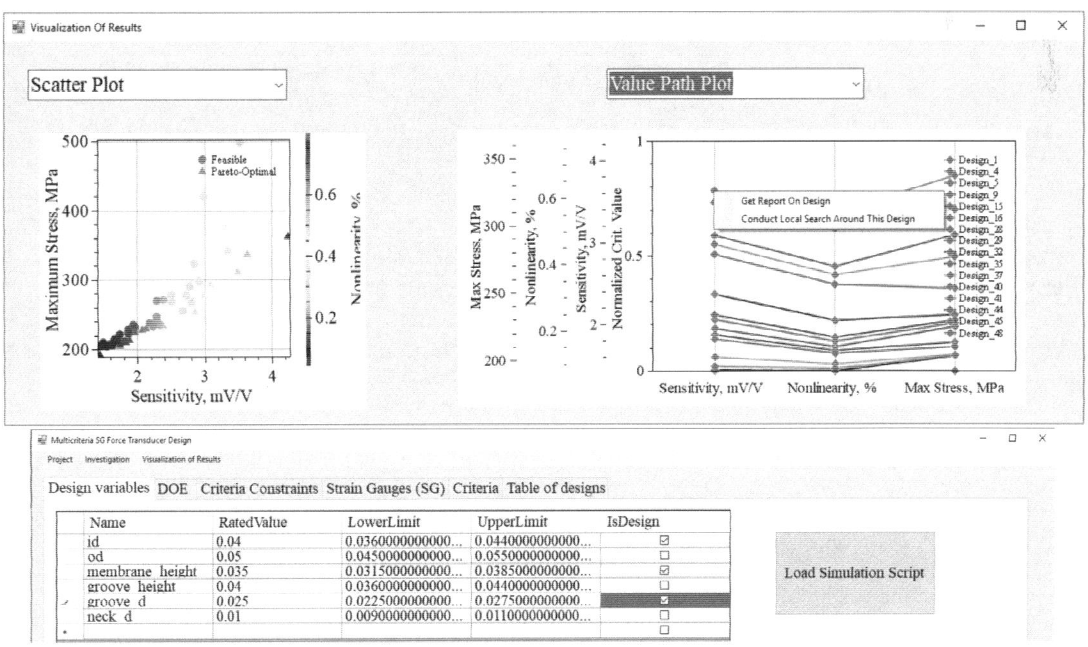

Figure 2. User interface of the developed software program.

The principle of interacting between the Finite Element software and the software is similar to the system presented in [10]. The information about feasible and Pareto-feasible designs is presented to the designer in the form of scatter plots, histograms, radar charts, etc. If the criteria constraints change, the sets of feasible and Pareto-optimal designs will be reconstructed automatically, as well as all plots visualizing this information. The design results are stored in the database.

4. Case study

The proposed approach will be demonstrated by designing a novel strain gauge force transducer. The measuring range of the transducer is 0...1000 kN. The topology of the transducer's elastic element is shown in figure 3 (a). The information about design criteria is shown in table 1. Let us elaborate on the criteria. The output signal of most force transducers is 2 mV/V at rated load. However, different compensation elements of the Wheatstone bridge can reduce the output signal by as much as 10%. Therefore, the minimum acceptable value of the output signal at rated load is 2.2 mV/V; this corresponds to the value of strain approximately ±1100 ppm measured using constantan strain gauges with the gauge factor of 2. The material of the elastic element is quenched stainless steel having hardness 40 HRC, the yield strength σ_y of 1000 MPa and the ultimate tensile strength σ_u of 1200 MPa. We utilized constantan strain gauges actively used in transducer manufacturing. The dimensions of the strain gauge are shown in figure 3 (b).

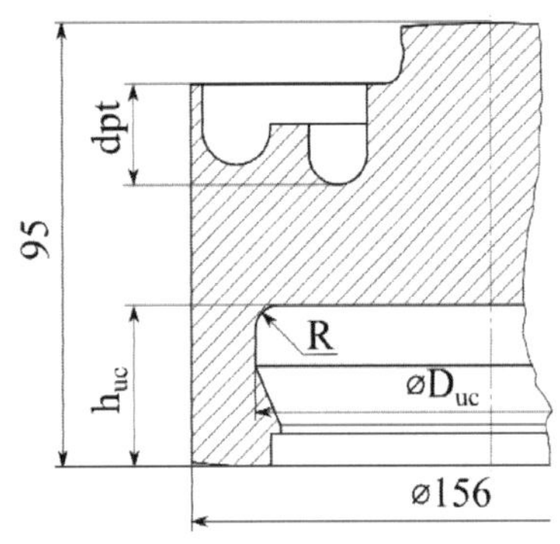

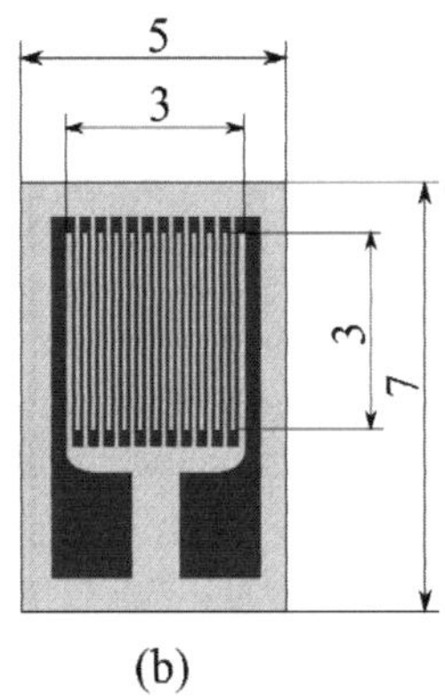

Figure 3. Topology of the elastic element (a) and the dimensions of the strain gauge (b). The dimensions of the strain gauge and the overall dimensions of the elastic element are in mm.

The explicit calculation of the transducer's hysteresis is not straightforward. It is not known whether accurate calculation of hysteresis is possible using FEA. The hysteresis simulation presented in [11] is limited to column compression force transducers. However, empirical data shows that transducers exhibiting large hysteresis error tend to have large values of slip at the contact interface between a force transducer and its loading pad. If a transducer slips on the loading pad, the forces of friction convert some of the strain energy into heat. This results in different strain values at the same load during loading and unloading. Therefore, if we minimize the slip value, the dry friction hysteresis is going to be small. The initial constraint imposed on the elastic element mass is arbitrary and can be altered in the design process.

Table 1. Design criteria and initial criteria constraints.

Criterion name	Initial criterion constraint
Output signal at the rated load	≥2.2 mV/V
Max. nonlinearity	≤0.02 % of full scale
Slip at the contact interface	≤1 μm
Mass	≤8 kg

The initial investigation included roughly 400 designs in the LP_τ sequence. The design process revealed that the constraint imposed on the maximum contact slip was too stringent; therefore, it was weakened to 2 µm. The designer was presented with the Pareto optimal set of solutions. The designer chose the design having the least mass and nonlinearity no more than 0.02% and conducted a local search to order to further minimize the mass. The criteria values for the initial and the final design are presented in table 2.

Table 2. Design criteria values for the initial and the final design.

Criterion name	Initial design	Final design
Output signal at the rated load, mV/V	2.3	2.2
Max. nonlinearity, % of full scale	0.02	0.01
Slip at the contact interface, µm	12.0	2.0
Mass, kg	8.0	7.6

The picture of the manufactured prototype is shown in figure 4.

Figure 4. Manufactured prototype of the designed force transducer.

The value of creep was not deal with in the scope of the current study. We applied the strain gauges used in “Tenso-M” for manufacturing conventional force transducers. The force transducers manufactured at “Tenso-M” have the accuracy class C3 per OIML R60 and have adequately low values of creep. We used the same material for the elastic element and the same type of gauges, so it would make sense that the creep error would be similar to the creep error of the force transducers manufactured at “Tenso-M”. We measured the creep error according to the procedure specified in OIML R60 using a 1000 kN hydraulic force machine. The creep error time history is shown in figure 5. The creep error does not exceed 0.01% after one hour.

As the hysteresis was evaluated indirectly, we tested the prototype using the 1000 kN hydraulic force machine. The measured values of hysteresis are shown in figure 6. As can be seen, the calculated value of the slip at the contact interface equal to 2 µm corresponds to the maximum hysteresis of ±0.02% of the full scale. However, it should be noted that hysteresis is not only affected by the friction losses at the contacts, but by the creep of the transducer as well and the internal friction hysteresis of the elastic element material [12]. For instance, strong negative creep coupled with nearly zero dry friction hysteresis can lead to the large overall negative hysteresis that can detrimentally affect the measurement accuracy of a force transducer.

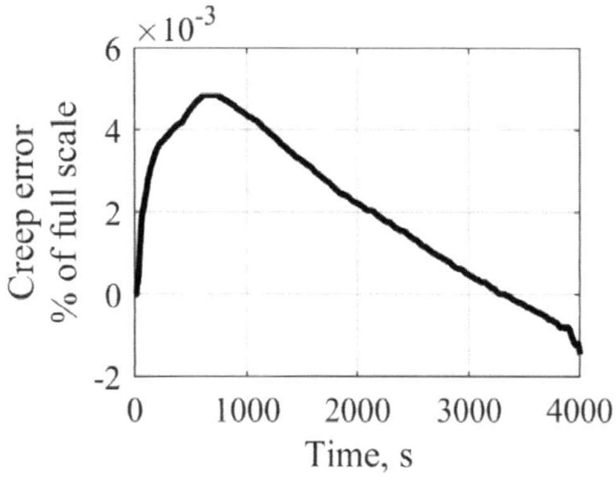

Figure 5. Creep error of the manufactured prototype.

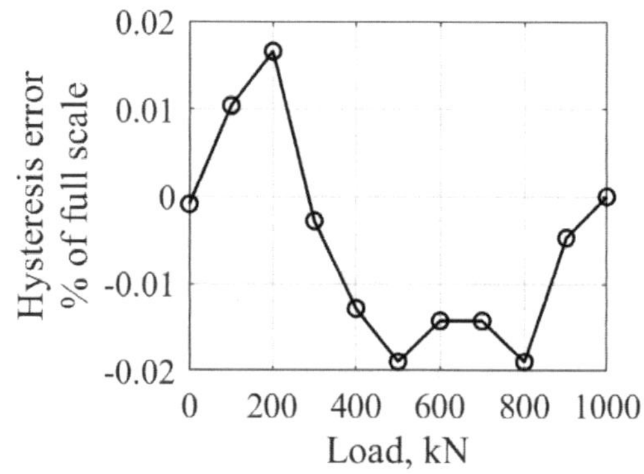

Figure 6. Hysteresis error of the manufactured prototype.

Thus, in order to make a low-hysteresis transducer it is necessary not only to minimize the slip at contact interface, but also select the proper material of the elastic element and the proper strain gauges.

The characteristics of the final design (nonlinearity and hysteresis errors ≤0.02%, creep error ≤0.01% and output signal at rated load 2.2 mV/V) are comparable to the characteristics with the conventional 1000 kN force transducers for industrial applications (the NHS model produced by Keli Sensing Technology and the RTN model produced by HBM), although hysteresis value can be further minimized by placing more stringent constraints on the maximum value of the contact slip. More superior designs can be obtained by subsequent, more thorough investigations of the parameter space and placing more challenging design criteria constraints.

5. Conclusions

In this paper, we proposed a novel multicriteria approach to designing strain gauge force transducers. The approach is based on the Parameter Space Investigation Method, methods of multicriteria optimization and special mathematical model of a strain gauge force transducer. The proposed approach was implemented in a software system that can be used to optimize new strain gauge force transducers and optimize the existing ones. A novel force transducer was designed using the system in order to demonstrate the efficacy of the system.

References

[1] Stefanescu D M 2011 *Handbook of Force Transducers: Principles and Components* (Berlin: Springer Science+Business Media) p 612

[2] Statnikov R B, Gavriushin S S, Dang M H and Statnikov A R 2014 *Int. J. of Multicriteria Decis. Making* **4** 252-78

[3] Gavryushin S S, Skvortsov P A and Skvortsov A A 2018 *Periodico Tche Quimica* **15** 679-87

[4] Karpenko A P 2014 *Modern Algorithms of Search Optimization. Algorithms Inspired by Nature* (Moscow: Bauman Moscow State Technical Univeristy Press) p 446 (in Russian)

[5] Statnikov R B and Matusov I B 2013 *J. of Mach. Manuf. Reliab.* **42** 289-92

[6] Gavryushin S S and Skvortsov P A 2017 *Solid State Phenom.* **269** 60-70

[7] Gavryushin S S and Nikolaeva A S 2016 *Mech. of Solids* **51** 339-348

[8] Andreev K A, Vlasov A I and Shakhnov V A 2016 *Autom. and Remote Control* **77** 1281-85

[9] Schajer G 1993 *Strain: an Int. J. for Exp. Mech.* **29** 9-13

[10] Eremeykin P A, Zhargalova A D and Gavriushin S S 2018 *Adv. in Int. Sys. and Comp.* **658** 259-65

[11] Robinson G M 1997 *Measurement* **20** 103-107

[12] Fank S and Demirkol M 2006 *Sens. Actuators A: Physical* **126** 25-32

Journal of Physics: Conference Series **1379** (2019) 012011 doi:10.1088/1742-6596/1379/1/012011

Solving interdisciplinary tasks: the challenge and the ways to surmount it

Vladimir Yu. Kneller, Alexandr M. Fayans

V.A. Trapeznikov Institute of Control Sciences of Russian Academy of Sciences, Moscow

E-mail: vkneller@ipu.ru; alfayans@mail.ru

Abstract. The paper outlines the first steps on the path to identification and systematization of the types of purposeful activity tasks and to the ways to solve them. The authors have been developing a non-traditional interdisciplinary approach and apply it here to the concept of a task as such. This research area shows a good promise, since it enables the development of a framework for a holistic objective interdisciplinary picture of knowledge in any scientific field. The result of the work is the classification of object types and their characteristics, as well as the classification of the latter in the aspect of their relationship to objective reality. This result points out key directions to the systematization of objects, their related task types, and solution methods.

1. Introduction

The increasing relevance of solving interdisciplinary tasks reflects the need to return to a holistic perception of the surrounding world. This perception was partially lost in the period of rapid differentiation of the areas of knowledge and the corresponding narrow specialization. These processes resulted in different development level of various areas. To overcome the challenge, a return to the integrity paradigm is required, because the research experience shows that breakthrough solutions can be found, as a rule, either with substantial use of the results of more developed areas of knowledge or at the junction of different areas.

When solving interdisciplinary tasks at the junctions of areas, it is necessary to agree on the concepts related to different areas of knowledge, whereas employing solutions from another area requires the appropriate formulation of the problem in terms of the area of interest. This implies the following difficulties: coordination of concepts and terminology from different areas, development of a unified view of apparently cardinal difference in solution methods traditionally inherent in different disciplines.

The difficulties mentioned are increasing by the occurring polysemy of a number of concepts in use (even within the same area) can be also referred to as an obstacle in eliminating the separation.

For example, "parameter" according to [1] is a quantity whose values distinguish the elements of a certain set from one another. The same concept, according to [2], means "generally, any characteristic that can help in defining or classifying a particular system (such as an event, project, object, situation, etc.). That is, a parameter is an element of a system that is useful, or critical, when identifying a system or evaluating its performance, status, condition, etc.". According to the first definition, parameter and value are synonyms because, on the one hand, a parameter is a type of value, and, on the other hand, because a value (as well as a parameter) is by definition a quantitative means of distinguishing private realizations of a value. According to the second definition, the parameter is identical to the attribute used in both qualitative and quantitative identification of a system.

The purpose of this work is to present a path which, according to the authors, will allow to overcome these challenges.

2. Possible ways to overcome the challenges of solving interdisciplinary tasks

An obvious way to overcome these challenges is a joint discussion and mutual understanding of experts in problematic areas resulting in the development of mutually acceptable solutions. A key drawback of this approach is the participation in the discussion of experts from a limited number of areas that certainly limits the scope of the solutions developed.

Another way is to rely on a holistic world view covering all possible areas of conscious activity and the tasks solved in them. By looking "from above", this approach potentially allows to identify and systematize the tasks and methods for solving them by the levels of their interdisciplinary community, starting from their place of origin, clearly separate interdisciplinary points, (i.e. the points common to different areas) from the ones that inherent in a particular knowledge area(s). On this way, it becomes extremely important to ensure the unambiguity of the interpretation of certain concepts. This interpretation should be achieved by a generally accepted logical conclusion rather than just by mutual agreement.

The existence and effectiveness of the second path is indicated by the interdisciplinary nature of information technologies and the underlying mathematical methods.

The research aimed at developing constructive tools for implementing the second way was initiated at the Institute of Control Sciences of Russian Academy of Sciences. These studies were based on the ideas expressed by V. Kneller in a number of works [3-7], the developed inductive-deductive interdisciplinary approach to revealing and systematizing methods for solving tasks of purposeful activity (hereinafter, the Approach) and the experience of its application to a number of key tasks, such as creating a conversion of physical quantities [8], ensuring the invariance of the transformation to the influence of disturbing factors [9, 10], analog-to-digital conversion [11]. The mentioned approach involves the following steps: (a) identify the essence and formulate the task in terms of unambiguously interpreted concepts, (b) generic generalization – repeated transition to a more general task, all solutions of which are potentially applicable to solving a generalized task, (c) consider the procedure for solving the problem at the highest level of generalization from various aspects in order to identify possible methods of solution, d) the top-down movement to the initial task – the transfer of solutions of a more general task to the level of less common tasks allowing for the specificity of the latter.

The authors propose to apply the Approach to the task as such: to reveal the deep sources of its origin and to trace the possibilities caused by these sources. The results obtained would be obviously applicable to any areas of knowledge and, just by the way of their construction, form an interdisciplinary communication language.

3. Fundamentals of the way to systematization of tasks and their solution methods based on the Approach

The result of executing a sequence of steps prescribed by the Approach makes it possible to identify and systematize the types of tasks arising from the constructed picture by operating only with extremely general attributes of the task as such. By virtue of the logic of the Approach, it is necessary to identify and systematize consistently the solution methods common to any tasks as well as methods that are unique to each of the tasks at the highest level of generalization. Next, it is necessary to investigate possible directions of specifying the attributes inherent in the task, as such, and its types at the highest level of generalization for problems. This specification engenders subsequent levels of the hierarchy of types of tasks and their inherent solution methods, forming a holistic picture of both the tasks solved in the process of human activity and the methods for their solution.

The main distinctive feature of the direction under consideration is the urgent need in introducing unambiguously interpreted generally accepted concepts and in harmonizing them with the concepts adopted in various scientific fields and schools.

The encouraging results were obtained at a number of stages of the presented path, some of which were presented at the XIII All-Russian Conference on Control Problems "RCCP-19" and will be briefly described below; other works are either in the process of research or are waiting for their enthusiasts. It is extremely important that the tasks solved at the same time be precisely positioned in the overall picture of the tasks and methods for their solution, which may require certain improvements of the above-mentioned holistic picture based on the results of these studies.

To apply the Approach in a new area, it is, first of all, necessary to provide an unambiguous definition of the task, as such, through extremely generalized concepts. *A task is a formulated by the subject(s), conscious and sensible necessity to satisfy the need for something under certain conditions*. Consistently generalizing the detailed attributes of comprehension, consciousness, and activity, we come to philosophical concepts of the special, the single and the Universal. But since it is known from the practice that the Universal is capable of development, it is necessary to indicate the reasons for such development. From the authors' viewpoint, this is the energetic fullness of the Universal and the possibility of changes in the manifestations of this fullness.

4. Key directions of task types specification

Since the concept of a task and the concepts involved in defining a task, such as need, formulations, conditions, etc., are singles, the specification of task types is entirely determined by the specification of the types of single ones. In turn, the species of the single are determined by the development of the Universal, starting from the single, as such.

The research [12] allowed to build a hierarchical picture of singles species up to the level of conscious comprehension activity, whose distinctive feature is the presence of tasks and methods for their solution. This picture in a simplified form, conditionally divided into layers, is shown in figure 1. The layers in the figure are separated by horizontal lines; attributes of layer separation are indicated on the right side. In the Figure 1 OS is the operating (active) single. In accordance with the picture construction procedure, the terms used in Figure 1 were selected as the best (from the authors' viewpoint) fits for the content of the revealed concept, rather than as a result of the generalization of various concepts denoted by one term. For example, the term "recognition" in figure 1 means establishing or denying ~~of~~ the fact that a single belongs to a certain set, the term "detection" means establishing the fact of the presence or absence of a single, the term "identification" means the collation of singles in order to establish or disprove their similarity.

The highest layer is the layer of the only single. The next layer focuses on a set of singles. This is followed by a layer that takes into account the possibility of connections between the singles, then a layer, which is characterized by the possibility of singles variability and, finally, a layer where additional possibilities open up associated with operating (active) singles, which may have the potential for consciousness and comprehension.

We would emphasize once again that the decomposition shown in figure 1 is extremely simplified; theoretically, a more detailed one can be proposed. The borders of these layers are blurred in reality. But in our case, the nature of the division into layers is not that important: the main thing is to illustrate the existence of a consistent logically based procedure for identifying and systematizing the types of objects, based on the possibilities that are consistently opened in the process of the development of the Universal.

Further specification directions also generate the types of single but already detailed objects, presented in figure 1. First of all, this detailing of objects appears through the need to describe the features of specific singles ones through their inherent characteristics. Since characteristics are the product of comprehension activity, they cannot be considered without employing the concept of subject.

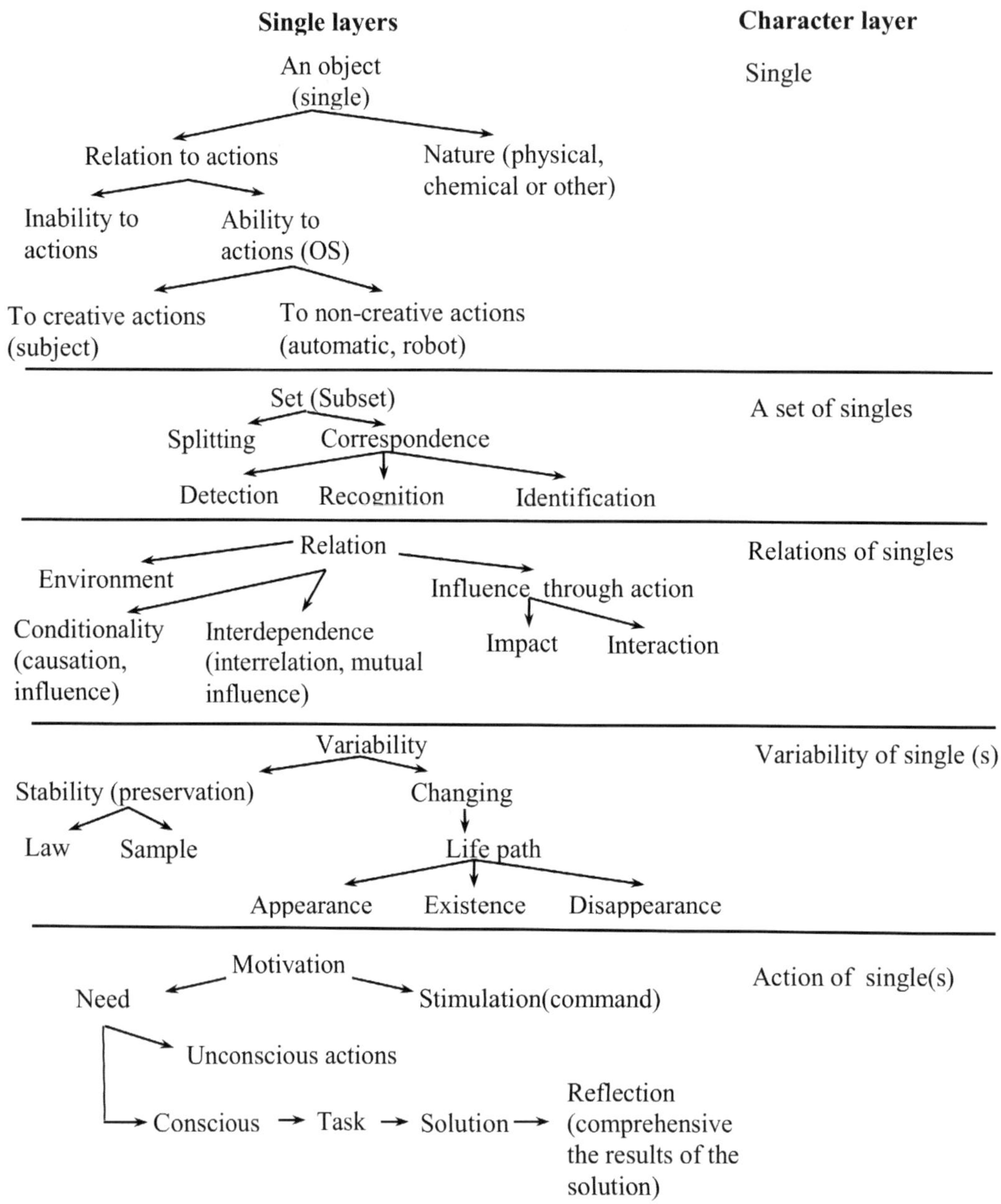

Figure 1. The hierarchy of key classes of objects at the highest level of consideration

By analyzing the characteristics in more detail, it is easy to see that they can also be systematized according to the generic principle, forming their own hierarchy. We propose to designate the characteristics corresponding to these levels as attributes, indicators, values, and constants. Selected terms are recommended; it is their semantic content that actually matters.

Here, attributes are the most common, qualitative characteristics irreducible to each another. Indicators are also qualitative characteristics, but they depend on attributes or other indicators. A value is a property that is common in a qualitative sense for a certain set of homogeneous (by the totality of attributes and indicators) singles, and in a quantitative (estimated, orderable) relation, single for each element of this set. Values are divided into intensive (estimated) and extensive, for which the axioms of addition are valid. And, finally, a constant is a characteristic permanent in its quantitative content.

The above hierarchy allows to define a complex object as an object with several independent (i.e., not reducible to each other) characteristics at the same hierarchy level. From this viewpoint, multidimensionality is a type of complexity that involves the presence of several independent scalar values characterizing a complex object.

Next direction of the origination of task types is the nature of the characteristics. From this viewpoint, the characteristics are divided into real (objectively existing) and phantom (purely subjective, unrelated to reality). The first, in turn, are subdivided into two classes. The characteristics of the first class are objective in their essence, the others are subjective, but, at the same time, they reflect the properties of real objects as well as the laws and regularities applicable to them. We conditionally designate them as physical and nonphysical, meaning in this case their attitude towards objective reality rather than their membership in a specific field of knowledge. Physical characteristics reflect directly the nature of things (e.g., resistance, valence), whereas non-physical ones are auxiliary and reflect the characteristics of real objects only indirectly (e.g., reliability, appearance and characteristics of a real object model). The description of objects through their characteristics within the framework of a task has a model nature which reflects the most significant characteristics of the objects from the viewpoint of the task under consideration. Thereby, variable quantitative characteristics of such descriptions not necessarily tied to their nature will be referred to as parameters.

The multiplicity of task types is determined not only by these directions, but also by their conjunctions, possible consistent combinations of tasks of different types, as well as by transitions to tasks with combined objects or the result of their interpenetration.

5. Development outlook and future work

Consideration of the variety of tasks that are logically derived from the picture briefly presented in this paper on the one hand, and the variety of practical tasks on the other, results in the need to establish a correspondence between these varieties. This requires the development of an extremely formalized procedure for determining the place of each real-life task in the overall picture and, at the same time, determining the sources of the origin of this task and the component tasks from which the initially considered task was formed. It is quite obvious that the availability of knowledge about the methods for solving the above mentioned component tasks and the rules for combining them opens up the way to considering potentially possible (in addition to the invented) methods for solving the original task. This proves the practical significance of the direction being developed.

From a theoretical viewpoint, it is critically important to identify the sources of origin of various key tasks, such as, for example, creating a transformation, ensuring the transformation invariance to the influence of interfering factors, measurement, monitoring, prediction, and diagnostics tasks. Answers to relevant questions will allow to specify unambiguously interpreted definitions of relevant concepts. The traditional way of defining such concepts, based on accumulated experience, due to its multidimensionality often leads to a number of definitions sometimes conflicting with each other.

To illustrate some opportunities opened up by the Approach, we consider a part of a logical chain arising from the need to organize the solution of a task regardless of its type. The solution of the task under complete or partial uncertainty of object's characteristics preventing the attainment of the desired result causes the need for developing control actions aimed at overcoming the specified obstacle. Under these conditions, it is required, first of all, to establish the fact of the need to perform control actions. This necessitates (i) the detection of monitored characteristics, which make it possible to indicate the necessity of the mentioned actions and to check the conformity of these characteristics to the norm, i.e., a state in which no control action is required. The statement of the need for impacts, in turn, raises the need (ii) for identifying events which cause the appearance of a nonconformity and which should be affected in order to eliminate the nonconformity. And, finally, there is a need (iii) to determine the characteristics of impacts (points of application and, if these characteristics are values, the intensity of impacts). By convention, we denote the selected needs, respectively, as needs in monitoring, diagnosing and determining control actions. If we take into account the aspect of localization of the state of characteristics in the process of object changes, then each of the monitoring (in relation to controlled characteristics) and the diagnostics (in terms of conditioning characteristics) tasks is detailed in three subtasks: per the current state, per the previous one, and per the state resulting from the current one.

It can be noted that the known definitions of diagnostics (see, for example, [13–15]) do not contradict the need for detecting events that cause the appearance of a discrepancy; this need is detected by logical inference rather than by generalizing various formulations of diagnostics.

Thus, following the Approach allows not only to establish the places of certain tasks in the overall picture, but also to specify their wording in the most accurate, unambiguous way due to the principle of establishing the independence of tasks once their solution methods are independent.

Given the exceptional practical significance of operating procedures with extensive characteristics, one of most important tasks also should be recognized as the task of establishing a relationship between quality and intensive characteristics on the one hand and extensive characteristics on the other. For physical extensive quantities, this task is reduced to the classical measurement task based on exemplary measures. Otherwise, we are talking about a generalized theory of measurement, which includes using the significance scores given by experts or as a result of testing procedures. These scores are accepted as the values of the corresponding extensive characteristics. The corresponding field of knowledge is discussed, in particular, in the works by L. Finkelstein (e.g., [16]).

At the same time, it is known [11] that a measurement task (even in its generalized version) requires solving a number of simpler tasks: dividing the set of singles into subsets, designating subsets (forming a scale), localizing the single to be measured on this scale, and designating the result localization.

The results of the application of the Approach indicate the existence of a small set of fundamental types of tasks arising at the highest level of generalization. It is these tasks that underlie all possible types of tasks, and the methods for their solution determine the methods for solving possible tasks. Their revealing and systematization is an important component of the theory of tasks requiring further research.

6. Conclusions

The material presented in this study corresponds to initial stages of building an interdisciplinary theory focused on identifying and systematizing the types of tasks of purposeful activity and methods for solving them, as well as formalizing the synthesis of methods for solving various tasks.

The study results outline possible directions of specifying the tasks of purposeful activity, and, hence, the methods of their solution, regardless of the related knowledge area.

Built general overview of the types of objects and, tasks underlie not only the development of a generally acceptable conceptual and terminological toolkit, but also the establishment of interdisciplinary bridges in the synthesis of highly efficient solutions at the junctions of various fields, as well as innovations in the learning processes of relevant disciplines, starting at the interdisciplinary level.

The material of this study claims to be new mainly in terms of the comprehension of logical connections arising from the types of objects, their characteristics and levels of generalization and, therefore, determining the species diversity of tasks and methods for their solution.

The authors hope that the study will stimulate discussion and further research aimed at harmonizing within the framework of a single whole multiple substantive views in various fields of knowledge.

References

[1] https://dic.academic.ru/dic.nsf/business/19277
[2] https://en.wikipedia.org/wiki/Parameter
[3] Kneller V 2003 Measurement, control and other processes: to the problem of knowledge systematization *Proc. XVII IMEKO World Congress* (June 22-27, 2003, Dubrovnic, Croatia) pp°1119–24
[4] Kneller V 2004 Quantities transformation: general techniques and accuracy improvement *Proc.*

IMEKO TC-4 XIII Int. Symp. on Measurements for Research and Industry Applications and IX European Workshop on ADC Modeling and Testing (29 September – 1 October, Athens, Greece) vol 2 pp 537–42
[5] Kneller V 2007 Physical values transformation: specificity, relationships with other processes, and the methods for basic problems solution *Sensors & Systems* 12 pp 58–67 (in Russian)
[6] Kneller V 2009 Reveal and systematization of quantities transformation methods *Proc. XIX IMEKO World Congr.* (6–11 September, 2009, Lisbon, Portugal) pp 1118–22
[7] Kneller V and Fayans A 2010 Revealing and systematization of methods for the implementation of purposeful processes *Proc. Russian Conf. with international participation "Technical and software of control, monitoring and measurement systems"* CMM'10 ISC RAS (18–20 October Moscow Russia) pp 112–26 (in Russian)
[8] Kneller V and Fayans A 2011 General methods of quantities transformation: revealing and systematization *Sensors & Systems* 11 pp 64–75 (in Russian)
[9] Kneller V and Fayans A 2015 Methods to ensure invariance transformation processes: revealing and systematization *Sensors & Systems* 4 pp 55–67 (in Russian)
[10] Kneller V and Fayans A 2015 Methods to Ensure the Invariance of Transformation Processes: Reveal and Systematization *Automation and Remote Control* **76** 12 pp 2266 - 79
[11] Fayans A 2018 Revealing and systematization of methods of analog-to digital conversion *Sensors & Systems* 1 pp 61–76 (in Russian)
[12] Kneller V, Fayans A 2019 About the approach to systematization of tasks and methods for their solution *Proc. XXIII All-Russian Conf. on Control Problems RCCP-2019* June 17-20 Moscow in print (in Russian)
[13] https://dal.academic.ru/dic.nsf/ruwiki/17181
[14] https://epistemology_of_science.academic.ru/158/диагностика
[15] 1997 Big Encyclopedic Dictionary *Moscow Big Russian Dictionary* (S-Pb: Norint) (in Russian)
[16] Finkelstein L 2008 Problems of widely-defined measurement *12th IMECO TC1 & TC7 Joint Symp. on Man Science & Measurement* (3-5 September, Annecy, France) pp 23-9

Journal of Physics: Conference Series **1379** (2019) 012012 doi:10.1088/1742-6596/1379/1/012012

Dissolved oxygen and pH detection system based on the fluorescence characteristics of coumarin-modified CdSe quantum dots

Chao Zheng, Wei Tao*, Quan Liu, Na Lv, Kan Wang, Hui Zhao

Department of Instrument Science and Engineering, Shanghai Jiao Tong University, Shanghai 200240, China

E-mail: taowei@sjtu.edu.cn

Abstract. In the field of biomedicine, the pH and dissolved oxygen (DO) content of the cell metabolism fluid can reflect the health of the cell and provide an auxiliary reference for the early detection of cancer. Based on the preliminary research results of the same group, a dual-parameter synchronous detection system for cell metabolic fluid was established. The system is based on the fluorescence characteristics of 7-Amino-4-methylcoumarin-modified CdSe quantum dots. Both of pH and DO are solved by the voltage signal of the photodetection module. Firstly, the characteristics of 7-Amino-4-methylcoumarin-modified CdSe quantum dots were introduced. Then the dual-parameter detection method was briefly proposed. Next, the hardware components and software functions of the system were introduced. Finally, the detection algorithm was proposed and the solution model was built according to the calibration experiment. The verification experiment showed that the sensitivity for detecting pH was 0.05 and the accuracy was 0.07. The sensitivity for detecting DO was 0.04 mg/L, and the accuracy was 0.07 mg/L. As the prototype machine, the system can basically realize the dual-parameter detection of the cell metabolism liquid with high accuracy, fast speed and small loss of the original liquid.

1. Introduction

Cancer is one of the most serious diseases threatening human health. “Cancer statistics, 2018” predicted that the number of new cancer cases in the United States would be 173,350 and the number of new cancer deaths would be 609,640 in 2018 [1]. Early screening for cancer is one way to reduce cancer mortality. The abnormal metabolic process of cancerous cells leads to different parameters of the cell metabolic fluid than normal cells. The most representative parameters are pH and DO. The pH of most cancer cell metabolites is 6.0-7.0, which is lower than that of normal cells by 7.3-7.4 [2]. The DO of abnormal cell metabolizing fluid is lower than that of normal cell [3]. Therefore, the pH and DO of the cell metabolic fluid can reflect the health of cells and provide a reliable basis for early diagnosis of cancer [4].

At present, the detection methods of the pH of the cell metabolism liquid mainly include MRI method, microelectrode method, fiber colour modulation method and fluorescence spectroscopy. The MRI method has low measurement accuracy and a large volume of equipment, which is not suitable for general laboratory. The microelectrode method has many sources of error, high measurement cost and poor reliability. The colour modulation of optical fiber has many research results. Tou et al. realized the optical fiber pH sensor by means of interferometry [5]. Fan S F used the dual LED for reference and produced a commercial fiber pH sensor with high measurement accuracy, but the production cost is high [6]. Fluorescence spectroscopy method is widely used, but the commonly used organic dyes have poor optical stability, wide and overlapping emission spectrum, so the signal is difficult to obtain [7].

The methods of measuring DO mainly include iodometry, electrochemical probes and optical sensors. The iodometry method requires much liquid to be tested, and the operation is very complicated. The electrochemical probe requires the test solution to ensure a certain flow rate. The optical DO sensor is widely used, but they must be stored in a humid environment when not working, which leads to some trouble in using it. The DO measurement chip proposed by Shao et al is effective for DO measurement in solution [8], but it is difficult to use for a small amount of solution. Boonya *et al.* developed a method for measuring DO under extremely acidic conditions [9], but not advisable to cell living environments.

However, simultaneous detection of DO and pH has not been reported. a method for detecting DO and pH is proposed in this paper, which is based on the fluorescence characteristics of 7-Amino-4-methylcoumarin-modified CdSe quantum dots. The fluorescence signal of two wavelengths can be obtained by excitation of one wavelength, and these can be calculated to obtain the pH and DO. Based on this, two photomultiplier tubes are used to construct a dual-parameter detection system.

2. Principle of Dual-parameter Detection

The coumarin-modified CdSe quantum dots have a large Stokes shift, a wide excitation wavelength range and a narrow emission wavelength range. The fluorescence intensity and stability are better than ordinary fluorescent dyes, and they have good biocompatibility. A method based on this for detecting pH and DO is proposed, and the principle is shown as figure 1. CdSe quantum dots containing ZnS nucleocapsid were modified by 7-Amino-4-methylcoumarin. Under the ultraviolet light with the wavelength of 365 nm, CdSe quantum dots will produce fluorescence with the wave peak of 625 nm, and 7-Amino-4-methylcoumarin will produce fluorescence with the wave peak of 485 nm. The fluorescence of CdSe quantum dots is sensitive to pH but not sensitive to DO. The fluorescence will be quenched to different degrees due to pH. The fluorescence of 7-Amino-4-methylcoumarin is sensitive to DO and pH. The pH and DO of the solution can be solved by measuring the fluorescence intensity of the 7-Amino-4-methylcoumarin-modified CdSe quantum dots at these two wavelengths.

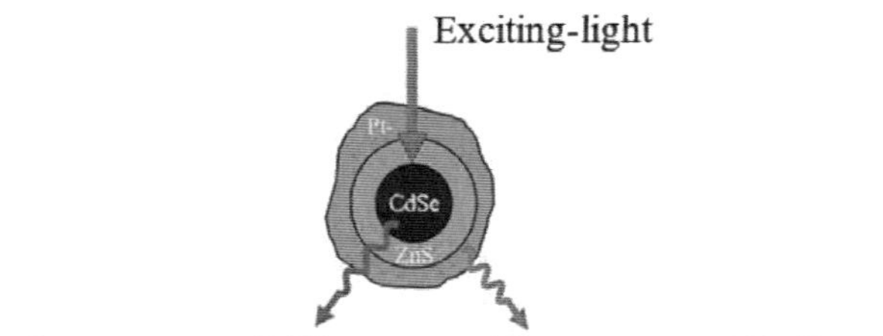

Figure 1. Principle of dual-parameter detection.

3. Dual-parameter Detection System

Figure 2 is a block diagram of the dual-parameter detection system. The system frame is made of black plexiglass and evenly painted with black matt lacquer. The light source is selected from CUN6GF1A UVLED353 violet lamp bead of Seoul SVC. The two photomultiplier tube modules are selected from the same batch of Hamamatsu H10493-013 photomultiplier tube module. The cuvette is 400 μL slit cuvette, and the data acquisition card is MP411L Module. The self-made constant voltage power supply supplies power to the light source and two photomultiplier modules. The light source uses 3.1 V, the two photomultiplier modules use ±15 V for power and 0.9 V for control voltage. The exit tube and the incident tube are in contrast to the cuvette, preventing the beam from diffusing into other areas and minimizing the effects of other light that may be present in the environment. A 365 nm filter is mounted at the rear end of the exit barrel. The rear ends of the two incident cylinders are symmetrically mounted with a 625 nm filter and a 485 nm filter respectively. Two photomultiplier tubes are symmetrically mounted at the rear of the two entrance tubes and connected to the data acquisition module. The data acquisition module is connected to the PC. The data is transmitted to the host computer and processed. Figure 3 is a physical diagram of the detection system.

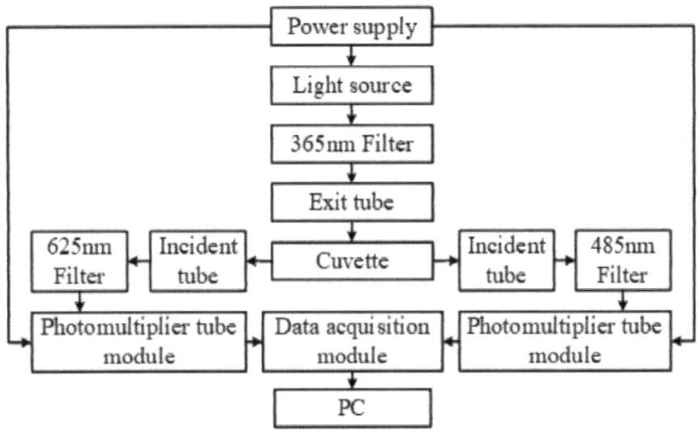

Figure 2. Block diagram of two-parameter measuring system.

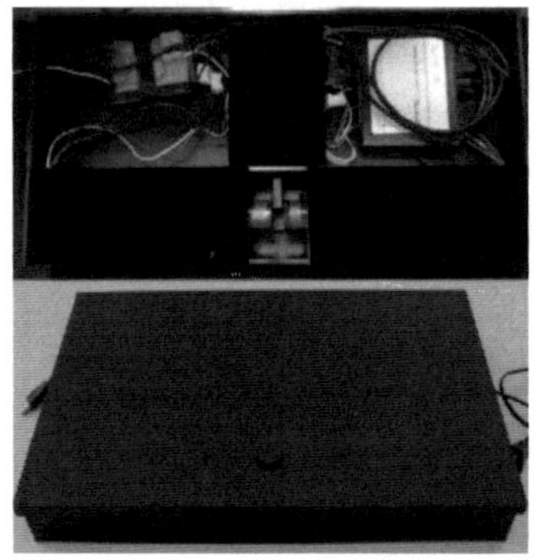

Figure 3. Image of two-parameter measuring system.

The software of dual-parameter detection system displays the two voltage values collected by the data acquisition module in real time, and separately filters and processes the two voltage signals for subsequent analysis. Figure 4 shows the interface of the host computer software.

The sensitive substance used in the dual-parameter detection system is the self-made 7-Amino-4-methylcoumarin-modified CdSe quantum dots, as shown in figure 5.

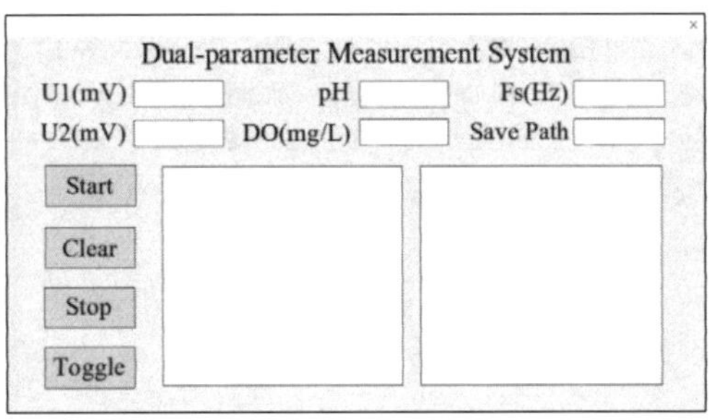

Figure 4. PC interface of two-parameter measuring system.

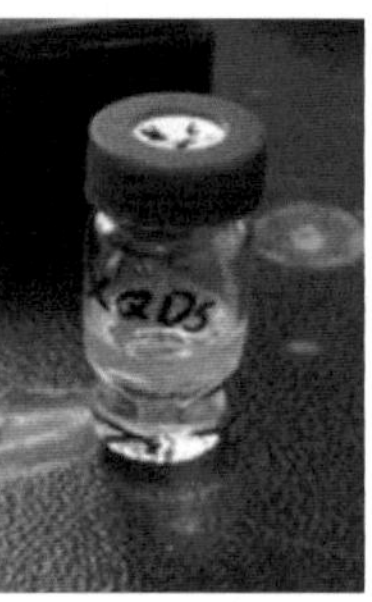

Figure 5. 7-Amino-4-methylcoumarin-modified CdSe quantum dots.

When the system works, a stable power supply will supply power to the light source and the two photomultiplier modules. The light source emits the light with a wavelength of about 365 nm, which passes through the 365 nm filter and is irradiated onto the cuvette side of the detection chamber through the exit tube. The solution in the cuvette will fluoresce when excited. Fluorescence is detected by two identical photomultiplier modules after passing through symmetric filters on both sides of the detection chamber. When the working condition is stable, the voltage signal output by the photomultiplier module is proportional to the collected light intensity, so the pH and DO of the solution can be calculated synchronously by the voltage signals of the two photomultiplier modules. The calculation process is: firstly, the pH value of the solution is calculated from the voltage corresponding to the wavelength of 625 nm, then the DO value of the solution is calculated according to the pH result and the voltage corresponding to 485 nm.

Journal of Physics: Conference Series **1379** (2019) 012012 doi:10.1088/1742-6596/1379/1/012012

4. Calibration and Verification

In the dual-parameter detection system, the photomultiplier module is used to detect the fluorescence intensity. It is necessary to establish a detection model by using the output voltage of the two photomultiplier modules and the two parameters of the solution. The fluorescence intensity of 625 nm is mainly related to the pH of the solution. It is necessary to establish a detection model of the pH and the output voltage corresponding to the 625 nm, which is called Model 1. The fluorescence intensity of 485 nm is mainly related to the pH and DO of the solution. Under the condition of knowing the pH, it is necessary to establish a detection model of pH, DO and the output voltage corresponding to the 485nm, which is called Model 2. By substituting Model 1 into Model 2, the final model for dual-parameter simultaneous detection can be obtained.

4.1. pH Detection Calibration

The experiment was carried out by selecting eight groups of PBS buffers having pH of 6.1, 6.3, 6.52, 6.72, 7.07, 7.33, 7.63, 7.87. The pH was measured by a pH meter as standard values. After extensive experiments, it was found that the best results were obtained by mixing 297 μL of PBS buffer with 3 μL of 7-Amino-4-methylcoumarin-modified CdSe quantum dots. 297 μL of the buffer was mixed with 3 μL of these quantum dots and placed in a cuvette of the detection system. The output voltage corresponding to the wavelength of 625 nm was recorded, and the relationship between the voltage and the pH at this wavelength was investigated. Figure 6 shows the test results for each set of experiments. The average of each set was taken as a stable output voltage, and the relationship with the pH of each group was established, as shown in figure 7.

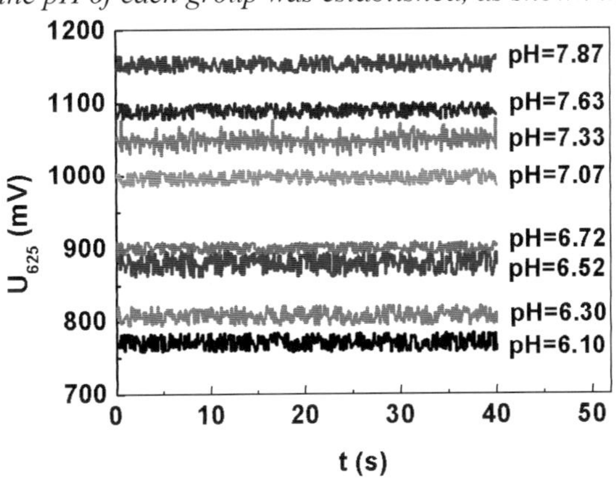

Figure 6. Experimental results of eight groups.

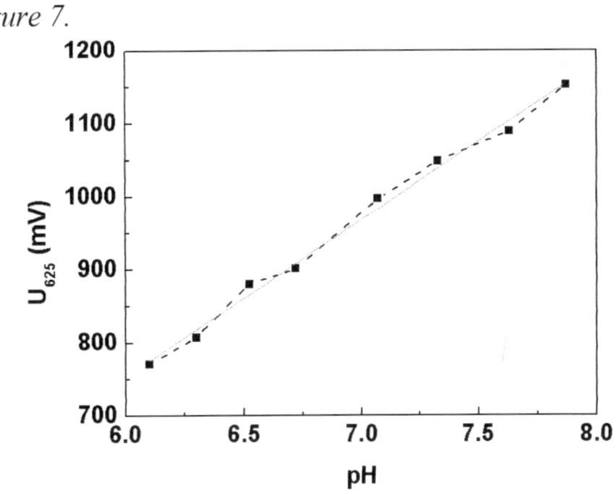

Figure 7. Relationship between pH and voltage.

The results of figure 7 can be fitted to obtain the linear relationship between the fluorescence intensity at 625 nm and the output voltage of the photomultiplier modules, which is Model 1 as follows.

$$U_{625} = 214.432x - 532.746 \quad (1)$$

Where x is pH, U_{625} is the voltage of photomultiplier module corresponding to 625nm, R^2 is 0.9922. The standard deviations of the eight experimental data were 8.463 mV, 7.763 mV, 10.740 mV, 5.097 mV, 6.447 mV, 8.815 mV, 6.476 mV, 7.164 mV. The maximum standard deviation was taken to calculate and the sensitivity limit for detecting pH was 0.05.

4.2. DO Detection Calibration

19 sets of PBS buffers of different pH were selected, and each set of buffers was taken out in four equal portions for deoxygenation to obtain different DO content. The deoxygenation operation is to introduce nitrogen into the solution for different time. 76 groups of liquid to be tested were obtained with different pH and DO. The DO content of each group of solutions was measured using a Hash Polymetron 9582

Dissolved Oxygen Analyzer and used as a standard value. For each set of experiments, 297 μL of buffer was mixed with 3 μL of such quantum dots and placed in the cuvette of the detection system. The relationship between the output voltage corresponding to the wavelength of 485 nm and two parameters of the solution was recorded. The output voltage of each set of experiments was averaged and shown in figure 8. It can be seen from the data that when the pH is constant, the output voltage has a good linear relationship with DO, but the slope and intercept of the linear relationship vary with pH. The data at each pH was linearly fitted and expanded. The measured DO maximum of 9.25 mg/L and the measured minimum of 8.18 mg/L were used as the endpoints, as shown in figure 9.

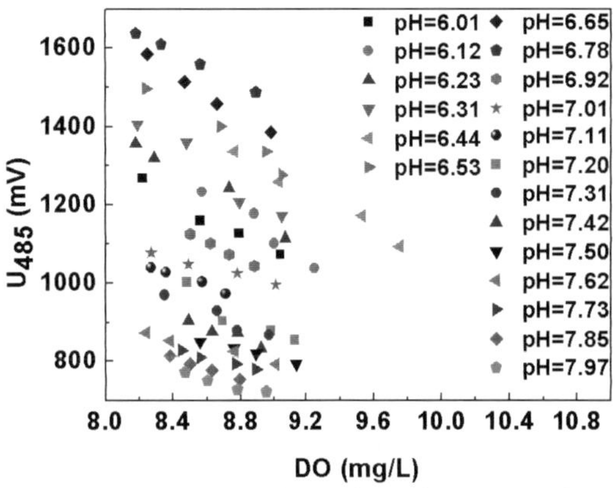

Figure 8. Experimental results of seventy-nine groups.

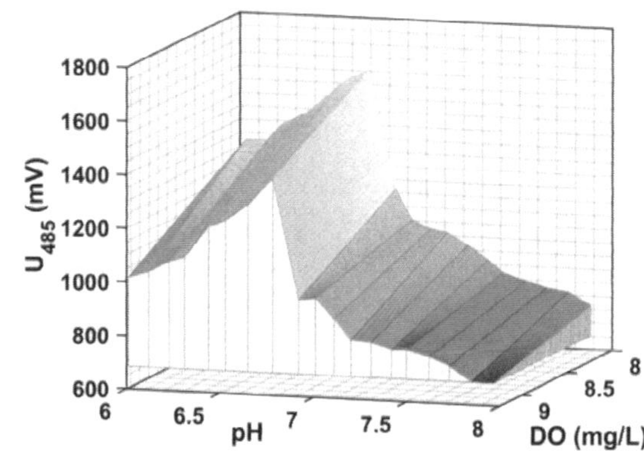

Figure 9. Fitting results of nineteen groups.

Taking the pH and the voltage as input and DO as output, a mathematical model was established, and the establishment process is as shown in figure 10. The detection model of the voltage, pH and DO was finally obtained as follows, which is Model 2.

$$y = 9.25 - \frac{1.07}{f_1(x) - f_2(x)}\left(3.95319U_{485} - 2349.47055 - f_2(x)\right) \tag{2}$$

$$f_1(x) = \begin{cases} -49668.50354x^4 + 1.26801\times10^6x^3 - 1.21316\times10^7x^2 \\ +5.15547\times10^7x - 8.21073\times10^7 (6.01 < x \le 6.78) \\ -32719.13809x^5 + 1.21945\times10^6x^4 - 1.81686\times10^7x^3 + 1.35268\times10^8x^2 \\ -5.03245\times10^8x + 7.48472\times10^8 (6.78 < x < 7.97) \end{cases} \tag{3}$$

$$f_2(x) = \begin{cases} -4.60869\times10^5x^6 + 1.77387\times10^7x^5 - 2.84367\times10^8x^4 + 2.4303\times10^8x^3 \\ -1.16785\times10^{10}x^2 + 2.9918\times10^{10}x - 3.19219\times10^{10} (6.01 < x \le 6.78) \\ -1206.43972x^3 + 27538.43289x^2 - 210258.72245x + 538467.76128 \\ (6.78 < x < 7.97) \end{cases} \tag{4}$$

Where x is pH, y is DO (mg/L), and U_{485} is the voltage corresponding to 485 nm (mV). In each set of data, the maximum voltage standard deviation is 12.823 mV, and the detection sensitivity limit is 0.04 mg/L. The pH x in this model was replaced by Model 1, and the whole model was obtained, whose input are U_{625} and U_{485}, and the output are pH and DO. Process is shown in figure 11

4.3. Dual-parameter Detection Verification

Eight groups of PBS buffers were selected for the verification. The pH values of the eight groups of solutions were respectively measured by a hand-held pH meter as standard values, and the DO values of the eight groups of solutions were respectively measured by a Hash Polymetron 9582 dissolved oxygen analyzer as a standard value. For each set of solutions, 297 μL of PBS buffer was mixed with 3 μL of 7-Amino-4-methylcoumarin-modified CdSe quantum dots and detected using the dual-parameter detection system. The output voltages were recorded and calculated according to the above model and the pH and DO measured by the system were obtained. Table 1 shows the system verification results.

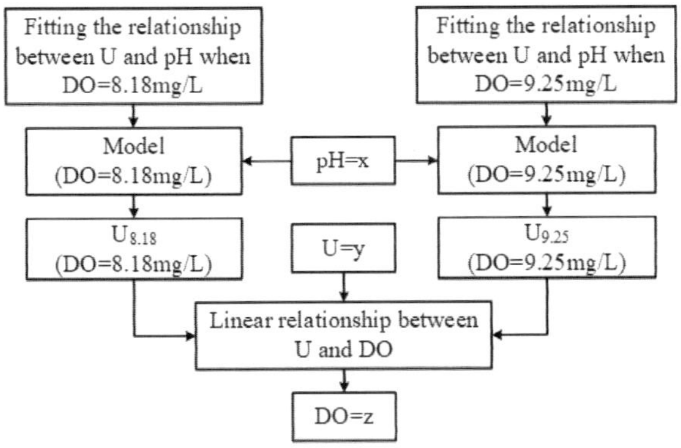

Figure 10. Established process of Model 2.

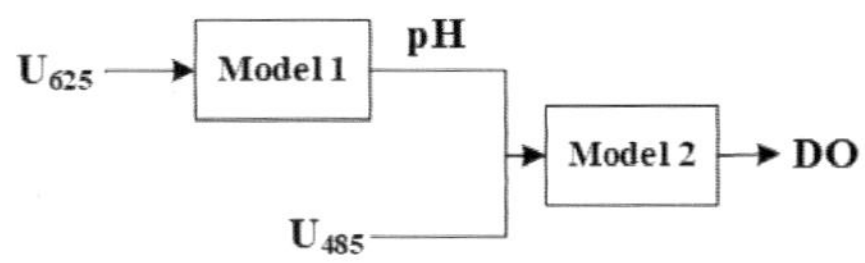

Figure 11. Established process of the whole model.

Table 1. Formatting sections, subsections and subsubsections

Group	pH standard value	pH measured value	Error	DO standard value (mg/L)	DO measured value	Error (mg/L)
1	6.10	6.079	-0.021	9.04	9.017	-0.023
2	6.30	6.251	-0.049	9.12	9.103	-0.017
3	6.52	6.589	0.069	8.97	9.004	0.034
4	6.72	6.686	-0.034	9.03	9.002	-0.028
5	7.07	7.135	0.065	8.82	8.755	-0.065
6	7.33	7.375	0.045	9.03	8.982	-0.048
7	7.63	7.566	-0.064	8.94	8.879	-0.061
8	7.87	7.860	-0.010	9.05	9.106	0.056

It can be seen from the above results that the dual-parameter detection system can realize the simultaneous measurement of pH and DO. The sensitivity of measuring pH satisfies 0.05, the accuracy satisfies 0.07, the sensitivity of measuring DO satisfies 0.04 mg/L, and the accuracy satisfies 0.07 mg/L.

5. Conclusion

In this paper, a dual-parameter detection system based on the analysis of fluorescence characteristics of 7-Amino-4-methylcoumarin-modified CdSe quantum dots was established. The hardware structure, software functions, and the work principle were introduced. Then the detection algorithms of pH and DO were proposed respectively, and the two parameters were decoupled. Finally, the mathematical calculation formula and detection sensitivity limit of dual-parameter detection were given. The experimental results showed that the sensitivity of the two-parameter detection system was 0.05, the accuracy was 0.07, the sensitivity of detecting DO was 0.04 mg/L, and the accuracy was 0.07 mg/L. As a prototype machine, the system basically realized the dual-parameter detection of liquid, with high accuracy, fast speed, small loss of the original liquid and high reference value.

References

[1] Siegel R L , Miller K D and Jemal A 2018 Cancer statistics, 2018. *Ca-Cancer J. Clin.* **60**(5) pp 277–300

[2] Felber A E, Dufresne Marie-Hélène and Leroux J C 2012 pH-sensitive vesicles, polymeric micelles and nanospheres prepared with polycarboxylates *J. Advanced Drug Delivery Reviews* **64**(11) pp 979–92
[3] Koppenol W H , Bounds P L and Dang C V 2011 Otto Warburg's contributions to current concepts of cancer metabolism *J. Nature Reviews Cancer* **11**(5) pp 325–37
[4] Vander Heiden MG, Cantley LC and Thompson CB 2009 Understanding the Warburg effect: the metabolic requirements of cell proliferation *J. Science,* 324(5930) pp 1029–33
[5] Tou ZQ, Chan CC, Hong J, Png S, Eddie KMT and Tan TAH 2014 Double-pass Mach–Zehnder fiber interferometer pH sensor *J.Biomed. Opt.* **19**(4) pp 047002-1 – 047002-7
[6] Gu B, Yin MJ, Zhang AP, et al. 2009 Low-cost high-performance fiber-optic pH sensor based on thin-core fiber modal interferometer *J. Optics Express* **17**(25) pp 22296–302
[7] Redy -Keisar O, Huth K, Vogel U *et al.* 2015 Enhancement of fluorescent properties of near - infrared dyes using clickable oligoglycerol dendrons *J. Org. Biomol. Chem.* **13** pp 4727–32
[8] Shao Q , Biel M A and Ashkenazi S 2014 Noninvasive tumor oxygen imaging by photoacoustic lifetime imaging integrated with photodynamic therapy[J]. *Proc. SPIE - The Int. Soc. Opt. Eng.* **8931**(1) pp 1115–21
[9] Charnnok B , Suksaroj T, Boonswang P and Chaiprapat S 2013 Oxidation of hydrogen sulfide in biogas using dissolved oxygen in the extreme acidic biofiltration operation[J]. *Bioresource Technol.* **131** pp 492–499

Reduction of the bias of measurement uncertainty estimates with significant non-linearity of a model equation

I P Zakharov[1,2], **P I Neyezhmakov**[1], **O A Botsiura**[2]

[1]NSC "Institute of Metrology", Kharkiv, Ukraine

[2]KhNURE, Kharkiv, Ukraine,

E-mail: newzip@ukr.net, pavel.neyezhmakov@metrology.kharkov.ua, olesia.botsiura@nure.ua

Abstract. The analysis of the nonlinear model equation is carried out. The nonlinear model equation in a Taylor series is expanded. It is shown that the bias in the estimation of the combined standard uncertainty by using the Law of Propagation of Uncertainty is due to the terms of the expansion of the second degree. To eliminate this bias, it is necessary to take into account the kurtosis of the input quantities distributions. A finite increments method for obtaining reliable estimates of the uncertainties contributions for nonlinear model equations is proposed. A practical example of calculation of comparison loss in microwave power meter calibration by various methods was considered. Estimate of the measurement uncertainty obtained with Law of Propagation of Uncertainty in this case has a bias. The result obtained with the finite increments method coincides with the results obtained using the Monte Carlo method.

1. Introduction

In general, the model of the measurand Y as a function of several input quantities $X_1, X_2, ..., X_N$, can be represented in the following form [1]:

$$Y = f(X_1, X_2, ..., X_N). \quad (1)$$

In the GUM [1] the value of the measurand y and its standard uncertainty $u(y)$ are calculated by the formulas:

$$y_{\mathrm{GUM}} = f(x_1, x_2, ..., x_N), \quad (2)$$

$$u(y) = \sqrt{\sum_{j=1}^{N} c_j^2 u_j^2 + 2\sum_{j=1}^{N-1}\sum_{i=j+1}^{N} c_j c_i \operatorname{cov}(x_j, x_i)}, \quad (3)$$

where $x_1, x_2, ..., x_N$ are estimates of the input quantities $X_1, X_2, ..., X_N$; u_j is standard uncertainty of the j-th input quantity; $\operatorname{cov}(x_j, x_i) = r(x_j, x_i) u_j u_i$ is covariance of the i-th and j-th input quantities; $r(x_j, x_i)$ is correlation coefficient; $c_j = \partial y / \partial x_j$ – sensitivity coefficient for the j-th input quantity.

In the nonlinear equation (1), the estimates (2) and (3) of y and $u(y)$ have a bias, the value of which depends on the type of the model equation and the values of the relative uncertainties of the input quantities [2]. The bias compensation of y in [1] is not carried out, and the bias compensation of $u(y)$ is realized in [1] by taking into account the terms of the second order of the expansion (1) in the Taylor series. In this case, it is necessary to have the values of the partial second-order derivatives function (1) with respect to the input quantities $X_1, X_2, ..., X_N$. It should be noted that the expression for an unbiased estimate of the standard uncertainty of the measurand is given in [1] only for the Gaussian probability density functions (PDF) of input quantities.

The purpose of this report is to obtain expressions that provide unbiased estimate of the standard uncertainty of the measurand for nonlinear model equations and for specified PDF of input quantities.

2. Expansion in a Taylor series of the model equation

Expression (1) can be represented as a Taylor series expansion:

$$Y = f(x_1, x_2, ..., x_N) + \sum_{j=1}^{N} c_j (X_j - x_j) + \frac{1}{2} \sum_{\substack{j=1 \\ i=1}}^{N} c_{ji} (X_j - x_j)(X_i - x_i) + R_0 , \quad (4)$$

in which $c_{ji} = \partial^2 f / (\partial x_j \partial x_i)$ – the mixed partial derivative (1) with respect to X_j and X_i; R_0 – the remainder term of the series.

The mathematical expectation of the expression (4) for $R_0 = 0$ will have the following form:

$$E(Y) = E[f(x_1, x_2, ..., x_N)] + \sum_{j=1}^{N} c_j E(X_j - x_j) + \frac{1}{2} \sum_{j,i=1}^{N} c_{ji} E[(X_j - x_j)(X_i - x_i)] . \quad (5)$$

Taking into account that $E(X_j - x_j) = 0$, and for uncorrelated input values

$$E[(X_j - x_j)(X_i - x_i)] = \begin{cases} u_j^2, & \text{when } j = i; \\ 0, & \text{when } j \neq i, \end{cases}$$

we get the following expression:

$$E(Y) = y = f(x_1, x_2, ..., x_N) + \frac{1}{2} \sum_{j=1}^{N} c_{jj} u_j^2 , \quad (6)$$

where $c_{jj} = \partial^2 f / \partial x_j^2$ – is the partial derivative of the second order of the measurand with respect to the j-th input quantity.

Thus, to reduce the bias of measurand estimate when the nonlinear model equation is used, it is necessary to take into account the standard uncertainties of input quantities.

Let's write down the expression for the difference between the measurand and its mathematical expectation (6):

$$Y - E(Y) = Y - f(x_1, x_2, ..., x_N) - \frac{1}{2} \sum_{j=1}^{N} c_{jj} u_j^2 =$$

$$= \sum_{j=1}^{N} c_j (X_j - x_j) + \frac{1}{2} \sum_{j=1}^{N} c_{jj} [(X_j - x_j)^2 - u_j^2] + \sum_{j=1}^{N-1} \sum_{i=j+1}^{N} c_{ji} (X_j - x_j)(X_i - x_i) . \quad (7)$$

The mathematical expectation of the square of the resulting expression will be an unbiased estimate of its variance $u_{ub}^2(y)$:

$$E[Y - E(Y)]^2 = u_{ub}^2(y) = \sum_{j=1}^{N} c_j^2 u_j^2 + \frac{1}{4} \sum_{j=1}^{N} c_{jj}^2 (\mu_j - 1) u_j^4 + \sum_{j=2}^{N} \sum_{i=1}^{j-1} c_{ji}^2 u_j^2 u_i^2 , \quad (8)$$

where $\mu_j = E[X_j - x_j]^4 / u_j^4$ – is the normalized fourth-order central moment of the *j*-th input quantity. The values of μ for different probability density function (PDF) are given in Table 1.

Table 1. The values of μ for different PDF of the input quantities.

PDF	μ
Arcsine	1.5
Uniform	1.8
Triangular	2.4
Gaussian	3
Student's with number of degrees of freedom ν	$6/(\nu-4)+3$

Thus, the value of the variance of the measurand will depend on the PDF of the input quantities. So, for example, for Gaussian PDF of all input quantities $\mu = 3$, therefore expression (8) takes the form known from the literature [3]:

$$u_{ub}^2(y) = \sum_{j=1}^{N} c_j^2 u_j^2 + \frac{1}{2}\sum_{j=1}^{N} c_{jj}^2 u_j^4 + \sum_{j=2}^{N}\sum_{i=1}^{j-1} c_{ji}^2 u_j^2 u_i^2 . \quad (9)$$

For a function of single variable, the value of the standard uncertainty of the measurand takes the following form:

$$u_{ub}(y) = \sqrt{c_x^2 u_x^2 + c_{xx}^2 u_x^4 (\mu_x - 1)/4} . \quad (10)$$

For a function of two variables, the value of the standard uncertainty of the measurand takes the following form:

$$u_{ub}(y) = \sqrt{c_1^2 u_1^2 + c_2^2 u_2^2 + \frac{1}{4} c_{11}^2 (\mu_1 - 1) u_1^4 + \frac{1}{4} c_{22}^2 (\mu_2 - 1) u_2^4 + c_{12}^2 u_1^2 u_2^2} . \quad (11)$$

It is seen from expressions (8) to (11) that in order to obtain an unbiased estimate of the standard uncertainty of the measurand, it is necessary to know the second partial derivatives of the measurand with respect to the corresponding input quantities, i.e. the model of the measurand should not only be fully known, but also twice differentiable.

3. The finite increments method

To calculate the unbiased estimate of the numerical value and the uncertainty of the measured quantity, the finite increment method [4] can be applied.

The difference partial derivative of the first order of the measurand with respect to the *j*-th input will be equal:

$$c_j^* = \frac{f[x_1,...,(x_j+u_j),...,x_N] - f[x_1,...,(x_j-u_j),...,x_N]}{2u_j} . \quad (12)$$

The difference partial derivative of the second order of the measurand with respect to the *j*-th input quantity will be equal:

$$c_{jj}^* = \frac{1}{u_j^2}\{ f[x_1,...,(x_j+u_j),...,x_N] - 2f(x_1,x_2,...,x_N) + f[x_1,...,(x_j-u_j),...,x_N]\} . \quad (13)$$

The difference partial derivative of the second order of the measurand with respect to the *j*-th and *i*-th input quantities will be equal:

$$c_{ji}^{*} = \frac{1}{4u_1u_2}\{f[x_1,...,(x_j+u_j),...,(x_i+u_i),...,x_N] - f[x_1,...,(x_j+u_j),...,(x_i-u_i),...,x_N] -$$

$$-f[x_1,...,(x_j-u_j),...,(x_i+u_i),...,x_N] + f[x_1,...,(x_j-u_j),...,(x_i-u_i),...,x_N]\}.. \quad (14)$$

For the function of N variables, the calculation of the partial difference derivatives (13) to (15) is easily realized by using a simple program executed in the MS Excel.

4. An example of calculation of comparison loss in the calibration of a microwave power meter

In subsection 9.4 of Appendix 1 to GUM [5], in the example of "Comparison loss in the calibration of a microwave power meter", the function $Y = X_1^2 + X_2^2$ for $x_1 = 0$, $x_2 = 0$ and standard uncertainties $u_1 = 0,005$, $u_2 = 0,005$ is considered. In this case, using GUM [1], we have:

$$u_{\text{GUM}}(y) = \sqrt{(2x_1u_1)^2 + (2x_2u_2)^2} = 0.$$

For the model equation under consideration, we get:

$$c_1 = 2x_1 = 0; \; c_1 = 2x_2 = 0; \; c_{11} = 2; \; c_{22} = 2; \; c_{12} = c_{21} = 0.$$

Similar results can be obtained using the formulas (12) to (14):

$$c_1^{*} = \frac{1}{2u_1}\{[(x_1+u_1)^2 + x_2^2] - [(x_1-u_1)^2 + x_2^2]\} = 0; \qquad c_2^{*} = \frac{1}{2u_2}\{[x_1^2 + (x_2+u_2)^2] - [x_1^2 + (x_2-u_2)^2]\} = 0;$$

$$c_{11}^{*} = \frac{[(x_1+u_1)^2 + x_2^2] - 2(x_1^2 + x_2^2) + [(x_1-u_1)^2 + x_2^2]}{u_j^2} = 2; \qquad c_{22}^{*} = \frac{[x_1^2 + (x_2+u_2)^2] - 2(x_1^2 + x_2^2) + [x_1^2 + (x_2-u_2)^2]}{u_j^2} = 2;$$

$$c_{12}^{*} = \frac{1}{4u_1u_2}\{[(x_1+u_1)^2 + (x_2+u_2)^2] - [(x_1-u_1)^2 + (x_2+u_2)^2] - [(x_1+u_1)^2 + (x_2-u_2)^2] + [(x_1-u_1)^2 + (x_2-u_2)^2] = 0.$$

Therefore, for uniform distributed input quantities, from equation (11), we have:

$$u_{ub}(y) = \sqrt{c_1^2u_1^2 + c_2^2u_2^2 + 0.8\frac{c_{11}^2u_1^4 + c_{22}^2u_2^4}{4} + c_{12}^2u_1^2u_2^2} = \sqrt{0.2(4u_1^4 + 4u_2^4)} = \sqrt{1.6 \cdot 0.005^4} = 0.005^2\sqrt{1.6} = 3.162 \cdot 10^{-5}.$$

A value of $u_{\text{MCM}}(y)$=3.159·10^{-5} was obtained by Monte Carlo method (MCM) with the NIST Uncertainty Machine [6], which coincides with the value $u_{ub}(y)$.

5. Conclusions

To reduce the bias of measurand estimate when the nonlinear model equation is used, it is necessary to take into account the standard uncertainties of input quantities. This approach, in contrast to GUM, is suitable in the presence of uncertainties in the input quantities, estimated by both statistical and nonstatistical methods.

It is shown that the unbiased estimate of standard uncertainty of the measurand is depending from normalized fourth-order central moments of the input quantities.

To simplify the calculation of the unbiased estimate of the numerical value and the uncertainty of the measurand, the finite increment method can be applied.

References

[1] JCGM 100 2008 *Evaluation of Measurement Data – Guide to the Expression of Uncertainty in Measurement* (GUM) (Joint Committee for Guides in Metrology)
[2] Zakharov I P, Ponomareva O V, Safaryan G G and Sergienko M P 2005 Investigation of the limits

of applicability of the linearization method for the processing of results of indirect measurements *Management Information System and Devices* (Kharkiv: Kharkiv national university of radiorlectronics) **130** pp 86-90 (in Russian)
[3] Granovsky V A and Syraia T N 1990 *Methods for Processing Experimental Data in Measurements* (Leningrad**:** Energoatomizdat) p 288 (in Russian)
[4] Botsiura O 2017 Application of the finite increments method to evaluating the results of indirect measurements *Standardization, certification, quality* (Kyiv: SE "UkrNDNC") **1**(104) pp 57-65 (In Ukrainian)
[5] JCGM 2008 *JCGM 101:2008 Evaluation of Measurement Data – Supplement 1 to the "Guide to the Expression of Uncertainty in Measurement" – Propagation of Distributions Using a Monte Carlo Method.* (Joint Committee for Guides in Metrology)
[6] https://uncertainty.nist.gov/

Journal of Physics: Conference Series **1379** (2019) 012014 doi:10.1088/1742-6596/1379/1/012014

Methods for quantitative evaluation of force and technique in competitive sport climbing

D Pandurevic, A Sutor and K Hochradel

Institute of Measurment and Sensor Technology, UMIT - Private University for Health Sciences, Medical Informatics and Technology GmbH, Hall in Tirol, Austria

E-mail: dominik.pandurevic@umit.at, klaus.hochradel@umit.at

Abstract. Increasing media attention of and progress in competitive climbing demand for novel solution in sport analysis to accompany and contribute to the evolution of climbing. We present a holistic approach to acquire movements in sport climbing. Knowledge on force and corresponding body motion helps to understand technique, performance and might even lead to injury prevention during training. Our system allows detection of peaks in force on each climbing hand and foot hold while recording the associated body position and motion. To achieve this, we equipped each hold with a three-axis force sensor and combined this with a three-dimensional markerless motion capture system based on a depth camera. Existing methods to determine finger strength and other parameters such as temperature are employed to ensure comparable and repeatable results. In addition, pre-assesement of athletes will be used to evaluate inter- and intra-individual variability among athletes.

1. Introduction

Rapid changes and growth of the climbing sport from an unknown niche towards mass and competition sport are clearly visible during the last 15 years. Artificial climbing gyms are nowadays widely spread and offer a save and defined environment for training. Climbing athletes are therefore capable to concentrate on complex motion sequences and aimed weight training to evolve the climbing sport. Other factors indicating the fast development are the increase from the native six to 12 degrees of difficulty established by the UIAA (Union Internationale des Associations d'Alpinisme) as well as the heavily increase of media attention and progress in competition. Starting with growth of participating nations since the first world cup in 1991, climbing as a competitive sport reaches its historical peak with the inclusion to the Olympic games in Tokyo 2020.

Thus, the requirements for athletes are increasing. In comparison to the initial competitive problems focusing on physical strength and stamina, athletes nowadays are required to add large amount of coordination and technique to their strength and stamina skill. This complexity presents trainer and athlete with new challenges. Hence, coaches try to support exercises by using video analysis tools and help with personal experience.

Except of the isolated way of determining the maximum force of fingers by hanging on a defined holds, their are no objective measurement methods up to date. Therefore, the following presents a new method for measuring and evaluating force and motion during climbing trials.

2. State of the art

Typical climbing athletes are distinguished from other sportsmen by a slim and light body structure with muscular torso, whereby the performance and related factors have been topics of various studies [1]. As mentioned before, finger strength is usually used as a limiting factor, though different studies showed inconsistent results regarding the correlation of maximum finger strength and degree of difficulty [2, 3, 4, 5]. Measurement of the maximal strength of fingers is mostly achieved by static, single-handed exercises (figure 1a) or, alternatively, using both hands and additional weights (figure 1b).

(a)

(b)

Figure 1. Typical methods for determining finger force; (a) measuring one-handed traction [6]; (b) both-handed with additional weight. [7]

Similar to this, the quantitative determination of the strength endurance is done by using the maximum time hanging on a defined hold as a parameter.

Compared to other mass sports like cycling or running, where measurements of critical parameters to increase the personal performance got indispensable, the development of the climbing sport in respect thereof is backward, but with rising trend. Accordingly, new methods to support evaluation and monitoring are important to improve this sport. Moreover, no quantitative methods exist to detect injury risks during training.

3. Method

The holistic evaluation of the performance of athletes in climbing sport is due to its complexity nearly impossible. Parameters like mental strength and its influence on the performance are not quantifiable. Compared with this, the force and technique as well as outside influences (temperature, humidity, etc.) are of today's standards measurable. Therefore, the presented method should support the development of the climbing sport and can be separated in following branches:

- pre-investigation of athletes
- determination of environmental impact
- direct measurement of forces during climbing
- tracking of joint motion during climbing.

3.1. Pre-Investigation

An independent dataset for a period of substantial length and different athletes is essential for the validation of the method and comparability and signification of the results. Thus, to quantify inter- as well as intra-athletes-variability, pre-investigations are necessary:

- *Defined and individual warm-up* is important to ensure a comparable degree of performance
- *Physiology* as size, weight, resting heart rate, etc.
- *Isolated maximum finger force* by existing methods
- *Environmental factors* like temperature or humidity for possible correlations with performance
- *Questionnaire* to find not measurable factors

3.2. Measurement of forces during climbing trials

Existing methods shown in figure 1 are based on isolated, climbing-unspecific trials. Therefore, we are presenting for the first time a direct method for measuring force while climbing. Three-axis force sensors (K3D120) from *ME-Meßsysteme* are used to determine forces on holds during climbing (figure 2a).

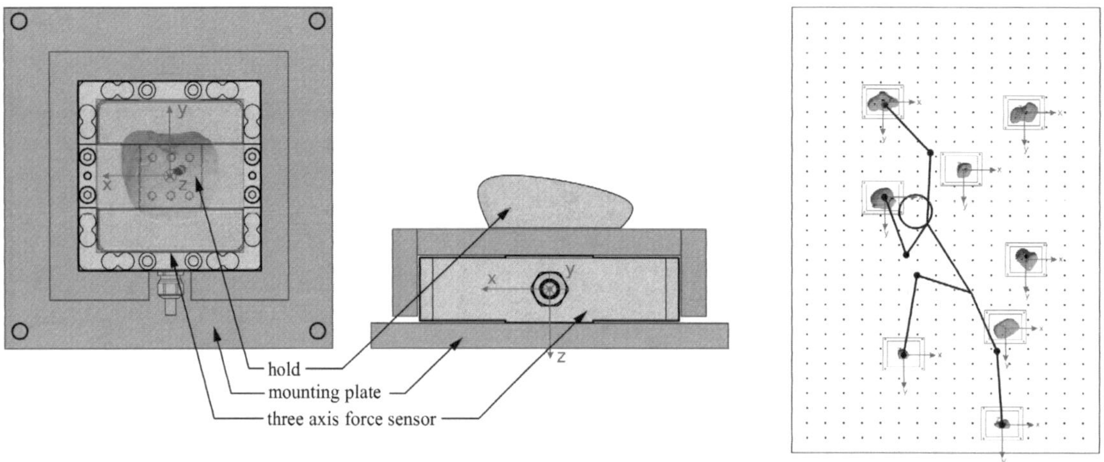

(a) (b)

Figure 2. (a) Three-axis force sensor (gray) with mounting plates (brown) and hold (blue); (b) Synchronous recording of motion of all joints and force distribution.

The outer plates are designed for an easy integration into existing artificial climbing walls. Another key part of the measurement system is the analog to digital conversion as well as the amplification of the output of the sensor. It is important to capture all three axis of every single sensor synchronously. High-precision amplifiers and AD-converters are employed, which allow the digitalization of all channels with 12-bit-resolution. For the case of integration and handling, the system is compact, battery powered and wireless. Thus, the system is equipped with microcontrollers featuring embedded WiFi module. According to a server/clients principle one of these microcontrollers serves as an access point and saves the incoming data of the clients off-site.

3.3. Determination of motion during climbing trials

Knowledge about the distribution of forces between single limbs is very important for the evaluation of strength as well as technique. To understand the cause of this distribution it is necessary to measure both force and current body position with all joints. For the first time a depth camera is employed, that enables the following:

- synchronous images and force measurements
- determination of position of body and joints
- measuring the climbers distance to the climbing wall

Journal of Physics: Conference Series **1379** (2019) 012014 doi:10.1088/1742-6596/1379/1/012014

This challenges are solved by modifying the so-called *Markerless Motion Capture* Method using a RealSense D435 modular unit from *Intel*. Figure 2b describes exemplary the process using *Markerless Motion Capture*. In our case, we use the Realsense depth camera to obtain a RGB and depth frame for every snapshot and evaluate the data using *OpenPose* [8] software to jointly detect human body keypoints. Hence, it is possible to record and analyse a motion pattern. The advantage over classic video analysis is the possibility to determine automatically joint angles and the distribution of force as a consequence thereof.

4. Results

During initial experiments, we evaluated the ability of a *Markerless Motion Capture* system and a prototype of a force sensor embedded climbing hold to track joints and measure forces during movement in three dimensions. Figure 3 shows a frame captured with the Realsense D435 during moving patterns. In a next step the position of the joints are calculated off-line by processing the taken RGB frames with *OpenPose*. As a result we get pixel coordinates of all relevant human body joints, which are used afterwards to get corresponding three-dimensional coordinates by analysing the related depth frames. Thus, we obtain after simple vector calculations, the joints angles over time (see figure 3). Concurrently,

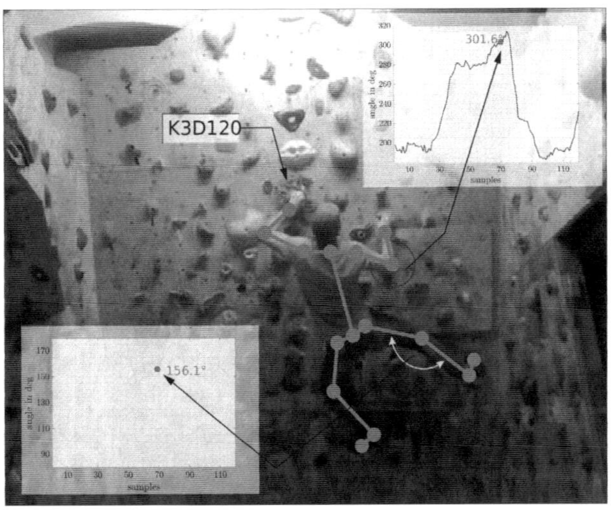

Figure 3. Rendered sample of the moving pattern including joints position and the time course of the right elbow and knee angles.

the integrated force sensor measures all three parts of the force vector simultaneously and sends the data leadless to the computing hardware.

Figure 4a illustrates the result of forces in all three directions during our first climbing trials. The hold was mounted to a inclined climbing wall at an horizontal angle of 45 degrees to strain all axes of the sensor. The force amplitudes corresponding to the momentary movement in figure 3 are marked with red dots.

We can determine the moment of first contact with an initial rise in force. Furthermore, the peaks illustrate among others the force necessary to readjust the hand position on the climbing hold. After readjusting the hand we see varying forces during the repositioning of the body until the moment of exiting the climbing hold.

To get comparable results, we repeated three different moving patterns several times in a row with the same athlete. Figure 4b represents for every axis the resulting differences for each time step and four trials. Due to the variation in climbing speed, the results were post-processed to compare measurement amplitudes. The remaining motion sequences were compared with the first try and time wrapped to recognize similarities.

Journal of Physics: Conference Series **1379** (2019) 012014 doi:10.1088/1742-6596/1379/1/012014

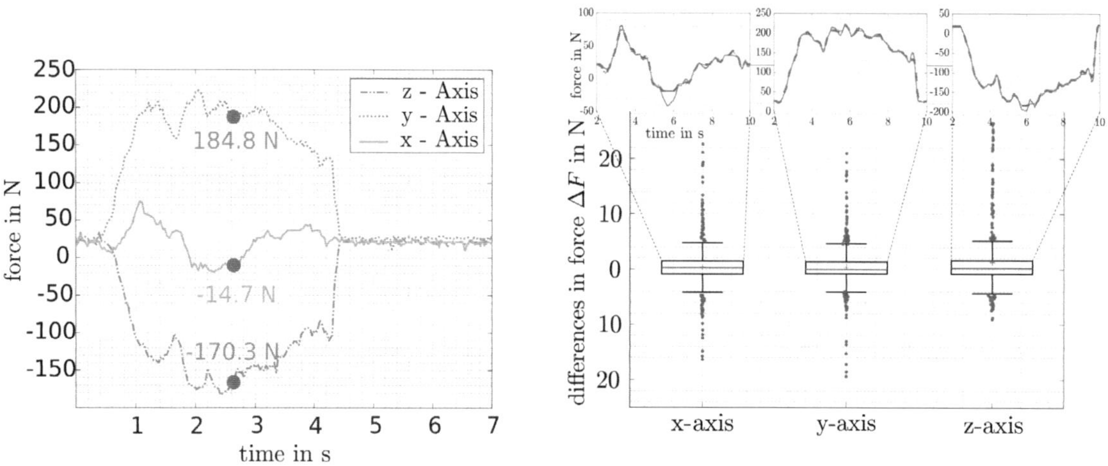

(a) (b)

Figure 4. (a) Force vector measured with the K3D120; (b) Time wrapped signals of first and third try (on top) and differences in force plotted for all three axes.

With respect to a few outliers, the outcome of this analysis shows that the climber was able to repeat nearly the same exercise for four times.

5. Conclusion

The introduced results illustrate the working principle of the in this paper presented method for simultaneous measurement of forces and motion capturing to detect human body keypoints. An evaluation of the outliers in figure 4b with the greatest deviation would be the next interesting step. Possible reasons for this significant variance could be different angles of joints while grabbing the sensor hold or signs of fatigue. To get an understanding of this changing factors, an extension of the existing system by adding more sensor holds and an enlargement of data sets by performing further iterations of the same moving pattern will be done for multiple athletes.

The received data sets of force sensor and motion capturing system will enable further studies on improving technique, performance and injury prevention of climbing athletes.

References

[1] Giles L V, Rhodes E C and Taunton J E 2006 *Sports Med. (Auckland, N.Z.)* **36** 529–545
[2] Ferguson R A and Brown M D 1997 *Eur. J. Appl. Physiol. Occup. Physiol.* **76** 174–180
[3] Grant S, Hasler T, Davies C, Aitchison T C, Wilson J and Whittaker A 2001 *J. Sports Sci.* **19** 499–505
[4] Grant S, Hynes V, Whittaker A and Aitchison T 1996 *J. Sports Sci.* **14** 301–309
[5] Mermier C M, Janot J M, Parker D L and Swan J G 2000 *Br. J. Sports. Med.* **34** 359–65; discussion 366
[6] Kraftolizer https://www.kraftolizer.com accessed: 2018-03-26
[7] Maximum weigt hang survey https://www.trainingbeta.com/concept-climbing-max-weight-hang-survey/ accessed: 2018-26-03
[8] Cao Z, Hidalgo G, Simon T, Wei S E and Sheikh Y 2018 *arXiv preprint arXiv:1812.08008*

Journal of Physics: Conference Series **1379** (2019) 012015 doi:10.1088/1742-6596/1379/1/012015

Comparison of experimental procedures of a method for continuous friction measurement in airport pavements

L Martins[1], A Ribeiro[1], J Alves e Sousa[2] and H Dimas[1]

[1]LNEC – National Laboratory for Civil Engineering, Lisbon, Portugal
[2]IPQ – Portuguese Institute for Quality, Caparica, Portugal

E-mail: lfmartins@lnec.pt

Abstract. This paper describes a comparison performed between two experimental procedures (with the Grip Tester and the Continuous Friction Measurement Equipment) of the fixed-slip method for longitudinal continuous friction measurement in airport pavements. The measurement framework is described, namely, the experimental testing and the related conformity assessment procedure supported by different friction thresholds. The experimental results of the comparison are presented (raw data, differences between friction estimates and corresponding experimental probability distributions) and its impact on the conformity assessment result is discussed. Conclusions are drawn based on the obtained difference dispersion, instrumental measurement uncertainties and expected difference between the two experimental procedures.

1. Introduction

In the framework of laboratories accreditation based on ISO/IEC 17025:2017 [1] method validation is considered a major requirement for the quality of experimental results and, in cases where conformity assessment is performed, it provides confidence to apply decision rules.

This is the case studied in this paper, regarding the continuous friction measurement in airport pavements, because measurement results are related to the pavement condition by scales establishing classes, being subsequently these used to take management decisions on the technical improvement of the tested pavement conditions.

The common cases for the use of proficiency tests [2, 3] as "quality assurance procedures" consider that the measurand is obtained from a reference method having experimental procedures that can be compared and evaluated using quality parameters (e.g., z-score). In this particular case, however, conventional proficiency testing comparison cannot be applied, as the experimental procedures are based on the same method but use two different numerical scales not directly comparable by the measurement results but only by the classes they establish (in both scales the output classes are the same).

In view of the above, the approach here considered was to develop a testing setup to evaluate the performance obtained by the two procedures, both having measurement traceability under comparable reference conditions (using same alignments on an airport runway). Validation of the relation of procedures with the method was achieved by the comparison and the analysis of the classification categories obtained.

2. Continuous friction measurement framework

The experimental testing activities performed in the context of airport pavements construction, maintenance and rehabilitation, have a significant impact on safety, characterization of surfaces and improved knowledge on operational performance. One major test regularly performed in airport pavements is the determination of the skid resistance in a wet condition, which can be achieved by the use of different continuous friction measurement equipment developed independently by different entities, in some cases, for over 50 years. Since different equipment can be used to determine friction in the same airport runway, a measurement uncertainty component related to the method reproducibility should be accounted for. In addition, friction traceability to the SI is indirectly obtained by the calibration of force (horizontal and vertical) measuring chains in a static mode, and the use of reference materials suitable for long distance measurement is unknown.

In this context, conformity assessment of the friction surface condition of airport pavements [4] is supported by three reference thresholds and the corresponding performance levels – design level (new pavement), operational level (used pavement), maintenance level (pavement requiring rehabilitation) and minimum level (non-admissible pavement). The magnitude of the friction thresholds depends on the adopted test speed and water depth (influence quantities) but also on the used equipment, as shown in figures 1 and 2.

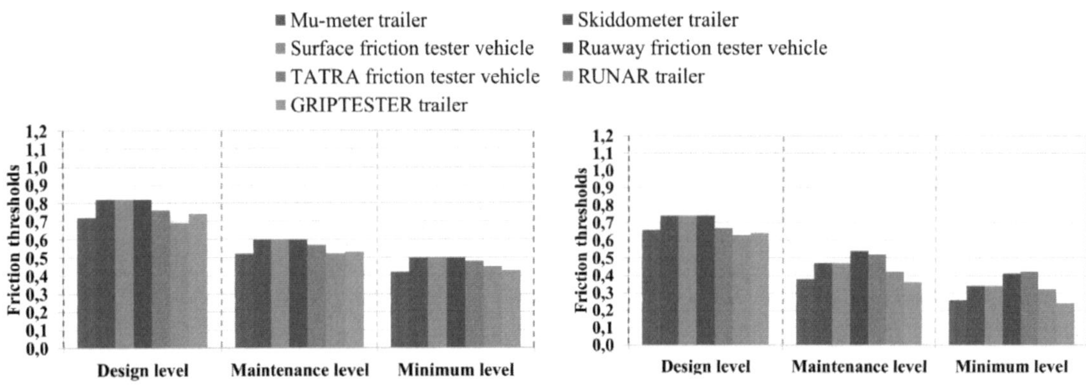

Figure 1. Friction thresholds for 65 km·h^{-1} test. **Figure 2.** Friction thresholds for 95 km·h^{-1} test.

Considering a certain test speed, the friction threshold differences can be interpreted as different measurement scales related to the applied equipment. This paper is focused on two particular measuring equipment – the Grip Tester (GT) [5] and the Continuous Friction Measurement Equipment (CFME) [6] – which were used for testing the same airport runway, aiming at the validation of the measurement method (fixed-slip method).

Both equipment allows the longitudinal friction measurement through the use of a standard tire [7, 8], mounted on a test wheel connected to a partial blocking system, which results in a lower tire circulation speed with respect to the test speed, thus creating a horizontal slip force on the tire. The action of this horizontal force and the vertical (load) force on the test wheel results in the elastic deformation of the axis where the wheel is installed. This axis is instrumented in order to measure the instantaneous longitudinal friction value, defined by ratio between the horizontal and vertical forces. Complementary measurement chains allow measuring both the travelled distance, the test speed and, in some cases, the water flow.

Different friction estimates given by these two equipment, in the same runway alignment, can be related to: (i) the standard tires used; (ii) the water projection systems of each equipment; (iii) the suspension system which mechanically connect the test tire to the surrounding structure; (iv) the dynamic stability of the test wheel attached to a trailer (in the case of the GT) or integrated in a vehicle (in the case of the CFME).

3. Experimental results

The continuous friction measurement tests were performed during the night period in an airport runway, between its initial and final markings, with a total dimension of 3100 m. The results shown are related to an alignment located 7.5 m away from the runway centerline. A nominal test speed of 65 km·h^{-1} and a 1 mm nominal water depth were defined and both equipment were subjected to prior calibration immediately before the test took place.

The following figures show the results obtained, namely: (i) the collected raw data of friction estimates measured in a 10 m spatial interval (Figure 3); (ii) the corresponding differences between friction estimates of the GT and CFME devices (Figure 4); (iii) the corresponding experimental probability distribution of differences (Figure 5); (iv) the order of changes in class (zero, one, two or three levels) in the surface condition due to friction estimate differences (Figure 6), considering compared results obtained by the two experimental procedures.

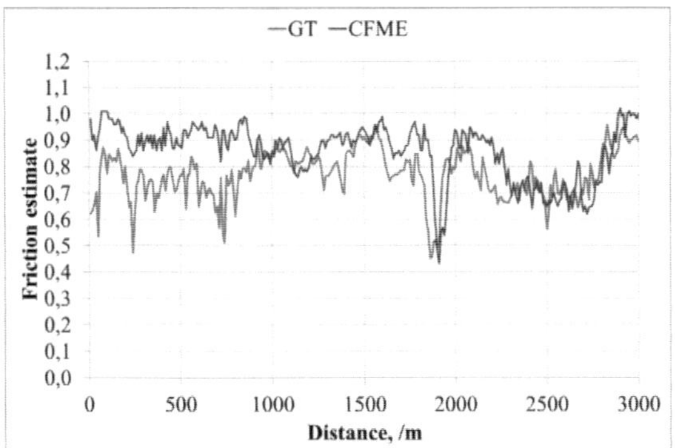

Figure 3. Raw data of the friction estimates.

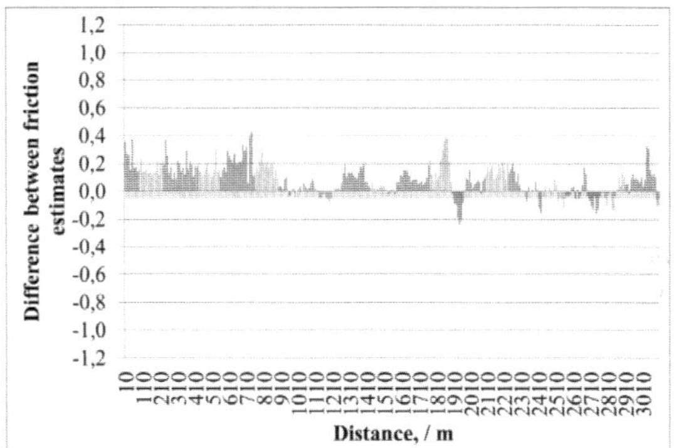

Figure 4. Differences between friction estimates.

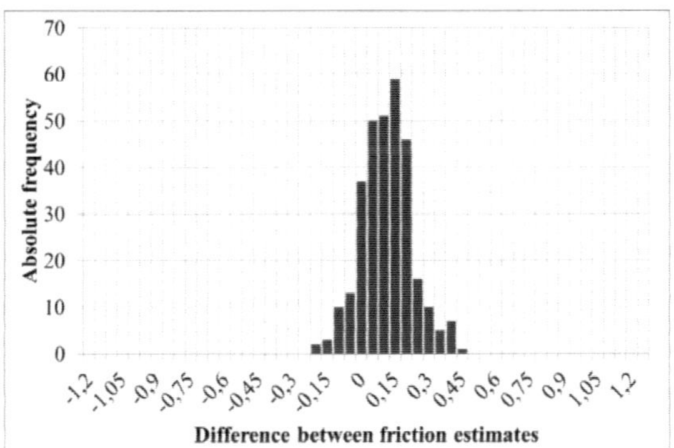

Figure 5. Histogram related to the differences between friction estimates.

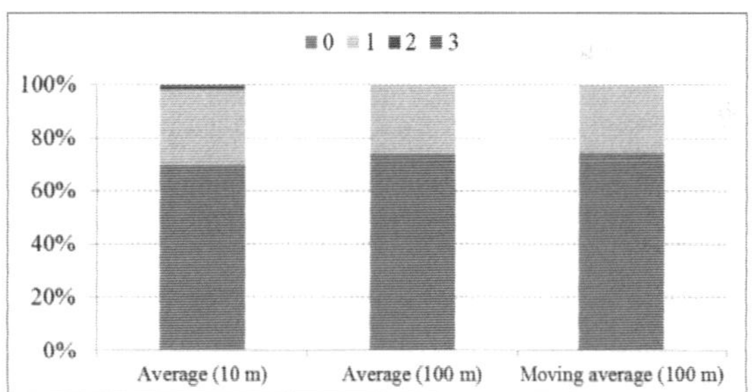

Figure 6. Changes in the classification due to friction differences between the two procedures.

The magnitude of the observed estimate differences is characterized by a dispersion of values mainly comprised between 0.00 and 0.20, which includes the expected difference related to the friction threshold values. In fact, for a test speed of 65 km·h^{-1} and a water depth equal to 1 mm, differences between GT and CFME threshold values correspond to 0.07 and 0.08 [4]. The obtained experimental probability distribution is centered around the 0.15 value and as a symmetrical shape, close to a Gaussian distribution. The instrumental measurement uncertainties of the equipment – between 0.01 and 0.08 [9] – also contributed for the obtained dispersion of values and took into account the following uncertainty components: calibration, resolution and systematic deviations of the measurement standard and the calibrated equipment; the interference between horizontal and vertical force measurement chains.

A residual number of cases were obtained where differences between estimates result in changes of two or three classification levels and only when the friction averaged values are related to a 10 m spatial interval (raw data). In general, from 70 % up to 80 % of the assigned classifications by both equipment match and between 20 % and 30 % differ only in one level, noticing that sometimes the measured values are quite close to the threshold value. The use of the moving average in a 100 m spatial interval originates a slightly lesser impact of the assigned levels (filtering effect).

The different spatial intervals (10 m or 100 m) and the type of average (fixed or moving) used for the estimate calculation does not have a significant impact on the evolution and on the obtained differences.

Additional alignments were performed in the same runway showing similar results. An exception was recorded concerning an alignment located 3.0 m away from the runway centerline, where the results shown in figures 7 to 10 were obtained.

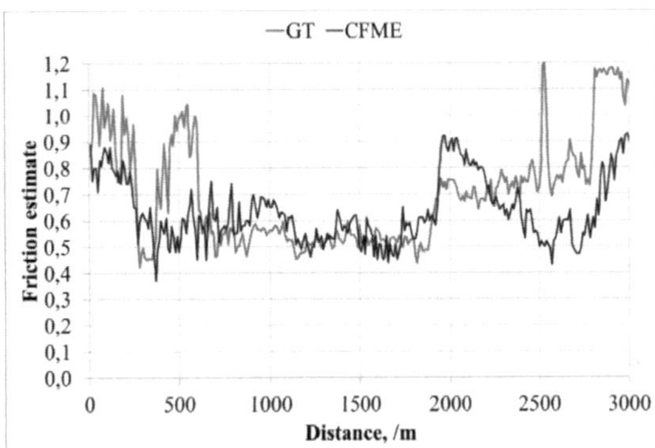

Figure 7. Raw data of the friction estimates.

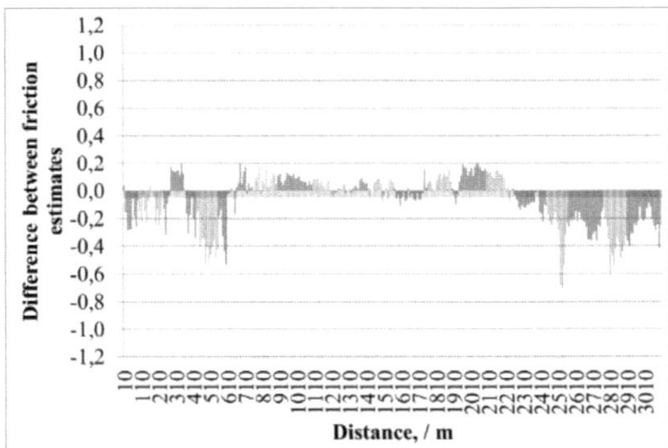

Figure 8. Differences between friction estimates.

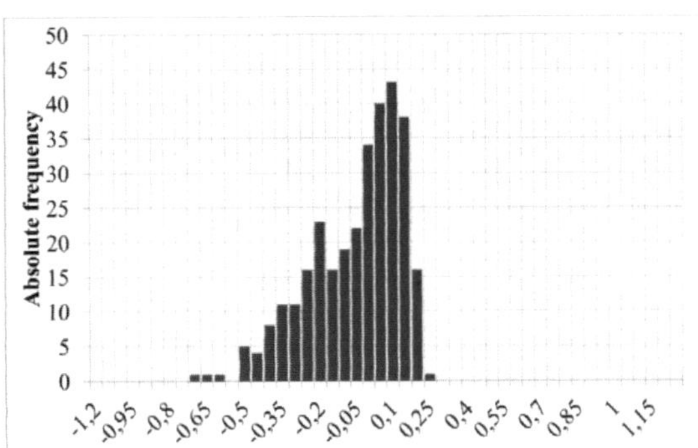

Figure 9. Histogram related to the differences between friction estimates.

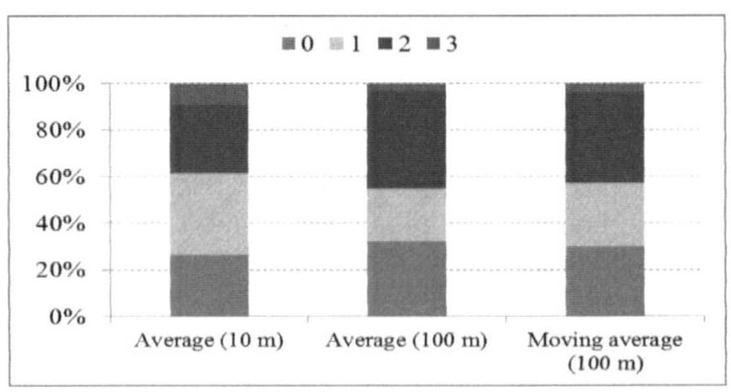

Figure 10. Changes in the classification due to friction differences between the two procedures.

Major differences were observed (Figures 7 and 8) in the initial and final segments of this alignment which can be justified by: (i) circulation of the equipment with a spatial gap between both passages; (ii) transient effects in the beginning and in the ending of the test due to the variation of influence quantities such as circulation speed (deviation from the 65 km·h^{-1} circulation speed nominal value) and water film thickness (deviation from the 1 mm nominal value).

Journal of Physics: Conference Series **1379** (2019) 012015 doi:10.1088/1742-6596/1379/1/012015

As a result of these differences, the obtained experimental probability distribution (Figure 9) shows a higher asymmetrical and non-Gaussian shape when compared with the previous one (Figure 5). This fact can be used as a quality assessment tool in the comparison of results produced by different testing equipment.

The dimension of these segments is reduced when compared with the total dimension of the alignment, noticing that in the central region of the runway, the results obtained are within the dispersion obtained in the previous studied alignment (located 7.5 m away from the runway centerline). However, the founded differences result in significant changes of the classification levels (Figure 10) noticing a high number of cases where a difference of one or two levels occurs and, in some residual cases, even three levels.

The impact of the different friction estimates obtained in the performed alignments can also be seen in a global perspective, considering the minimum values (conservative safety approach) of the moving averages related to each testing equipment in partial (the middle region B and the two ends A and C of the runway) and total dimension of the runway, as shown in tables 1 and 2.

Table 1. Minimum friction values of the moving averages in the first studied alignment.

Testing equipment		GT		CFME	
Runway	Region A	0.65			0.85
	Region B	0.55	0.55	0.61	0.61
	Region C	0.68			0.67

Table 2. Minimum friction values of the moving averages in the second studied alignment.

Testing equipment		GT		CFME	
Runway	Region A	0.47			0.52
	Region B	0.48	0.47	0.48	0.48
	Region C	0.68			0.50

In the first alignment, the global and partial classifications assigned by both equipment agree (green operational level) with only one exception, in region C, where the GT assigns a blue design level (high friction in the runway) while the CFME classifies this region with a green operational level (normal friction in the runway), although the proximity of the friction estimates.

An interesting case is noticed in the second alignment, namely, in the global classification. A minimum friction value of 0.47 (yellow maintenance level) is assigned by the GT, while the CFME assigns a higher friction estimate, however, corresponding to unacceptable friction (red level).

4. Conclusions

The experimental tests performed with different equipment, aiming at the comparison analysis of the airport pavements continuous friction measurement, shows results within a difference dispersion between 0.00 and 0.20, being of the same magnitude of the instrumental measurement uncertainty and of the expected difference between GT and CFME results.

The best obtained dispersion has a direct impact between 20 % and 30 % of the assigned classifications for the surface condition of the tested runway. The risk of changes in the classification can be mitigated with the improvement of the spatial sampling procedure (higher number and proximity between alignments) and the knowledge of the test repeatability (successive repetitions of the performed alignment in a given direction).

Future work to be developed intends to evaluate the influence of repeatability and to perform a sensitivity study of other influence quantities [10] in order to obtain a more robust analysis of the comparison evaluation.

References

[1] ISO/IEC 2017 *ISO/IEC 17025:2017 General Requirements for the Competence of Testing and*

Calibration Laboratories

[2] ISO 2010 *ISO 17043:2010 Conformity Assessment. General Requirements for Proficiency Testing*

[3] ILAC 2014 *ILAC-P9-06 ILAC Policy forParticipation in Proficiency Testing Activities*

[4] ICAO 1999 *ICAO – International Standards and Recommendation Pratices. Aerodromes. Annex 14 to the International Civil Aviation vol I: Aerodrome Design and Operations* 3rd edition

[5] CEN 2009 *CEN/TS 15901-7:2009 Road and Airfield Surface Characteristics - Part 7: Procedure for Determining the Skid Resistance of a Pavement Surface Using a Device with Longitudinal Fixed Slip Ratio (LFCG): the Grip Tester®*

[6] ASTM 2011 *ASTM E2340/E1240M Standard Test Method for Measuring the Skid Resistance of Pavement and Other Trafficked Surfaces Using a Continuous Reading, Fixed-Slip Technique*

[7] ASTM 2016 *ASTM E1551-16 Specification for Special Purpose. Smooth-Tread Tire,Operated on Fixed Braking Slip Continuous Friction Measuring Equipment*

[8] ASTM 1996 *ASTM E1844-96 Specification for A Size 10 x 4-5 Smooth-Tread Fiction Test Tire*

[9] Martins L, Ribeiro A, Batista F, Freire A and Maia A 2016 Conformity assessment study and measurement requirements for the Grip Tester test *Proc. the 1st Conf. on Testing and Experimentations in Civil Engineering* (Lisbon: RELACRE) 379-86

[10] White G 2012 Pavement Surface Friction Testting for Airports *Proc. the Australian Airports Association National Technical Convention* (Melbourne) 1-14

Journal of Physics: Conference Series **1379** (2019) 012016 doi:10.1088/1742-6596/1379/1/012016

Elastography of vocal folds

R Lamprecht[1], M Maghzinajafabadi[1], M Döllinger[2], M Semmler[2] and A Sutor[1]

[1] Institute of Measurement and Sensor Technology, UMIT - Private University for Health Sciences, Medical Informatics and Technology, 6060 Hall in Tirol, Austria

[2] Department of Otorhinolaryngology, Division of Phoniatrics and Pediatric Audiology, University Hospital Erlangen, Friedrich-Alexander-University Erlangen-Nürnberg, 91054 Erlangen, Germany

E-mail: raphael.lamprecht@umit.at

Abstract. The diagnosis of changes in the structure of the vocal folds is an important clinical task. There is no approach available, which is capable of getting synchronous spatial and biomechanical properties of them. Whereas biomechanical changes, particularly the elasticity, are a good maker for malignant and benign conditions. To gain the elastic characteristics of the vocal folds we are proposing an endoscopic approach using ultrasound elastography.

1. Introduction

The voice is one of the most important way of communication in everyday life. Consequently, disorders of the voice forming organs cause a big impact on patients. Approximately 10% to 30% of the population are affected by problems of their voice in their lifetime, professional speakers or singers up to 50% [1, 2]. Voice disorders have many reasons: the most common ones are vocal strain/excessive tension, reflux, vocal fold paralysis, functional dysphonia, nodules, cysts, presbyphonia and Reinke's edema [1, 3]. Diagnosing and assessing those pathologies is a wide clinical- and research-field. Especially imaging techniques to gain qualitative data have been studied recently.

A fundamental part of voice production are the vocal folds (VFs). The biomechanical properties of the VFs change if pathologies occur [4]. To separate the nondysphonic cases from dysphonic, it was shown that the measurement of VFs stiffness is highly sensitive [5]. Different methods are used *in vivo* to determine the stiffness of the VFs: aspiration method [6], tension meter, optical measurements and ultrasound [7].

But so far only one group [8, 9] used (transcutaneous) ultrasound elastography to assess the VFs. We are proposing a laryngoscopic, semi-static-elastography approach to examine the stiffness of the VFs.

2. Anatomical and biomechanical basics of the vocal folds

The true vocal folds are located in the larynx. They are surrounded by the false vocal folds, also called the vestibular fold (Figure 1a).

The oscillation of the true VFs is responsible for a major part of phonation, whereas the vestibular folds play a minor role in the phonation process. The VFs are 3-10 mm thick and measure 10-20 mm anterior-posterior and 8-12 mm in the medial-lateral direction [7].

The human VFs (Figure 1b) are formed by five different layers: epithelium, lamina propria and the vocalis muscle. The lamina is divided into three sublayers (superficial, intermediate and deep lamina

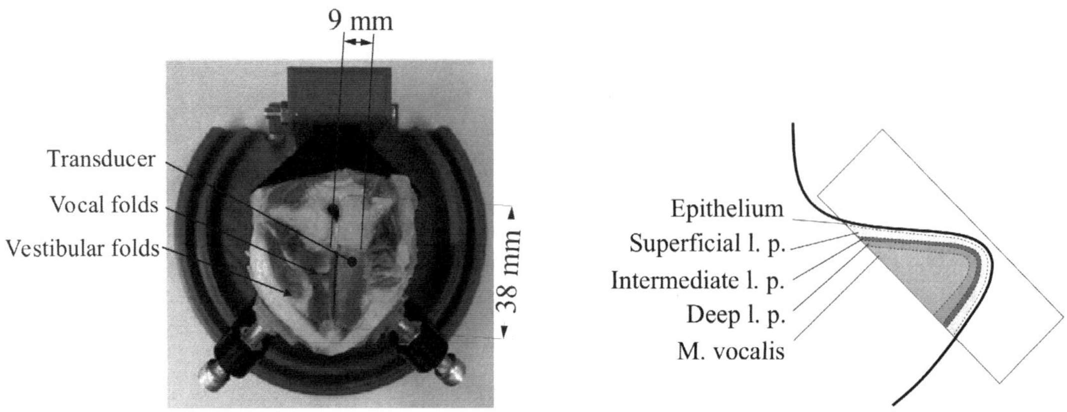

(a) Porcine vocal folds.

(b) Layers of the human VFs: Epithelium, lamina propia (l.p.) and Musculus vocalis [7].

Figure 1. Anatomical properties of the vocal folds.

propria) [10]. The outer layers primarily consist of an extra cellular matrix, whereas the inner layers consist mainly of cells [7]. Likewise, they fulfil different tasks: the epithelium maintains the shape and protects the underlying tissue whereas the lamina propria mainly keeps the elastic properties. The muscle has two functions: passive as a stiff rubber band and active as a muscle [10].

Due to the complex structure of the VFs it is clear that their elastic behaviour is nonlinear [7]. The VFs are primarily responsible for the phonation process. They oscillate self-sustained with a high frequency and small amplitude [7]. Different models have been developed to describe the biomechanical properties and the oscillation of the VFs [11].

3. State-of-the-art methods

3.1. Diagnosing vocal fold disorders

For a clinical assessment of voice disorders endoscopic imaging techniques are widely used, mostly videostroboscopy or high-speed videoendoscopy [12]. Stroboscopy still remains the standard examination because it is cost-efficient, easy to use and it provides synchronized audio and visual data. The stroboscopic approach is limited by its relatively low sample rate. If the dysyphonia exceeds a moderate level, the stroboscopy cannot visualize the changes because it does not capture cycle-to-cycle changes [13].

To differentiate between malignant and benign it is necessary to take a biopsy of the affected tissue [14]. Like every invasive examination a biopsy does have some risk and potentially leaves scars on the VFs, which could lead to a decrease in the patient's voice quality [15]. New endoscopic techniques - like narrow band imaging, optical coherence tomography (OCT), three-dimensional imaging and iScan technology - are being examined and show promising results [15, 16, 17]. The narrow band imaging and iScan cannot detect the depth of the infiltration that limitates it to superficial lesions [16]. Whereas the OCT can visualize the tissue in depth [15]. The three-dimensional laser imaging is capable of imaging the surface of the VFs in three dimension during phonation [17].

Other medical imaging techniques like ultrasound, computer tomography and magnetic resonance imaging [18] are also used to examine the larynx and the vocal folds. Those approaches are capable of gaining spatial information. New approaches - especially using ultrasound [19] - are being developed, but still barley used.

An elastographic approach could combine the spatial image and biomechanical information, which possibly leads to a higher sensitivity in discovering lesions.

3.2. Measuring the elastic properties of the vocal folds

Pathologies of the VFs affect the biomechanical properties [20]. For example previous studies indicate that healthy VFs are stiffer, compared to the ones with polyps [21]. Measuring the biomechanical properties of the VFs has always been a challenge because of not only their size but also their anatomical accessibility, their anisotropies and their broad range of elasticity. It is desirable to measure the elasticity of human VFs *in vivo*, but for an approximation also *ex-* and *in-situ* and animal studies have been made [7]. To mechanically measure the elasticity of a tissue it is necessary to apply known stress and measure the displacement (e.g. aspiration technique) [12]. Other approaches observe shear wave propagation in the tissue and measure their group velocity (e.g. shear-wave-elastography) [22].

The *in vivo* approaches can be separated into three groups: mechanical, optical and medical imaging techniques [7]. One approach to mechanically examining the elastic properties is an aspiration tool, which sucks the tissue into a tube and measures the largest displacement [23]. Other aspiration tools excite the tissue, using air-pressure oscillations and determine the vibration [24]. All mechanical approaches need direct contact to the VFs, which could lead to tampering with the results.

Optical methods, like stroboscopic or high-speed imaging, focus on observing the mucosal waves and gain qualitative information about the movement of the VFs. Using high-speed images, quantitative methods have been developed. Those are capable of visualizing the time dependent distances between the VFs. Variations in mucosal wave propagation indicate biomechanical changes [21].

In addition, an optical method, but here summarized with the medical image techniques, is the OCT. It was shown that OCT is a proper approach to measure the histological layers of the vocal folds [21]. The optical coherence elastography was successfully used to assess the biomechanical properties of porcine trachea [25].

Another medical imaging technique is the ultrasound. Two approaches are known: transcutaneous and direct imaging. Ultrasound imaging of the VFs has been tested and it showed usable results [26]. The transcutaneous access is easy to use because a common clinical ultrasound device is used. Identifying the VFs on the ultrasound image could be challenging for the examiner. Especially the true vocal folds are hard to find [27].

Various ultrasound methods have been proposed examining the VFs: B-mode- [28], Doppler- [29], Nakagami-imaging [26] and elastography [8, 9]. Except B-mode-imaging, all of them are capable of measuring the biomechanical properties. Ultrasound elastography is a widely used assessment method to examine the elasticity of tissues [22]. All the transcutaneous approaches have the limitation that identifying the VFs is difficult. This limitation could be resolved by a direct measurement on the VFs.

4. Methods

To overcome the discussed limitations, a semi-static-elastography approach to examine the stiffness of the VFs is proposed here. The measurement directly on the VFs enables the examiner to gain results that are more exact. It is difficult to identify the VFs with transcutaneous approaches and the overlying tissues could distort the measurement. To show the feasibility, porcine larynxes are examined with a 8-17 MHz intraoperative transducer. The ultrasound transducer is directly placed on the VFs. Establishing a small displacement, an internal strain is applied. The elastography package of Alpinion E-CUBE 15 EX is used to get qualitative results. To quantify the results the B-Mode images are registered with a Bspline-registration with *elastix* [30, 31]. *elastix* is an established tool for image registration in medicine. The strain in the tissue was determined according to Lai et al. [32].

Journal of Physics: Conference Series **1379** (2019) 012016 doi:10.1088/1742-6596/1379/1/012016

5. Results

The two measurements, which are shown in figure 2 & 3, were done on the same porcine VFs. The transducer was placed on the position shown in figure 1a. In both measurements, the results were comparable.

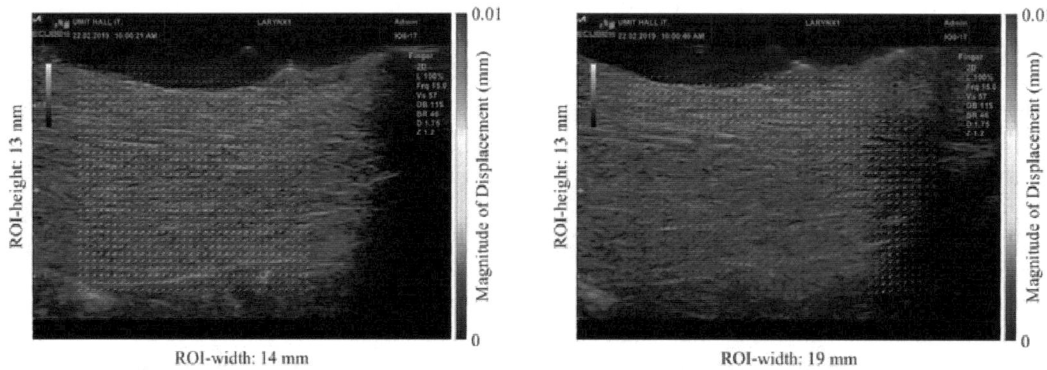

(a) Measurement A (b) Measurement B

Figure 2. Displacement field in the tissue of a porcine VFs.

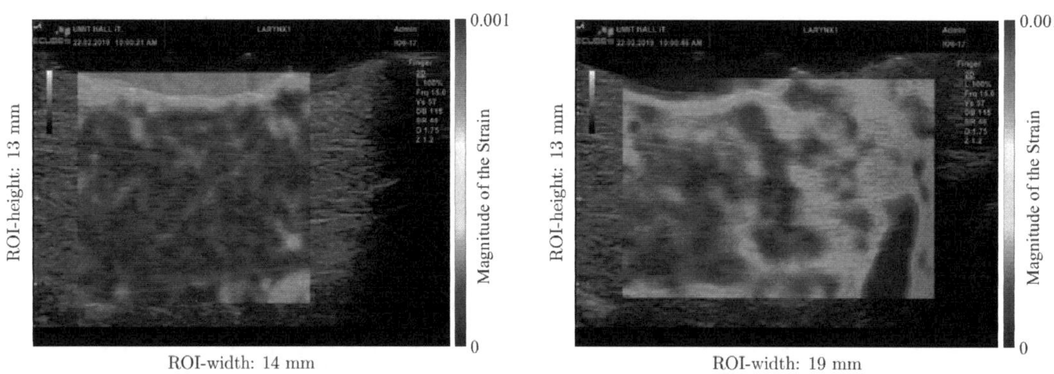

(a) Measurement A (b) Measurement B

Figure 3. Absolute value of the strain field.

The visualization of the internal strains is a new assessment method. Therefore, no comparable data was found. Different studies were made to examine the surface strains of the VFs [33, 34]. The strain in material depends on its elastic properties. The stiffer a material under constant stress the lower the strain and vice versa. The tissue of the vocal folds has different layers that have different elastic properties. In the evaluated porcine larynx, it was not yet possible to distinguish between those layers. The image registration process needs to be refined to differentiate them.

6. Conclusion & Outlook

We showed an assessment method, which is capable of displaying the strain distribution in the VFs. Our approach is based on a conventional ultrasound and image registration. The strain distribution gives us an idea of the elasticity in the tissue.

To get quantitative results it is necessary to measure the applied forces. To determine the applied force field, we are using a stress sensor fixed on the transducer (Figure 4). The stress sensor has well known elastic properties. Measuring the deformation of the sensor, it is possible to determine the force field on the surface [35]. This approach is still under development. To evaluate the registration process and get

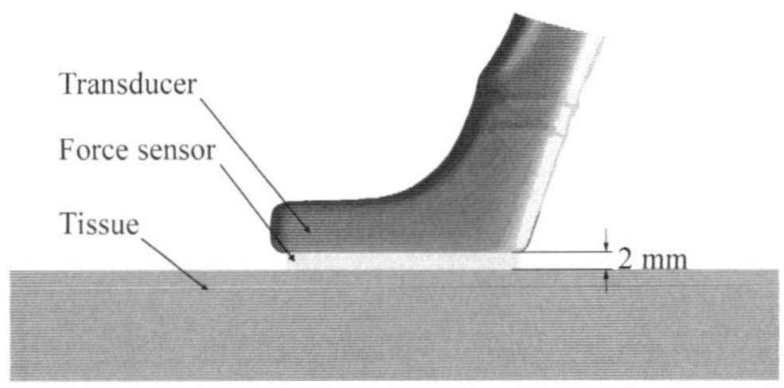

Figure 4. Measurement set up to determine the surface forces on the tissue.

quantitative material results a finite-element body-cover-model [4] should be implemented in COMSOL. The simulation results are used to update the material properties [36].

The advantages of our approach is the direct measurement on the VFs, which should provide better results than transcutaneous approaches. Disadvantages are that the patient needs to be anesthetized and possible irritations of the VFs. For a clinical use, it would be necessary to develop a new transducer, which has the required size and properties.

Acknowledgments
This work was supported by Austrian Science Fund Grant no. I 3806 - B28 and by the German research foundation (DFG) no. DO1247/7-1.

References

[1] Martins R H G, do Amaral H A, Tavares E L M, Martins M G, Gonc¸alves T M and Dias N H 2016 *J. Voice* **30** 761.e1–e9
[2] Paul B C, Chen S, Sridharan S, Fang Y, Amin M R and Branski R C 2013 *The Laryngoscope* **123** 215–9
[3] Misono S, Marmor S, Roy N, Mau T and Cohen S M 2016 *Otolaryng. Head Neck* **155** 33–41
[4] Zhang Z 2009 *J. Acoust. Soc. Am.* **125** 1091–102
[5] Gómez-Vilda P, Fernández-Baillo R, Nieto A, Díaz F, Fernández-Camacho F J, Rodellar V, Alvarez A and Martínez R 2007 *J. Voice* **21** 450–76
[6] Weiss S, Sutor A, Ilg J, Rupitsch S J and Lerch R 2016 *Acta Acust. United Ac.* **102** 214–29
[7] Miri A K 2014 *J. Voice* **28** 657–67
[8] Gao J, Zhu Q, Xia C X, Shin J, Shih G and Min R 2018 *J. Ultras. Med.* **37** 2537–44
[9] Gao J, Xia C X, Zhu Q, Hamilton J and Min R 2018 *Clin. Imag.* **51** 292–9
[10] Hirano M, Kakita Y, Ohmaru K and Kurita S 1982 *Speech and Language* **7** 271–97
[11] Cook D D, Nauman E and Mongeau L 2009 *J. Acoust. Soc. Am.* **126** 2002–10
[12] Goodyer E, J Jiang J, Devine E, Sutor A, Rupitsch S, Zorner S, Stingl M and Schmidt B 2011 *Curr. Bioinform.* **6** 344–61
[13] Mehta D D and Hillman R E 2012 *Curr. Opin. Otolaryngo.* **20** 429–36
[14] Ahn A, Wang L, Slaughter J C, Nguyen A M, Ossoff R H and Francis D O 2016 *The Laryngoscope* **126** 923–7
[15] Hawkshaw M J, Sataloff J B and Sataloff R T 2013 *J. Voice* **27** 738–43
[16] Rzepakowska A, Sielska-Badurek E, Cruz R, Sobol M, Osuch-Wójcikiewicz E and Niemczyk K 2018 *Head Neck* **40** 927–36

[17] Semmler M, Döllinger M, Patel R R, Ziethe A and Sch ützenberger A 2018 *The Laryngoscope* **128** 2367–74
[18] Ruffing S, Struffert T, Grgic A and Reith W 2005 *Der Radiologe* **45** 828–37
[19] Fukuhara T, Donishi R, Matsuda E, Koyama S, Fujiwara K and Takeuchi H 2018 *World J. Surg.* **42** 130–6
[20] Dion G R, Jeswani S, Roof S, Fritz M, Coelho P G, Sobieraj M, Amin M R and Branski R C 2016 *Mat. Sci. Eng. C-Mater.* **64** 444–53
[21] Goodyer E, Muller F, Bramer B, Chauhan D and Hess M 2006 *Eur. Arch. Oto-Rhino-L* **263** 455–62
[22] Sigrist R M, Liau J, Kaffas A E, Chammas M C and Willmann J K 2017 *Theranostics* **7** 1303–29
[23] Tanaka S and Hirano M 1990 *Arch. Otolaryngol.* **116** 721–4
[24] Weiß S, Thomson S L, Lerch R, Döllinger M and Sutor A 2013 *J. Mech. Behav. Biomed.* **17** 137–51
[25] Bu R, Price H, Mitran S, Zdanski C and Oldenburg A L 2016 *Proc. SPIE–Int. Soc. Opt. Eng.* **9689** 968923
[26] Tsui P H, Wan Y L and Chen C K 2012 *Curr. Opin. Otolaryngo.* **20** 437–42
[27] Kandil E, Deniwar A, Noureldine S I, Hammad A Y, Mohamed H, Al-Qurayshi Z and Tufano R P 2016 *JAMA Otolaryngol.* **142** 74–8
[28] Hu Q, Zhu S Y, Luo F, Gao Y and Yang X Y 2010 *J. Ultras. Med.* **29** 1023–30
[29] Hsiao T Y, Wang C L, Chen C N, Hsieh F J and Shau Y W 2002 *Ultrasound Med. Biol.* **28** 1145–52
[30] Klein S, Staring M, Murphy K, Viergever M A and Pluim J P W 2010 *IEEE T. Med. Imaging* **29** 196–205
[31] Shamonin D P, Bron E E, Lelieveldt B P F, Smits M, Klein S and Staring M 2013 *Front. Neuroinform.* **7** 50:1–50:15
[32] Lai W M, Rubin D and Krempl E 1993 *Introduction to continuum mechanics* 3rd ed (Oxford, New York, USA: Pergamon Press)
[33] Alipour F and Vigmostad S 2012 *J. Voice* **26** 816.e21–816.e29
[34] Bakhshaee H, Young J, Yang J C W, Mongeau L and Miri A K 2013 *J. Voice* **27** 659–67
[35] Kennedy K M, Chin L, McLaughlin R A, Latham B, Saunders C M, Sampson D D and Kennedy B F 2015 *Scientific Reports* **5** 15538
[36] Pheiffer T S, Ou J J, Ong R E and Miga M I 2011 *IEEE T. Bio-med. Eng.* **58** 2607–16

Journal of Physics: Conference Series 1379 (2019) 012017 doi:10.1088/1742-6596/1379/1/012017

An acoustic pressure pipette aspiration method for isotropic materials

M Maghzinajafabadi[1], R Lamprecht[1], M Döllinger[2], M. Semmler[2] and Sutor[1]

[1] Sensor technology institute, University for Health Sciences, Medical Informatics and Technology, Hall in Tirol, Austria
[2] University Hospital Erlangen, Division of Phoniatrics and Pediatric Audiology at the Department of Otorhinolaryngology Head & Neck Surgery, Erlangen, Germany

E-mail: Mohammadali.maghzinajafabadi@umit.at

Abstract: A measurement setup is presented as a method to determine the elasticity modulus of both artificial and real tissue as a function of frequency. The future goal at the end is to develop and validate the method to measure the elasticity modulus of ex vivo cadaver animal vocal folds over the human phonation frequency range. In the present study, we investigate an innovative acoustic pressure pipette aspiration technique to measure the material characteristic of an isotropic silicone specimen with similar characteristics as human vocal folds. With the aim of ensuring reliable results, we test our method on different mixtures of silicone rubber. Four silicone samples with different consistencies were tested over a frequency range of 50 –500Hz. Influence of vibrometer laser spot location in an enclosed area on the sample surface was investigated by repeating the measurements ten times. The results show only small standard deviations demonstrating the potential of the acoustic pressure pipette aspiration technique for material characterization.

1. Introduction

The determination of mechanical material properties of synthetic as well as natural vocal fold material is an important task. The information is necessary for the verification of synthetic models, for numerical simulations and finally for the further understanding of the phonation process. In particular, silicone or polyurethane mixtures are used as substitute for vocal folds tissue on which the research of this study is aimed at. This work presents a method to measure the elasticity modulus of silicone specimen with different mixtures as a function of frequency. We focus is on the pipette aspiration method which was originally introduced to measure the stiffness and intracellular pressure of red blood cell membranes by Rand and Burton [1]. Therein, a pipette was placed onto the material sample and excited via a static low pressure, sucking it into the pipette. Aoki [2] applied it to the local elasticity measurement of soft tissue. In a recent investigation, Weiss [3-6] and Zörner [7] developed a new pipette aspiration setup to measure silicone sample elasticity. In the both experimental setups, the silicone sample was placed under the pipette and by means of a pistonphone, which was mounted on an electromechanical shaker and connected to a small gauge head by a flexible tube, a fluctuating pressure was generated within the pipette resulting in a vibration of the area enclosed by the pipette. The out-of-plane velocity at several measurement points on the specimen's surface was measured by the laser scanning vibrometer. We enhanced the aspiration pipette method to a miniaturized pipette with acoustic pressure excitation which allows to dynamically apply an acoustic pressure excitation to the surface of the material at frequencies in the human phonatory range.

2. Experimental setup of the enhanced aspiration pipette

For experimental investigations, four mixing ratio samples 1:1:0, 1:1:2, 1:1:3 and 1:1:4 were made of a two-part silicone composite, Ecoflex TM Platinum Cure Silicone Rubber, Part-A: Part-B: silicone thinner. The measurement of the four samples have been performed by the pipette aspiration with the experimental setup shown in (figure 1). In the original setup as reported in Aoki [2] a static low pressure was used to suck the material into the pipette and the resulting deformation was measured. By a simple analytical formula, the elasticity modulus is determined. We replace the static low pressure by using a loudspeaker to produce an acoustic pressure at a predefined frequency as displayed in figure 1.

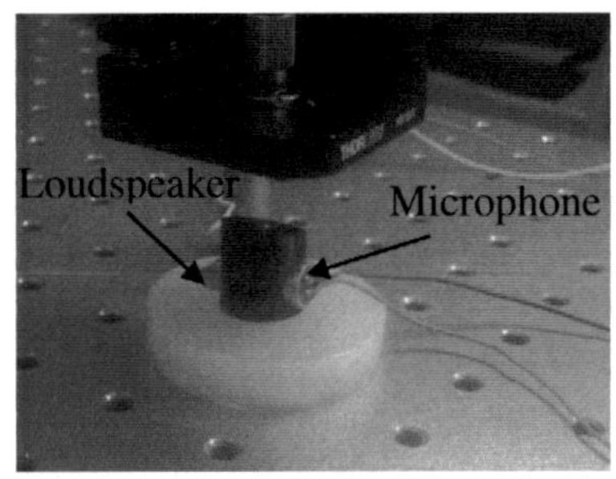

Figure 1. Picture of the measurement setup.

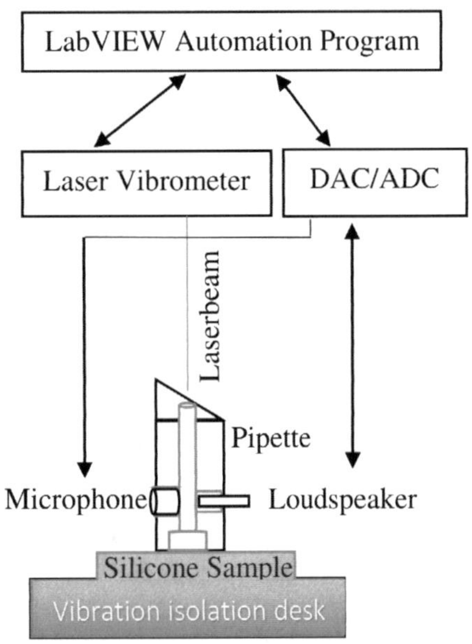

(a)

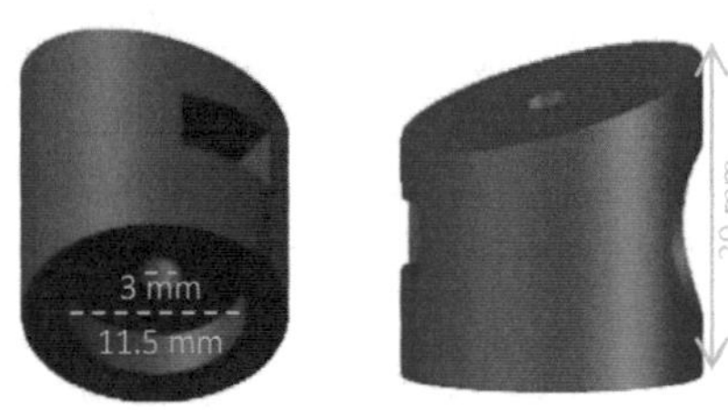

(b)

Figure 2. Setup: (a) Schematic picture of the measurement setup; (b) Drawing of the miniaturized pipette.

The loudspeaker is mounted in 10 mm distance from the base of the pipette wall with a pipette fixed to it that is placed upon the sample. The loudspeaker is driven at ten frequency steps in the range of 50 – 500Hz with 50Hz intervals to cover the frequency range of normal vocal fold oscillations to obtain excitation pressure (sound) on the specimen surface. The acoustic pressure applied to the inner volume of the pipette enables to have an enclosed pressure on the surface of the silicone sample. The displacement of the vibration induced by the sinusoid fluctuating force generated by the acoustic pressure on the silicone sample surface inside the pipette walls is measured by a laser vibrometer. Thereby, the exciting pressure on the sample surface can be measured by a microphone and the resulting mechanical vibrations by a laser vibrometer. Since the wavelength of the sine wave with 500Hz maximum frequency is much longer than the dimension of the pipette, the pressure considered constant in all the pipette volume, therefore, the pressure on the surface of specimen is equal with the pressure sensed by the microphone positioned in the pipette. The sound pressure received by the microphone is the exciting pressure to calculate the module of elasticity. The displacement of the sample due to the acoustic pressure is in the range of one micrometer which demands a very accurate measurement method. Therefore, an OVF-5000 Xtra, the newest Polytec modular laser vibrometer is applied. A helium-neon laser exploits the Doppler Effect to derive the velocity. The displacement is a sinusoidal movement and obtained by integrating over velocity, which is done by a Labview automation program. A sketch of the whole setup is depicted in (figure 2 (a)). Due to the small displacement, external influences must be kept to a minimum or remedied. Hence, the measurement setup and laser vibrometer sensor head were placed on a vibration isolated table to damp external vibrations and interferences.

3. Results

The bottom end of the pipette was placed on the surface of the specimen perpendicularly (figure 3 (a)). Due to the sinusoidal sound pressure in the pipette, the surface area enclosed by the pipette walls is aspirated into the pipette and pushed in (–z) direction on specimen (figure 3 (b)). The displacement wave (L) at the center of the pipette cross-section is optically measured as a function of the acoustic pressure (ΔP). In this study, we assumed that the silicone samples used are isotropic and homogenous and has linear elasticity. Based on these assumptions and using the pipette aspiration technique, the linear equation (1) taken from Aoki [2] used to calculate Elasticity modulus E. The coefficient C is corresponding to the effect of the variations in pipette wall thickness (*b/a*), specimen thickness (*h/a*) and specimen radius (*R/a*) which form Aoki [2] plots (figure 4) we consider the coefficient 1.1 according to our pipette and silicone sample dimensions.

$$\frac{L}{a} = C\frac{\Delta P}{E} \quad \Rightarrow \quad E = C\frac{\Delta P}{\frac{L}{a}} \qquad (1)$$

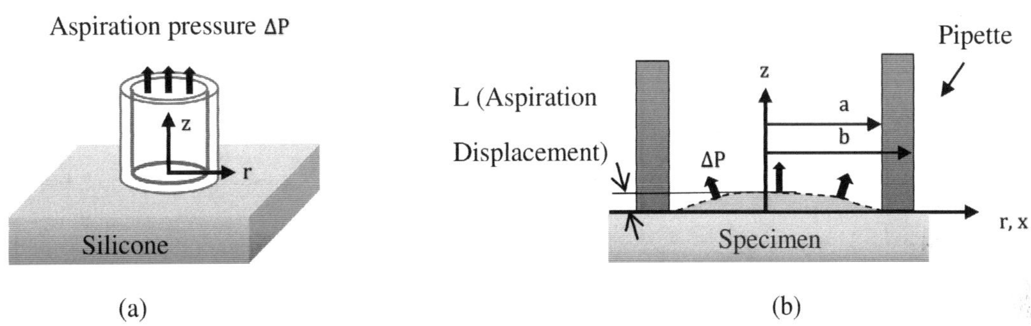

Figure 3. Pipette and sample: (a) Pipette Aspiration; (b) Schematic of the deformed sample under pipette.

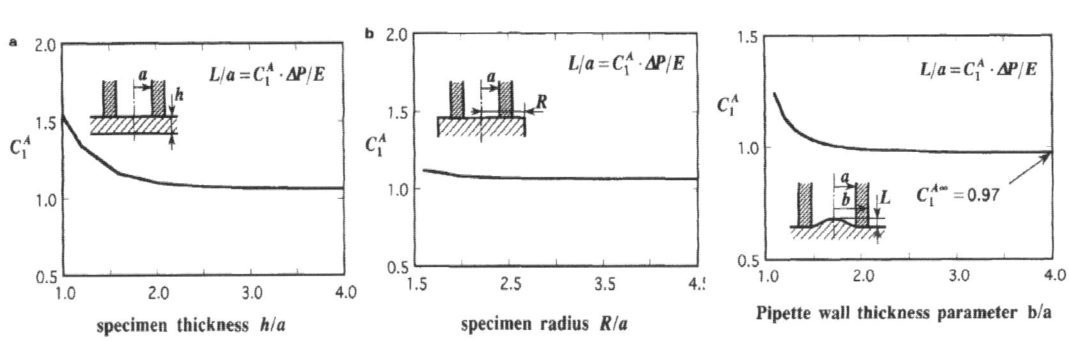

Figure 4. Effect of pipette wall thickness, specimen size and radius extracted from Aoki [2].

The elasticity modulus of all four samples summarized with respect to the frequencies at which they were excited are shown in figure 5. The results clearly show the change of the elasticity modulus over the considered frequency steps. For the specimen with 1:1:0 mixing ratio, an elasticity modulus of about 37 kPa is determined at an excitation frequency of 50 Hz (step 0) while in the same step, 25 kPa, 11 kPa and 8 kPa are shown by the 1:1:2, 1:1:3 and 1:1:4 sample respectively. The elasticity modulus raise to 210 kPa at an excitation frequency of 500 Hz (step 9) in the 1:1:0 specimen. The highest and the lowest elasticity modulus belongs to the 1:1:4 sample with maximum 450 kPa at 500 Hz and minimum 8 kPa at 50 Hz.

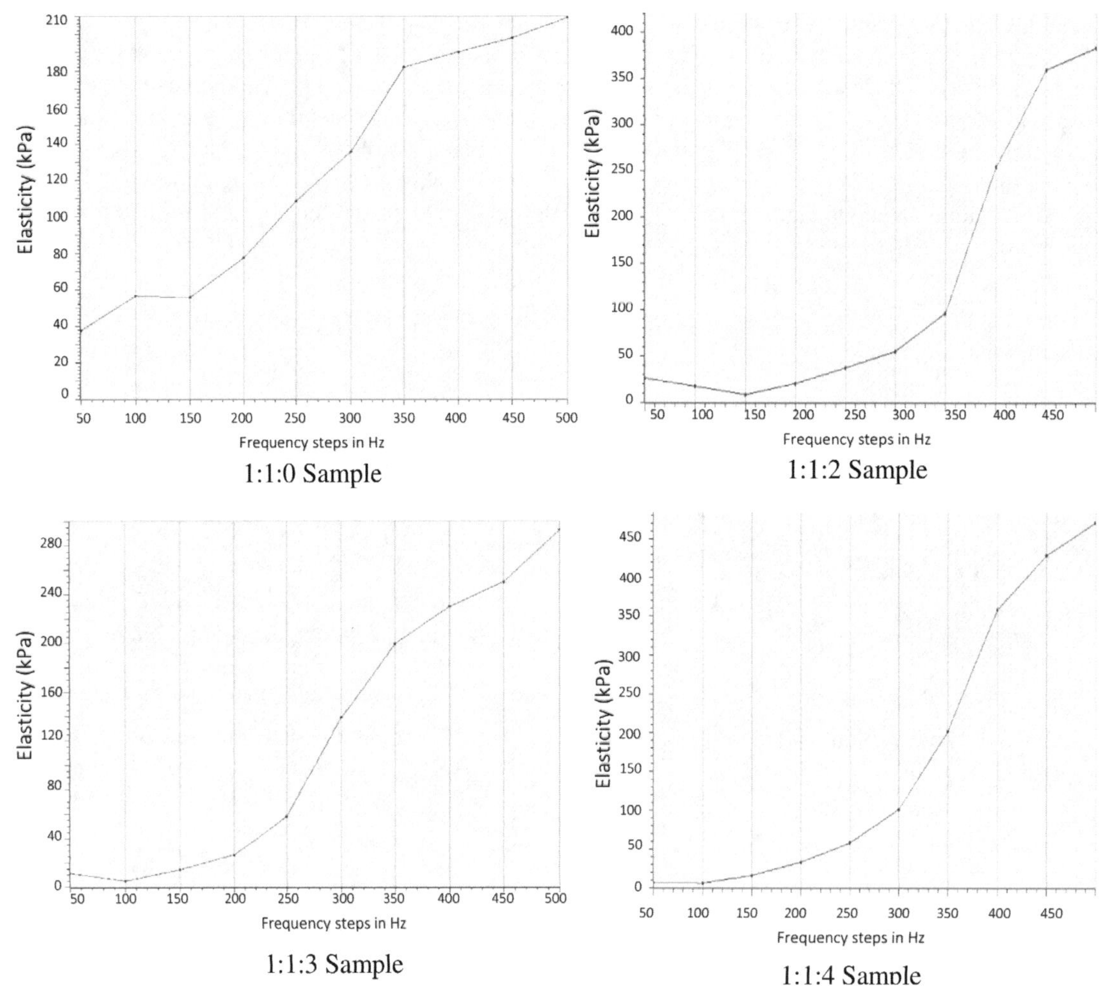

Figure 5. The Elasticity moduli of the four samples.

We measured the modulus of elasticity with various pressure amplitudes and it remained independent from the pressure amplitude diversity as it demonstrated in figure 6. In the chart (a), the elasticity modulus of the 1:1:3 specimen calculated based on disparate pressure amplitudes from around 1 Pa to 12 Pa for each frequency steps. While the chart (b) illustrated module of elasticity with unique pressure amplitude, 3 Pa for all frequency steps.

Like the previous studies which took into account, rather than the thickness and radius of the pipette walls which we assumed their effects by the coefficient (C), the elasticity modulus shows very high dependency to the force imposed on the silicone surface by the pipette walls. Moreover, since the displacement is in range of nanometers or micrometers depend on the excitation pressure amplitude, and the sound pressure measured by the microphone should not be distorted, the environmental factors like as vibrations, temperature, noise and humidity are essential to get precise results.

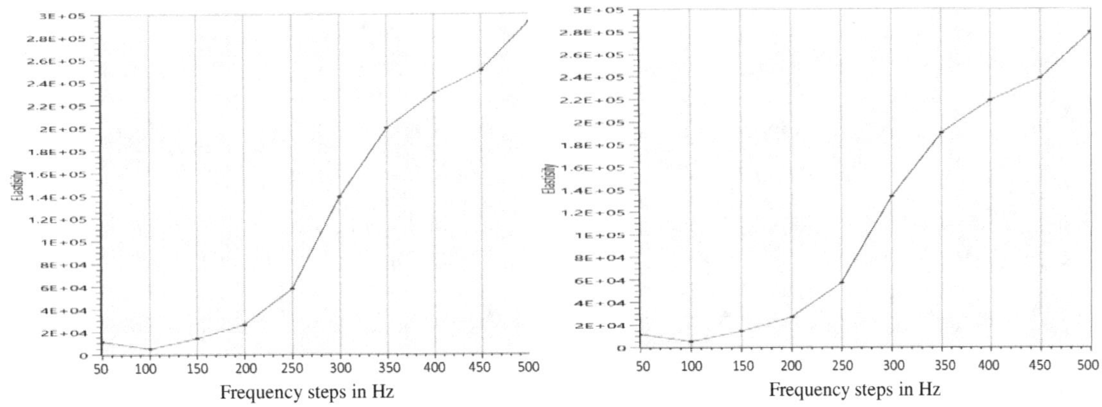

Figure 6. Elasticity modulus with (a) disparate pressure amplitudes (b) unique pressure amplitude.

We changed the position of the laser spot between ten different locations in the area of a circle with a diameter of 3 mm on the center of the silicone surface in the pipette-enclosed area and measured the elasticity module for each of the laser spot positions for the ten frequency steps. Elasticity modules for the ten different laser spot positions are shown in (figure 7). The mean value and the standard deviation of the values in the first step and the last step of the 1:1:0 graph is (34.203 kPa, 2.75925 kPa) and (214.053 kPa, 5.525 kPa) respectively. The mean value and standard deviation of the 1:1:2, 1:1:3 and 1:1:4 elasticity modulus graphs over the 500 Hz frequency are (316.405 kPa, 37.795.5 kPa), (344.162 kPa, 53.585 kPa) and (562.654 kPa, 52.291 kPa) respectively. Since in this case, we are measuring off the center, the producing measurement error is very small except in the first step of 1:1:0 specimen and in some of the ending steps in all the other samples. Therefore, this method has a good quality of the measurement.

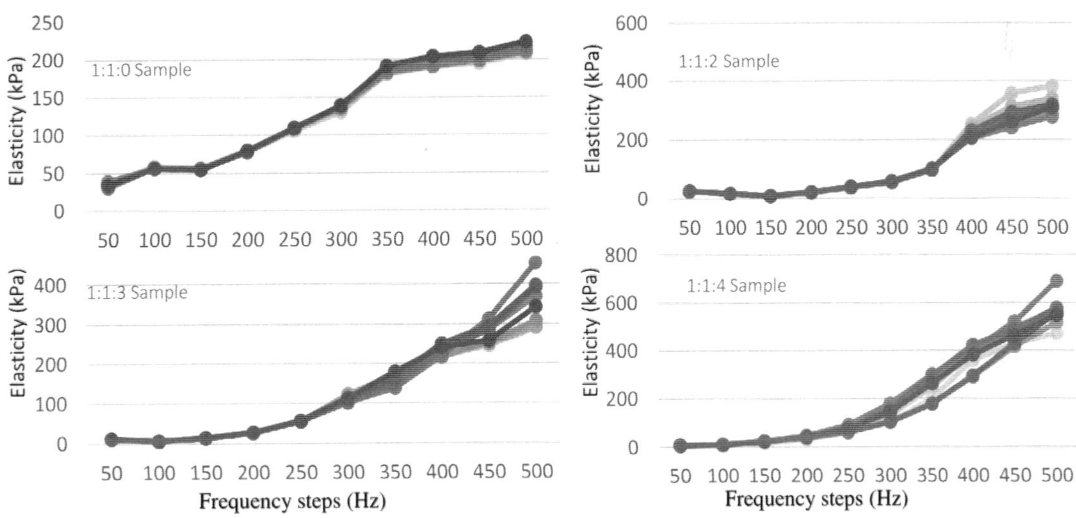

Figure 7. Measured elasticity module for 10 different laser center points of the for silicone .sample.

It is clear that they represent a higher deviation at the higher elasticity modulus, in these steps we calculate high elasticity modulus values because of a small excitation in the first step of 1:1:0 specimen and a small displacement amplitude in the ending steps of the rest of the specimens. Moreover, the equipment, measurement errors or the environmental conditions like as vibrations, noises, temperature, etc. might be interfere in the results and be the cause of the high deviations.

4. Conclusion

We have presented a measurement method to determine the elasticity modulus at specific frequency, for silicone materials as used for artificial vocal folds. Therewith, a sample is placed under a pipette, which is used to channel the acoustic pressure produced by a loudspeaker. With the help of a Labview program, where the measured excitation pressure as well as the displacement of the silicone surface is used as input data, the elasticity modulus is determined. The obtained elasticity modulus clearly shows an increase over frequency in all the three silicone mixtures. The elasticity moduli were found for the four different silicone mixing ratios in the range of 8kPa to 450 kPa. Extreme miniaturization and simplicity of the used measurement technique are the big advantages of this method with compare to the previously studied pipette aspiration method using pistonphone. Further work to develop this method needs to be done by comparing the results with numerical simulation and the silicone sample static elasticity measurement.

Acknowledgment

This work was supported by Austrian Science Fund Grant no. I 3806 - B28 and by the German research foundation (DFG) no. DO1247/7-1.

References

[1] Rand R P and Burton A C 1964 Mechanical properties of the red cell membrane i. membrane stiffness and intracellular pressure *Biophys. J.* **4** pp 115–35

[2] Aoki T, Ohashi T, Matsumoto T and Sato M 1997 The pipette aspiration applied to the local stiffness measurement of soft tissues *Ann. Biomed. Eng.* **25** pp 581–7

[3] Weiss S, Sutor A, Ilg J, Rupitsch S J and Lerch R 2016 Measurement and analysis of the material properties and oscillation characteristics of synthetic vocal folds *Acta Acust. United Ac.* **102** (2) pp 214-29

[4] Weiss S, Thomson S L, Sutor A, Rupitsch S J and Lerch R 2016 Influence of pipette geometry on the displacement profile of isotropic materials used for vocal fold modeling *Int. Conf. on Biomedical Electronics and Devices (BIODEVICES)* pp 108-13

[5] Ilg J, Rupitsch S J, Sutor A and Lerch R 2012 Determination of dynamic material properties of silicone rubber using one-point measurements and finite element simulations *IEEE T. Instrum. Meas.* **61** pp 3031–8

[6] Weiss S, Thomson S L, Lerch R, Döllinger M and Sutor A 2013 Pipette aspiration applied to the characterization of nonhomogeneous, transversely-isotropic materials used for vocal fold modeling *J. Mech. Behav. Biomed.* **17** pp 137–51

[7] Zörner S, Kaltenbacher M, Lerch R, Sutor A and Döllinger M 2010 Measurement of the elasticity modulus of soft tissues *J. Biomech.* **43** pp 1540–5

Journal of Physics: Conference Series **1379** (2019) 012018 doi:10.1088/1742-6596/1379/1/012018

Conformity assessment of glare in buildings considering the uncertainty of the parameter UGR_L

A S Ribeiro, L Martins and A Santos
LNEC – National Laboratory for Civil Engineering, Lisbon, Portugal

E-mail: asribeiro@lnec.pt, lfmartins@lnec.pt, asantos@lnec.pt

Abstract. The evaluation of discomfort related to glare is a common need in buildings, requiring conformity assessment defined in the standard EN 12464-1:2011, using the parameter UGR_L index and a related scale with threshold levels associated with perceptible behaviour transitions. The scale is defined for a dimensionless parameter related to the degree of psychological (discomfort) glare of an electric lighting installation in an indoor space. The aim of this paper is to provide a comprehensive understanding of the role of the UGR_L uncertainty in the conformity assessment procedure, being a critical quantity used in the definition of a decision rule under the new edition of the ISO / IEC 17025 standard. The impact of the uncertainty of the UGR_L quantity is shown using examples of decision rules in measurements near the transition limits of the scale.

1. Introduction

In buildings, an adequate lighting should allow individuals to perform different visual tasks efficiently and accurately. The degree of visibility and comfort required for a wide range of visual activities is determined by the type and duration of the activity [1]. A good lighting design should take into account not only the required illuminances but also other qualitative and quantitative requirements. To comply with these requirements, satisfaction of three basic human needs should be achieved: i) visual comfort, where individuals have a feeling of well-being; (ii) visual performance, where individuals are able to perform their visual tasks, even under difficult circumstances and during longer periods and (iii) safety.

The main parameters determining the luminous environment with respect to electric lighting and daylighting are: (i) luminance distribution; (ii) illuminance; (iii) directionality of light and lighting in the interior space; (iv) variability of light (levels and colour light); (v) colour rendering and colour appearance of the light; (vi) glare and (vii) flicker. With the present paper, the authors will address specifically the effect of glare, and in particular, the discomfort glare [1].

Glare is the result of the presence of undesirable light in the visual field of an observer and generally originates from one or more excessively bright light sources. The CIE International Lighting Vocabulary (ILV) defines glare as "*the conditions of vision in which one experiences discomfort or a reduction in ability to see details or objects, due to an unsuitable distribution or range of luminance, or to extreme contrasts*" [2]. Glare can be experienced either as *discomfort glare* (causes discomfort without necessarily impairing the vision of objects) or as *disability glare* (impairs the vision of objects without necessarily causing discomfort). In interior spaces disability glare is not a major problem if discomfort glare limits are met.

The rating of discomfort glare caused directly from the luminaires of an indoor lighting installation is calculated using the CIE Unified Glare Rating (UGR_L) tabular method, based on the following formula [1, 2]:

$$UGR_L = 8\log_{10}\left(\frac{0.25}{L_B}\sum\frac{L^2\omega}{p^2}\right), \tag{1}$$

where L_B is the *background luminance*, calculated as $E_{\mathrm{ind}} \cdot \pi^{-1}$, in which E_{ind} is the *vertical indirect illuminance* at the observer's eye, in lux L is the *luminance* of the luminous parts of each luminaire in the direction of the observer's eye, in cd/m^2; ω is the *solid angle*, in steradian, of the luminous parts of each luminaire at the observer's eye; p is the *Guth position index* for each individual luminaire which relates to its displacement from the line of sight [3-5]. The use of this parameter in the design of electric lighting installations implies the definition of upper limits applied to different degrees of reflectance of materials, walls, ceilings, floors and other working planes, considering specific lighting design solutions, such as the positioning and characteristics of luminaires.

The UGR_L estimated value of the lighting installation shall not exceed the values referred in [1]. The recommended limiting values of the UGR_L form a series whose steps indicate noticeable changes in glare. The series of UGR_L is: 10 (imperceptible), 13 (just perceptible), 16 (perceptible), 19 (just acceptable), 22 (unacceptable), 25 (just uncomfortable) and 28 (uncomfortable). This quantitative scale of visual discomfort is based on the classification proposed initially by Hopkinson [6], supported on the statistical analysis of human qualitative perception of representative samples.

The conformity assessment requires the comparison of the glare rating estimate (in this case, UGR_L) and its uncertainty with the transition values of the scale of visual discomfort showed above. The decision of compliance should be supported by a decision rule.

2. Glare index measurement uncertainty evaluation

The knowledge about the measurement uncertainty is essential for a rigorous conformity assessment. This section presents the measurement uncertainty evaluation related to the glare index, based on the example presented in [5], which describes a simplified calculation approach for this quantity.

The studied example considers a square room with length X and width Y equal to 6 m and a ceiling height of 3 m. The height of eye level for a seated person is 1.2 m above the floor, therefore, the distance H between the height of eye level for a seated person and the ceiling is estimated to be 1.8 m. The reflectance is assumed to be 0.7 for the ceiling, 0.3 for the walls and 0.2 for the floor. A "worst case scenario" is considered with respect to glare, by assuming that the luminaires – with a luminous flux equal to 6 500 lm and 1 650 mm in length – are viewed crosswise. In this simplified approach, the first calculation step consists in an initial glare index estimate obtained by linear interpolation from reference tabular glare data (Table 7.10 [5]). Based on the relation between X, Y and H ($X = Y = 3.3H$), an initial glare index estimate of 11.06 is obtained from the interpolation of four tabular estimates.

The second calculation step involves the application of corrections related to: (i) a mounting height above 1.2 m eye level different than 2 m; (ii) a total downward luminous flux different than 1000 lm; (iii) extra correction terms for different luminaire sizes or lamp types than the ones related to the reference tabular glare data. In the studied example, these corrections are once again obtained by linear interpolation from reference tabular data (Tables 7.10 and 7.11 [5]).

The quantification of the corresponding uncertainty components is given by the differences between the interpolation estimates and the closest tabular estimates, described by triangular probability distribution functions based on the assumption that the estimates are the most probable values of the distribution. The reduced size of the experimental data used for the interpolation, does not allow to assume a Gaussian or a t-Student probability distribution function. No information regarding the measurement uncertainty of the tabular estimates is found in [5] and, therefore, it is not considered in the evaluation.

In this example, a linear mathematical model (addition of initial glare index and corresponding corrections) of the simplified approach was considered suitable, allowing to use the GUM conventional method [7] for the determination of the glare index measurement uncertainty. The measurement uncertainty budget (Table 1) includes the main contributions for the calculation. Regarding the degrees of freedom, the assigned values were estimated using the GUM approach for type B sources of uncertainty (equation G.3 of Annex G of the GUM, being the value of uncertainty reliable to 10 %).

Table 1. Glare index measurement uncertainty budget.

Input quantity	Estimate	Probability function	Standard uncertainty	Degrees of freedom
Initial glare index	11.06	Triangular	0.29/√6=0.12	50
Correction due to lamp luminous flux	4.88	Triangular	0.22/√6=0.090	50
Correction due to height above eye level	-0.20	Triangular	0.20/√6=0.082	50
Correction due to luminaire length	0.32	Triangular	0.31/√6=0.13	50

The final glare index estimate was equal to 16.06 with a combined standard uncertainty of 0.21 and 178 effective degrees of freedom (obtained using the Welch-Satterthwaite formula), for which a coverage factor of 1.97 is obtained for 95 % of confidence interval. Thus, the 95 % expanded uncertainty is 0.42.

3. Decision rule for conformity assessment using glare index measurement uncertainty

The growing need to assure that performance of buildings meets all the applicable requirements, namely, to conditions related to comfort, brings the interest to understand how conformity assessment should be developed and how decision rules are applied for measurement quantities such as glare index and, also, how its measurement uncertainty is accounted for.

Currently, the definition of decision rules to support conformity assessment, taking into account the uncertainty associated with the estimate of parameters is subject of reference documents based on statistical approaches of the problem [8]. This is the subject of particular importance nowadays, as it is one of the major changes introduced in the recently revised ISO/IEC 17025:2017 [9] supporting the reference scheme for the accreditation of testing and calibration laboratories.

The approach to establish the decision rule (criterion) can have two options (with single upper limit): to consider an approach making a direct comparison between the estimate and its expanded measurement uncertainty with the upper tolerance limit (the adopted approach); or the definition of a guard band based on the expanded measurement uncertainty interval.

In the case studied, experimental values were obtained for the estimate of the UGR_L parameter, its standard uncertainty, $u(UGR_L)$ and the significance level expected (usually, of 95 % confidence), being the decision rule (for upper tolerance limit proximity T_U – a similar approach should be considered for the lower tolerance limit T_L) with a significance level of α the following.

$$\text{Acceptance, if is true the hypothesis H0, } P(UGR_L \leq T_U) \geq (1-\alpha), \tag{2}$$

$$\text{Rejection if the hypothesis H0 is false, } P(UGR_L \leq T_U) < (1-\alpha). \tag{3}$$

The expression to test corresponds to

$$P_C = P(\eta \leq T_U) = \Phi\left(\frac{T_U - UGR_L}{u(UGR_L)}\right). \tag{4}$$

This process is illustrated in figure 1 for the upper tolerance limit.

Journal of Physics: Conference Series **1379** (2019) 012018 doi:10.1088/1742-6596/1379/1/012018

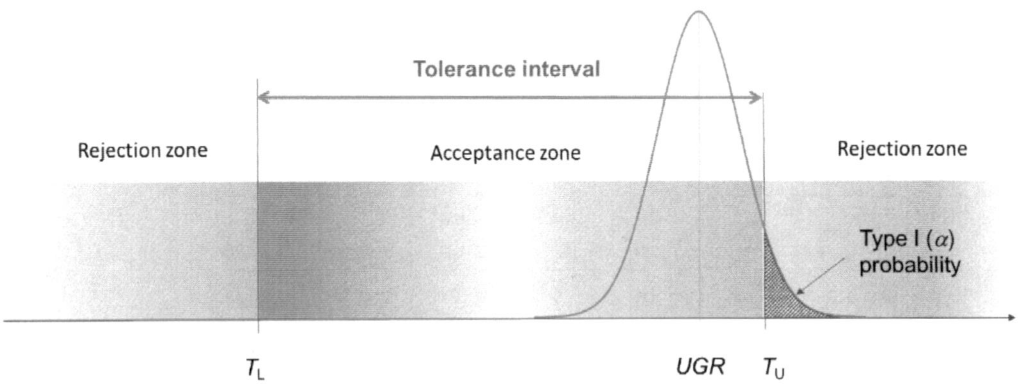

Figure 1. UGR_L value and its uncertainty near the upper limit of a tolerance interval.

An alternative approach (Figure 2) is based on the definition of a guard band, G_u, for a defined confidence interval of $(1 - \alpha)$, requiring that the measurement uncertainty applied to the estimates of the quantity is fixed. In this case, comparison can be made directly between the estimate and the guard band limit, which can be practical for engineering purposes. The guard band lower limit is obtained using:

$$G_u = T_U - u(UGR_L) \cdot [\Phi^{-1}(1 - \alpha)]. \quad (5)$$

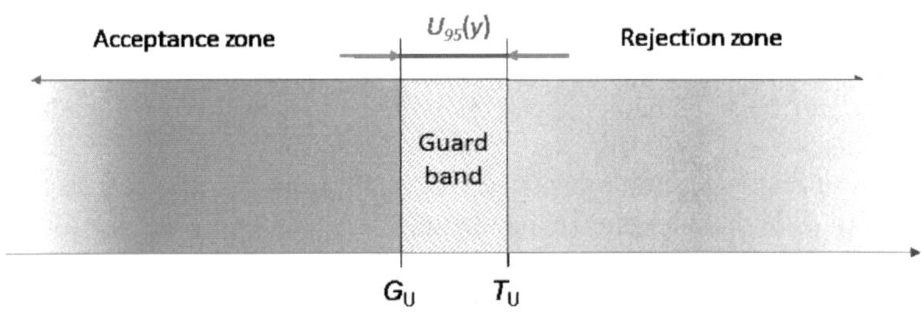

Figure 2. Guard band for upper tolerance and guarded acceptance defined with an expanded uncertainty of 95 %.

4. Conclusions

The role of uncertainty in the conformity assessment procedure is relevant due to the fact that a larger value of uncertainty implies higher risk of rejection of the conditions evaluated with technical and economic impact.

To illustrate the need of using the criteria presented in order to make a decision, consider an estimate of 16.06 and uncertainty of 0.21, being the uncertainty probability distribution partially inside two scale categories [13, 16] and [16, 19]. In this case, the comparison with the transition value 16 gives the amount of probability above the transition value,

$$P_C = P(\eta \geq T_L) = 1 - P(\eta \leq T_L) = 1 - \Phi\left(\frac{T_L - y}{u(y)}\right) = 1 - \Phi\left(\frac{16 - 16.06}{0.21}\right) = 0{,}612. \quad (6)$$

In this case, the probability of being lower than 16 is 38.8% and higher of 16 is 61.2%, thus the conformity assessment for 95% of confidence, related to the scale categories [13, 16] and [16, 19], is not conclusive.

Another illustrative example following the same approach, using an estimate of UGR_L of 15.6, an uncertainty of 0.3 and a confidence interval of 95% would have two distinct decisions, based on equation (6): if the decision rule is based only on the estimate, it will be accepted for the scale category of [13, 16]; otherwise, accounting the uncertainty in the same test, it would be rejected.

$$P_C = P(\eta \leq 16) = \Phi\left(\frac{16-15.6}{0.3}\right) = \Phi(1.33) = 0.909. \quad (7)$$

This result shows that, despite the estimate is lower than 16, the probability of being higher than 16 is (1 – 0.909), equal to 9.1% (higher than 5%) thus being rejected. As the uncertainty value increases, the probability of rejection of measurement conditions also increase. When the measurement uncertainty has the same value for the range of the scale, is possible to introduce an alternative approach based on the definition of guard bands near the tolerance limits, allowing to simplify the analysis by making the comparison of estimates directly with the reduced acceptance region [7].

The second approach, based on the definition of a guard band, allows an analysis of the relation between the UGR_L uncertainty and the guard band near a transition limit of the scale. Using eq. (5) and the transition limit of 16 (meaning the scale classification of *perceptible*),

Figure 3 shows the functional relation, including the areas of compliance and non-compliance (in this case, for the interval between 13 – just perceptible – and 16 – perceptible and uncertainties between 0 and 1). This type of analysis, can be useful in engineering processes, as it can give a quick indication of the guard band amplitude for an uncertainty within the interval selected, providing a simple way to apply a decision rule.

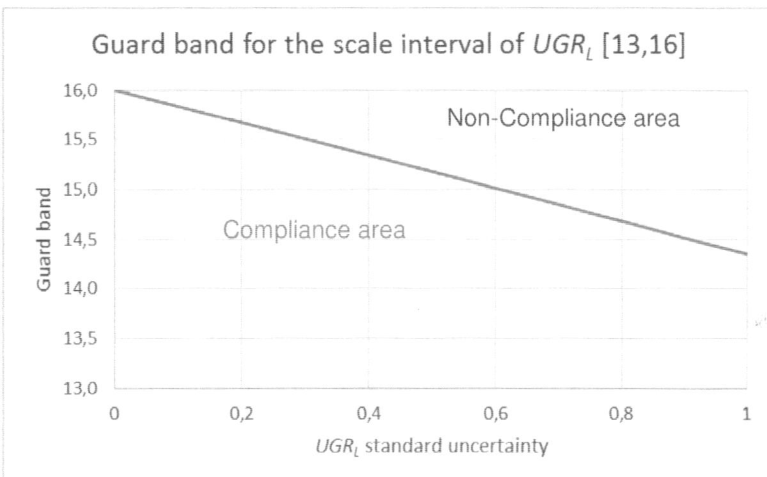

Figure 3. Guard band variation obtained for a measurement standard uncertainty interval.

The use of the parameter UGR_L in the decision making process relies on its comparison with the transition values of the quantitative scale of visual discomfort. This scale defines a classification based on a discrete set of transition values, for which conformity assessment is made, by the comparison of estimates and related uncertainties with scale transition values, being the uncertainty the key for the decision process. The proper use of statistics in the criteria adopted is, therefore, a relevant issue for the quality of the assessment and decision.

Future research work is intended to be made for the evaluation of the UGR_L measurement quality applying other calculation methods and models including a sensitivity analysis of input quantities. Other relevant daylighting and electric lighting quantities and parameters will also be addressed in the future.

References

[1] CEN 2011 *EN 12464-1: Light and Lighting - Lighting of Work Places - Part 1: Indoor Work Places* (European Committee for Standardization)

[2] CIE 2016 *ILV - International Lighting Vocabulary* 2nd edition

Journal of Physics: Conference Series **1379** (2019) 012018 doi:10.1088/1742-6596/1379/1/012018

[3] CIE 2011 *Discomfort Glare In Interior Lighting*
[4] Guth S 1963 A method for the evaluation of discomfort glare *Illuminating Engineering* 58
[5] Pritchard D C 1999 *Lighting* 6th edition (Essex: Longman)
[6] Hopkinson R G 1963 *Architectural Physics: Lighting* (London: BRS, HMSO)
[7] JCGM 100 2008 *Evaluation of Measurement Data – Guide to the Expression of Uncertainty in Measurement* (Joint Committee for Guides in Metrology)
[8] JCGM 106 2012 *Evaluation of Measurement Data – The Role of Measurement Uncertainty in Conformity Assessment* (Joint Committee for Guides in Metrology)
[9] ISO/IEC 2017 *ISO/IEC 17025:2017 General Requirements for the Competence of Testing and Calibration Laboratories* (International Organisation for Standardization)

Journal of Physics: Conference Series **1379** (2019) 012019 doi:10.1088/1742-6596/1379/1/012019

Natural Semantic Metalanguage as an approach to measuring meaning

R F Cavanagh[1], **Y Asano-Cavanagh**[1] **and W P Fisher Jr**[2]

[1] School of Education, Curtin University, GPO Box U1987, Perth, Western Australia
[2] BEAR Center, Graduate School of Education, University of California, Berkeley, Berkeley California 94702 USA

E-mail: R.Cavanagh@curtin.edu.au

Abstract. Do any interesting or valuable fundamental commonalities arise when measurement and linguistic methods are used to understand the same phenomena? A basis for such commonality resides in the human desire for meaning, a need manifest in all cultures and languages. However, the notion of meaning is rarely associated with measurement; it resides more comfortably in the study of language and linguistics. This exploration commences with an examination of measurement theories, principally Rasch Measurement Theory (RMT), with the aim of identifying opportunities for elucidating the meaning of objects of measurement. A brief overview of the discipline of linguistics then reveals the importance of semantics in expressing meaning. An explanation of how the Natural Semantic Metalanguage (NSM) can enable explication of linguistic meaning follows. The paper concludes with a proposal for ecological applications of invariant measurement and the NSM.

1. The meaning of measurement in the human sciences

Contemporary conceptions of measurement in the human, social, behavioral, and health sciences, are varied:

- "... the process of locating both persons and items on a line that represents a latent variable" [1];
- measures "... must be objective abstractions of equal units" [2]; and
- as being "... concerned with properties considered in terms of more or less, and measured by instruments that operate in some relevant range" [3].

Andrich [3] identified three related definitions of measurement applicable to the social sciences: classical measurement, representational measurement, and additive conjoint measurement. Classical definitions are based on the test-score tradition of measurement comprising classical test theory and true score theory [1]. Classical test theory uses the statistical methodology of correlation coefficients to solve psychometric issues related to the estimation of reliability coefficients. The function of instruments is dependent on the sample being measured, and hypothesis testing focuses on estimates of statistical significance levels. According to true score theory, the observed score for a person is the sum of a true score and an error score. The focus is on the sum of observed scores, not on item characteristics, and the error value is for the test, overall; error values for individual items or individual persons are not estimated.

The second, representational, definition of measurement in Andrich's classification states "... the properties of objects can be assigned numbers in such a way that the relationship among the properties is the same as the relationships among the numbers" [3]. For example, concatenation in which the extent of more or less in a property of an object is represented numerically.

Third, in additive conjoint measurement, the consistent relation between two observable variables is taken as evidence of a third, but latent, variable that is not directly measured. For instance, the property of density is not observable, but can be inferred by the relation between the observable properties of mass and volume.

A detailed examination of contemporary measurement theories, particularly Rasch Measurement Theory (RMT), reveals the following more nuanced characteristics conceived in terms amenable to the broad metrological sense of measurement [4-5].

Sufficiency: The requirement that all the information needed for measurement is available in the observed response data is articulated as a testable hypothesis in models positing statistically sufficient estimators of separable parameters. The meaning of the unit of measurement is qualified by the *a priori* development of a construct model or variable map tested empirically against data.

Objectivity and separability: "Objectivity requires that agent calibrations (e.g. item difficulties) be independent of the sample of objects (e.g. persons) used in the calibration procedure. Object measures must also be independent of the particular agents used to obtain them" [6]. Specific objectivity and separability of parameters in RMT meet this requirement.

Invariance: The comprehensive class of all invariant rules is essentially identical with that subclass of invariant rules based on sufficient statistics [7, 8]. In mathematics and the physical sciences, the principle of invariance refers to the property of remaining unchanged under pressure to transform, or in the case of measurement, from changes in the conditions of measurement. The principle is fundamental to understanding the application of scientific laws in different frames of reference such as stationary and moving systems. Invariance in measures developed from RMT requires invariant, linear item difficulty calibration and person measurement.

Differential item functioning: Each item is tested for differential functioning to minimize bias due to confounding or extraneous person factors. At the item level, persons or objects with a similar level of the attribute being assessed should receive equivalent item measures irrespective of differences in person factors including location, time of assessment, and gender.

Discrimination and anomalies: Implicit in the definitions of measurement was the idea of comparison of the properties of persons or objects. The metrological systems in scientific measurement use interval and ratio scales to measure properties of objects and make comparisons between objects. According to Kuhn [9], discovering and investigating anomalies arising in these comparisons is a primary aspect of scientific progress.

Philosophy: The hallmark of meaningfulness is a truth value obtained independent of the particulars of the choice of representation, within certain limits, such that formal analysis in the context of measurement theory leads to a focus on invariance under changes of scale [10, 11]. Additional philosophical justifications for measurement theory can be derived from hermeneutical phenomenology and Actor-Network Theory [12, 13].

Non-deterministic (stochastic) constructs: The object of measurement is not precisely defined but is described probabilistically in terms of self-organized, complex adaptive systems [14]. The essence of the object of measurement is systematically refined and reproduced through multiple iterations of data collection and analysis.

These seven characteristics are fundamental principles for the construction of invariant measures, a lens into the meaning of measurement. Now we take up the second major construct of this project, language.

2. Language

Having language distinguishes humans from other animals. Chomsky (1972), asserted "When we study human language we are approaching what some might call the 'human essence' the distinctive qualities of mind that are so far as we know unique to man" [15]. The study of human language is linguistics. Linguistics provides structure for understanding the nature of language. Linguists investigate morphology, syntax, phonetics, phonology, semantics and pragmatics. *Morphology* is the study of the structure of words. *Syntax* is the arrangement of words and phrases to create sentences, and to create the structure within sentences. The syntactical branch of language structure is also presented as prescriptive grammars. *Phonetics* concerns the production of speech sounds, *phonology* is the system and patterns of the speech sounds. *Semantics* is the branch of linguistics and logic concerned with the meaning of words and their meaning within sentences. *Pragmatics* is the study of language from the perspective of usage and context.

The thesis of the current project focuses on the fact that measurement and language have a common purpose, the expression of meaning individually and collectively. Accordingly, the following examination of meaning in language will centre on *semantics*. Pragmatic issues in the proliferation of proficiency testing, particularly in second language teaching and assessment, will be covered elsewhere.

Wierzbicka [16] makes clear distinctions between language, meaning, and communication. Language is essentially a tool for expressing meaning and communicating; both apply language. But in contrast to communication, meaning can be expressed without the need for an audience, as, for example, when we talk to ourselves or think aloud. However, the expression of meaning is complicated by some words having language-specific meanings and other words having a common meaning across languages. Resolution of a potential conundrum is possible by employing a two-parameter model of linguistic meaning in which meaning can be separated from language; the meaning of words is separated from their morphology. Some words will retain the same meaning cross-linguistically and cross-culturally; their meaning is invariant. Other words' meanings will be variably dependent on cultural factors. The presence of words with invariant meanings in different languages provides an anchor for cross-linguistic comparisons and semantic explication. These are features of the NSM [17].

3. Natural Semantic Metalanguage (NSM)

The philosophical orientation of the NSM was influenced by the pre-Enlightenment rational philosophers who believed all people share some simple ideas, the basic elements of thought. For example, Goddard [18] recounted Leibniz acknowledging the occurrence of concepts which could be understood in isolation but could be combined to describe more complex concepts. Similar bottom-up patterns of emergent meaning lead to the spontaneous introduction of new concept-word-thing semantic triangles as cultural systems change. Reciprocally, these complex concepts can be understood by reduction, as when complex concepts can be decomposed into a small number of simpler concepts.

Wierzbicka [16] applied a linguistic perspective to the decomposition notion and defined the semantic prime. A semantic prime is a word-meaning or concept which cannot be paraphrased (decomposed) into any simpler terms; and semantic primes are expected to be lexical universals. Universality requires that, within one language, a prime cannot be paraphrased, and that semantic primes have equivalents in other languages, there should be exponents of the prime in many languages. Every semantic prime has its own grammar of relationships involving the other primes with which it can be combined. For every prime, the other primes with which it can be combined have been identified.

The NSM currently employs 65 semantic primes; for example, substantives, determiners, speech, time and space, with equivalents in most languages. The following semantic analysis (a partial explication), illustrates how the NSM is applied to clarify the meaning of the Japanese word "kawaii" (dictionary translation "cute") [19]. "Kawaii" was chosen because although Japanese girls and women extensively use it to express positive emotions towards clothes and feminine paraphernalia, there is no equivalent in English. The corpora for the analysis were publicly available web-based information found in 50 postings.

First, Japanese dictionary definitions of "kawaii" were obtained and examined in conjunction with previously proposed definitions. Next, the reductive paraphrase methodology was applied to the corpora. A textual analysis of the web-based discourse of the native speakers was conducted to reveal embedded emotion, thoughts, cultural significance, and physical attributes. For example, in the case of "kawaii", these included "small" and "positive emotion". Simultaneously, the repertoire of 65 NSM semantic primes was consulted to tentatively specify the semantic components of "kawaii", the constituent semantic primes.

The process was iterative and proceeded until the components of meaning could not be further reduced. Finally, the resulting explication included only semantic primes and their respective English components for the following categories of primes; substantives (e.g. "thing"), evaluators (e.g. "good") and descriptors (e.g. "small") [19]. The analysis of "kawaii" illustrates how the NSM enables meaning-based cross-lingual understanding of a word, in contrast to literal translation. In literal translation, the expression of meaning is limited by reliance on synonyms and resulting circularity of argument.

4. Issues in the application of semantics in measurement

As was noted earlier, both measurement and linguistic approaches can be applied to express meaning. For example, applying RMT constructs meaningful invariant measures, and applying the NSM elucidates linguistic meaning. This commonality of intent provides a rationale for exploring the possibility of complementarity and synergy. In keeping with the remit of the project, emphasis now shifts to the opportunities afforded by the NSM for expressing meaning when constructing invariant measures in the human, social, behavioral, and health sciences. The opportunities are in response to problems that arise through the use of language.

The first problem is deficiencies in the procedures and the resources used to compare the semantics of different languages. In the case of native language instruments used in social and psychological measurement, the availability of existing definitions can enable reliable analysis of the semantics. However, when instruments use vocabulary from multiple languages, the availability of bi-lingual definitions may be limited, particularly from bilingual dictionaries. Bilingual dictionaries have been criticized for insensitivity to cultural values, an ethnocentric view of the lexis of other cultures, and circularity between definitions.

A second problem derives from the treatment of substantive matters in the design of instruments, particularly multi-language instruments, where semantic analyses are omitted when developing multi-linguistic construct models. The choice of vocabulary in multi-language instrument construction is a semantic issue and use of semantic analyses during instrument development and administration is warranted.

A third problem is exemplified by large-scale international surveys. Measuring a common construct using multiple vernaculars requires attention to the semantics of each native language.

A fourth problem centers on the use of quantification in the absence of semantic analyses. Quantitative procedures and instruments do not necessarily provide meaningful results notwithstanding the selection of large samples and statistical data analyses.

A fifth problem is myopia in the choice of linguistic and measurement approaches, not canvassing multiple theories and methodologies, in particular across qualitative and quantitative divides often wrongly assumed to make opposing philosophical commitments.

5. Applying semantic analysis for expressing meaning in Rasch measurement

The following procedure describes how NSM analyses could deal with linguistic issues in the development of a bi-lingual measure. The vernacular of a language can be analysed by the NSM approach to identify native language semantic primes that define the phenomenon of interest, as, for example, in the analysis of "kawaii". Similarly, the meaning of the construct in another language can also be defined by an NSM analysis.

Although the NSM definitions use exponents specific to the respective native languages, the exponents have universal meaning and equivalents in most languages. Examination of the exponents enables comparison of the meaning of the phenomenon in the different languages. Depending on cultural and linguistic peculiarities, the comparison might or might not reveal common meaning. When there is common meaning, the equivalents in the common NSM definitions can comprise a construct model of the phenomenon with meaning in both languages and is applicable for the construction of instruments in accord with RMT.

The NSM approach has features analogous to aspects of invariant measurement. The reductive paraphrasing in the NSM analysis is typically conducted in the native language using examples from that language. This ensures fidelity of the NSM explication in expressing the meaning of attributes of the local culture including identifying anomalies to the NSM explication. Another analogy concerns the universality of the primes, as the function of the primes is independent of the language being analysed. Similarly, the universality of the sematic primes exemplifies specific objectivity because the development of the NSM and the testing of primes were independent of the languages being tested. These features of semantic analysis could be viewed as linguistic invariance.

6. Applying semantic analysis in ecologising

When meaning at the semantic level of word usage can be shared with fidelity in different communities, cultures, and languages, metrological networks become viable for the management of human and natural ecosystems [20-22]. A semantic approach to ecologising focusses on both the substance of what is being expressed and the mechanism for expressing meaning. The substance of expression, or the essence of the meaning, can be revealed by reductive semantic analyses that complete a return dialectic back to the patterns of phenomena that originally cohered into classes that could be named. Significantly, the notion of semantics can be extended beyond a linguistic conception to include any phenomenon or construct assigned meaning.

From an ecological perspective, a semantic analysis is a tool for understanding the constituents and dynamics within ecosystems and between ecosystems. Using semantic primes to explicate ecosystems enables cross-ecosystem system analysis. It also provides construct models for constructing invariant measures (Rasch Measurement Theory), as described in Section 5. These measures are the foundation for building metrological networks and complement the semantic approach to expressing meaning.

7. Summary

The core activity of the paper was a comparison between Rasch Measurement Theory and the Natural Semantic Metalanguage, an approach to semantic analysis. The impetus for the comparison was the possibility of synergy between the measurement theory and the linguistic theory. After a brief overview of measurement, the notion of invariant measurement was explicated in terms of seven distinguishing characteristics compatible with RMT:

- Sufficiency;
- Objectivity and separability;
- Invariance;
- Lack of differential item functioning;
- Discrimination and anomalies;
- Philosophy;
- Non-deterministic (stochastic) constructs.

The features were advocated as the principles for expressing meaning when constructing measures.

Attention was then given to linguistics and the role of semantics. The Natural Semantic Metalanguage (NSM) approach was introduced and the process for semantic analysis using the NSM was explained. The principal analytic procedure is defining the construct of interest by reductive paraphrasing and then using semantic primes to define the construct. Semantic primes (65) have universal meaning, are independent of language and culture, and enable stable cross-linguistic and cross-cultural comparisons. The invariant nature of the NSM framework of primes was described as "linguistic invariance".

Linguistic issues impacting measurement stem from limitations in the attention given to semantics in instrument development, particularly in cross-language studies. NSM could be used to ameliorate these limitations and provide construct models for the design of measures.

Applying the NSM to define constructs by explications comprising semantic primes appears to be a potentially powerful method for understanding the functioning of ecosystems and the design of ecologising, particularly when complemented by application of Rasch Measurement Theory.

In conclusion, we argue for more use of sematic analysis in human science measurement. While the NSM and RMT theoretical approaches appear dissimilar, there is significant similarity and important complementarity to warrant more investigation. Future work will design and implement a study to test the proposed methods.

References

[1] Engelhard G Jr 2012 *Invariant Measurement* (New York: Routledge Academic) p 28
[2] Bond T G and Fox C M 2001 *Applying the Rasch Model: Fundamental Measurement in the Human Sciences* (New Jersey: Lawrence Erlbaum) p 2
[3] Andrich D 2018 Advances in social measurement: A Rasch measurement theory In F Guillemin, A Leplege, S Briancon, E Spitz and J Coste (Eds) *Perceived Health and Adaptation in Chronic Disease* pp 66- 91 (New York: Routledge) pp 66-68
[4] Mari L and Wilson M 2014 An introduction to the Rasch measurement approach for metrologists *Measurement* **51** 315-327
[5] Pendrill L and Fisher W P Jr. 2015 Counting and quantification: Comparing psychometric and metrological perspectives on visual perceptions of number *Measurement* **71** 46-55
[6] Stenner A J 1990 Objectivity: specific and general *Rasch Measurement Transactions* **4** 111
[7] Hall W J, Wijsman R A and Ghosh J K 1965 The relationship between sufficiency and invariance with applications in sequential analysis *Ann. Math. Stat.* **36** 575-614
[8] Arnold S F 1985 Sufficiency and invariance *Stat. Probabil. Lett.* **3** 275-279
[9] Kuhn T 1970 *The Structure of Scientific Revolutions* (Chicago: University of Chicago Press)
[10] Mundy B 1986 On the general theory of meaningful representation *Synthese* **67** 391-437
[11] Roberts F S 1994 Limitations on conclusions using scales of measurement. In A. Barnett, S. Pollock and M. Rothkopf (Eds.) *Operations research and the public sector* (Amsterdam, The Netherlands: Elsevier) pp 621-671
[12] Fisher W P Jr 2004. Meaning and method in the social sciences *Hum. Stud.: A Journal for Philosophy and the Social Sciences* **27** 429-454
[13] Fisher W P Jr and Cavanagh R 2016 Measurement as a medium for communication and social action, I & II. In Q Zhang and H Yang (Eds.) *Pacific Rim Objective Measurement Symp. (PROMS) 2015 Conf. Proc.* pp 153-182 (Berlin: Springer-Verlag)
[14] Fisher W P Jr and Stenner A J 2017 Towards an alignment of engineering and psychometric approaches to uncertainty in measurement: Consequences for the future *18th Int. Congr. of Metrology* **12004** 1-9
[15] Chomsky N 1972 Language and mind https://www.sk.com.br/sk-chom.html
[16] Wierzbicka A 1992 *Semantics, Cultures and Cognition* (New York: Oxford University Press)
[17] Goddard C and Wierzbicka A 2014 *Words & meaning* (Oxford: Oxford University Press)
[18] Goddard C 2017 *Ten Lectures on Natural Semantic Metalanguage* (Leiden: Brill)
[19] Asano-Cavanagh Y 2014 Linguistic manifestation of gender reinforcement through the use of the Japanese term kawaii *Gender Lang.* **8** 341–359
[20] Fisher W P Jr and Stenner A J 2016 Theory-based metrological traceability in education: A reading measurement approach *Measurement* **92** 489-496
[21] Fisher W P Jr 2009 Invariance and traceability for measures of human, social, and natural capital: Theory and application *Measurement* **42** 1278-87
[22] Fisher W P Jr and Stenner A J 2015 The role of metrology in mediating and mobilizing the language and culture of scientific facts *J. Phys.: Conf. Ser.* **588** 012043

Journal of Physics: Conference Series **1379** (2019) 012020 doi:10.1088/1742-6596/1379/1/012020

Density measurements of viscoelastic samples with oscillation-type density meters

A Furtado[1], J Gavina[2], A Napoleão[2], J Pereira[1], M.T. Cidade[3] and J. Sousa[1]

[1] DMET-LNM, Instituto Português da Qualidade, I.P., Caparica, Portugal
[2] CQE-DQB, Faculdade de Ciências, Universidade de Lisboa, Lisboa, Portugal
[3] Departamento de Ciência dos Materiais e CENIMAT-I3N, Faculdade de Ciências e Tecnologia, Universidade Nova de Lisboa, Caparica, Portugal

E-mail: afurtado@ipq.pt

Abstract. The oscillation frequencies produced by oscillation-type density meters during the density measurements suffer damping due to the viscosity of newtonian liquids. The effect of viscoelastic behaviour of non-newtonian liquids in the damping of these oscillations is still not known. So, 5 viscoelastic liquids were rheologically characterized, resorting to a rotational rheometer, and their density measured in order to provide a deeper insight into the damping effects produced by these types of fluids. To do so, oscillatory data was related with the obtained density deviations. The results of this study are one of the first insights for the knowledge of the measuring behaviour of these density meters when measuring viscoelastic fluids, one of the scopes of the EMPIR Project 17RPT02-*rho*Liq.

1. Introduction

1.1. The metrology of liquids density

Density, ρ, is one of the major driving elements in economic transactions with important goods, ranging from fuels to liquid food products. The actual consumption of high-value liquids, such as wine, olive oil and fuel, can be considered of major importance to the European economy [1-3]. Liquids' physical properties such as: density, surface tension, viscosity and elasticity (i.e. non-Newtonian behaviour due to dissolved gases like carbon dioxide in beverages, suspended particles like in the pulp in a natural juice, etc.), among others, cover a wide spectrum of variation. Additionally, these liquids are handled in a wide range of temperature and pressure. So, the existent methods for density measurement are influenced by one or more of these physical properties as well as by temperature and pressure, meaning that their robustness need to be study. However, currently there are no EURAMET guides on liquid density measurements and the existing international standards [4, 5] and the reference documents used in Legal Metrology [6, 7] are either outdated or incomplete. Improvements in the quality of liquid density measurements will have a direct impact at industrial sites, leading to more competitive national industries and it will allow a free circulation within the EU market.

1.2. 17RPT02-rhoLiq: a new EMPIR Joint Research Project

In order to bridge this gap in the scientific knowledge of liquid density metrology a 3 years EMPIR project, the 17RPT02-*rho*Liq, counting with a Consortium of 11 National Metrology Institutes (NMI) (IPQ – Portugal, BEV-PTP - Austria; BRML- Romania, CMI – Czech Republic, DMDM – Serbia, GUM – Poland, IMBiH – Bosnia and Herzegovina, JV – Norway, PTB – Germany, TUBITAK – Turkey, INM - Moldova) and one density meters manufacturer (Anton Paar – Austria) has started last May 2019 [8]. One of the main focus of this project is to develop high-level measurements and calibration services, and to produce density reference materials for national stakeholders, e.g. food, chemical, pharmaceutical and petroleum industries. The international recognition of these NMI in this metrological field will indirectly lead to the reinforcement of mutual confidence and cooperation at regional and international levels. This project will facilitate compliance with economically relevant EU Directives, e.g. Directive 2007/45/EC [9], and it will further reinforce the competitiveness of production industries. These NMI will be able to develop strategies for accurately measuring the density of liquids with non-classical physical properties, i.e. with physical properties that differ from water or hydrocarbons which are normally used as standard liquids. This will include robustness studies using liquids with high viscosity, with viscoelastic behavior, and with dissolved gases or suspended particles. These kinds of liquids are the most commonly measured by the end-users (food, chemical, pharmaceutical, petroleum, biofuels, etc.), therefore, the knowledge gained about possible interferences and corrections, will be crucial to obtain accurate and traceable density measurement results. These kinds of measurements can be performed under limited conditions and will often result in larger uncertainties. This knowledge will also be disseminated in international guides and standards for scientific, applied (via 3 new EURAMET guides and the revision of existing ISO standards [4, 5]) and legal (via revision of existing OIML [6] and WELMEC [7] documents) metrology and this will address the lack of documentation on this issue.

1.3. Damping effect on oscillation-type density meters

The working principle of an oscillation-type density meter is based on the law of harmonic oscillation. In this kind of measuring instrument, a U-shaped tube is filled with the sample. This measuring cell containing the sample will act like a flexural oscillator; the density of the liquid is determined by measuring the period of oscillation. It is well known that the viscous component of a liquid during the oscillation leads to a damping and by this to a lower resonance frequency of the measuring system and, thus, to an incorrect density indication. Damping detection is performed by analyzing the frequency of harmonics of the oscillation. From these data, the density meter determines the non-viscosity-corrected density value, ρ_{nc}, and a viscosity-corrected density value, ρ_c [10]. However, the effect of viscoelastic behaviour of non-Newtonian liquids in the damping of these oscillations is still not known. So, the rheological behaviour, in oscillation mode, of 5 viscoelastic liquids was studied and the results correlated with the obtained density deviations, in order to provide a deeper insight into the damping effects produced by these types of liquids. The results of this study are one of the first insights to the knowledge of the measuring behavior of these density meters when measuring viscoelastic liquids, one of the scopes of the EMPIR Project 17RPT02-*rho*Liq.

2. Method and Materials

2.1. Test samples

To carry out the study of the influence of the viscoelasticity of the samples on the result of the measurement of the density by oscillation-type density meters, 3 aqueous solutions of different polymers (Carbopol 940 (Fagron) at 0,15 $cg{\cdot}g^{-1}$; polyvinyl alcohol (PVA, Sigma-Aldrich) and sodium borate (Borax, Dimor), both at 0.04 $g{\cdot}ml^{-1}$; hydroxymethylcellulose (Sigma-Aldrich) at 0.01 $g{\cdot}ml^{-1}$; all in ultrapure water) and 2 food liquids (mayonnaise and ketchup, both commercial formulations) were chosen.

2.2. *Density measurements*
The oscillation-type density meter used in this study was a DMA 5000 from Anton Paar (figure 1). The density indication errors were obtained by calibration of this density meter with newtonian certified reference liquids (CRM) (from H&D Fitzgerald) within the dynamic viscosity, η, interval from 1 to 795 mPa·s. The reference density values of these liquids were determined by hydrostatic weighing. The density indication errors of the density meter were calculated by the difference between the density measured value and the reference density value [11].

The density of the 5 viscoelastic samples was measured by gravimetric method with a 50 mL Gay-Lussac pycnometer (figure 1), according to ISO 2811-1 [12]. The filling of the pycnometer with these samples was performed with a peristaltic pump (ICC, ISMATEC) in order to avoid the formation of air bubbles. The density of these 5 samples was also measured using the oscillation-type density meter at the same temperature of the pycnometer test, for comparison of results.

Additionally, the density values of the 5 viscoelastic samples obtained by the oscillation-type density meter (i.e., ρ_{nc} - non-viscosity-corrected density value, ρ_c - viscosity-corrected density value and ρ'_c – density value corrected with the calibration curve obtained with the newtonian liquids) were compared against the density results obtained by gravimetric method using a Gay-Lussac pycnometer (ρ_{pic}).

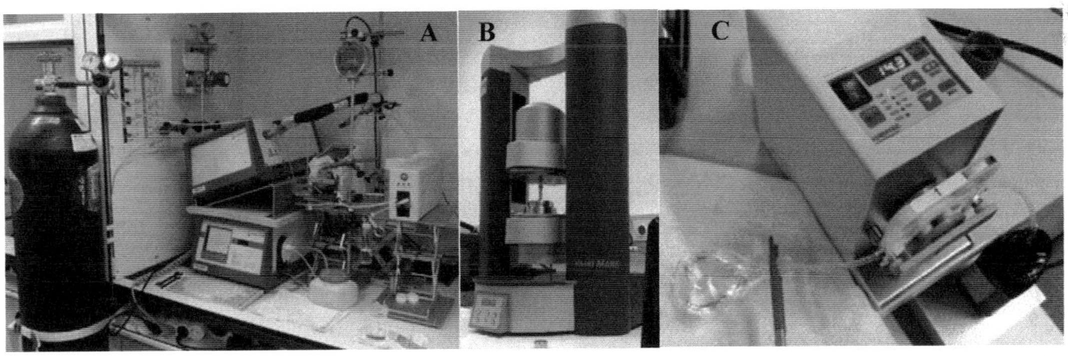

Figure 1. Photos of the measuring apparatus used. **A**: Oscillation-type density meter (DMA 5000, Anton Paar); **B**: Rheometer (Mars III, HAAKE ThermoScientific); **C**: Filling procedure of Gay-Lussac pycnometer with a peristaltic pump (ISMATEC).

2.2.1. *Uncertainty budget.*
The uncertainty of density values obtained in both measuring methods, i.e. with the oscillation-type density meter (1) and with the pycnometer (2), was obtained according to GUM methodology [13], having into account the following major contributions for the uncertainty budget: (1) density meter (resolution, drift and calibration including CRM used), measurements repeatability; (2) calibration of pycnometer volume, balance (resolution, drift and calibration), air buoyancy (and instruments used to measure air temperature, relative humidity and pressure), mass standards used, measurements repeatability, temperature, temperature coefficient of the liquid. In this paper the reported expanded uncertainty, U, is stated as the standard measurement uncertainty multiplied by the coverage factor $k = 2$, which for a t-distribution corresponds to a coverage probability of approximately 95 %.

2.3. *Rheological determinations*
Dynamic viscosity, η', storage modulus, G', and loss modulus, G'', dependence on shear strain and on frequency, of the non-newtonian samples tested were determined by oscillation tests in a rotational rheometer (Mars III, HAAKE ThermoScientific) (Figure 1) using a plate-plate (PP35 TiL) measuring geometry. Two different oscillation mode tests were performed for each sample: (1) amplitude sweep

at a constant angular frequency, ω, of 10 rad/s in the interval of shear strain, γ, from 0.01 to 100 % (log. ramp) for the determination of the linear viscoelastic regime (LVR) and (2) a frequency sweep at a constant shear strain, γ, of 0.1 to 1 % inside the LVR, for an angular frequency interval, ω, of 100 to 1 rad/s (from 0.016 to 16 Hz) (log. ramp). From the amplitude sweep tests it was possible to identify the prevalent behaviour of the sample (viscous or elastic) regarding the shear strain applied and also to check for yield stress value, τ_y, and for the flow stress value, τ_f. The data from frequency sweeps ran for a non-destructive deformation range, will give information regarding the dependence of the dynamic viscosity, η', with the oscillation frequency, *f*, up to 16 Hz, the upper limit of the PP35 measuring geometry.

3. Results and Discussion

3.1. Comparison of density results

Because it is a static method of measuring the density of liquids, pycnometry has been chosen to be used in comparison with vibrating tube densimetry. As a static method it is expected that the viscoelastic properties of the samples would not influence the measurement result, as in oscillation-type density meters. However, phenomena related to these properties were observed during the filling of the pycnometer which may have influenced the measurement results, and which cannot be easily accounted for. Table 1 summarizes the deviations of the bulk density results of the test liquids obtained by oscillation-type density meter compared to the results obtained by gravimetric method using the Gay-Lussac pycnometer. A maximum expanded uncertainty of $5.4\cdot10^{-5}$ g·cm^{-3} was obtained for the ρ_{nc}, ρ_c and ρ'_c values, with exception of the liquid foods (mayonnaise and ketchup) with a maximum uncertainty value of $3.3\cdot10^{-4}$ g·cm^{-3}, possible related to the heterogeneity of this kind of samples, leading to a consequent low repeatability of the density measurements. On the other hand, the uncertainty obtained for the pycnometric method was $5.0\cdot10^{-3}$ g·cm^{-3}. Given this, and considering the results from Table 1, the only significant deviations are those obtained for the mayonnaise sample, with a positive deviation of $1.4\cdot10^{-2}$ g·cm^{-3}. All the other deviations are lower than the $U\rho_{pic}$ values, therefore are not considered significant.

Table 1. Summary of the density deviations, δρ, of the results obtained by the oscillation-type density meter (ρ_{nc}, ρ_c and ρ'_c) and the results obtained by gravimetric method using a Gay-Lussac pycnometer (ρ_{pic}) of the 5 viscoelastic samples.

Samples		δρ (g·cm^{-3})	
	ρ_{nc} - ρ_{pic}	ρ_c - ρ_{pic}	ρ'_c - ρ_{pic}
Carbopol	$3.7\cdot10^{-3}$	$3.7\cdot10^{-3}$	$3.7\cdot10^{-3}$
PVA and Borax	$2.1\cdot10^{-5}$	$-1.8\cdot10^{-5}$	$-5.4\cdot10^{-5}$
Hydroxymethylcellulose	$-6.9\cdot10^{-4}$	$-8.6\cdot10^{-4}$	$-1.3\cdot10^{-3}$
Mayonnaise	$1.5\cdot10^{-2}$*	$1.4\cdot10^{-2}$*	$1.4\cdot10^{-2}$*
Ketchup	$-8.1\cdot10^{-4}$*	$-1.6\cdot10^{-3}$*	$-1.4\cdot10^{-3}$*

Notes: ρ_{nc} - non-viscosity-corrected density value, ρ_c - viscosity-corrected density value and ρ'_c – density value corrected with the calibration curve obtained with the newtonian liquids; *U* – expanded uncertainty stated as the standard measurement uncertainty multiplied by the coverage factor $k = 2$, which for a *t*-distribution corresponds to a coverage probability of approximately 95 %, obtained according to GUM methodology [13]; *U* for ρ_{nc}, ρ_c and ρ'_c values: $5.4\cdot10^{-5}$ g·cm^{-3} and *$3.3\cdot10^{-4}$ g·cm^{-3}; *U* for ρ_{pic} values: $5.0\cdot10^{-3}$ g·cm^{-3}.

3.2. Rheological determinations

The extrapolated value of the dynamic viscosity, $\eta'_{ext.}$, of each viscoelastic sample at the same oscillation frequency, f_ρ, produced in the oscillation-type density meter during the density measurement, was obtained by extrapolating the experimental lines obtained from the linear regression of $\log(\eta') = a \log(f) + b$ for the interval of frequencies, f, analysed in the rheometer (Table 2). The viscosity values estimated by the results of the oscillation-type density meter (DMA 5000, Anton Paar), $\eta_{est.\ \rho}$, are also given in Table 2, together with the description of the viscoelastic character predicted for each sample for f_ρ and the damping factor, tanδ, for the maximum frequency tested.

The hydroxymethylcellulose sample presented a predominant viscous behavior, G'' > G', in the entire range of frequencies tested. PVA/Borax and mayonnaise samples showed a predominant elastic behavior, with G' > G'', for frequencies higher than their crossover frequency, i.e. for f > 1.5 Hz, and for f > 2.3 Hz, correspondently. Ketchup sample showed a predominant elastic behavior, with G' > G'', in the entire range of frequencies tested. At last Carbopol sample showed a G' ≅ G''.

Earlier studies [10] showed a way of estimate the dynamic viscosity, $\eta_{est.\ \rho}$, of newtonian fluids by using the density results given by an oscillation-type density meter, with an uncertainty of 18 %. The same approach was now applied to estimate $\eta_{est.\ \rho}$ for each non-newtonian fluid tested. The $\eta_{est.\ \rho}$ were compared with the dynamic viscosity values extrapolated η'_{ext} from the curve $\log(\eta') = a \log(f) + b$, with an uncertainty of 10 %, and the relative deviation of the $\eta_{est.\ \rho}$ in relation to $\eta'_{ext.}$, $\delta\eta_\rho$ calculated (Table 2). Despite Carbopol and PVA/Borax samples presented $\delta\eta_\rho$ values below the $U\eta_{est.}$, the $\delta\eta_\rho$ results for the other samples showed that this approach seems not to be suitable to be use for non-newtonian fluids (Table 2).

Table 2. Summary of viscosity values estimated by the results of the oscillation-type density meter (DMA 5000, Anton Paar), $\eta_{est.\ \rho}$, and the dynamic viscosity values, $\eta'_{ext.}$, estimated by extrapolation of the log (η') curves as a function of log (f) determined by the rheometer (Mars III, ThermoScientific) with a plate-plate (PP35TiL) measurement geometry, for the frequency values, f_ρ, measured by the oscillation-type density meter for each sample.

Samples	Density meter		Frequency sweep tests				
	f_ρ (Hz)	$\eta_{est.\ \rho}$ (mPa·s)	Viscoelastic character predicted for f_ρ	tan δ (1)	$\eta'_{ext.}$ (mPa·s)	$\lvert\eta^*\rvert_{ext.}$ (mPa·s)	$\delta\eta_\rho$ (%)
Carbopol	278.62	< 6	G' ≅ G''	0.89	5.3	8119	13
PVA/ Borax	277.94	10.3	G' > G''	0.23	10.7	375	-4
Hydroxymethylcellulose	278.34	48.5	G'' > G'	2.16	29.1	34.2	67
Mayonnaise	282.13	> 795	G' > G''	0.44	88.5	329	798
Ketchup	270.52	> 795	G' > G''	0.60	245.5	313	224

Legend: f_ρ – oscillation frequency produced in the oscillation-type density meter (DMA 5000, Anton Paar) during the density measurement (Hz); $\eta_{est.\ \rho}$ – viscosity value estimated by the results of the oscillation-type density meter; $\log(\eta') = a \log(f) + b$ – linear fitting of the dynamic viscosity η' results against frequency, f, obtained in the frequency tests in the rheometer (Mars III, ThermoScientific); $\eta'_{ext.}$ - dynamic viscosity value extrapolated from the curve; $\lvert\eta^*\rvert_{ext.}$ – complex viscosity value extrapolated for f_ρ ; $\delta\eta_\rho$ – relative deviation of the $\eta_{est.\ \rho}$ in relation to $\eta'_{ext.}$

4. Conclusions

Oscillation-type density meters are a very robust, reliable and convenient instruments to measure the density of newtonian liquids in wide range of density, viscosity and temperature with an expanded uncertainty from $1.0 \cdot 10^{-5}$ g·cm^{-3} to $3.0 \cdot 10^{-5}$ g·cm^{-3}, by using a proper calibration curve, since the deviations due to viscosity may lead to a maximum density deviation of $6.2 \cdot 10^{-4}$ g·cm^{-3} in the viscosity interval up to 795 mPa·s [10]. This study showed that the knowledge of the effect of samples' viscoelasticity on the density measurements results using this kind of density meter is limited by the uncertainty of the pycnometer method ($5.0 \cdot 10^{-3}$ g·cm^{-3}), since these density meters are able to produce density results with lower uncertainty ($1.0 \cdot 10^{-5}$ g·cm^{-3}). Only the mayonnaise sample, with a gel-like structure, showed a density deviation of $1.4 \cdot 10^{-2}$ g·cm^{-3}, one order of magnitude higher than the uncertainty of the pycnometer method. With this, and to be safe, one can say that the density of viscoelastic samples may be determinate by using oscillation-type density meters with an expected maximum deviation of $1.4 \cdot 10^{-2}$ g·cm^{-3}. Additionally, one should have in consideration that these results may be instrument-dependent. Despite the obtained data one cannot conclude about how the rheological properties of a fluid affect the density measurements and viscosity predictions with an oscillation-type density meter. Additionally, the results showed that the approach used to estimate the dynamic viscosity of newtonian fluids by using the density results given by an oscillation-type density meter [10] cannot be used for non-newtonian fluids.

As already suspected in earlier studies [10], these results are other indication that oscillation of the density meter cell during density measurements can cause modifications in the internal structure and arrangement of the molecules of the non-newtonian samples, leading to a non-well described density deviation trend, which may be essentially due to the elastic portion of the viscoelastic behaviour of these samples. As planned in the 17RPT02-*rhoLiq* EMPIR Project the viscoelasticity effect on density results need to be study by a measurement method with a low uncertainty such as the hydrostatic weighing. This may lead to means of comparison that will be able to use the oscillation-type density meters in their maximum metrological capability also with non-newtonian samples. Or even to know the real limitation of this measuring instrument, to give the most accurate insights for reference documents, such as standards and guides.

Acknowledgments
The EMPIR project "17RPT02-rhoLiq" is carried out with funding of European Union under the EMPIR. The EMPIR is jointly funded by the EMPIR participating countries within EURAMET and the European Union.

References

[1] https://ec.europa.eu/agriculture/sites/agriculture/files/wine/statistics/price-evolution_en.pdf
[2] https://ec.europa.eu/agriculture/sites/agriculture/files/olive-oil/prices/market-situation_en_0.pdf
[3] https://www.eea.europa.eu/data-and-maps/indicators/fuel-prices-and-taxes/assessment-6
[4] ISO 1998 *ISO 15212-1:1998 Oscillation-type Density Meters - Part 1: Laboratory Instruments* (International Organisation for Standardization)
[5] ISO 2002 *ISO 15212-2:2002 Oscillation-type Density Meters- Part 2: Process Instruments for Homogeneous Liquids* (International Organisation for Standardization)
[6] OIML 2011 *OIML G 14:2011 (E) Density Measurement* (International Organization of Legal Metrology)
[7] WELMEC 2015 *Guide 6.4. Guide for Packers and Importers of E-marked Prepacked Products* (European Cooperation in Legal Metrology, WELMEC Secretariat)
[8] https://www.euramet.org/research-innovation/search-research-projects/details/?eurametCtcp_project_show%5Bproject%5D=1551&eurametCtcp_project%5Bback%5D=1081&cHash=515ddac86e38ea4ac36c4d8a2e0a6969
[9] Directive 2007/45/EC of 5 September 2007 *OJ L* 247 21 September 2007 pp 17-20
[10] Furtado A, Pagel R, Lorenz F, Godinho I and Wolf H 2017 Estimation of nominal viscosity of Newtonian liquids from data obtained by an oscillation-type density meter *Ann. T. Nord. Rheol. Soc.* **25** pp 321-8
[11] JCGM 2012 *JCGM 200:2012 International Vocabulary of Metrology—Basic and General Concepts and Associated Terms (VIM)* 3rd ed (2008 version with minor corrections) (Sèvres: Joint Committee for Guides in Metrology)

[12] ISO 2016 *ISO 2811-1:2016 Paints and Varnishes - Determination of Density - Part 1: Pycnometer Method* (International Organisation for Standardization)
[13] JCGM 2008 *JCGM 100:2008 Evaluation of Measurement Data - Guide to the Expression of Uncertainty in Measurement* (GUM 1995 with minor corrections) (Joint Committee for Guides in Metrology)

TCal: Contribution to metrology for Industry 4.0

Sasho Andonov[1] **and Marija Cundeva-Blajer**[2]

[1] PhD student, Ss. Cyril and Methodius University in Skopje, Faculty of Electrical Engineering and Information Technologies-Skopje, Ul. Ruger Boskovic, br. 18, POB 574, 1000 Skopje, R. North Macedonia
[2] Ss. Cyril and Methodius University in Skopje, Faculty of Electrical Engineering and Information Technologies-Skopje, Ul. Ruger Boskovic, br. 18, POB 574, 1000 Skopje, R. North Macedonia

E-mail: sasho.andonov@t.mk; mcundeva@feit.ukim.edu.mk

Abstract. Touchless Calibration (TCal) is introduced as a future development in calibration metrology. TCal is dealing mostly with decreasing the steps necessary to provide calibration traceability. Missing steps actually contribute to considerable decreasing of time and costs of calibration keeping the calibration traceability on high level. In the scope of Metrology for Industry 4.0, it could be one of the important improvements bringing increased effectiveness and efficiency for manufacturing industry. In this paper, the detailed explanation of overall process is provided.

1. Introduction

Touchless Calibration (TCal) is the manufacturing answer of the question: Why to send an instrument to laboratory for calibration? TCal is a concept in which the manufacturers can just be digitally connected to the calibration laboratory, receive a calibration data from the calibration lab and do the calibration in own premises. All this is achieved under umbrella of Industry 4.0. As a basis of metrology, calibration must follow these initiatives and TCal is a contribution to the efforts to provide effective and efficient Metrology for Industry 4.0. In the paper a detailed explanation of the benefits of TCal and the way how to achieve them is elaborated.

2. Concept of Touchless Calibration

TCal [1] is a concept which applies the Industry 4.0 requirements for fast, safe, accurate and flexible operations. It allows to do calibration of measurement instruments (Units Under Tests – UUT) in the manufacturing facilities, without the traditional necessary presence of the calibration reference standard and UUT in the same premises. In the concept of TCal, the calibration reference standard remains in the calibration laboratory and the UUT stays in its company's premises (which could be thousands of kilometres away) "without touching" each other. The overall concept is decribed in figure 1.

As it can be seen from figure 1, there is network connection between two laboratories which is used to communicate digitally the "calibration data" from calibration laboratory to the place where UUT is situated. In TCal, the term "calibration data" has wider meaning than in classical calibration and it comprises the calibration and environmental data. TCal calibration data (containg the figcalibration and environmental data!) are used for simulating a "same room" calibration. The link between laboratories is one-way oriented with direction from the calibration laboratory and the reference standard

(Transmitter) to the clients premises and the UUT (Receiver), but it could also be a two-way communication. The communication link is Industry 4.0 Network, but it also strictly depends on the process of transforming the ordinary calibration data into "numbers".

Similar concept as TCal is already implemented in Global Navigation Positioning System (GNSS). GNNS is consisting of few satellite systems launched for the purpose to provide navigation services to military, aviation, marine, civil societies, etc. The GNSS receivers are using trilaterations to calculate distance (PR - pseudoranges) from different GNSS satellites using speed of radio waves (speed of light). By comparing this position with Earth model, the receivers can calculate real position of the object on or above the Earth surface.

Each of the GNSS satellites (there are approximately around 60 already launched and orbiting) has few atomic (cesium and/or rubidium) clocks which provide to the GNSS receivers a highly accurate time. The GPS can provide time accuracies measured in nanoseconds (10^{-9} or billionths of a second) [2]. From the customer side, each GNSS receiver has its own clock which has no accuracy of the atomic clocks.

The information regarding the time from atomic clocks is embedded into GNSS navigational message and it is used to "calibrate" GNSS receiver's clock for a short period of time. This short period of time is enough for receiver to calculate the position of the subject. Considering that the GNSS signals are transmitted continuously and navigation messages are periodically part of the GNSS signals, the time information of atomic clocks is periodically presented at the receiver. So, in GNSS receivers, the receiver's clock is calibrated by an atomic clock from GNSS system. And this is very much similar to TCal: The information of time (in a format of a number!) is sent to reciver to calibrate its own time.

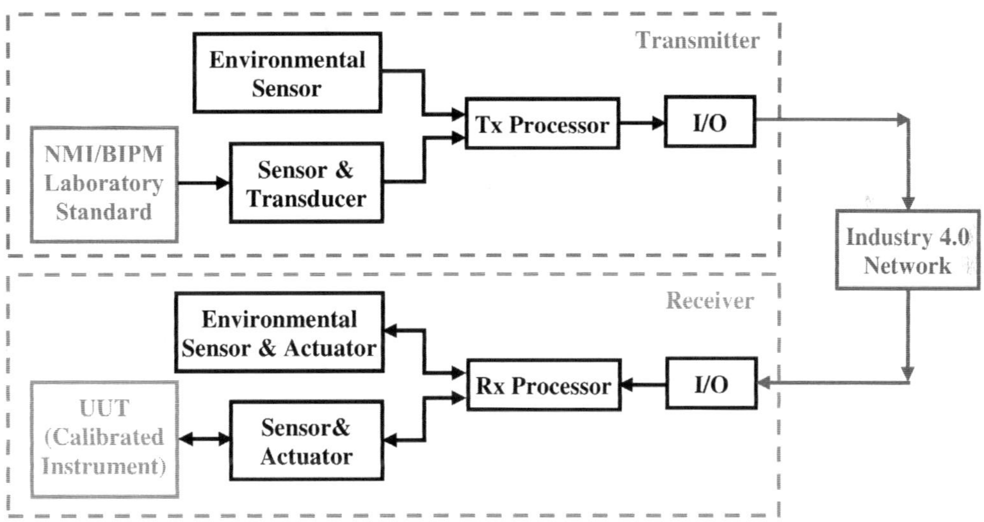

Figure 1. Flow-chart for TCal.

The main point in TCal is conversion of calibration data into "numbers" which can be communicated through the Industry 4.0 digital network. Previously it was presented example for transforming the weight into voltage using the paired piezoelectric sensors made from one piece of crystal.

The selection of the sensor which will transform the calibration data into "number" (embedded into calibration message) is depending from the innovative approach to the particular type of calibration. It is qualified as eligible if the Sensor&Transducer can produce "number" (which is transmitted from Transmitter to Receiver) in the form of time or frequency. This is justified by the following reason: time and frequency transmitted through Industry 4.0 Network can easily be recovered with great accuracy from the calibration message on Receiver side. Another benefit from transmitting time or frequency is the fact that their accuracy is superior compared to others quantities.

3. Traceability

Calibration is of no value if the requested traceability of measurements is not maintained. Traceability is defined as a requirement to relate individual measurement results through an unbroken chain of calibrations to any of accepted primary (NMI/BIMP) standards or to the fundamental (or natural) physical constants which values are accepted (or correlated) by renowned metrology institutions [2].
The traceability is one of the calibration characteristics where the benefit of TCal compared to classical calibration can be emphasized. The process of traceability establishement through classical (traditional) and TCal calibrations between the primary calibration laboratory (NMI/BIMP primary reference standard) and the manufacturing company (UUT) is given in figure 2.

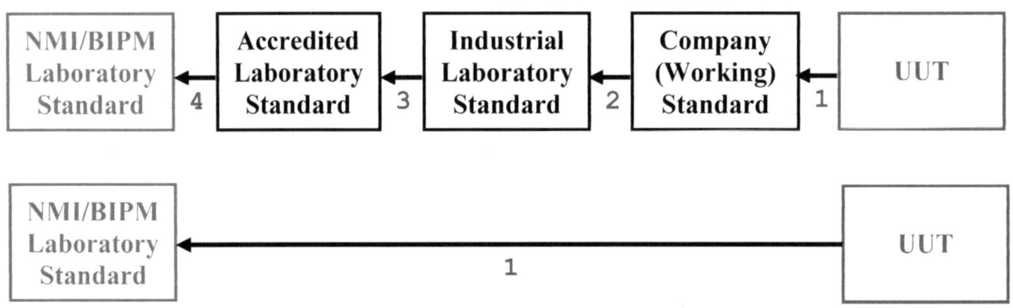

Figure 2. Traceability through Classical Calibration (above) and TCal (below).

From figure 2 it can be noticed that instead of the usual four steps to maintain the traceability chain during calibrations of company instruments (UUT), the TCal offers skipping of few steps. With TCal the traceability with NMI/BIPM primary reference standards is achieved directly in one direct step.

4. Total uncertainty with TCal

The all steps in classical calibration brought particular value of Type A and Type B uncertainty in the calibration process [3]. The total uncertainty (expressed as total variance - σ_{tot}^2) for both calibrations (TCal and classical) in general, can be expressed by the formula:

$$\sigma_{tot}^2 = \sum_{n=1}^{k} \left(\sigma_{Ak}^2 + \sigma_{Bk}^2\right) \tag{1}$$

σ_{Ak}^2 is Type A uncertainty of k-th step and σ_{Bk}^2 is Type B uncertainty of k-th step. In the steps presented in figure 2, for classical calibration, k = 4 and for TCal, k = 1.

It must be said here that Type A and Type B uncertainties exressed by variances, will add each other, but Type A uncertainty will be more than 4 times bigger for classical calibration than for TCal. This is reasonable to assume because for classical calibrations there will be 4 steps and each of these steps will have repeated measurements. It is also reasonable to assume that each of the steps will produce 10% bigger Type A uncertainty from previous one. So, if the Type A uncertainty (expressed as variance) for one-step TCal is $\boldsymbol{\sigma_{TCalA}^2}$, than uncertainty for 4-steps classical calibration ($\boldsymbol{\sigma_{CCA}^2}$) will be:

$$\boldsymbol{\sigma^2_{CCA}} = \boldsymbol{\sigma^2_{TCalA}} + (\boldsymbol{\sigma_{TCalA}} \times \mathbf{1.1})^2 + (\boldsymbol{\sigma_{TCalA}} \times \mathbf{1.1^2})^2 + (\boldsymbol{\sigma_{TCalA}} \times \mathbf{1.1^3})^2 \quad (2)$$
$$= \mathbf{5.446} \times \boldsymbol{\sigma^2_{TCalA}}$$

The equation (2) is valid under assumption that range of $\boldsymbol{\sigma^2_{TCalA}}$ and $\boldsymbol{\sigma^2_{CCA}}$ is same. It is clear that TCal will bring improvement in Type A uncertainties (expressed as variances) approximately more than 5 times compared to classical calibration. If we express difference of uncertainties as standard deviations than taking square root of both sides of (2) will result in:

$$\boldsymbol{\sigma_{CCA}} = \mathbf{2,33} \times \boldsymbol{\sigma_{TCalA}} \quad (3)$$

The TCal will keep some of the uncertainties of classical calibration in Type B uncertainty, but it will also introduce new uncertainties, which will show up due to the sensing and conversion of the sensors used. The total new Type B uncertainty ($\sigma^2_{\mathrm{TCalB}}$) of TCal may be expressed by sum of variances of new uncertainties:

$$\sigma^2_{TCalB} = \sigma^2_{ds} + \sigma^2_{env} + \sigma^2_{sp} + \sigma^2_{AD} \quad (4)$$

The σ^2_{ds} is uncertainty caused by difference of sensors characteristics, the σ^2_{env} is the uncertainty due to (possible) difference of data from environmental sensor on both sides, uncertainty σ^2_{sp} comes from the speed of communication (valid only for dynamic calibrations!) and σ^2_{AD} is caused by eventual data errors which A/D and D/C devices involved in the process could introduce to the TCal process (if they are used for digital communication).

The most critical is uncertainty caused by difference of sensors characteristics ($\boldsymbol{\sigma^2_{ds}}$), because all other uncertainties can be strictly controlled by TCal concept. Actualy, the $\boldsymbol{\sigma^2_{ds}}$ uncertainty makes the difference between classical calibration and TCal.

5. Uncertainty caused by difference in sensors characteristics ($\boldsymbol{\sigma^2_{\mathrm{ds}}}$)

An example, the Voltage to Frequency Convertor (VFC) and Frequency to Voltage Convertor (FVC) circuit will be presented for general illustration of the process of estimation sensor uncertainty ($\boldsymbol{\sigma^2_{ds}}$) in TCal for voltage calibrations. The TCal voltage calibration is conducted by using a device (Microchip circuit TC 9401) which can be configured as VFC on NMI/BIPM side and the same type of device can be configured as FVC on UUT side.

To keep the uncertainty Type B caused by difference in sensors characteristics($\boldsymbol{\sigma^2_{\mathrm{ds}}}$) as small as possible, it is necessary the sensors implemented on both sides to be paired. Whatever convertors are used on both sides, they must transfer "Mirror Image" where Industry 4.0 Network is the "Mirror". The simple example of "Mirror Image" for voltage calibration by using voltage to frequency conversions (and vice versa) is presented in figure 3.

In figure 3 the VFC is Voltage-to-Frequency Convertor and FVC is Frequency-to-Voltage Convertor. The "Mirror Image" is presented by formula:

$$\boldsymbol{V_1 = V_2} \quad (5)$$

It necessary means that:

$$\boldsymbol{f_1 = f_2} \quad (6)$$

because the f_1 frequency (communicated through Industry 4.0 Network as a number) and f_2 (which will be generated on UUT side) must be the same (6). Only than the FVC will produce $V_1 = V_2$ (5).

To check in advance the possibility of TCal, the theoretical analysis was conducted for VFC and FVC circuits built by using Microchip circuit TC 9401. The theoretical analysis was based on rough manufacturing data for this curcuit and pretty pessimistic scenario was executed only for Type B uncertainties.

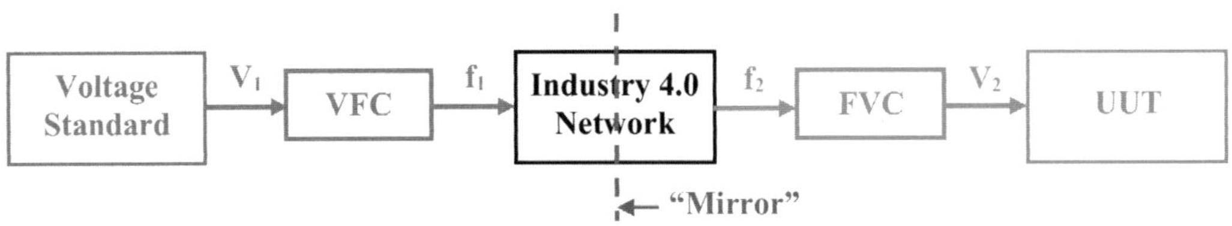

Figure 3. "Mirror Image" in TCal with VFC and FVC.

The Sensor&Transducer on the Transmitter side is not critical. One of the most critical contributing factors to the uncertainty is deriving from ability of Sensor$Actuator on the Receiver side to produce $V_2 = V_1$. So, based on the information regarding the frequency f_1, a frequency generator which will produce frequency $f_2 = f_1$ which will be used by FVC to produce the same voltage $V_2 = V_1$, has to be on disposal. The degree of sensors impairment produces the particular uncertainty which is expected and it can be decreased only by calibration of Sensor&Actuator. The main point is that $\boldsymbol{\sigma}^2_{\mathbf{ds}}$ will be actually uncertainty obtained by this calibration. FVC (Sensor&Actuator) must be calibrated "relative" to the used VFC (Sensor&Transducer) circuit. "Relative" means that there is no standard used for calibration. Having a standard is totally unnecessary to achieve "Mirror Image" with "relative" calibration method given in figure 4.

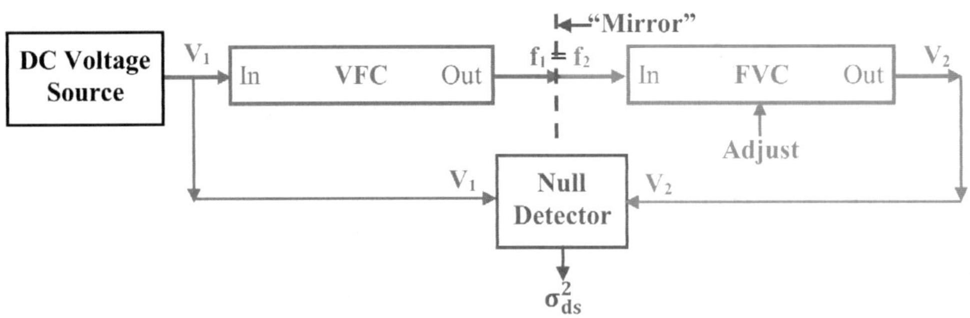

Figure 4. "Relative" calibration of FVC with VFC.

The DC Voltage Source (V_1) connected on VFC input is not of interest because its value is compared to V_2 ("relative" calibration!) and only the output of Null Detector ($\boldsymbol{\sigma}^2_{\mathbf{ds}}$) is point of interest. It is significant that the used DC Voltage Source must be with value in the same range as the voltage which will be calibrated by TCal. In figure 4, the adjustment of FVC must produce such a V_2, so the difference with V_1 to be such a small, that output of Null Detektor is as small as possible

The output of Null Detector (properly processed and calculated) will be the Type B uncertainty sensors contributing factor ($\boldsymbol{\sigma}^2_{\mathbf{ds}}$).

6. Conclusion

To conduct a comparison analysis of the uncertainties between the classical calibration and TCal as in figure 2 is of utmost importance [4]. It is noticeable that, in case of classical calibration, the contributing uncertainties (expressed as variances) are cumulative and add each other. From figure 2, it can be noticed that TCal could bring benefit only if the total Type A uncertainty introduced by TCal is better than uncertainties declared by the accredited laboratory reference standard, the industrial laboratory reference standard and the company working standard.

Assuming that each step has same range of contributing uncertainties and the company UUT (production instrument) must satisfy the traceability steps from figure 2, the total Type A uncertainty of classical calibration (expressed through the variance) will be many times bigger than Type A uncertainty of TCal.

The main Type B uncertainty factor in TCal is σ_{ds}^2. This factor is result of imperfection of the "Mirror" image which must to be provided. Whatever type or method of conversion is used, the accuracy of sensors and convertors must be very high, which means that the Sensors must be calibrated "relative" to each other.

Maybe someone will think that it is a problem to calibrate sensors on both sides. However, it is much less complicated, faster and cheaper to calibrate sensors in manufacturing premises and to send it into NMI/BIMP Laboratory than instrument (UUT). It means that instrument (UUT) will be still available for use in the company premises. Actually, it will not leave the company premises at all.

So, in the case presented here, providing small Type B uncertainty caused by used sensors (σ_{ds}^2) by "relative" calibration between VFC and FVC will increase TCal calibration accuracy. As one of the big benefits for companies will be the fact that using TCal, the time for UUT calibration will be transferred from weeks into hours.

References

[1] Andonov S M and Cundeva-Blajer M 2018 Calibration for Industry 4.0 Metrology: Touchless Calibration *J. Phys.: Conf. Ser.* 1065 *Measurement Science*

[2] Blewitt G 1997 Basics of GPS Technique: Observation Equations *Geodetic Applications of GPS* (Stockholm, Sweden: Swedish Land Survey) http://www.nbmg.unr.edu/staff/pdfs/blewitt%20basics%20of%20gps.pdf

[3] JCGM 100 2008 *Evaluation of Measurement Data - Guide to the Expression of Uncertainty in Measurement* (GUM 1995 with minor corrections) (Joint Committee for Guides in Metrology)

[4] Rabinovich S G 2005 *Measurement Errors and Uncertainties: Theory and Practice*, 3rd ed (New York, USA: Springer)

Journal of Physics: Conference Series **1379** (2019) 012022 doi:10.1088/1742-6596/1379/1/012022

Can there be one meaning of "measurement" across the sciences?

Luca Mari[1], **Andrew Maul**[2], **Mark Wilson**[3]

[1] School of Industrial Engineering, Università Cattaneo, Castellanza, VA, Italy
[2] Gevirtz Graduate School of Education, Univ. of California, Santa Barbara, CA, USA
[3] Graduate School of Education, University of California, Berkeley, CA, USA

E-mail: lmari@liuc.it

Abstract. The *International Vocabulary of Metrology* (VIM) gives the following definition of 'measurement': "process of experimentally obtaining one or more quantity values that can reasonably be attributed to a quantity". In pursuit of a common understanding of measurement, we explore this condition of *reasonableness* across the sciences, asking in particular whether there are common features of the structure of measurements in different fields of study that could meet this criterion. We propose that a general sufficient condition for measurement is that it is an experimental process of evaluation of empirical properties that produces *explicitly justifiable information*, which, we argue, is consistent with characterizations of science and epistemology more generally.

1. Introduction

The concept of measurement, along with associated concepts and terminology, is frequently invoked in many different scientific fields as well as throughout society at large, and the range of activities referred to as measurements or as being related to measurements continues to expand and diversify. In many ways this trend is simply a corollary to the continued expansion of the scope of science itself, given how integral measurement is, and is perceived to be, to many aspects of scientific inquiry [1]. Throughout the twentieth century and up to the present moment, a great deal of time and attention has been given by philosophers of science and the general public to the question of *what science is*, as distinct from other types of human activities (see, e.g., [2]); as could be expected, the similar and more local question of *what measurement is* has also received significant attention, particularly in light of claims made by psychologists and other human scientists starting around the turn of the twentieth century regarding the measurability of psychological and social properties (see, e.g., [3]).

Although an observer might note that identifying such distinct endeavors with the same terms ("measurement", "measurability", etc.) could be merely a lexical peculiarity based on some homonymy, we contend that significantly more is at stake: just as debates regarding the demarcation between science and pseudo-science were given urgency insofar as they bore on the degree of public trust that should be afforded to fields such as psychoanalysis and Marxist economic theory, debates regarding the legitimacy of claims about measurement similarly bear on the status of the human sciences, with some going so far as to assert that such claims are not only illegitimate but "pathological" [4] insofar as they misunderstand the necessary and sufficient conditions for an activity to properly be characterized as a measurement. More positively, we hypothesize that an exploration of the interdisciplinary similarities and differences in approaches to measurement (and discourse about measurement) may shed new light on the question of what measurement is, and what features of the measurement process serve to warrant its commonly acknowledged trustworthiness.

2. Can there be one meaning of "measurement" across the sciences?

The *International Vocabulary of Metrology* (VIM) [5], possibly the most authoritative existing source regarding measurement terms and concepts, gives the following definition of 'measurement': "process of experimentally obtaining one or more quantity values that can reasonably be attributed to a quantity". This is, in essence, a black-box characterization of measurement: measurement is defined as an experimental process whose input is a quantity, and whose output is one or more quantity values. (Two notes. First, the term "experimental" plausibly should be understood in a generic sense here, in contrast to "theoretical" (e.g., as based on a thought experiment); in other words, measurements are processes that interact with empirical features of the world. Second, we do not consider here the possible generalization to non-quantitative properties of objects, i.e., whether non-quantitative properties are also measurable – though see [6].)

Read at face value, this definition would equally apply to processes such as guessing or statement of personal opinion, if these were phrased in quantitative terms (e.g., "I think it will be three degrees warmer tomorrow than it is today"; "I'm ten percent happier today than I was yesterday"), with only the condition that some experimental activity was performed for obtaining the outcome. Given that such processes are not generally considered measurements, the validity of the VIM's definition critically relies on the adverb "reasonably".

Clearly reasonableness is not a formal condition; rather, it can be considered as a means to provide an encompassing definition, acceptable by the broadest set of researchers, independently of their field of activity (indeed, who would refuse to accept that measurement should produce "reasonable" results?). But this condition is not sufficiently specific: it could be that, for example, some guesses about empirical quantities could be considered reasonable, but nevertheless we might refuse to accept them as measurements. Hence the problem we explore here is: *can there be one criterion of "reasonableness" that is sufficient to characterize all processes that across the sciences are commonly considered measurements?*

This key problem should be interpreted as related to a descriptive, not normative, characterization of measurement: "measurement" is not a trademarked term, and in principle everyone is free to call "measurement" whatever (s)he likes. But this does not take into account the special epistemic role commonly acknowledged to measurement: were it the case that measurement and, e.g., opinion-making were equally valuable, why should money be spent for buying, maintaining, and operating expensive measuring instruments?

It is a critical condition for the criterion of "reasonableness" we are seeking that it be one that applies across the sciences. A *specific* sufficient condition could be, in particular, that the process is realized by means of a properly operated calibrated physical transducer that is sensitive to the property intended to be measured – which is indeed the sufficient condition usually applied to characterize physical measurements: it works for physical properties, and may also provide some guidelines for designing measurement systems for psycho-social properties. But treating it as a general sufficient condition would correspond to an old-fashioned physicalism, which assumes that only physical properties are measurable, thus considering non-physical properties as unmeasureable simply *by fiat*. Hence any such candidate criterion should be independent of domain-specific strategies of design of measuring instruments and implementation of measurement processes.

3. Different subject matters, different processes...

In the physical sciences, the properties intended to be measured are often already embedded in a well-established body of knowledge, and the work is usually aimed at the refinement of existing instruments or mathematical techniques. The term "instrumentation and measurement" (as found, for example, in the title of the IEEE journal *Transactions on Instrumentation and Measurement*) clearly conveys this message, and the books on measurement systems in these fields are largely devoted to presentation of

Journal of Physics: Conference Series **1379** (2019) 012022 doi:10.1088/1742-6596/1379/1/012022

the features and modes of operation of measuring instruments (e.g., [7, 8]). In the human sciences, such quantitative knowledge is less developed, and thus activities aimed at measurement necessitate first of all the task of formally characterizing the properties of interest, in such a way that quantitative information on them can be acquired and then mathematically processed. The key issues in these cases are the meaningfulness (e.g., [9]) and more generally the validity (e.g., [10]) of the hypotheses upon which the formalization is based, and such issues are the focus of measurement sub-disciplines within specific human sciences such as psychometrics and econometrics, as components of psychology and economics respectively.

This multiplicity of perspectives reflects the fact that researchers operating in different fields face quite different problems and challenges, and will thus have different associations with the concept of measurement. However, this does not preclude the possibility that there are common elements of the structures of the processes referred to as measurements in different fields, and that identifying these common elements may suggest paths towards the formulation of conception of measurement applicable across the sciences.

4. ... with some structural commonalities...

In this analysis we draw from previous works, in which we have discussed the very definition of 'measurement' (e.g., [11]) and its epistemology (e.g., [12]), the stereotypes that surround measurement (e.g., [13]), and in particular the mistaken assumption that measurement is identical to quantification [6]. A background commonality that emerges from these studies is that measurement (1) is an experimental process that (2) produces information on empirical properties of objects (where objects can be bodies, systems, phenomena, events, processes, individuals, organizations, etc.), such information being (3) in the form of values of property. This is consistent with the VIM's definition quoted previously and, as with the VIM's definition, it is an abstract, black-box (i.e., input-output) characterization of measurement, that nevertheless serves to delimit the scope, in particular by distinguishing measurement and computation (where the latter produces information from other information).

Even without "opening the box", such a characterization already includes several key conditions, related to

(i) *the procedural nature of the process* (highlighting that measurement is designed on purpose and therefore justifying the focus on the experimental and mathematical instrumentation as a tool to implement the procedure, as mentioned in the first category of cases above, more commonly emphasized in physical measurement), and to

(ii) *the definition of the property* intended to be measured and to which the produced information is attributed (highlighting that measurement requires modeling stages that are preliminary to data acquisition, as mentioned in the second category of cases above, more commonly emphasized in psycho-social measurement).

These can be interpreted as *necessary* conditions for a process of evaluation of an empirical property to be considered a measurement (an additional condition sometimes considered necessary is that the measured properties are quantities – this is in particular the VIM's position, reflecting a long tradition in physical sciences; as previously mentioned, we do not develop this subject here). But of course the critical issue is about the *sufficient* conditions.

5. ... and a common emphasis on trustworthiness...

As was previously observed, measurement is usually associated with trustworthiness. Elsewhere (e.g., [14]) we have argued that the trustworthiness of measurement results (more or less equivalently for this concept we used also the term "dependability") requires that they convey information that is *object-related* and *subject-independent*, or, more explicitly:

1) information that is specific to the measurand and, therefore, to a given property of the object under measurement. This means that the provided information should be independent of any other property of the object or the surrounding environment, which includes both the measuring system and the subject

who is measuring. This corresponds to guaranteeing that measurement results actually provide information about the measurand and not some other property. This condition is about the appropriate attribution of information to its claimed object: hence, it is a requirement of object-relatedness, or "objectivity" for short;

2) information interpretable in the same way by different users in different places and times. This corresponds to guaranteeing that measurement results are expressed in a form independent of the specific context and only referring to entities which are universally accessible, so that the meaning of a measurement result is unambiguous and can be easily reconstructed in principle by anyone, possibly on the basis of suitable conventions: hence, it is a requirement of subject-independence, or "intersubjectivity" for short.

According to this characterization, objectivity and intersubjectivity are independent features – something can be objective but not intersubjective, and vice versa –, that identify the two dimensions of measurement: the claim of the possibility *of obtaining information about empirical properties*, and the claim of the possibility *of socially reporting such information* (interestingly, these dimensions correspond to the two main stages of a measurement process, which Roman Morawski calls "conversion" and "reconstruction" [15] and Giovanni Battista Rossi and Francesco Crenna call "observation" and "restitution" [16]). It is thus through their objectivity and intersubjectivity that measurement results are considered trustworthy.

On the other hand, as intended here objectivity and intersubjectivity are not Boolean (i.e., yes-no) conditions: something can be more or less objective and more or less intersubjective. Hence for objectivity and intersubjectivity to be the sufficient conditions that we are looking for to characterize measurement as specific kind of evaluation, a threshold of minimum objectivity and intersubjectivity should be set. Thus the term "reasonable" in the VIM's definition of 'measurement' once again does the heavy lifting: a given attribution of values would be considered reasonable (and therefore, would be considered a measurement) if both its objectivity and intersubjectivity were sufficient for the intended purpose for which the instrument was designed and performed. This highlights the pragmatic nature of measurement: the same measurement may be considered good for some purposes and bad for some others.

Emphasis on sufficient objectivity and intersubjectivity for a given purpose is then operatively useful, for the general guidelines it provides regarding the design and performance of measurements (e.g., in [17]) – though more work should be done to develop better guidelines – but it is still too specific at least in one respect: it would assume that measurement is always *good* measurement. While pragmatically this is sound – if we know that what we are doing is a bad measurement (e.g., with measurement uncertainty greater than target uncertainty, so that the produced information is useless) we avoid doing it – the concept 'bad measurement' as such is not contradictory, and bad measurements do not fulfill the condition of sufficient objectivity and intersubjectivity. In other words, in order to maintain the VIM's characterization of "reasonableness", objectivity and intersubjectivity are useful but still not sufficient: an even more general condition has to be identified.

6. ... and a focus on producing explicitly justifiable information

Science aims at the development of knowledge, where, following Plato, knowledge is commonly understood as *justified true belief*. But although science *aims* at truth, it cannot *guarantee* it, as plainly illustrated by its history, and thus the truth or falsity (and more generally the quality, were a criterion of quality given) of any given scientific theory cannot be a definitional component of what makes it scientific: a theory can be of low quality, and even eventually admitted to be false, and nevertheless can be scientific. Although discussions of the definition and essential features of science are still ongoing, it is generally agreed that *scientific theories must be explicitly justifiable*, in that their logical and evidentiary grounding must be clear and publicly evaluable [2]. Thus, *science is characterized by its structure rather than its outcomes*; it is no contradiction to say that a theory is both scientific and false, but it would be a contradiction to say that a theory is both scientific and untestable, even if it were true. Similarly, the trustworthiness of measurement results is not earned solely by their objectivity and intersubjectivity, but by the *justification* that is provided for such, in the sense that an explanation can be given for how the structure of the measurement process serves to secure the trustworthiness of its results. The availability of such an explanation should, at least in principle, make the results reproducible, thus making possible their social critical analysis. Again, it is no contradiction to say that a given process is both a measurement and produces results that are of low quality – as in the case of a bad measurement – but it would be a contradiction to say that a process is both a measurement and cannot be justified, even if the results were (perhaps accidentally) accurate.

Of course, in many cases, particularly of non-scientific measurements, measuring systems remain black boxes and no justifications of their results are reported: what is important is that a justification *can* be reported, whenever required. Thus the idea of the general sufficient condition that we are seeking, applicable across the sciences, is that measurement is an experimental process of evaluation of empirical properties that produces *explicitly justifiable information*. In this way, the expectations of good measurement processes are continuous with the expectations of other sources of knowledge, in Plato's sense of knowledge as justified true belief. This is the case regardless of the subject matter and the details of the procedures involved in measurement.

The requirement that measurement results be explicitly justifiable helps explain why measurement cannot be adequately characterized solely using a black-box model: if a given attribution of value(s) to a property is claimed to be a measurement result (instead of, e.g., once again, a guess), it must be possible to explain how it was obtained, by "opening the box" and identifying the features of the process that secure the quality of the results. It is, of course, no simple matter to describe how such justification should take place in any given instance of measurement, and it is here in which we find significant differences across disciplines. As previously noted, the key activities associated with designing and using measuring instruments in the physical and human sciences are different, and could give the impression to a casual observer that there can be no shared meaning of what measurement is in such diverse fields. However, a closer look reveals a common condition – the possibility of explicit justification – and thus helps clarify how the "reasonableness" criterion of the VIM's definition of measurement is applicable across the sciences.

References

[1] Tal E 2015 Measurement in science *The Stanford Encyclopedia of Philosophy* Zalta E N (ed.)
[2] Hansson S O 2017 Science and Pseudo-Science *The Stanford Encyclopedia of Philosophy* Zalta E N (ed.)
[3] Duncan O D 1984 *Notes on Social Measurement – Historical and Critical* (New York: Russell Sage Foundation)
[4] Michell J 1999 *Measurement in Psychology: A Critical History of a Methodological Concept* (Cambridge: Cambridge University Press)
[5] JCGM 2012 *JCGM 200:2012 International Vocabulary of Metrology—Basic and General Concepts and Associated Terms (VIM)* 3rd ed (2008 version with minor corrections) (Sèvres: Joint Committee for Guides in Metrology) (www.bipm.org/en/publications/guides/vim.html)
[6] Mari L, Maul A, Torres Irribarra D and Wilson M 2017 Quantities, quantification, and the necessary and sufficient conditions for measurement *Measurement* **100** 115-121
[7] Bentley J P 2005 *Principles of Measurement Systems* (Harlow: Pearson)
[8] Doebelin E 2003 *Measurement Systems: Application and Design* (New York: McGraw-Hill)
[9] Narens L 2012 *Theories of Meaningfulness* (New York: Routledge)
[10] Markus K and Borsboom D 2013 *Frontiers of Validity Theory: Measurement, Causation, and Meaning* (New York: Routledge)
[11] Mari L 2013 A quest for the definition of measurement *Measurement* **46** 2889-95
[12] Mari L 2003 Epistemology of measurement *Measurement* **34** 17-30
[13] Mari L, Carbone P, Giordani A and Petri D 2017 A structural interpretation of measurement and some related epistemological issues *Stud. Hist. Philos. Sci.* **65-66** 46-56
[14] Mari L, Carbone P and Petri D 2012 Measurement fundamentals: A pragmatic view *IEEE Trans. Instr. Meas.* **61** 2107–2115

[15] Morawski R Z 2013 An application-oriented mathematical meta-model of measurement *Measurement* **46** 3753-3765
[16] Rossi G B and Crenna F 2018 A formal theory of the measurement system *Measurement* **116** 644-651
[17] Petri D, Mari L and Carbone P 2015 A structured methodology for measurement development *IEEE Trans. Instr. Meas.* **64** 2367-2379

Why metrology? Fair dealing and efficient markets for the United Nations' Sustainable Development Goals

W P Fisher Jr[1], L Pendrill[2], A Lips da Cruz[3], A Felin[2]

[1] Living Capital Metrics LLC, Sausalito CA 94965 USA, and BEAR Center, Graduate School of Education, University of California, Berkeley, 94702 USA
[2] RISE Research Institutes of Sweden, Gothenburg, Sweden
[3] ALDC Partnership, Stockholm, Sweden

E-mail: wfisher@berkeley.edu

Abstract. Why metrology? What special value does it offer to the design and use of sustainability impact metrics, and so to the realization of the United Nations Sustainable Development Goals? One answer in short, is, "For justice." The balance scale is a universal symbol of fairness, development and equality. A second brief reply is, "For efficient markets." Common currencies for the exchange of value are well known as a key factor in lowering transaction costs and supporting property rights, both essential to market formation. The repeatedly encountered dilemma, of course, is how to design, calibrate, and use measures of economic, environmental, human, and social value that lower costs and are just in more than name alone, serving some but not others. Longer answers justify metrological traceability to consensus unit standards as a key factor in realizing the SDGs by showing how it contributes to (a) the cultivation of shared meanings and improved communication; (b) the grounding of pragmatic expressions of verified trust; (c) making real, reproducible developmental trajectories visible and manageable; and (d) creating the multilevel systems of systems required for managing and governing complex, adaptive, self-organizing forms of social, political, legislative and economic life. A more complete answer to the question, "Why metrology?," then focuses on how the SDGs' diverse stakeholder interests may be most effectively fulfilled via collective actions facilitated on the basis of shared values.

1. Metrology: What, why, and how

Initiated in January, 2016, the United Nations' (UN) Sustainable Development Goals (SDGs) offer the nations and people of the world a forum in which to coordinate efforts at eliminating poverty, protecting the environment, and ensuring peace and prosperity. The 17 goals overlap in ways that will build synergies across interrelated domains ranging from democratic governance, climate change, economic inequality, education, and health care, to innovation, sustainable consumption, reduced violence, and social justice. The SDGs are key to UN Development Programme (UNDP) efforts in 170 countries, and will guide planning and funding until 2030. The UNDP is actively partnering with governments, the private sector, civil society, and citizens to integrate the SDGs into policies and practices at every level.

1.1. What exactly is metrology?

Metrology is "the science of measurement and its application" [1], especially as concerns the definition and reproduction of mathematically rigorous, experimentally validated quantities. Measurement science focuses on models, quantities and uncertainty, estimation, and construct theory. National metrology institutes work with this science and more broadly with traceability in support of product comparability, instrument conformity assessment, consensus standards formation, legal and regulatory issues in the wider quality infrastructure. A matter of essential concern is traceability, which is a "property of a measurement result whereby the result can be related to a reference through a documented unbroken chain of calibrations, each contributing to the measurement uncertainty" [1].

Though it may come as a surprise to many, social and cultural expectations and norms are likely a bigger barrier to the achievement of metrological traceability in the SDG domains than technical problems are.

Despite a wealth of evidence supporting the feasibility, viability, and desirability of metrological traceability, and despite the many advantages to be gained from creating traceability systems, instruments measuring the same human, natural and social constructs are not usually calibrated in common units in either research or practice. These constructs are very different from the economic and financial key ratio information used in accounting. Precision measurement of abilities, health, trust, and environmental quality provides actionable detail that is otherwise unavailable, and that is essential to the achievement of the SDGs.

Even though these systems have not been referred to as metrological until recently, they not only incorporate the measurement science and theory characteristic of metrology, they also serve as models for new initiatives at national metrology institutes [8-10]. Systematic presentations of philosophical and methodological issues are available in multiple forums [2-16].

Human abilities, attitudes, and behaviors have in fact long been measured in ways respective of individual differences while also being read from instruments calibrated in accord with models mathematically analogous to the balance scale [2-5]. Thousands of well-calibrated individual instruments and hundreds of millions of individual measures combining coherent continuity with local improvisations have been produced and used in social and psychological research and practice [6-7].

1.2. Why metrology?

The question as to why metrology should be important to the realization of these global goals can be answered quite easily: for justice. The idea that everyone should be treated equally, irrespective of their differences, makes the balance scale a universal symbol of global fairness development and equality opportunities. The metric system, the *Systeme Internationale des Unites*, or SI, is itself a product of the French Revolution's call for "one law, one weight, and one measure" [17] (omitting the reference to "one king"). That said, arriving at just relations in an even-handed way that does not reduce individual differences to false forms of homogenized sameness has proven so elusive many consider it impossible. The idea that a universal measure representing every individual human, organisational and natural resource could be found in globally shared human, organisational and natural characteristics seems antithetical to valuing unique differences, however much more easily managed such measures might be for bureaucracies [18], and however much more they would offer in the way of growth opportunities and profits [19].

In accord with the focus on justice, more complex answers to the question of "Why metrology?" speak to the need for a common language of sustainable development. Communicating what is learned about what works and how to use it demands fair, equitable, reproducible, verifiable, meaningful and efficient representation. Significant difficulties are associated with defining, estimating, and implementing shared standards for fair and equitable measures in an era of increasing global information where the Internet and digitalisation enable sometimes vapid and destructive connectivity.

1.3. The role of metrology in commerce and markets

Huge resources, nationally and internationally, are invested in metrological traceability to the SI reference standards. Those investments are however worth it, since they pay more than proportionate returns by lowering transaction costs, clarifying communications, and supporting property rights in efficient markets [20]. High transaction costs often make it difficult or impossible to determine the quantity and/or quality of a product or service, which then obscures communication and ownership. Metrology plays a key role in lowering the costs of market transactions, in supporting unambiguous communications, and in establishing legally defensible title. So a second answer to the question "Why metrology?" is "For efficient markets" [19, 13, 20-21].

This role of precision measurement science and metrological traceability in making markets has to date been almost completely ignored in the domain of sustainability. The achievement of the UN SDGs is fundamentally obstructed by widespread use of incommensurable metrics. Because communication is obscured, neither management efforts nor market comparisons are as well coordinated within shared frames of reference as they should be. This comes about even though the economic value of common currencies in trade is widely recognized and implemented in policy and practice, and even though the SI unit standards are universally acclaimed for their role in technological and entrepreneurial innovation.

Efficient markets for the exchange of scientific and economic value plainly require a wide array of stakeholder interests (legal, financial, economic, technical, aesthetic, moral, social, etc.) that are, like musical instruments and scales, both (a) coordinated and aligned relative to shared standards, and (b) open to and expressive of unique individuality and creative improvisation [11, 12, 22]. Metrological research and practice provides the strategic forum in which diverse stakeholders in sustainable policies and practices assert their interests in evaluating and managing the uncertainty, precision, meaningfulness, and use of sustainability information. Historically, metrology is the means by which groups with varied interests (consumers, lawyers, economists, accountants, managers, investors, regulators, financiers, etc.) realize their interests more fully and equitably via collective action than they can in isolation. Conformity assessment, including legal metrology, is an established sphere of activities national metrology institutes use to ensure stakeholder confidence in products and services.

Instrument calibrations, data analyses, estimation algorithms, and reporting mechanisms all require close attention to ensure that everyone is talking about the same thing in the same terms. Trust and confidence are based in verified inspections and accountability demonstrating that the requirements specified in consensus and sustainable standards are met. Conformity assessment is often essential to meeting the needs of the public interest, public health, safety and order, and protections of the environment, consumers, and fair trade.

1.4. How to extend metrology in new directions

And so the question arises as to how we might extend science productively into the new domains of innovation defined by the United Nations Sustainable Development Goals. Four areas of especially urgent interest can be identified. These involve meaning and relevance, trust based in reproducible objectivity, the manageability that follows from measurability, and the complexity required for adaptation to changing circumstances.

In each of these four areas, illustrative examples can be drawn from the target included in UN SDG 3 on Good Health and Well-Being concerning the achievement of "universal health coverage, including financial risk protection, access to quality essential health-care services and access to safe, effective, quality and affordable essential medicines and vaccines for all." Differences across global variations in how quality essential health care services are defined must be addressed in new ways if metrological unit standards are to be feasible and viable. Of paramount importance is the need for an inclusive system where equity in health care is assured for all global citizens independently of economic terms and individual countries' forms of constitutional governance.

2. Meaning and relevance

Scales that are mathematically rigorous and that effectively concatenate an abstract ideal unit are often in Health and Well-being contexts estimated from nonlinear data using algorithms that are also nonlinear. Individuals do not typically progress and grow along the dimension measured in a strictly linear, incremental fashion, so it is important to understand that change and performance occurs within a range bound by uncertainty. Accepting these limits, meaningful measures stand for things supported by theory and evidence as adding up in linear interval ways. Thus, the truth of meaningless statements:

Journal of Physics: Conference Series **1379** (2019) 012023 doi:10.1088/1742-6596/1379/1/012023

...depends on what scale or coordinate system is employed, whereas meaningful propositions have truth-value independent of the choice of representation, within certain limits. ...analysis of this distinction leads...to a...technical apparatus focusing upon invariance under changes of scale [23].

Evidence evaluated in sustainability terms ranges from numeric data like economic growth, energy consumption, water usage and the preservation of species to more qualitative perceptions, often scored on categorical scales. Meaningfulness is obtained, however, not only from the relevance of the phenomena observed, but also from the extent to which those phenomena can be communicated effectively, transparently, and efficiently. The metrology of categorical properties scaled in meaningful units of measurement is still in its infancy, despite more than 90 years of applications in many fields.

The meaningfulness of calibrations that remain invariant across the particular choice of representation contrasts with the kind of criteria for universal health coverage specified in the UN SDG 3 target, to take one example. Different countries and different care providers can reasonably be expected to offer services that vary markedly in their quality and terms, such that many higher percentages of enrollees will be included by some, while others with perhaps more stringent, expensive, and higher quality features will enroll lower percentages. A more rigorous approach will specify the operational terms of what services constitute a minimum standard of coverage, so that the experiences of consumers can be meaningfully documented, modelled, estimated, calibrated, and compared.

3. Trust based in demonstrated objectivity

Meaningfulness cannot, however, be created in contexts devoid of trust. When we give our word, we make a promise of trustworthiness that conveys an intention of objective verifiability. That is, multiple independent observers have to be able to reproduce invariant experimental results, just as multiple independent approaches to designing and deploying measuring instruments must reproduce the unit of measurement to within a tolerable range of uncertainty, and just as multiple independent approaches to explaining variation in the calibrations of questions and indicators must also reproduce the same predictive model, also to within a tolerable range of uncertainty. These factors together contribute to the objectivity of results capable of inspiring trust. In this sense, there is an important convergence of moral values concerning just relations with the quantitative values of precision measurement [24].

Trust in a metric for universal health coverage, as is implied in the UN SDG 3 target, cannot, then, be established on a basis of the percentages of enrollees. Taken at face value, these numbers vary meaninglessly across providers, and so cannot support the reproduction of an invariant connection between the numeric expression of quantitative amounts and a stable phenomenon of consumer engagement with health coverage services. Dealing with these numbers alone, as though they automatically meet the need for a quantitative metric, is insufficient to the task.

4. Manageability facilitated by measurability

Metrology facilitates manageability in specific and essential ways that otherwise remain inaccessible. It does so by providing a navigable map of the development terrain to be traversed [3, 6, 11-12, 22]. This map is based in the evidence of observed variation, it is calibrated in the shared language of a standardized unit, and it is validated by the explanatory model of a theory predicting the observed variation. The measured construct, no matter if it involves education, health care, human resource training, organizational learning, community service, or any other form of sustainable development, lays out the sequences through which development, growth, or change progresses. In this context, the maxim, "you manage what you measure," becomes a practical expression of meaningful formative engagement with what matters. Instead of meaningless scores, measures stand for reproducible structures of change and growth that are individually interpretable and actionable.

In this vein, a critical appraisal of the health-related SDGs calls for "a robust formative evaluation system that scrutinises accountability mechanisms and possible double-duty paradoxes embedded in the operationalisation and implementation of the SDGs" [26]. Formative evaluations of just this kind have been developed and used in education and health care for decades [3, 6-7, 11-12, 14, 22]. In the context of the UN SDG 3 target on universal health coverage, a calibrated developmental sequence mapping a continuum of increasing quality of care lays out a roadmap to the management of improved access. The most easily provided and commonest features of universally accessible care will be rated as present most often by consumers, and the rarest, least easily provided features will be rated as present least often. Later features and programs will rely and build on earlier ones. Not all features and programs will be relevant or needed in different locations; sequences will not be rigidly followed everywhere. Thus we need to deal with complexity to be able to nimbly adapt to varying circumstances and change over time.

5. Complexity

The most fundamentally important answer to the question "Why metrology?" concerns discontinuities between the varying levels of complexity that structure information. This is of special importance today, as digital information increasingly influences the allocation of resources and investments towards SDG targets. We are so habituated to the flexibility language affords us in communication that we tend to ignore, accommodate, or work around instances in which the complexity of information becomes problematic. This happened, for instance, in the formulation of the SDGs, where the word "we" refers to humanity in general when demand for sustainable solutions are the topic of discussion, but refers to a different and unidentified class of people when speaking to issues of accountability [25]. Another disconnection of responsibility from established channels of governance occurs in sustainability impact investing when producers accountable to investors circumvent citizens' reliance on their elected officials, effectively turning people into faceless commodities [26].

This kind of restricted market definition follows closely from the quality of the measurement information used to track and evaluate productivity. Metrics tied to specific indicators are not commensurable across (or even within) providers, and so cannot be the medium of fair trade or efficient markets, since they do not lower transaction costs, enhance communication, or support property rights [13]. But social studies of metrological science show its objects of study "are both plastic enough to adapt to local needs and the constraints of several parties employing them, yet robust enough to maintain a common identity across sites" [27 p 392]. That is, metrological principles and practice can facilitate both local improvisations and general navigable continuity in measurement and management [11]. This capacity is essential to meeting the practical demands for meaning, trust, manageability, and complexity involved in the management of the SDGs.

6. Conclusion

The established viability of metrological traceability for the instruments of human, social, and natural capital and new initiatives set in motion by national metrology institutes [2-16] set the stage for fundamental changes in the organization of approaches to realizing the SDGs.

Sustainability impacts plainly and obviously at present lack common product definitions, objective measures, efficient markets, and associated capacities for competing on improved quality. The absence of these landmarks to date in the domain of sustainability interests is a result of inattention and cultural biases far more than it is a result of the inherent characteristics or nature of sustainability itself. Given the economic importance of these kinds of capacities and the urgent need for new innovations supporting sustainable development, it is curious how even those most stridently advocating new ways of thinking seem to systematically ignore well-established opportunities for advancing their cause. The wealth of historical examples of rapidly emerging, transformative, disruptive, and highly profitable innovations would seem to motivate massive interest in how extend those successes in new directions.

Economists have long noted how common currencies reduce transaction costs, support property rights, and promote market efficiencies. Language itself is well known for functioning as an economical labor-saving device in the way that useful concepts representing things in the world as words need not be re-invented by everyone for themselves, but can simply be copied. In the same ways that common languages ease communication, and common currencies facilitate trade, so, too, do standards for common product definitions contribute to making markets and to innovative quality improvements [21].

Metrologically traceable measurements allow everyone everywhere to know how much of something there is. This is important, first of all, because things have to be identifiable in shared ways if we are to be able to include them in our lives, socially. Anyone interested in obtaining or producing that kind of thing has to be able to know it and share information about it as something in particular. Common languages capable of communicating specifically what a thing is, and how much of it there is, support claims to ownership and to the fruits of investments in entrepreneurial innovations. Metrology ought, then, to be a top-level priority in SDG programs and policies. How else will sustainable impacts be made universally identifiable, individually owned, efficiently exchanged, and profitable?

The electronics, computer, and telecommunications industries provide ample evidence of metrology's role in reducing transaction costs, establishing common product definitions, securing property rights, and reaping huge profits [21]. The music industry's use of these technologies combines the science of precision measurement with the artistic creativity of intensive improvisations. Instruments tuned to standardized scales are used to achieve wholly unique levels of individual innovation. Much stands to be learned, and even more to be gained, in focusing sustainable development on harnessing the economic power of the profit motive and individuals' capacities for creativity.

The potential for embodying shared moral, economic, and scientific values in common metrological languages [24] should not be judged impossible without thorough investigating its viability, feasibility, and desirability. This task ought to be the defining mission of this generation. The time to act is now.

References

[1] Joint Committee for Guides in Metrology 2012 *International vocabulary of metrology: Basic and general concepts and associated terms* 3rd ed (Sevres, France: BIPM)
[2] Rasch G 1980 *Probabilistic Models* (Chicago: University of Chicago Press)
[3] Wilson M 2005 *Constructing measures* (Mahwah, New Jersey: Lawrence Erlbaum)
[4] Pendrill L and Fisher W P Jr 2015 *Measurement* **71** 46-55
[5] Mari L and Wilson M 2014 *Measurement* **51** 315-27
[6] Fisher W P Jr and Stenner A J 2016 *Measurement* **92** 489-96
[7] He W and Kingsbury G G 2016 *J. Phys.: Conf. Ser.* **772** 012022
[8] Cano S, Pendrill L, Barbic S and Fisher W P Jr 2018 *J. Phys.: Conf. Ser.* **1044** 012057
[9] Cano S, Melin J, Fisher W P Jr, Stenner A J, Pendrill L and EMPIR NeuroMet 15HLT04 Consortium 2018 *J. Phys.: Conf. Ser.* **1065** 072033
[10] Cano S, Pendrill L, Melin J and Fisher W P Jr 2019 *Measurement* in press
[11] Fisher W P Jr and Oon E P-T 2019 *Kybernetes* in review
[12] Fisher W P Jr and Wilson M 2015 *Pensamiento Educativo* **52** 55-78
[13] Fisher W P Jr 2009 *Measurement* **42** 1278-87
[14] Wilson M 2013 *Psychometrika* **78** 211-36
[15] Maul A 2013 *Theor. Psychol.* **23** 752-69
[16] Pendrill L R 2018 *Meas. Sci. Technol.* **29** 034003
[17] Alder K 2004 *The measure of All Things* (London: Abacus)
[18] Scott J C 1998 *Seeing Like a State* (New Haven: Yale University Press)
[19] Ashworth W J 2004 *Science* **306** 1314-17
[20] Barzel Y 1982 *J. Law Econ.* **25** 27-48
[21] Miller P and O'Leary T 2007 *Account. Org. Soc.* **32** 701-34
[22] Wright B D, Mead R J and Ludlow L H 1980 KIDMAP: person-by-item interaction mapping *MESA Memorandum #29* (Chicago: MESA Press)
[23] Mundy B 1986 *Synthese* **67** 391-437
[24] Wise M N 1995 *The values of Precision* (Princeton, New Jersey: Princeton University Press)
[25] Engebretsen E, Heggen K and Ottersen O P 2017 *The Lancet* **389** 365
[26] Sinclair S, McHugh N, Huckfield L, Roy M and Donaldson C 2014 *Social Policy Rev.* **26** 119-36
[27] Star S L and Griesemer J R 1989 *Soc. Stud. Sci.* **19** 387-420

Addressing traceability in social measurement: establishing a common metric for dependence

T Salzberger[1], S Cano[2], L Abetz-Webb[3], E Afolalu[4], C Chrea[4], R Weitkunat[4], K Fagerström[5], J Rose[6]

[1] WU Wien, University of Economics and Business, Vienna, Austria, thomas.salzberger@wu.ac.at
[2] Modus Outcomes, Spirella Building, Letchworth Garden City, UK, stefan.cano@modusoutcomes.com
[3] Patient-Centered Outcomes Assessments, LI, Macclesfield Cheshire, UK, linda.abetz-webb@p-coa.com
[4] PMI R&D, Philip Morris Products S.A., Neuchâtel, Switzerland, Esther.Afolalu@pmi.com, Christelle.Chrea@pmi.com, Rolf.Weitkunat@pmi.com
[5] K. Fagerström Consulting AB, Vaxholm, Sweden, karl.fagerstrom@hotmail.com
[6] Rose Research Center, Raleigh, NC, USA, jed.rose@roseresearchcenter.com

E-mail: thomas.salzberger@wu.ac.at

Abstract. Measurement in the social sciences is characterised by a multitude of incompatible paradigms, most of which fail to adhere to fundamental principles of measurement explicated in metrology. This has led to a fragmentation of instruments that are hard to interpret and lack a common reference. Rasch measurement theory, the metrological framework of the Rasch model, combined with strong substantive theories of the measurand, has the potential to advance measurement in the social sciences significantly. The Rasch model establishes a reference standard with a common unit of measurement against which different instruments can be calibrated ensuring traceability in social measurement. Uncertainty in the calibration process is addressed by a standard error of measurement for a specific estimate taking into account that precision varies over the continuum of the measurand. A strong substantive theory of the measurand allows for concrete predictions that can be tested empirically with positive evidence supporting accuracy. We exemplify the potential of the Rasch model by developing the ABOUT-Dependence, a new instrument for the measurement of dependence on tobacco- and nicotine-containing products (TNP). The instrument provides comparable measurements of dependence on different TNPs as well as dependence on multiple TNPs used concurrently. A co-calibration of the new instrument and existing, product-specific instruments allows for a conversion of scores of existing instruments into scores on the new instrument, and vice versa, via the established equal-unit reference scale. The score comparability adds to traceability in the measurement of dependence, is likely to facilitate the acceptance of the new instrument, and helps avoid a discontinuity in the research of dependence.

1. Introduction

Measurement is a precisely defined concept, with traceability and a stated range of uncertainty being its core properties [1, 2]. While these principles apply to all disciplines alike [3], measurement in the social

Journal of Physics: Conference Series **1379** (2019) 012024 doi:10.1088/1742-6596/1379/1/012024

sciences (also referred to as social, or human, measurement) has been plagued by a multitude of incompatible paradigms [4, 5] bearing on very different definitions of measurement [6]. These paradigms have contributed to the widespread view that social measurement and its properties are the subject of negotiation and pragmatism. Numbers resulting from coded observations (such as ticking a box in a questionnaire) have been interpreted as measurement with little, if any, justification. Most paradigms fail to reveal anomalies disproving successful measurement implying that the scientific mechanism of self-correction by replication stops functioning. Measurement models are supposed to best describe the data rather than expose anomalies revealing weaknesses in the measurement process. Furthermore, concepts to be measured are often vaguely defined, leaving plenty of room to substantiate them after the fact, essentially defining them in terms of the items retained in the instrument, quite often according to arbitrary statistical criteria. As a consequence, instruments tend to be fragmented and weakly linked to one another. To attain the same level of scientific rigour as the natural sciences with respect to measurement, the social sciences need to reconsider the underpinnings of measurement. The paradigm of Rasch measurement [7], or Rasch measurement theory (RMT [8, 9]), lends itself as a conceptualisation, or philosophy, of measurement that takes traceability and uncertainty adequately into account. The measurement of dependence on tobacco- and nicotine-containing products (TNP) illustrates how RMT contributes to concept-driven measurement in the social sciences, providing interpretable measures with a specified level of uncertainty. In the social sciences, the latter is typically addressed by the standard error of measurement based on reliability.

2. A measurement theory for the social sciences

Measurement instruments in the social sciences typically consist of a set of items (questions or statements) collectively supposed to represent the measurand. Each item is a reference to which subjects can be compared. Each response of a subject to an item is the manifest outcome of such a comparison as the measurement method. The probability of a particular response is modelled as a function of the item's quantity and the subject's quantity, which are represented by an item location and a person location parameter, respectively. Item and subject quantities are estimated from a series of observed responses to a set of items forming a scale representing the same measurand. In order to provide measurement, items need to be like stable markings on a ruler that are invariant with respect to the respondents. The independence of item locations from the specific respondents, instrumental in the item location estimation, is achieved by the Rasch model for measurement [7] by virtue of its defining characteristic of specific objectivity [10, 11]. In the case of binary response formats (dichotomous response scales, such as yes versus no or agree versus disagree, scored 0 and 1, respectively), one item location parameter is estimated. Ordered multi-categorical formats (polytomous response scales, such as rating scales) require the estimation of a set of threshold locations marking the boundaries between adjacent response categories. The expected score on an item as a function of the item location and the person location forms an s-shaped curve, represented by a logistic function. In the Rasch model, there is no additional parameter describing variation in the slope of this curve. Thus, in the dichotomous case, the Rasch model requires the slope to be the same for all items to retain an equal unit of measurement across the continuum. In the polytomous case, the slope at each threshold has to be the same across all thresholds and items. This implies that respondents discriminate equally between each pair of adjacent response categories.

Valid measurement requires two conditions to hold. First, the observed responses need to correspond to the responses expected under the model. Second, the item location estimates need to form a meaningful hierarchy in terms of the measurand and its conceptualisation. The first requirement is addressed by various statistical tests of fit assessing whether the observed responses match the responses expected under the model. Another approach to test whether the data fit the model is based on specific objectivity. It implies that item parameters are invariant regardless of the subjects used to estimate them. Thus, item location estimates must not differ between different respondent groups. All of these tests involve formal statistical procedures that, by their very nature, do not take item content into account. Furthermore, they are based on empirically estimated item location parameters with no reference to predictions by the conceptual theory of the measurand. Thus, mere model fit is a necessary, but not a sufficient condition of measurement.

Therefore, to substantiate accuracy of measurement, the estimated item parameters have to be linked to the conceptual theory. To begin with, the item hierarchy in terms of item locations needs to be meaningful and interpretable. It demonstrates what more or less of the property assessed actually means, allowing for an unambiguous interpretation of person measures. If the conceptual theory proposes a particular item hierarchy, comparing the empirical item hierarchy and the expected hierarchy tests whether theoretical predictions apply. Ideally, the theory predicts a concrete location for each item based on some exposed mechanism driving the item's location [12]. While successful implementations of such elaborate construct theories are still rare, the focus in RMT has shifted towards more powerful construct theories.

In the case of less well-developed conceptual theories, empirical insights in terms of the actual item hierarchy revealed by the Rasch model may help advance the conceptual theory. The contribution of applications of the Rasch model to theory extension is particularly relevant in the transition phase, which has to cope with two obstacles. First, an abrupt change from firmly established but inadequate rules to new principles is likely to face resistance [13], and existing instruments will continue to be defended and will not fall into sudden disuse. Second, an abrupt shift may lead to a discontinuity in substantive research, implying unjustified devaluation of existing research findings. Rather, legacy instruments should be scrutinised, re-evaluated, improved when needed, and linked to new measures and to one another. In this regard, RMT is particularly powerful, allowing for establishment of a common metric for different measures of the same measurand, which is an important contribution to traceability.

3. The measurement of dependence on TNPs

Nicotine dependence is a key driver underlying maintenance of use of TNPs. The measurement of nicotine dependence has primarily focused on assessing product-specific dependence (notably cigarettes) rather than overall dependence on TNPs [14]. A broad range of alternative TNPs have become available (e.g. electronic cigarettes, heat-not-burn tobacco products, smokeless tobacco), and some of these products are currently under investigation by regulatory bodies as part of a population harm reduction strategy [15-17]. The measurement of dependence is currently hampered by two obstacles: Measures for dependence on different TNPs are not comparable; and no measure for dependence on multiple TNPs is available [18]. To overcome these limitations, a new instrument (the ABOUT-Dependence, as part of the ABOUT Toolbox initiative [19]) was developed.

First, a conceptual model of dependence was generated based on a literature review, expert opinions, and concept elicitation interviews with TNP users (n=40 [20, 21]). Second, a draft version of the instrument comprising 19 potential items was developed, with items best representing the aspects of the concept of interest, response categories meaningful to intended respondents (TNP users), and appropriate recall periods. The items were grouped into seven aspects of dependence (such as compulsion to use, automaticity of using, or self-awareness of dependence). In the cognitive interviews, some of the items were identified as being conceptually redundant (alternative phrasing of the same issue), while one item did seem to be relevant, but only to some of the participants. However, all 19 items were tested psychometrically to identify the best-working items when multiple items capturing the same issue were generated. Third, in a field test, data from a sample of 2,434 respondents in the U.S. comprising 1,253 poly-users and 1,181 exclusive users (seven different TNPs considered: conventional cigarettes, e-cigarettes, smokeless tobacco, cigars, pipes, waterpipes, and nicotine replacement therapy) were collected to assess the proposed instrument. As strict unidimensionality was not tenable, the conceptualisation was reconsidered, and eventually, three domains of dependence were proposed (*extent-of-use*, covering the urgency or pervasiveness with which the product(s) is/are used; *doing*, comprising aspects of how dependence impacts on daily activities; and *feeling*, related to symptoms and signs of dependence experienced by TNP users). With regard to the *doing* domain, a partial ordering of the items was to be expected, with *using more of the product than intended* being a basic symptom of dependence that should be endorsed most easily. Endorsing *stopping an activity in order to use a TNP* was supposed to require less dependence compared to *avoiding an activity altogether*. For the *feeling* domain, *a strong desire* was supposed to be more readily endorsed than *finding it hard to control the urge*, as the latter implies the former but not vice versa.

The application of the polytomous Rasch model confirmed the three domains of the revised conceptualisation, revealed redundancy where it had been expected, and identified one item suspected to misfit as inadequate. The reduction of the number of items to 12 by omitting six redundant items and one misfitting item did not result in any lack of content coverage. The partially expected ordering of items was confirmed, and the full item hierarchy turned out to be conceptually meaningful. As the three domain measures were correlated in the order of 0.5 to 0.8 in the entire sample, a composite score capturing overall dependence was proposed covering all three domains. The *extent-of-use* domain required an adjustment for differential item functioning by the type of TNP used.

Technically, the composite scores were analysed by applying the Rasch model to domain sub-scores as super-items (or subtests). The Rasch model transformed the total sum score (raw score) into a linear measure on a logit scale, which was then linearly transformed into a more convenient and accessible metric of 0 to 100. By ordering all threshold locations across all three domains, for each observed total score, the most likely score on each domain was inferred. Subsequently, each domain score was characterised by the most likely response pattern based on the previous application of the Rasch model to the domain items. By doing so, a comprehensive interpretation of each total score on the composite measure was achieved.

After successfully establishing a meaningful and interpretable structure of items forming composite scores, various existing instruments can be linked to the newly established metric. This is done by co-calibrating the new instrument (three domains as subtests) and existing instruments (represented by another subtest), whereby the subtest parameter estimates for the new instrument were anchored to values estimated in preceding analysis of the new instrument only. This procedure provides the opportunity to transform raw scores on various existing product-specific instruments to a common metric established by the new instrument. Figure 1 shows how ABOUT-Dependence measurements among exclusive cigarette users are converted to measurements by the very frequently used Fagerström Test for Nicotine Dependence (FTND) [22], recently renamed the Fagerström Test for Cigarette Dependence [23], and vice versa. The item locations for each instrument are represented by squared dots on the common linear metric. Each raw score on the FTND matches a score on the ABOUT-Dependence. For example, a score of 2 on the FTND is equivalent to a score of 12 on the ABOUT-Dependence. Conversely, scores of 10 to 14 on the ABOUT-Dependence all map onto a score of 2 on the FTND. This shows that the new instrument has a better resolution compared to the FTND.

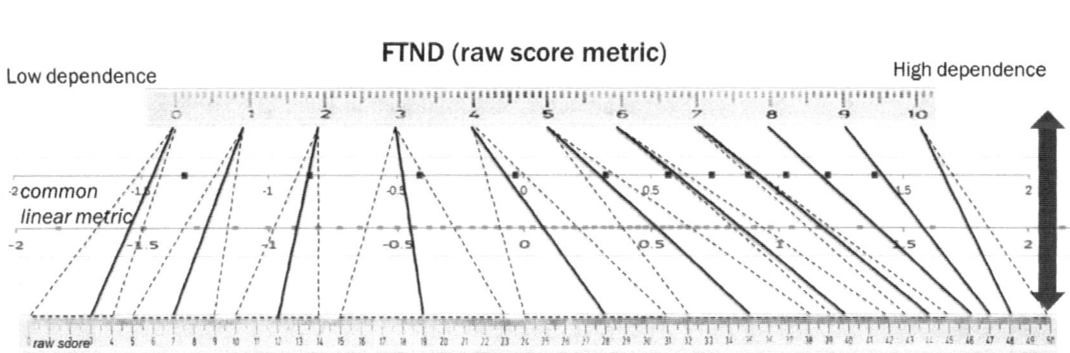

Figure 1. Conversion of the ABOUT-Dependence raw score (composite score) into the FTND raw score via a common linear metric.

Figure 2 illustrates the conversion of ABOUT-Dependence raw scores to linear measurements. Each raw score from 0 to 50, on the x-axis, is transformed into a linear measurement of dependence, on the y-axis. The transformation is non-linear demonstrating the raw score is not a linear measure. The error bars represent 95 % confidence intervals for the estimated measures illustrating measurement uncertainty. Towards the extremes, measurement uncertainty increases, while most precise measures are provided in the middle of the scale.

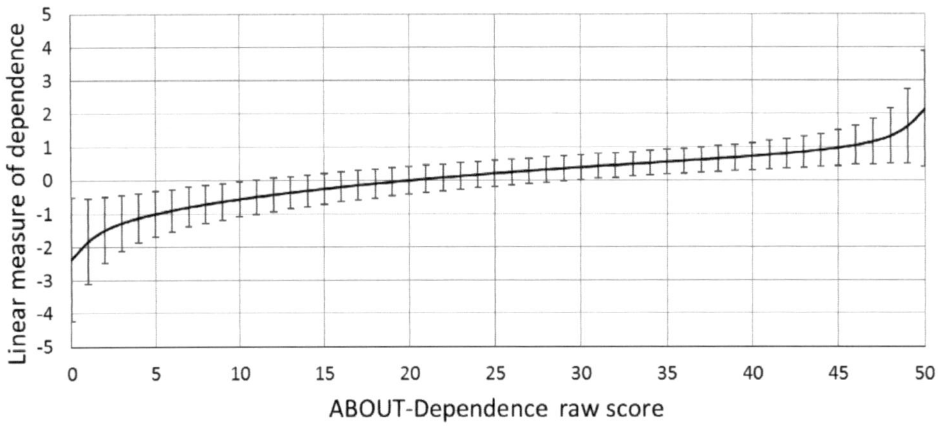

Figure 2. Conversion of the ABOUT-Dependence raw score (composite score) into a linear measurement with 95 % confidence intervals for measurement uncertainty.

4. Conclusions

The measurement of TNP dependence has focused on specific tobacco products, most notably cigarettes. Today, such instruments are no longer fit-for-purpose due to the diversity of products available and the prevalent habits of using multiple products concurrently. This is an obstacle for health policy makers, regulatory bodies, and researchers in the field of tobacco regulatory science as dependence across different TNPs is hard to compare. Based on a conceptual model of dependence, the interplay of insights from the psychometric analysis applying the Rasch model and further conceptual considerations, a proposed three-domain structure of dependence could be empirically supported. The high correlations of these three domain measures justified the formation of a composite score summarising dependence by one overall measure. Compared to existing measures, the ABOUT-Dependence provides a better interpretable measure of dependence by referring to individual item and response category locations. In addition, the ABOUT-Dependence provides measures with better resolution (Figure 1) compared to existing instruments, such as the FTND. The composite score allows for a consolidation of existing instruments, and the ABOUT-Dependence by providing crosswalks from the metric of one instrument to the metric of another. The transformation of scores from existing instruments to the newly established metric should be interpreted with caution, however, as existing measures do not necessarily meet the requirements of the Rasch model. What is more, the transformed scores retain the level of the original uncertainty, which is typically considerably higher than uncertainty associated with ABOUT-Dependence measurements. When comparing dependence measures across products, it is important to realise that what the instrument measures is the existing dependence on a specific TNP or, in case of poly-users, a bundle of TNPs. In a cross-sectional study, the measurements do not allow for inferring a TNP's potential to induce nicotine dependence. Applied research, in particular longitudinal studies, will reveal how dependence develops over time and the predictive validity of the instrument.

References

[1] JCGM 2012 *JCGM 200:2012 International Vocabulary of Metrology—Basic and General Concepts and Associated Terms (VIM)* 3rd ed (2008 version with minor corrections) (Sèvres: Joint Committee for Guides in Metrology) (www.bipm.org/en/publications/guides/vim.html)

[2] JCGM 2008 *JCGM 100:2008 Evaluation of Measurement Data - Guide to the Expression of Uncertainty in Measurement* (GUM 1995 with minor corrections) (Joint Committee for Guides in Metrology)
[3] Pendrill L R and Fisher Jr W P 2013 Quantifying human response: linking metrological and psychometric characterisations of man as a measurement instrument *J. Phys.: Conf. Ser.* **459** 1 01205
[4] Cano S et al 2016 *On Trial: the Compatibility of Measurement in the Physical and Social Sciences J. Phys.: Conf. Ser.* **772** 012025
[5] Andrich D 2004 Controversy and the Rasch model: a characteristic of incompatible paradigms? *Med. Care* I7-I16
[6] Michell J 1997 Quantitative Science and the Definition of Measurement in Psychology *Brit. J. Psychol.* **88** 3 355-383
[7] Rasch G 1960/1980 *Probabilistic Models for Some Intelligence and Attainment Tests* (Copenhagen: Danish Institute for Educational Research; expanded edition 1980 Chicago, IL: The University of Chicago Press)
[8] Andrich D 2018 Advances in social measurement: A Rasch measurement theory. In F Guillemin, A Leplège, S Briançon, E Spitz, and J Coste (Eds) *Perceived Health and Adaptation in Chronic Disease* (London: Routledge) pp 66–91
[9] Cano S, Pendrill L R, Melin J, Köbe T and Fisher Jr W P 2018 *Metrology for the Social Sciences: A case for Rasch Measurement Theory (not Rasch Analysis)* (New York: IOMW)
[10] Rasch G 1977 On specific objectivity: an attempt at formalizing the request for generality and validity of scientific statements *Danish Yearbook of Philosophy* **14** 58-93
[11] Fischer G H 1995 Derivations of the Rasch Model *Rasch Models, Foundations Recent Developments, and Applications,* ed G H Fischer and I W Molenaar (New York: Springer) 15-38
[12] Stenner A J, Fisher Jr W P, Stone M H and Burdick D S 2013 Causal Rasch models *Front. Psychol.* **4** 1-14
[13] Andrich D 2002 Implications and applications of modern test theory in the context of outcomes based education *Stud. Educ. Eval.* **28** 2 103-121
[14] Strong D R *et al.* 2017 Indicators of dependence for different types of tobacco product users: Descriptive findings from Wave 1 (2013–2014) of the Population Assessment of Tobacco and Health (PATH) study *Drug Alcohol Depen.* **178** 257-266
[15] Gostin L O 2009 FDA regulation of tobacco: politics, law, and the public's health *Jama* **302** 13 1459-60
[16] Warner K E and Schroeder S A 2017 FDA's innovative plan to address the enormous toll of smoking *Jama* **318** 18 1755-56
[17] Fagerström K O and Bridgman K 2014 Tobacco harm reduction: the need for new products that can compete with cigarettes *Addict. Behav.* **39** 3 507-511
[18] Kasza K A *et al.* 2017 Tobacco-product use by adults and youths in the United States in 2013 and 2014 *New Engl. J. Med.* **376**(4) 342-353
[19] Chrea C *et al.* 2018 Developing fit-for-purpose self-report instruments for assessing consumer responses to tobacco and nicotine products: the ABOUT™ Toolbox initiative [version 1; peer review: 2 approved] *F1000Research* 7:1878 (https://doi.org/10.12688/f1000research.16810.1)
[20] Chrea C *et al.* 2018 Development of a tobacco and nicotine products dependence instrument. *24th Annual Meeting of the Society for Research on Nicotine and Tobacco (SRNT) (Baltimore, USA)*
[21] Chrea C *et al.* 2018 Development of a fit-for-purpose tobacco and nicotine products dependence instrument ISPOR (The International Society for Pharmacoeconomics and Outcomes Research) (Europe Barcelona, Spain)

[22] Heatherton F, Kozlowski L T, Frecker R C and Fagerström K O 1991 The Fagerström Test for Nicotine Dependence: A revision of the Fagerström Tolerance Questionnaire *Brit. J. Addict.* **86** 1119–27
[23] Fagerström K O 2012 Determinants of tobacco use and renaming the FTND to the Fagerström Test for Cigarette Dependence *Nicotine Tob. Res.* **14** 75–78

The status of the concept of reference object in measurement in the human sciences compared to the physical sciences

M Wilson[1], **L Mari**[2] **and A Maul**[3]

[1] Graduate School of Education, University of California, Berkeley, CA, USA
[2] School of Industrial Engineering, Università Cattaneo, Castellanza, VA, Italy
[3] Gevirtz Graduate School of Education, Univ. of California, Santa Barbara, CA, USA

E-mail: MWilson@berkeley.edu

Abstract. For measurement in the human sciences, one barrier to being able to see the commonalities with measurement in the physical sciences has been the common prominence that is often given to the important place that "reference objects" are given in introductory accounts in metrology. The obvious difficulties in having such objects, say, "reference humans," makes it difficult to build connections. In this paper, we outline a way to broaden the concept of a reference object in way that addresses this issue. In this light we then discuss the recent developments in the International System of Units (SI), and comment on certain similarities.

1. Introduction

Many basic introductory account of measurement feature centrally the idea of 'reference object' (also known as "measurement standards" in the tradition of physical metrology). For example, in [1], the initial step in his proposed formalization of measurement is to create a "standard reference set" (of objects) $\boldsymbol{A}$, by associating them with a set of symbols $\boldsymbol{\Theta}$. The next step is then to carry out *data acquisition*: through empirical interaction, select an element of $\boldsymbol{A}$ that is indistinguishable from the object under measurement with respect to the considered property, and assign it the element of $\boldsymbol{\Theta}$ that corresponds to the element of $\boldsymbol{A}$.

When we look to the human sciences, this would seem to imply that we need to have accessible, stable and reproducible humans who we can (a) certify as reference objects, and therefore (b) use for comparisons in the practice of measurement. A technique for extending this logic to avoid the use of a set of human beings as reference objects has been discussed in [2]—the argument presented there does not attempt to detach from the primacy of the concept of reference objects, but instead builds on that concept to develop a different basis for measurement. Hence, this still assumes that at least in principle one could indeed find such "reference humans," even though they are not required to be exploited in practice.

Given the common assumption that such reference objects are empirical entities such as rods, gauge blocks, standard cell electrochemical batteries, standard resistors, and so on, the typical pragmatic ploy in the human sciences is to replace the human reference object with a reference instrument. But, this ploy then leaves the reference object as missing in the context of measurement in the human sciences.

Thus, we address the question – what if there were a more direct relationship between the instrument and the human property that is under measurement? To see a different perspective than the typical one mentioned above, think about what we mean by "human being." A human being is many-fold—incorporating a very large number of features, so many that they might (in practical terms) be thought of as infinite. This is one of the ways in which measurement in the human sciences is seen to differ from the stereotype of measurement in the physical sciences. In comparison, a rod used as a reference object for the property of length (say being a reference for a certain length—the meter etc.) would seem to have just one principal characteristic, the property of length. However, in fact, this is NOT true, for any such rod is actually also a cluster of a very large number of features. For example, each such rod is not mathematically straight, it has not the same width at every point, it does not have truly plane and parallel ends, it may not have the same electric charge, it does not weigh exactly the same as other "meter rods," and it is not composed of the same atoms as other such rods, nor even the same mixture of types of atoms. Each meter rod used as a reference object is in fact uniquely itself, and different in many ways from every other reference object used to reference the meter. When this sort of example is being used in an introductory exegesis of measurement, these other properties are generally ignored, and hence a reader unfamiliar with physical measurement (such as even experts in measurement in the human sciences) can easily take away a stereotype that the reference objects are somehow "pure" examples of the property in question. Of course, it is an important point that, while physical reference objects can be *designed* to have one "principal" feature, that is not the case for human beings.

2. Our contentions

It is our first contention that we are held back from understanding the way in which instruments are indeed one aspect of reference objects by this underestimation of the individual diversity of the reference objects that are used in traditional measurement in the physical sciences. Or, put another way, we are over-appreciative of the individual diversity of human beings. This seems to us to be a forgivable weakness, as we are indeed all human beings, and we value our uniqueness both as individuals and as a species (at least until the artificial intelligences catch up with us!). Thus, a more humble perspective might well be that indeed a mere "meter rod" could be seen as having near-to the uniqueness of a human being. So, why do we not see a similar problem in creating reference objects for length as we do for human properties?

This introduces our second contention, that what we consider to be a "human property" is, in our thinking, overburdened with metaphysical complexities. For example, take reading comprehension ability as an example of a human property. When thinking about a reference object, we immediately imagine a person reading a text (an observable event), and, unobservably, forming a "comprehension" of the meaning of the text in their minds (this of course, is the erstwhile human property). This "comprehension" an observer might ascertain by asking certain questions of the reader, and interpreting/judging the responses. In fact, is it not a more straightforward stance to consider that the human property of reading comprehension is in fact inherent in the event of being asked questions (of a certain type, about the text and its content) and responding with (correct or incorrect) answers? From this standpoint, the property is seen as being an aspect of an event—the event of responding to the question(s).

3. Beyond reference objects

With this perspective in mind, one can then see that a reading comprehension test can indeed constitute a reference object (or, perhaps, more accurately, a system of reference objects). But the essence of such an insight lies in conceiving of the test as more than just the surface-level "object"—the test is not just the letters that form the words that make up the questions, or command, that constitute the items that make up the test. Each such item must be seen as the combination of (a) the prompt, (b) the response(s), and (c) the methods of interpreting those responses, in the sense of indicating the extent of the property revealed by those responses. When an item is seen as being composed of these parts, it makes more sense to see it as (part of) a "reference object" for a human property that is wrought in a deeper way than would usually be conveyed by the term "object." Hence, instead of relating measurement to "reference objects," the contention here is that we can broaden the concept of an object, beyond being a "body" so that we could think of a "reference event." In the context of reading comprehension, the event would be a reader reading a text passage, and responding to a series of questions posed about that reading passage. For each item, there would be a protocol for coding the responses to give a response category—the set of these for all the items would constitute the standard set of reference events, $\boldsymbol{A}$. And this would be used to generate a symbol from $\boldsymbol{\Theta}$ (as above).

Consider, for example a classic example of traditional measurement, the Kansas Silent Reading Test [2]. An example item from this test is:

> *I have red, green and yellow papers in my hand. If I place red and green papers on the chair, which color do I still have in my hand?* _______________

This one might refer to as "nude" item—it stands alone, without reference to what it is purported to be measuring (apart from the title of the test, "Silent Reading", it does not have specifications for scoring the item responses; it is assumed this is self-evident), and it does not come with a complicated procedure for accumulating the scores from the items to an overall test score (the scores are simply added up—as is common in education). This traditional style of item design encourages the conceptualization of each item as a stand-alone prompt, without the connections to a measurement system that would make clear the connection to the human event of engaging in certain type of situation designed to invoke a certain human characteristic, and responding in ways that can be then observed, coded and accumulated to make an inference about a person in terms of that property.

This can then be contrasted to an alternative example of an item that is indeed embedded in a context that scaffolds its status as a "reference event." In this case, the theory about the person's cognition takes a particularly simple form: the property is modeled as a "construct" consisting of a linear succession of discrete segments of a continuum, and when these are illustrated in a figure, it is termed a "construct map". An example of this is shown in figure 1, which was designed as part of a test of middle school students' practice of designing and critiquing displays of data.

DaD6: Integrate case with aggregate perspectives.
DaD5: Consider the data in aggregate when interpreting or creating displays.
DaD4: Recognize or apply scale properties to the data.
DaD3: Notice or construct groups of similar values.
DaD2: Focus on individual values when constructing or interpreting displays of data.
DaD1: Create displays or interpret displays without reference to goals of data creation.

Figure 1. The levels of the Data Display construct map.

Each item is developed to match to one or more levels of the construct map. An example item for the Data Display construct map is shown in figure 2. Figure 3 then shows sample student responses to this item that have been matched to certain levels of the DaD construct map. Gathering sample student responses can lead to a revision of the construct map and/or the hypothesized link, as well as modification/deletion of the item.

Once an item set is established, all items having the characteristics described in the previous paragraph, "banding" or "standard setting" takes place—this is the analogue of the "calibration" described by Mari in [1]: (a) locations for each item is estimated via a Rasch scaling, and then (b) the delineation of segments ("bands") of the Rasch logit scale. The upper and lower boundaries of these bands are determined by the locations of the lowest and uppermost of the items linked to each level of the construct map. Quite often, these item locations do not result immediately in a "clean" segmentation of the logit scale—hence an expert judgment process [3] is needed to gauge reasonable locations for the band boundaries of each band. This may require further decisions regarding the suitability of certain items, and may also involve demographic and other information about the persons, when that is available.

These successive bands can then be used as a basis for criterion-referenced interpretations of measurement results, enhancing and deepening the interpretations. At the same time, the existence of the underlying Rasch scale means that (a) technical aspects are available, such as estimates of instrumental uncertainty (standard errors etc.), and (b) technical advantages of item response scales are still available, such as flexibility in item choice, ability to link forms through items, and the possibility of computerized adaptive item administration.

Shimmering Candle Company claims that their candles burn longer, on average, than candles made by the Brilliant Candle Company. Testers for Consumer News burned 15 Shimmering candles and 15 Brilliant candles and recorded the number of minutes that each candle burned. The plot below shows the burning time and the middle-50 percent of the data around the median.

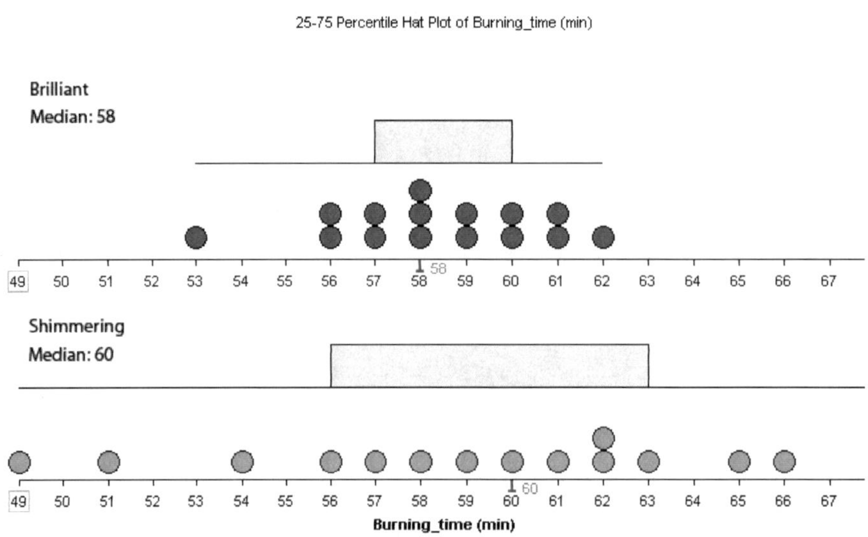

1. Is the claim made by the Shimmering Candle Company, that their candles burn longer, supported by the test results? Circle your answer.

Yes **No**

Why do you think so?

Figure 2. An example item ("Shimmering Candle.") related to the DaD construct.

Returning to the meter rod, we can see, from this broader perspective, that the rod itself, although it is indeed an object, is by and of itself insufficient for measurement. What is required is actually a reference *event*, which would be the comparison of that reference rod to the rod under measurement. And this must be carried out in certain ways (e.g., carefully aligning the "bottom" of the two rods, etc.), and under certain conditions (e.g., both being at the same temperature, etc.). And this would be coded by noting which rod was longer, or the same, and so on for the set of rods. And, as for reading comprehension, the assignment to $\boldsymbol{\Theta}$ would be made on the basis of this set of codes.

Description of Level	Examples of Responses
DaD5: Student quantifies aggregate property of the display using one or more of the following: ratio or proportion or percent. *OR* Student recognizes that a display provides information about the data as a collective.	• "50% of Brilliant candles are between 57 and 60, and 50% of Shimmering candles are between 56 and 63."* • "No, b/c the majority of their burning times overlap." • "No, b/c even though their media was larger, they were also more spread out."
DaD3 Notice or construct groups of similar values from distinct values.	• "Yes, b/c the most of the Brilliant candles were between 57-59, while most of the shimmering candles were between 61-63."
DaD2 Concentrate on specific data points without relating these to any structure in data.	• "Because the last candle burned 6 min. more than Brilliant's last candle." • "They have more with larger numbers." • "The median is 58 for the other company, but Shimmering's median is 60." • "Because Shimmering candle has more candles that burnt 58 long." • "I think this because I added them up and got 875 minutes for Brilliant Candles and 953 for Shimmering Candles." "Because I calculated their means."
DaD1 Create or interpret data displays without relating to the goals of the inquiry. *OR* Makes a numerical assertion that is present in the data but not related to assertions about burning time.	• "There are 30 measurements."

Figure 3. Examples of responses to the DaD construct levels for the Shimmering Candles item.

Thus, this broadening of the specification of measurement basics not only makes the inclusion of typical human sciences practices more straightforward, but also makes the delineation of physical science measurement more comprehensive—making clear that it is not just the existence of a set of reference objects that matters, but also a set of informational specifications about how those objects must be used to produce the measurement result (i.e., part of the "measurement procedure").

A person's response to a set of items is certainly LESS than a human, but that can also be said of a rod—when we measure the rod by comparing it to a meter rod, we are ignoring all of the characteristics and features that make it an individual rod—we are only looking at one property—its length. And, symmetrically, even though the meter rod is itself a rod, we are ignoring all of the characteristics and features that still pertain to even this "reference object." It could as well be LESS than a rod, and it would still suffice for the purpose of being a reference object. In fact, one could say that this is a way to present measurement, and more generally science: the art of selecting something, and neglecting everything else in complex systems (i.e., "abstraction").

4. Developments in physical sciences

In fact, this development is exactly what has happened, and continues to happen, in the advancement of measurement in the physical sciences—classical reference objects (even though they remain as a centerpiece of basic textbooks in measurement) have been superseded by events and even computations from theory—see for example [4]. Today the situation in physical sciences is radically different, even though we might still say that on the scene there are reference properties and reference objects. Now the SI units (a particular but important case of reference properties!) are defined as functions of properties of physical phenomena: the metre is the length of the path of light in vacuum in a given fraction of one second, and so on. By adopting this semiotic, encompassing concept of 'object', we may still consider units as properties of objects, but such objects are not physical bodies anymore: they are *phenomena that are assumed to be invariant* according to the best available theories (and the revised SI -- just approved by the CGPM -- makes this strategy of definition even more radical).

This situation in physical measurement is acknowledged in the VIM [5], where, as part of the definition of measurement standard—the "realization of the definition of a given quantity, with stated quantity value and associated measurement uncertainty, used as a reference"—the following is noted: "A "realization of the definition of a given quantity" can be provided by a measuring system, a material measure, or a reference material." Thus, the extension of the possible basis for measurement beyond the traditional measurement object (referred to as "measurement materials") is encompassed within this broader definition. The case of a reference instrument, as discussed above would be included in the term "measurement system."

5. Conclusion

The ultimate need remains to be able to calibrate measuring systems and therefore to guarantee the metrological traceability of measurement results, and since calibrating a measuring system against a phenomenon such as, say, light in vacuum, is not exactly is trivial thing to do, the definition of a unit is realized by producing (more traditional) objects against which (more easily) measuring systems can be calibrated.

Similarly, the extension of the conceptualization of measurement outlined here, replacing the basic concept of the reference object with a reference event, a practice common in the human sciences, can be seen as pointing the need for a broader conceptualization of measurement. One could speculate that what has happened is that the cementing of the stereotypical reference objects into the whole account of measurement has affected the way it is thought about and the way it is expressed. Then, when we come to a situation (such as measurement in human sciences) where the erstwhile reference object is less amenable to being used as such, the account has to be "turned around", and what is seen as being the first step may no longer be so--what is important is that we get to the same place in the end (after calibration).

From this perspective, the use of reference instruments in the human sciences should not seen as a sort of idiosyncratic step sideways due to the theoretical weakness of those sciences, but rather can be seen as a prescient advancement in the fundamentals of measurement that has been taken by certain of the human sciences to overcome the ontic complexities of human properties.

References

[1] Mari L 2000 Beyond the representational viewpoint: a new formalization of measurement *Measurement* **27** 71-84
[2] Kelly E J 1916 The Kansas silent reading tests *J. of Educ. Psychol.* **7** 63–80
[3] Wilson M and Draney K 2002 A technique for setting standards and maintaining them over time *Measurement and Multivariate Analysis* ed S Nishisato Y Baba H Bozdogan and K Kanefugi (Tokyo: Springer-Verlag) pp 325-332
[4] Zim X-Z 2018 The kilogram is dead. Long live the kilogram! *New York Times* November 16
[5] JCGM 2012 *JCGM 200:2012 International Vocabulary of Metrology—Basic and General Concepts and Associated Terms (VIM)* 3rd ed (2008 version with minor corrections) (Sèvres: Joint Committee for Guides in Metrology)

Interpreting the relationship between item difficulty and DIF: examples from educational testing

M Wilson[1] **and M Dulhunty**[2]

[1] Graduate School of Education, University of California, Berkeley, USA
[2] Melbourne Graduate School of Education, University of Melbourne, Australia

E-mail: MarkW@berkeley.edu

Abstract. The investigation of differential item functioning (DIF) is commonplace in educational testing. Most often it results in no more than the discarding of a proportion of the developed items. Apart from the potential threat to validity of the instrument due to differential loss of certain types of items, this also is inefficient in terms of item development costs. This paper discusses certain situations where particular patterns of non-uniform DIF can be related to item difficulty, and modelled by enhanced Rasch models. This allows for a more efficient usage of items, as well as, potentially, an improved understanding of the complexities of how educational tests function across different sub-populations.

1. Introduction

In educational testing, as in other measurement contexts, the validity of interpretation of the results of statistical analysis of measurement data depends critically on the reasonableness of the assumptions behind the statistical models used to analyze the data. If these assumptions are violated, then serious mistakes may result from the uncritical usage of the measurements. One particular family of such threats to validity is differential item functioning (DIF), which occurs whenever the probability of a particular response to a specific item is influenced by factors other than the respondent's underlying ability [1-3].

While this may occur at the individual student level (as it almost certainly does), the existence of such effects is most readily detected at the group level—standard cases would be differentials between ethnic and racial groups, socio-economic groups, or gender effects. For example, in US contexts, comparing African-American students with White students is quite common, and where certain items are found that display evidence of DIF, the test developers systematically delete them from the test. This is done usually without any recourse to investigating why there was DIF for that particular item for that particular group comparison, as the existence of DIF for an item is seen as indicating that the item is in some way unfair, and hence must be eliminated. However, this systematic deletion of items from the item sample may constitute a threat to test validity in itself (in the sense of effectively altering the definition of the underlying property under measurement), especially where the deletions were concentrated on specific types of items etc.

Thus, it is of interest to try and find patterns in DIF results, and then relate these patterns to substantive aspects of the underlying property and/or features of the measurement instrument. The aim of this paper is to examine one particular way that DIF may be systematically occurring in tests, to relate that to potential patterns of DIF results, to describe some statistical models that could be used in this situation, and to recommend next steps in a research agenda focused on this issue.

2. Relating DIF to item difficulty

One prominent example of systematic DIF was brought to attention by Freedle [4] who showed how many of the more difficult SAT items displayed DIF effects which favored African American students while easier SAT items had DIF effects favoring White students. Freedle hypothesized that this relationship between item difficulty and DIF estimates could be explained by cultural familiarity and semantic ambiguity. He postulated that, on the one hand, "Easy" Verbal items are contextualized within a more culturally specific content and hence can be hypothesized to be perceived differently, depending on the student's own cultural and socioeconomic background. On the other hand, "Hard" Verbal items will tend to be expressed using rarely used words that are less likely to have differences in interpretation across ethnic communities because these are only familiar to those with higher levels of education, which is more uniform than the home environment.

This explanation leads then to an idea that, in this case, for the comparison of African-American students and White students, there might be a monotonic relationship, or possibly a linear one, between DIF and item difficulty, with the DIF effect decreasing as the difficulty increases. Indeed, this is similar to what Freedle in [4] found in his analyses, and he reported a correlation approximately of 0.5. Similar results in similar situations have been reported by Kulick and Hu [5], Scherbaum & Goldstein [6] and Santelices and Wilson [7]. Interestingly, these have all been found to be approximately linear relationships.

There has been considerable criticism of the interpretation of these results. Some researchers have claimed that the results are due to one of the following explanations: (i) the confounding of DIF and impact by the shortcomings of one particular technique called the Standardization Approach and/or (ii) by random guessing. These researchers were all associated with testing companies however [8-12]. An investigation by researchers not associated with a testing company found that the results persisted even when these effects were accounted for [13].

3. Statistical models related to DIF

In a standard dichotomous Rasch model,

$$\text{logit}\,(X_{pi} = 1 \mid \theta_p, \delta_i) = \theta_p - \delta_i\,, \qquad (1)$$

where X_{pi} is the response of person p to item i (which is assumed to be dichotomous, so that X_{pi} has two values 0 and 1), θ_p is the person location, and δ_i is the item difficulty for item i.

Now, assume there is a student covariate, C_p, which could be, for example, a student's gender or ethnicity. For simplicity's sake, assume that C is dichotomous—i.e., it takes the values of either 1 or 0 — hence, we can form the labeling function $c(p)$, which takes on the value 1 when p is in group C_p, and 0 otherwise. In the psychometric literature the group with $c(p) = 1$ are termed the *focal* group, and the group with $c(p) = 0$ are termed the *reference* group. The standard Rasch DIF model [14] is given by

$$\text{logit}\,(X_{pi} = 1 \mid \theta_p, \delta_i, \gamma_i) = \theta_p - \delta_i + c(p)\gamma_i\,, \qquad (2)$$

where γ_i is the DIF effect for item i. This model is not estimable unless we impose a constraint, for example, that $\gamma_0 = 0$. This constraint is often expressed in terms of the reference group—the difficulty parameters are seen to be those for the reference group, and the DIF parameter expresses how much the item difficulty for the focal group differs from this. This is referred to as having *uniform* DIF.

A linear relationship between item difficulty and the DIF effect, as postulated above, could then be expressed as:

$$\text{logit}\,(X_{pi} = 1 \mid \theta_p, \delta_i, \gamma_i) = \theta_p - \delta_i + \theta_p\, c(p)\gamma_i\,. \qquad (3)$$

Or, alternatively, using a function that is monotonically-related to person ability, such as the sufficient statistic for the person ability under the Rasch model, the person score[1], denoted as $t(p)$,

[1] Under an assumption of no missing data.

$$\text{logit}\,(X_{pi} = 1 \mid \theta_p, \delta_i, \gamma_i) = \theta_p - \delta_i + t(p)c(p)\gamma_i. \tag{4}$$

Models such as the one in equation 4, that go beyond the uniform DIF specification in equation 2, are referred to as having *non-uniform* DIF.

A non-linear (but functional) relationship could also be modeled using a slightly more complicated equation:

$$\text{logit}\,(X_{pi} = 1 \mid \theta_p, \delta_i, \gamma_i) = \theta_p - \delta_i + f(\theta_p)c(p)\gamma_i, \tag{5}$$

where $f(.)$ could take on various shapes such as quadratic, etc. where that was suggested by substantive theory.

Estimation of these models cannot be carried out by most of the standard Item Response Theory (IRT) software estimation packages due to the complexities of having a non-uniform DIF as a part of the model. However, because it incorporates the possibility of using a design matrix as part of its model specification, the *ConQuest* software [15] can be used to estimate the models in equations 4 and 5. More general statistical estimation techniques, such as those used in *gllamm*, WinBugs and STAN can also be used, though they will likely be slower, and each has their own complexities.

4. Examples

For example, recent work by Dulhunty [16], illustrated in figure 1, shows evidence for a systematic relationship between DIF related to indigenous status among Australian students and item difficulty in the national reading test for Grades 3 and 5 in 2017. These results are *expressed* in terms of item-wise DIF estimated using equation 2 with non-Indigenous students used as the reference group. The reported results were obtained by comparing an item calibration using only the reference group (i.e. non-Indigenous students) to a separate item calibration using only the focal group (i.e. Indigenous students). Each point on the plot refers to a specific test question. Negative y-axis values indicate that the item is relatively easier for Indigenous students (after adjusting for average ability). Positive values indicate Indigenous students find the item relatively difficult compared to non-Indigenous students. These figures display a similar pattern as was observed by Freedle for the comparison of African-American with White students in the US. The interpretation by Freedle could be applicable here also—Australian Indigenous students at the lower end of the ability spectrum find the items more difficult than do their equal-ability non-Indigenous students, while the effect reverses for higher ability Indigenous students.

When we look at the DIF effects for gender in the same test, similar results are also found for Grades 3 and 5 (see figure 2). In this case, consistent with the comments above, the interpretation would be that Australian Male students at the lower end of the ability spectrum find the items more difficult than do their equal-ability Female students, while the effect reverses for higher ability Male students. However, it would seem less likely that this could be due to a similar "cultural familiarity" effect as for the indigenous students, as the phenomenon of gender bias in educational tests has been known for many years, and the items in the NAPLAN test have been screened for this. Thus, there needs to be investigation of an alternative explanation in the case of Gender—a possibility would be that there was differential shrinkage of the item difficulties in the separate calibrations carried out by NAPLAN for each of the reference and focal groups. Note that this effect would not be present in analyses carried out using equation 2 above, but is a possible confounding effect when two separate calibrations are used (and this is one reason why equation 2 was used to define the DIF effect in the account above).

Now, it is important to examine instances of linear DIF effects as well as non-instances. One example is sown in figure 3, which plots DIF effects against item difficulties in the Spelling test for Indigenous versus non-Indigenous students in the same grades and year as in figures 1 and 2. Here we see an example of an absence of a linear DIF effect—there are indeed largish DIF effects for individual items (up to approximately 0.5 logit)—but the plots are all centered around a DIF effect of 0.0. In fact, most of the "DIF plots" in Dulhunty [16] were similar to those in figure 3.

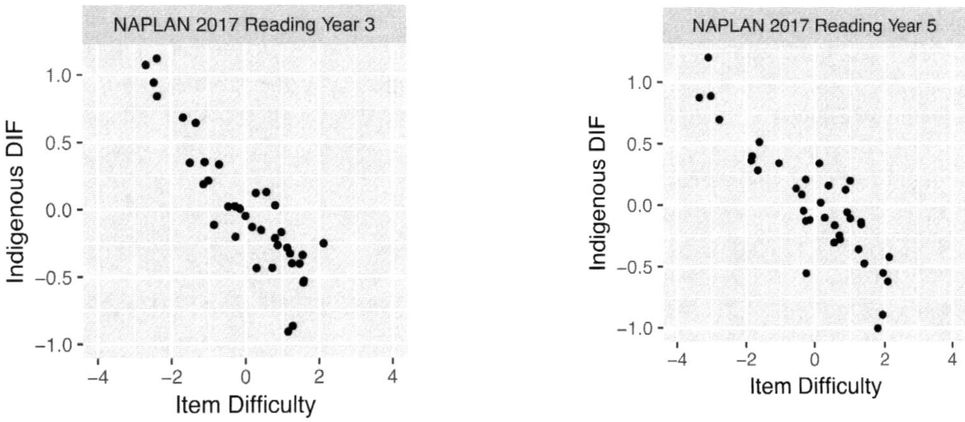

Figure 1. Plot of DIF effects with item difficulties when comparing Australian Indigenous and non-Indigenous students for Grades 3 and 5 Reading.

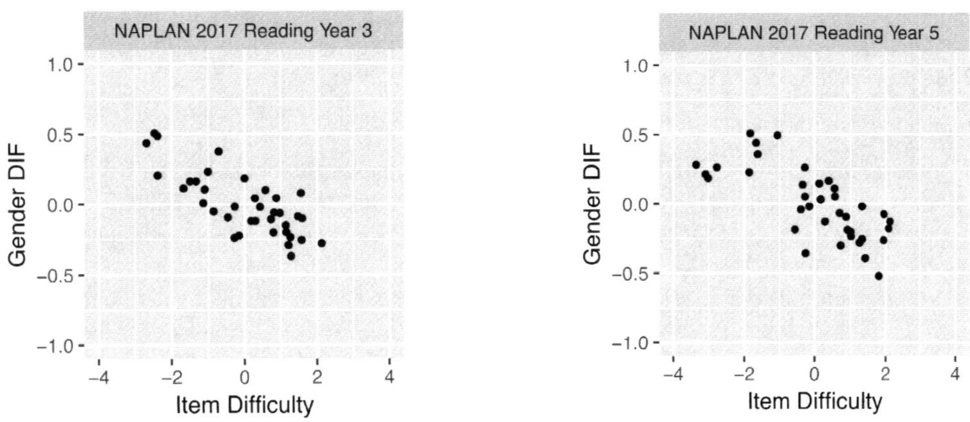

Figure 2. Plot of DIF effects with item difficulties when comparing Australian Male and Female students for Grades 3 and 5 Reading.

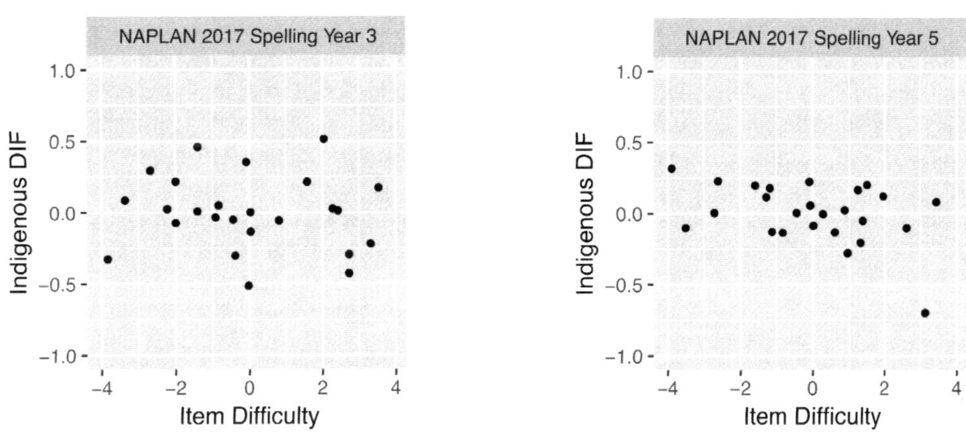

Figure 3. Plot of DIF effects with item difficulties when comparing Australian Indigenous and Non-indigenous students for Grades 3 and 5 Spelling.

5. Discussion

The analysis of publicly reported results for 240 NAPLAN test forms [16] shows DIF estimates can at times be found to be correlated with item difficulty. Furthermore, the strength of the relationship depends on the assessment domain for which DIF was examined and the population subgroups for which the DIF analyses were performed. Some relationships appear to be stable year after year despite the use of different test forms administered to different student cohorts. For example, there appears to be a linear relationship between Rasch estimated item difficulty and Indigenous DIF on the Reading assessments. The association persists across year levels (from Year 3 to Year 9) and has been observed annually from 2013 to 2017 (which are all the technical reports publicly available to date).

The results reported in [16] are based on the assumption inherent in the NAPLAN analyses that the metrics resulting from the two Rasch model calibrations (i.e., for the reference group and the focal group) are the same. Preliminary investigations suggest this assumption is not always valid: When calibrated separately the standard deviations of the item parameters are sometimes not the same. When this occurs we end up with the familiar pattern of one group finding easy items relatively hard and, at the same time, finding hard items relatively easy. This phenomenon can be considered a case of test equating when a 'shift and scale' adjustment is required to equate tests rather than just a 'shift' (e.g., [17]). This type of issue has been recognized as an issue for measurement in general [18]. Further investigation of the NAPLAN results (say, using equations 2 to 5) are needed to eliminate the possibility that some observed linear DIF effects may be the result of a scaling artifact.

6. Conclusion

The phenomenon of item-wise DIF is a common occurrence in educational testing: it mainly constitutes a headache for test developers, who typically develop extra items beyond those that are expected to be needed specifically to account for loss of items due to DIF. As mentioned above, although this practice does indeed result in a set of items that are not subject to DIF effects for the groupings that have been tested, the elimination of particular types of items could be a source of bias in the item pool. Fitting models such as those in equations 3 to 5 can allow the possibility that items with these specific DIF patterns can be maintained within the item pool, and thus avoid such threats to validity.

The accumulation of results such as these leads one to seek more robust theoretical understanding of the phenomenon. The models in equations 4 and 5 allow an efficient and unbiased estimation of the underlying relationship (when the assumptions are correct), and also the estimation of bias-corrected measurements of students in both groups. Each time this phenomenon is found, it potentially contributes to a deeper understanding of the complications of testing across different sub-populations.

Further work along these lines is needed. Of course, one important step is to go beyond the examination of reported DIF effects as described above, to the direct estimation of linear DIF effects utilizing equations 3 to 5. This will require a careful re-analysis of the data at the item-level, and, as mentioned above, the investigation of appropriate estimation software for that purpose (and investigation for any scaling artifacts using, say, simulation studies). Beyond that, interpretation of what it would mean if linear DIF effects were indeed confirmed, then further investigations into the possible causes for such effects would also be needed—these could include (a) deep dives into the relevant literature (which would vary, of course, depending on the reference/focal group pair), and (b) simulation studies to eliminate possible statistical artifacts, as well as (c) qualitative investigation of student response patterns.

References

[1] Lord F M 1980 *Applications of Item Response Theory to Practical Testing Problems* (Hillsdale, New Jersey: Lawrence Erlbaum Associates)

[2] Hambleton R, Swaminathan H and Rogers J 1991 *Fundamentals of Item Response Theory* (Thousand Oaks, California: Sage Publications)

[3] Thissen D, Steinberg L and Wainer H 1988 Use of item response theory in the study of group differences in trace lines *Test Validity* ed H Wainer and H Braun (Hillsdale, New Jersey: Lawrence Erlbaum Associates)
[4] Freedle R 2003 Correcting the SAT's ethnic and social-class bias: A method for re-estimating SAT Scores *Harvard Educ. Rev.* **73** 1–43
[5] Kulick E and Hu P G 1989 *Examining the Relationship Between Differential Item Functioning and Item Difficulty (College Board Report No. 89-5; ETS RR-89-18)* (New York: College Entrance Examination Board)
[6] Scherbaum C and Goldstein H 2008 Examining the relationship between race-based differential item functioning and item difficulty *Educ. Psychol. Meas.* **68** 537– 553
[7] Santelices M V and Wilson M 2010 Unfair treatment? The case of Freedle, the SAT and the standardization approach to differential item functioning *Harvard Educ. Rev.* **80** 106– 133
[8] Dorans N 2004 Freedle's Table 2: Fact or fiction *Harvard Educ. Rev.* **74** 62–73
[9] Dorans N and Zeller K 2004 *Examining Freedle's Claims and His Proposed Solution: Dated Data, Inappropriate Measurement, and Incorrect and Unfair Scoring (Educational Testing Service Report No. RR-04-26)* (Princeton, New Jersey: Educational Testing Service)
[10] Bridgeman B and Burton N 2005 Does scoring only the hard questions on the SAT make it fairer? *The Annual Meeting of the American Educational Research Association (Montreal)*
[11] Wainer H 2009 Ethnic bias or statistical artifact? Freedle´s folly *Picturing the Uncertain World. How to Understand, Communicate and Control Uncertainty through Graphical Display* ed H Wainer (New Jersey: Princeton University Press)
[12] Wainer H and Skorupski W 2005 Was it ethnic and social-class bias or statistical artifact? Logical and empirical evidence against Freedle´s method for reestimating SAT scores *Chance* **18** 17–24
[13] Santelices M V and Wilson M 2012 On the relationship between differential item functioning and item difficulty: An issue of methods? Item response theory approach to differential item functioning. *Educ. Psychol. Meas.* **72** 5–36
[14] Meulders M and Xie Y 2004 Person-by-item predictors. *Explanatory Item Response Models. A Generalized Linear and Nonlinear Approach* ed P De Boeck and M Wilson (New York: Springer)
[15] Wu M, Adams R J and Wilson M 1998 *ACER-ConQuest* (Hawthorn, Victoria, Australia: ACER Press)
[16] Dulhunty M 2019 *Item Difficulty Association with Reported Gender, Indigenous and LBOTE DIF in NAPLAN 2013 to 2017* (Melbourne, Australia: MGSE, University of Melbourne)
[17] Humphry S M 2011 The role of the unit in physics and psychometrics *Measurement* **9** 1–24
[18] Wu M, Hak P T and Tsung-Hau J 2016 *Educational Measurement for Applied Researchers: Theory into Practice* (Singapore: Springer)

Journal of Physics: Conference Series **1379** (2019) 012027 doi:10.1088/1742-6596/1379/1/012027

Upper limbs biomechanics: drinking evaluation protocol in hemiparetic patients

Francesco Crenna[1], Camilla Bergamino[1], Chiara Ponte[2], Marina Simonini[3] and Giovanni B Rossi[1]

[1] Università degli Studi Genova – DIME/MEC, Laboratorio di Misure, Genova, Italy
[2] Università degli Studi Genova – DINOGMI, Genova, Italy
[3] SC Recupero e Rieducazione Funzionale Osp. La Colletta Arenzano, ITALY

E-mail: francesco.crenna @unige.it

Abstract. Biomechanical measurements applied to upper limbs gesture, characterisation and clinical evaluation are less diffused than lower limbs evaluation where gait analysis is rather common. When dealing with upper limbs a set of critical problems have to be afforded. In walking analysis we know that the movement is cycled and repeatable, instead upper limb gestures are more complex principally because of lack of the periodicity. In this paper we present a protocol to face such problems and obtain an objective biomechanics evaluation of the upper limbs, with the particularity to be focused on a daily life gesture such as drinking. The experimental protocol is presented together with the parameter identified to characterise the movement. Some preliminary results are discussed with reference to a set of healthy and clinical subjects.

1. Introduction

Stroke is the most common neurological disease in the adult population worldwide [1]. Since often brain area affected by stroke is the motor control section, one of the most common consequences is hemiparesis, defined as muscular weakness or partial paralysis restricted to one side of the body. For this reason, beside standing and walking, also grasp and manipulation movements are critical in stroke patients. However, the ability to reach, grasp, transport and release objects is essential for daily activities, such as feeding and grooming. Rehabilitation protocols are applied as soon as possible after the injury to maximise recover possibilities.

As regards the evaluation of patients' biomechanics before, during or at the end of rehabilitation protocol, some biomechanical evaluation protocols are available for gait and walking analysis [2], and a few are available for upper limbs [3]. Beside that while gait and walking are natural daily life gestures, in general the evaluation of upper limb biomechanics is based on reference gestures such as pointing, grasping or hand to mouth [4]. Such movements are not the usual ones, they just mimic common daily life activities, so the patient may be less motivated and the evaluation is not directly related to subject autonomous capabilities in daily life [5].

The focus of this work is the design and test of a three-dimensional upper limbs biomechanical assessment system to quantify the injury and the progress during stroke rehabilitation. The reference gesture we have selected is based on a Reach to Grasp Cycle, but it is modified by also considering the drinking gesture. This motion sequence is very important in daily activities and incorporates all major

Journal of Physics: Conference Series **1379** (2019) 012027 doi:10.1088/1742-6596/1379/1/012027

joints off the upper limb and simulates a functional task that is feasible, yet challenging enough to reveal key motor deficits in individuals with movement disorders [6]. The experimental protocol includes kinematics measurements by a three dimensional infrared video system, referred to a biomechanical model. The protocol includes the possibility to introduce surface electromyography measurements to evaluate muscular activations of the subject. Gesture performance can be evaluated both trough time histories and spatial or temporal features. Moreover, an interesting approach for grasp evaluation is proposed: the hand area time history. It can give information on hand opening - closing subject's capabilities [7]. The proposed protocol has been tested on a group of healthy subjects in order to construct a healthy reference range, useful to evaluate patients' results.

2. Experimental protocol

2.1. The reference cycle

The subject is initially asked to sit comfortably on a chair placed in front of a table, as in figure 1. Hips and knees are flexed 90° and both feet lay flat on the ground. Both arms are at rest on the table so that the shoulder is in a neutral position, elbows are flexed, forearms are pronated, and wrist is held in a neutral position, with hand palms flat on the table. A water bottle (half a litre, only 1/3 filled) is placed on the table in front of the subject at 80% of the maximum reach distance.

Figure 1. Initial position of the reference cycle.

The cycle starts with the subject moving one limb to reach the bottle, grasp and bring it to mouth to simulate drinking; then back to table to release bottle in the same point of table and return to the initial position. Soon after the same cycle is repeated with the other limb. The healthy subjects were instructed to start with the dominant limb while patients start with the healthy limb.

The task is paradigmatic since it includes two relevant phases: a first open-chain movement (reaching for the bottle) and a second movement (bringing the glass to the mouth) where a strict coordination among upper limb, trunk and head movements is required, thus challenging the motor control system even in presence of upper limb minimal disability.

2.2. Biomechanical Model

We have designed a new biomechanical model, based upon Rab's upper limb protocol [8], with some modifications due to the specificity of the task we are going to study. It requires 23 retroreflective skin markers: 4 on the head, 3 on the trunk, 4 for each forearm, 4 for each hand. Two additional markers are positioned on bottle top and bottom.

A procedure was defined in order to guarantee proper and reproducible marker positioning. A graphical representation of the biomechanical model is presented in figure 2. The red marker is positioned on subject's chin in order to evaluate the accuracy of the drinking gesture, while on the hand there are markers on thumb, index and little finger nails to evaluate the hand's opening and closing movements.

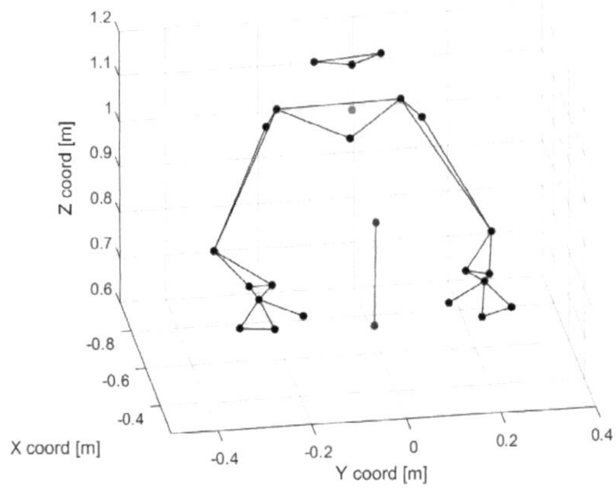

Figure 2. Biomechanical upper limbs model.

2.3. Participants

We experimented the protocol on 10 healthy subjects and 8 hemiparetic patients, of which two with right and six with left hemiparesis, at different injury levels, but always capable to conduct the test. All of them were informed and agreed to participate to the experimentation. Initially the protocol was checked on healthy subjects as a control group following a dual purpose: on one hand to have data on healthy subjects on the other to validate the protocol and the parameters set obtained. This preliminary data set was useful to identify movement strategies and associated joint angles histories to create a normality band.

2.4. Acquisition sequence

As already mentioned, the two limbs perform the gesture alternatively. Moreover, in order to evaluate subject's repeatability, a set of five consecutive repetitions is required. According to subject's availability and fatigue state, another set of sequences might be acquired, up to an overall of 5 sets. A short pause of approximately 2 minutes is planned between each set trial. The complete recording session lasts about half an hour for each subject. This is not a difficult task for a healthy subject but it may be complex for an injured one, according to the stroke consequences importance.

Table 1. Dirking cycle characterisation parameters.

Parameter		Unit	Description
Temporal	Cycle time	s	Cycle duration in seconds
	Mean velocity	m/s	Cycle mean velocity
Spatial	Trajectory length	m	Length of the curve tracked on the index finger
	ROM	deg	Range of Motion of involved articulations, for each degree of freedom
	Compensation	deg	Degrees of compensation movements of head and trunk
Spatial - temporal	Grasp adjustment	Cycle fraction Δ%	Fraction spent in adjusting the hand approaching the bottle
	Target adjustment	Cycle fraction Δ%	Fraction spent in adjusting the bottle on the target (mouth)
	Grasp	Cycle moment %	Cycle instant at which the bottle is grasped
	On target	Cycle moment %	Cycle instant at which the bottle is properly on the target
Hand	Max aperture	%	Ratio of maximum hand area on hand area at rest
	Carrying	%	Ratio of hand area during carrying to mouth on hand area at rest

2.5. Data processing

Three-dimensional kinematic data are acquired by a motion capture optoelectronic system, BTS Bioengineering GaitLab System, able to automatically record 3D trajectories of passive markers by means of stereo photogrammetric methods using 6 IR cameras. Marker's positions measurements and their tracking is carried out in BTS environment with manual correction of eventual error or missing. Kinematic data enables the biomechanical analysis, that starts with the identification of a standard cycle, in order to be able to compare gestured carried out with different speeds, for example by healthy and injured subjects.

The velocity of the index finger marker defines the beginning and the end of the drinking cycle. The onset of movement from the resting start position is identified as the first instant when the normalized velocity of the index finger marker exceeds 5% of peak reaching velocity [9], while the end of the cycle is signified by a decrease in the same marker velocity to less than 5% of the maximum velocity upon returning to the initial position.

The biomechanical model consists of 11 segments (head, trunk, right and left upper arm, right and left lower arm, right and left hand, right and left fingers and bottle) whose local coordinate systems are used to calculate upper limbs motion [10]. A standard biomechanical analysis is applied to each segment obtaining Euler angles useful to determine the three-dimensional joint angles. Articular angles are used to express the joint angles of the proximal segment respect to the distal segment, following the biomechanical usual convention for rotation succession: flexion/extension, adduction/abduction and axial rotation. The segments of the head, the trunk and the bottle are described with references to the lab coordinate system. The shoulder joint is modelled as a ball and socket joint with three-degrees-of-freedom, located in the centre of the humeral head. Movement is calculated between the trunk and the humerus. The elbow joint is modelled as a rotating-hinge with two-degrees-of-freedom like the wrist, but this last one is modelled as a universal (saddle) joint. Moreover we have also defined a fingers' joint as a hinge joint in order to calculate fingers' flexion/extension movements.

A set of time or cycle histories is available for detail studies, while a set of overall spatial and temporal parameter is computed to obtain a simpler characterisation of the gesture and easier comparison between subjects. The parameters available at present moment are presented in table 1.

2.6. Hand's surface area analysis

Among gesture parameters an interesting and rather new one is based on the idea that hand opening could be described by the surface area [7] defined by interconnecting the markers positioned on subject's hand. In particular, a quadrilateral is obtained by the sum of two triangles with a common vertex, as shown in figure 3. Since the value of this parameter clearly depends on the hand size, the trajectories have been normalised to the subject's hand size, obtained as the initial hand surface area in resting position.

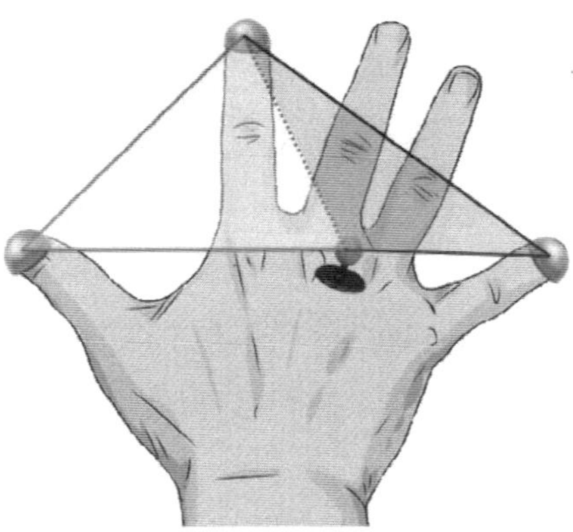

Figure 3. Hand's surface area definition. A quadrilateral is obtained by the sum of the yellow triangle - defined by thumb, index and hand markers - and the light blue one -defined by index finger, little finger and hand markers.

3. Results

In Section 2.4 we have presented the overall set of parameters that can be obtained from the test. As preliminary examples, we focus on two of them: elbow's range of motion and hand's area variability during the cycle task as shown in figure 4. These two cycle histories are useful to understand where the movement is compromised. Through elbow's joint analysis, figure 4a, it is evident that both left and right limbs are over-flexed during task cycle. This could suggest problems in proximal side of the upper limb, possibly in the shoulder, that cannot flex and support the arm in the gesture in a physiological way. On the other hand, in figure 4b, it's possible to analyze the hand area of another patient during a cycle. In this case also both sides present difficulties in managing hand aperture, but in particular the injured left side presents not only low area levels but also a different evolution during the cycle, suggesting distal problems in the limb.

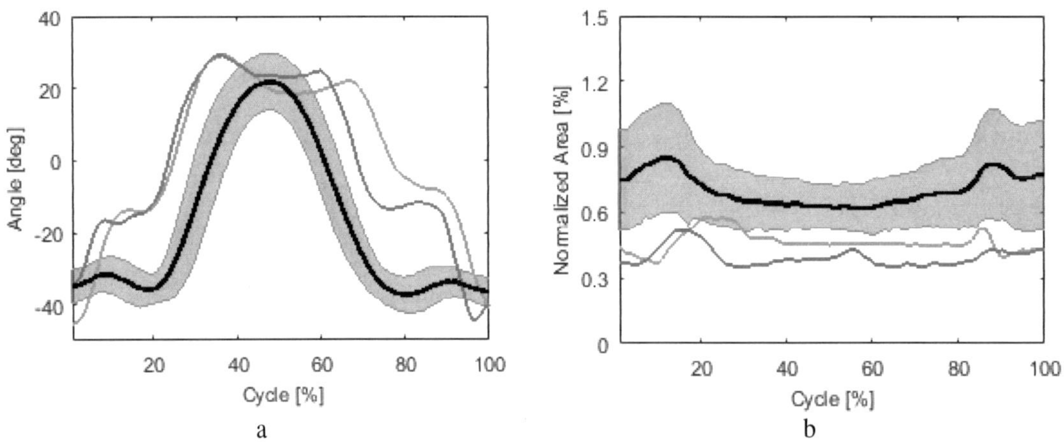

Figure 4. Elbow range of motion (a) and hand's area (b) during the task cycle. Patient's right arm (green) and injured left arm (red). Normality range (grey) and mean (black).

In addition to the detailed cycle analysis, as presented before, it is possible to evaluate cycle overall parameters. Let's consider as an example spatial temporal parameters expressed as a fraction of the cycle, as shown in figure 5. As before, these parameters allow to define if the stroke compromise mostly proximal or distal segment and quantify the quality of gesture of drinking. Yet this is not an exhaustive set since new parameters con be imagined as the experimentation proceeds.

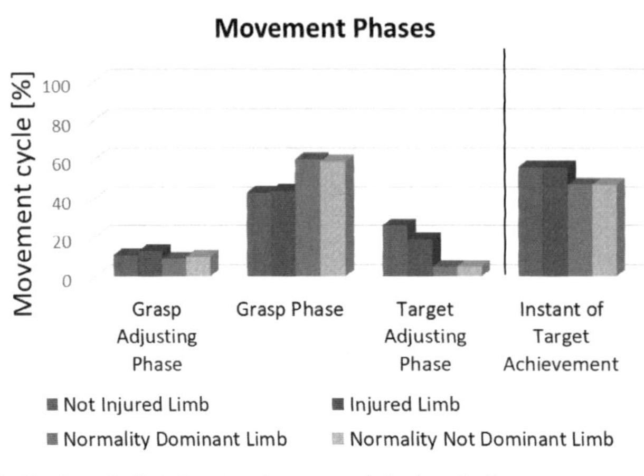

Figure 5. Cycle subdivision in phases and their relative percentage. Healthy arm (green), injured arm (red) of an injured subject and normality (grey).

4. Conclusions

A novel approach to upper limbs biomechanical objective evaluation is proposed. The method is promising for the evaluation of patients with after stroke injured upper limb or with upper limbs orthopaedic problems. The protocol is based on a sequence of tasks that incorporates all major joints of the upper limb and simulates a functional task that is feasible yet challenging enough to reveal key motor deficits in individuals with movement disorders. We have chosen this daily life gesture to obtain as much motivation as possible from the subject and to focus the rehabilitation on the recovery of subject's daily life autonomy. The information available from the proposed biomechanical model applied to the experimental protocol is huge and a set of possible evaluation parameters is presented in the paper. Preliminary results on healthy subjects and patients are promising and enable the localisation of the biomechanical problem is specific regions, where the rehabilitation will be focused. Application to a larger set of subjects and patients is going on in collaboration with the rehabilitation unit at La Colletta Hospital. It would be interesting to use this method in order to support upper limb treatment, e.g. botulinum toxin injections, constraint-induced movement therapy, and corrective surgery. The drinking cycle can offer effective and standardized quantitative evaluation pre- and post-intervention. In addition, these quantitative measures of upper limb functions may help delineate diagnoses of different movement disorders, and thus, may be important in determining structure–function relationships between brain and motor abnormalities.

References

[1] Mackay J and Mensah G 2004 *The Atlas of Heart Disease and Stroke, Part Three: Global Burden of Stroke* World Health Organization, WHO
[2] Datta Gupta A, Tucker G, Koblar S, Visvanathan R and Cameron I D 2018 Spatiotemporal gait analysis and lower limb functioning in foot dystonia treated with botulinum *Toxins* **10**(12) 532
[3] Rau G, Disselhorst-Klug C and Schmidt R 2002 Movement biomechanics goes upwards: from the leg to the arm *J. Biomech.* **24** pp 969-977
[4] Merdler T, Liebermann DG, Levin MF and Berman S 2013 Arm-plane representation of shoulder compensation during pointing movements in patients with stroke. *J. Electromyogr. Kinesiol.* **23**(4) pp 938-47
[5] Petuskey K, Bagley A, Abdala E, James M A and Rab G 2007 Upper extremity kinematics during functional activities: three-dimensional studies in a normal pediatric population *Gait & Posture* **25**(4) pp 573-579
[6] Butler E E, Ladd A L, Louie S A and LaMont L A 2010 Three-dimensional kinematics of the upper limb during a Reach and Grasp Cycle for children *Gait & Posture* **32**(1) pp 72-77
[7] Supuk T, Bajd T and Kurillo G 2011 Assessment of reach-to-grasp trajectories toward stationary objects *Clin. Biomech.* **26**(8) pp 811-818
[8] Rab G, Petuskey K and Bagley A 2002 A method for determination of upper extremity kinematics *Gait & Posture* **15**(2) pp 113-9.
[9] Coluccini M, Maini E S, Martelloni C, Sgandurra G and Cioni G 2007 Kinematic characterization of functional reach to grasp in normal and in motor disabled children *Gait & Posture* **25**(4) pp 493-501
[10] Hingtgen B, McGuire J R, Wang M and Harris G F 2006 An upper extremity Kinematic model for evaluation of hemiparetic stroke *J. Biomech.* **39**(4) pp 681-688

Journal of Physics: Conference Series **1379** (2019) 012028 doi:10.1088/1742-6596/1379/1/012028

Squat biomechanics in weightlifting: foot attitude effects

Francesco Crenna and Giovanni Battista Rossi

Università degli Studi Genova –DIME/MEC, Laboratorio di Misure, Genova, Italy

E-mail: gb.rossi@dimec.unige.it.

Abstract. Squat weight lift is a common exercise in physical training used both in weightlifters specific training and in general athletes for leg force increment. When weight load on subject shoulders is rather heavy as compared with full body weight, gesture biomechanics becomes critical, since injuries may occur when gesture is not controlled or performed in a wrong way. Experienced trainers are aware of such issues and give proper instructions to obtain best performance at a minimum injury risk. In this work a quantitative approach is presented, supported by biomechanical measurements to determine the effects on main articulations of different postures during the exercise. Preliminary results confirm good practices giving a quantitative indication of joints solicitations.

1. Introduction

Squat weight lift is a widely diffused practice in force training programs applied to several sports such as athletics soccer, rugby and, specifically, in Olympic weightlifting. The focus is on upper leg and low back muscle conditioning [1-3]. Specific literature is available in sport related journals, but it is not common to find a quantitative movement description [4-6, 7]. A brief qualitative gesture overview is given hereafter.

Barbell loaded squat movement can be executed with barbell in front or back positions. In the following we are going to consider the most common execution, the high back load squat with barbell positioned upon shoulders, over the acromyon, as showed in Figure 1. The movement starts with the subject in a static upright position. A continuous leg compression, including hip, knee and ankle movements, brings the body in the final position as deep as desired. A rising movement follows with the subject returning in upright position. Different leg compressions are possible, usually distinguished according to the amount of knee flexion, denoted by partial, half or complete squat. In this communication, complete squat with a knee flexion above 100° is considered. Weight lift purpose in training is to develop muscular force and in the literature papers are available that propose the use of EMG to describe muscle activation as a function of gesture parameters such as squat depth. As regards rehabilitation some references focus on the effect on overall squat movement, of low efficiency or low force developed by some specific muscles involved in the movement, or articular mobility problems mainly in the ankle.

During squat weightlift exercise, injuries are possible. Muscular injuries are mainly due to overloads, improper warm up or sudden unbalance. On the other hand, a wrong or un-correct execution of the exercise might procure, during time, injuries involving articulations [8]. These problems are mainly due to the improper loads on the articulations procuring mechanical effects on the compliant elements such as ligaments and cartilages. Papers are available regarding muscle activation studies by surface electromyography in various conditions [9]. On the other hand, as far as we know, a quantitative approach considering movement deficiencies and improper gesture execution effects on articulations stresses, based upon a set of biomechanical measurements and models is not available in literature, and here we propose a first approach with some preliminary results [10-13]. In particular, we are interested in investigating the effect on articular moments of foot anterior-posterior attitude during the gesture. In fact, it is possible to maintain the load in the middle of the foot sole, or force a frontal loading moving the force toward foot tip, or toward the heel moving the load on the back. Correct gesture procedure requires a median loading, as recommended by athletics trainers, but due to heavy loads and little unbalances it is rather easy to move the load under the foot sole, with consequences on articular load. Besides that, it is interesting to note that specific weightlifting shoes are available [14, 15]. They present a rigid sole to increase foot stability and a few centimeters wedge so that the heel is just a little higher than the foot tip [16]. This seems to be in contrast with the recommended good practice, since the wedge implies the tendency to lead the foot in front, so shoes effect on articular load is worth to be investigated. In the following we will define the experimental protocol, introduce the experimental setup, and final we will present and discuss some preliminary results.

2. Methods

2.1. Experimental protocol

As mentioned in the Introduction, our goal is to understand the effects of foot attitude on joints internal forces, during gesture execution. A protocol has been thus designed, to guarantee repeatability evaluation and reproducibility of the experiment. The test sequence was defined according to table 1, considering the different factors under investigations and controlling subject fatigue during the experiment. A preliminary static measurement is necessary to establish marker placement positions on the subject. Then subject performs the sequence of weightlifting according to table 1, introducing pauses as required.

Table 1. Experimental protocol.

		Test #		Test #		Test #		Test #
Load weight [kg]	8		44		8		44	
Tech. Shoes	NO		NO		YES		YES	
Foot Attitude	Front	1	Middle	7	Middle	13	Middle	19
	Rear	2	Rear	8	Rear	14	Rear	20
	Middle	3	Front	9	Front	15	Front	21
	Front	4	Middle	10	Middle	16		
	Rear	5	Rear	11	Rear	17		
	Middle	6	Front	12	Front	18		

2.2. Biomechanical model

A planar biomechanical model has been considered, since we are mainly interested in studying the gesture in the sagittal plane, with a special focus on the anterior posterior direction. In fact, the gesture is almost planar and such approximation is in general good to describe gesture biomechanics unless medio-lateral asymmetries have to be investigated. The model we are considering is schematised in figure 1 with the most important anatomical landmarks describing the movement reported in the table. Eight active markers have been placed on the subjects in such anatomical points. Markers misalignment from the reference sagittal plane is small and during the gesture they move very little out of the plane – less than $\pm$ 30 mm. Markers are placed on the subject only since we are interested in subject's lower limbs biomechanics, and we are considering rear barbell squat only so during the gesture barbell does not move in relation to subject's arms. Some model parameters can be measured directly on the subjects such as anthropometric segments lengths, while others such as mass anthropometry, centres of mass positions and inertia are derived from standard anthropometric tables such as those reported in [17]. The biomechanical model enables inverse kinetics analysis to determine internal forces and moments at the articulation joints, as it will be presented in the next subsection.

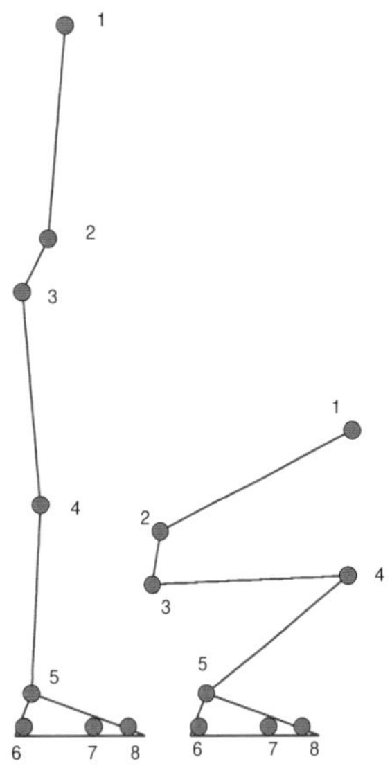

Marker	Position
1	acromion
2	iliac crest
3	trochanter
4	knee
5	ankle
6	heel
7	metatarsus
8	foot tip

Figure 1. Squat weightlift biomechanical model and markers positions.

2.3. Measurement system

Kinematic measurements are carried out with a video system based on a Dalsa Falcon camera – 1400x1024, 100Hz frame rate, Camera Link connection -. The video system is properly aligned and calibrated with a reference length (1,2 m - 0,1 m steps) placed on the measurement plane the to obtain the sensitivity constant (about 1,4 mm/pixel) and minimise prospective compensations and residual distortions. In these conditions markers movement out of the plane implies a deviation of less than ± 2% in the sensitivity. Acquisition and processing MatLab® code gives as output markers coordinates in the sagittal plane in time. Kinetic measurement are based on two force platforms - P6000 BTS Bioengineering, 600 x 400 mm, 2kN full scale, 1kHz sampling rate -. Data acquisition is managed separately for kinematic and kinetics systems, and they are hardware synchronised.

3. Results

Two subjects – male, 22 years old– underwent the protocol presented in par. 2.1. In order to be able to compare different gesture executions by the same or different subjects it is necessary to introduce a standard gesture cycle [18]. This is for some gestures straightforward, such as in the case of walking or running, for example considering successive heel contacts on the ground. In weightlifting it is a little bit tricky since there is no specific event at execution start or stop. A standard cycle has been defined according to hip position – marker 3 in figure 1: by setting a proper threshold it is possible to identify the initial downward movement and gesture conclusion when it returns in the initial position. Once identified the time instants gesture evolution is normalised on a 0-100 % squat cycle, so executions different speed can be compared. Note that we experienced limited cycle time duration in the gesture, so we consider an analysis based on the reference cycle suitable for our purposes. In the following results will refer to such normalised execution cycle.

3.1. Kinematics

Kinematic measurements give as output marker's positions during the gesture. A first result is presented in Figure 2 where kinematics results depending on foot attitudes are compared. Heel and knee anterior-posterior markers positions are presented for a squat cycle. When attitude is placed on foot anterior part, knee's position goes in the forward direction toward foot-tip. On the contrary, when attitude is on the rear of the foot, knee remain on the back also and its position is always behind the foot-tip. This experimental evidence confirms the proper execution of the gesture as regards subject's control of foot attitude as requested by the test protocol.

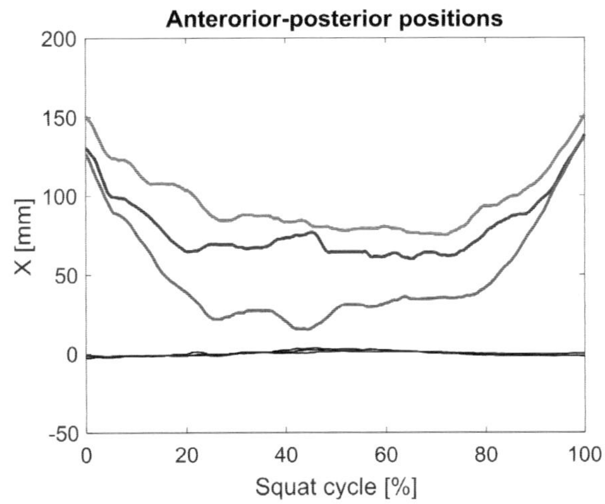

Figure 2. Heel and knee anterior-posterior positions: red front, blue middle and green rear foot attitudes.

The planar model described in 2.2 identifies articular, or relative, angles at ankle, knee and hip. Figure 3 presents an example of angular motion at the ankle for executions with and without technical shoes. Generally, graphs suggest that shoes increase the ankle range of motion, probably due to the wedge inclination under foot sole.

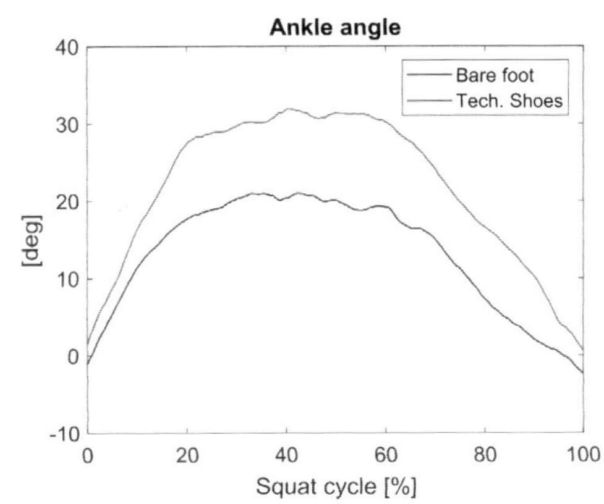

Figure 3. Angles at the ankle during executions with and without technical shoes.

3.2. Kinetics

Figure 4 presents a typical Ground Reaction Force (GRF) time history measured during an execution.

As already mentioned, gesture dynamics is subjective and for this reason a normalisation is necessary. Nevertheless, dynamics affects the forces exchanged with the ground of course. When dealing with a rather slow execution vertical GRF can be considered constant, neglecting continuous little movements due to subject asset control. In figure 4 a rather fast execution is presented in which at about the middle of the flexion-extension cycle (50 % on the graph) there is a clear peak, about 200 N higher than the mean. This is typical in a fast dynamic weightlift movement at maximum flexion, where we have the contribution of a fast deceleration-acceleration in the dynamics and the sudden change in muscle behaviour from extension to contraction with the contribution of muscles elasticity – *myotatic reflex*.

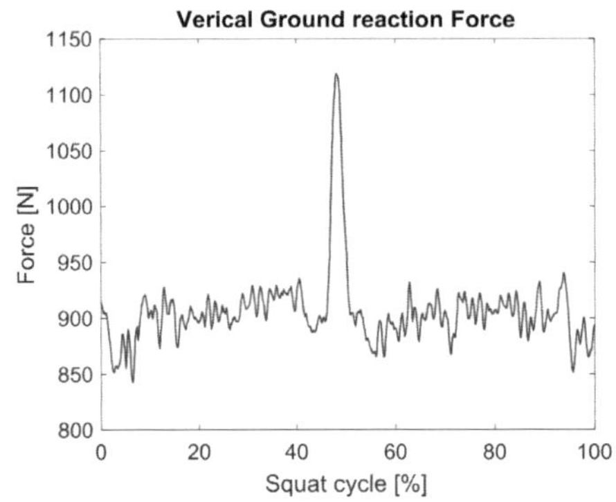

Figure 4. Vertical GRF during a fast dynamic movement cycle

3.3. Internal moments and foot attitude

Internal moment at articular joints due to muscle action can be measured by inverse kinetic analysis, based on the biomechanical model. Dynamic equations define the behaviour of each model segment and starting from the distal one, the foot, where the external force is exchanged with the ground, it is possible to obtain articular moments. In figure 5 we consider two examples to compare results with and without technical shoes and with different foot attitude.

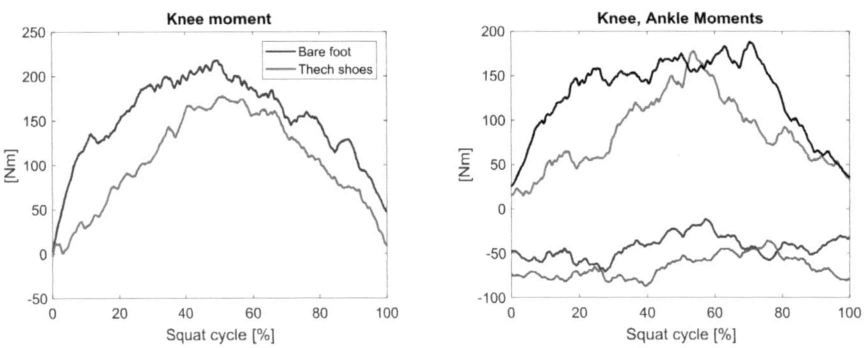

(a) (b)

Figure 5. Example of moments at knee and ankle in different conditions: (a) - knee moment with and without technical shoes; (b) - ankle moment: rear (blue) and front (green) attitudes; knee moment rear (red) and front (black) attitudes.

4. Conclusions

A quantitative biomechanical analysis of squat weightlifting has been carried out to investigate the effect of foot attitude and technical shoes on weight lift kinematics and articular moments. A protocol has been developed to investigate the influence of technical shoes and foot attitude. A standard gesture cycle has been defined to compare different executions by different subjects or with different dynamics. Preliminary results quantitatively indicate that foot attitude on the front of the foot implies a higher articulation load as compared with a balanced middle sole attitude, as it was expected. Technical shoes with a low height wedge under foot sole mainly stabilise the movement and produce a slight load reduction on knee and ankle. Further work has to be done to analyse the overall set of data to verify results repeatability and statistical significance of the effects we have investigated. An increase in subjects' number will give a more robust set of experimental evidences.

References

[1] Chandler J T and Stone M H 1991 The squat exercise in athletic conditioning: a review of the literature, *Natl. Strength Condit. Assoc. J.* **3** (5) pp 51– 60

[2] Wretenberg P, Feng Y and Arborelius U P 1996 High-and low- bar squatting techniques during weight-training. *Med. Sci. Sports Exerc.* **28** pp 218 –24

[3] Mc Caw S T and Melrose D R 1999 Stance width and bar load effects on leg muscle activity during the parallel squat. *Med. Sci. Sports Exerc.* **31** pp 428 –36

[4] Shoenfeld B J 2010 Squatting Kinematics and kinetics and their application to exercise performance *J. Strength Cond. Res.* **24** (12) pp 3497-506

[5] Dioniso V C, Almeida G L, Duarte M and Hirata R P 2008 Kinematic, kinetic and EMG patterns during downward squatting. *J. Electromyogr. Kines.* **18** pp 134–43

[6] Escamilla R F, Fleisig G S, Lowry T M, Barrentine S W and Andrews J R 2001 A three-dimensional biomechanical analysis of the squat during varying stance widths. *Med. Sci. Sports Exerc.* **33** pp 984–98

[7] Braidot A A, Brusa M H, Lestussi F E and Parera G P 2007 Biomechanics of front and back squat exercises, *J. Phys.: Conf. Ser.* **90**(1) 012009

[8] Senter C and Hame S L 2006 Biomechanical analysis of tibial torque and knee flexion angle: Implications for understanding knee injury. *Sports Med.* **36** pp 635–41

[9] Donnelly D V, Berg W P and Fiske D M 2006 The effect of the direction of gaze on the kinematics of the squat exercise. *J. Strength Cond. Res.* **20** pp 145–50

[10] Signorile J F, Kwiatkowksi K, Caruso J F and Robertson B 1995 Effect of foot position on the electromyographical activity of the superficial quadriceps muscles during the parallel squat and knee extension *J. Strength Cond. Res.* **9** pp 182–7

[11] Hung Y J and Gross M T 1999 Effect of foot position on electromyo-graphic activity of the vastus medialis oblique and vastus lateralis during lower-extremity weight-bearing activities *J. Orthop. Sports Phys. Ther.* **29** pp 93–102

[12] Caterisano A, Moss R F, Pellinger T K *et al.* 2002 The effect of back squat depth on the EMG activity of 4 superficial hip and thigh muscles *J. Strength Cond. Res.* **16** pp 428–32

[13] Dooley E, Carr J, Carson E, and Russell S 2019 The effects of knee support on the sagittal lower-body joint kinematics and kinetics of deep squats *J. Biomech.* **82** pp 164-70

[14] Southwell D J, Petersen S A, Beach T A C and Graham R B 2016 The effects of squatting footwear on three-dimensional lower limb and spine kinetics, *J. Electromyogr. Kines.* **31** pp 111-8

[15] Legg H S *et al.* 2017 The effect of weightlifting shoes on the kinetics and kinematics of the back squat *J. Sports Sci.*, **35**(5) pp 508-15

[16] Fortenbaugh D, Sato K and Hitt J K 2010 The effect of weightlifting shoes on squat kinematics *Proc. 28 Int. Conf. on Biomechanics in Sports (Marquette, Michigan, USA, July 19 – 23, 2010)*

[17] Winter D A. 2009 *Biomechanics and Motor Control of Human Movement* 3rd ed., Wiley

[18] Stevens W R, Kokoszka A Y, Anderson A M, and Tulchin-Francis K 2018 Automated event detection algorithm for two squatting protocols *Gait & Posture* **59** pp 253-7

Towards patient-centred cognition metrics

J Melin[1], LR Pendrill[1], S J Cano[2], and the EMPIR NeuroMET 15HLT04 consortium[3]

[1] RI.SE, Eklandagatan 86, S-412 61, Göteborg, Sweden, jeanette.melin@ri.se, leslie.pendrill@ri.se
[2] Modus Outcomes, Spirella Building, Letchworth Garden City, SG6 4ET, UK, stefan.cano@modusoutcomes.com
[3] NeuroMET coordinator Milena Quaglia, LGC, Milena.Quaglia@lgcgroup.com

E-mail: jeanette.melin@ri.se

Abstract. Metrological quality assurance is essential if reliable decisions about diagnosis, treatment and rehabilitation are to be made consistently throughout the healthcare system. However, many observations in healthcare are considered to lie 'off the scale' of quantitative measurement. We have previously argued that a fundamental reappraisal of metrology is needed if such ordinal or nominal properties are to be included in an extended quantity calculus on which the SI could be based. As just one example, metrological quality assurance in person-centred care now seems possible in a novel way for partners in care; physician ability (e.g., to diagnose cognitive impairment correctly) and patient ability (e.g., cognitive performance of patients with Alzheimer 's disease) using novel formulations of causal Rasch models articulated through construct specification equations (CSEs). Here, as part of our research in the EMPIR NeuroMET 15HLT04 project, we illustrate how construct specification equations and new metrological references for task difficulty for block tests, such as the Knox Cube Test, can be established in terms of entropy, reversals and average distance.

1. Introduction

Comparability through SI (System of International Units) traceability and uncertainty analysis is an, as yet, unmet requirement for regulatory approval of biomarkers, patient centred outcome measures (PCOMs), clinical thresholds and new therapeutic drugs. However, we have previously argued that: a) measurement, whether physical or social, should have the same definition (i.e., the ratio of two magnitudes of the same attribute, where the denominator is the unit, providing for invariant comparisons) and the same broad goal (i.e., quantification of meaningful variables) [1]; b) to ensure decision makers within health systems have access to the necessary objective evidence, we require the same kind of quality-assurance of PCOMs as is established in physics and engineering [2]; and c) Rasch Measurement Theory [3], and especially causal Rasch models [4], have a central role to enable references for traceability and means of evaluating measurement uncertainty to be established [5].

2. Patient centred cognition measurement

The European EMPIR NeuroMET 15HLT04 project [6] combines expertise from clinical, academic, and National Metrology Institutes to provide a better understanding of how to improve, combine and analyse measurements in Alzheimer's disease (AD) diagnosis and treatment. As part of the project,

PCOMs of cognitive performance (e.g., recall, language and praxis) are examined. The ultimate goal is to improve the analysis of correlations between cognitive function and volumes of AD-related brain structures and neurometabolite concentrations. Rasch Measurement Theory is being used to develop a 'NeuroMET Memory Metric' based on results from administering an extensive battery of legacy cognitive performance PCOMs (e.g., Mini-Mental State Examination, Corsi Block Test, Digital Span Test). This is being achieved through the formulation of novel construct specification equations (CSEs) using multivariate principal component regression [7].

CSEs are defined as the highest level of construct theory in social measurement [8, 9]. At the lowest level 1, social measurement systems have no explicit construct theory and items are simply calibrated empirically from data analysis. At the highest level 5, measurement systems have a construct theory and CSE; a conjoint model that measures task difficulty and patient ability in a common unit. This is known as theory-referenced measurement. It affords three central benefits [7]: a) an observational outcome (i.e., a response); b) a causal mechanism that transmits variation in the intended attribute (e.g., cognitive test item difficulty), via an instrument to the observed response (e.g., the test person); and c) attribute measures (i.e., the measurand) expressed in some unit. Consequently, CSEs resemble 'recipes for certified reference materials' for traceability in chemistry [7]. They provide a comprehensive empirical understanding of: how a collection of items works together; what is being measured; and how validity is ensured.

3. A CSE for the Knox Cube Test – a worked example

A CSE breaks the relationship between intention (e.g., task difficulty) and attainment (e.g., person ability) by independently formalizing intent as an equation that can produce item calibrations (or ensemble means) that are in close correspondence with empirical estimates [4]. A construct-specification equation for task difficulty, $\delta = f[x_1,\ldots x_m]$, is developed by regressing object construct values δ on selected characteristics, X_i, ('explanatory' variables). The results from this multiple regression analysis yield two important sources of information: a) the R^2 index indicates the amount of variance in construct values δ accounted for by the model; b) the analysis provides a regression equation that indicates which characteristics of the items are important in predicting construct values δ [8].

The CSE approach provides a predictive tool for the design of, for instance, new cognitive tasks complementing existing tests and scales and demonstrating item equivalence. These sorts of cognitive performance PCOMs can often rely on the concept of entropy. Entropy can be described as a measure of order. The higher the order, the lower the entropy, and the other way around. Task difficulty should be expected to decrease with entropy, since more ordered tasks are more 'informative' and should be easier to perform. Information theory draws analogies between thermodynamic entropy in physics and the measures of information content in communication systems [10]. Information content can range from basic examples (e.g., the number of elementary symbols) to increasingly sophisticated information, through syntax, semantic and pragmatic aspects of meaningful information in many contexts.

Development of cognitive performance PCOMs includes considerations of the item factors that can explain the level of difficulty of remembering, say, a particular sequence (e.g., tapped blocks or a series of numbers or words). A basic idea is that task difficulty is proportional to the entropy, as follows. The Shannon entropy is proportional to ln(P), where P is the probability of a message. The less expected a message is (i.e., smaller P), the greater the amount of information conveyed ('surprisal'). Taking the logarithm facilitates addition and subtraction of different amounts of information. Consider in general a message in which a number, N_j, (j = 1, …, M) of symbols of M different types can be distributed in a number, G, of categories (or cells) $G = \sum_{j=1}^{M} N_j$. The probability of encountering the jth symbol is $p_j = \frac{N_j}{G}$, which can be summed to unity. The total number, P, of messages that can be obtained by distributing the symbols at random over the G cells (with never more than one symbol per cell) is $P = \frac{G!}{\prod_{j=1}^{M} N_j!}$ [11]. The information theoretical entropy, which is a measure of the amount of information in these messages, is then:

$$I = K \cdot lnP = K \cdot \left[ln(G!) - \sum_{j=1}^{M} ln(N_j!)\right] \cong K \cdot \left[G \cdot ln(G) - \sum_{j=1}^{M} N_j \cdot ln(N_j)\right] \quad (1)$$

where K is an arbitrary constant.

A CSE for task difficulty as a function of entropy can be developed for sequence memory tests, such as the Knox Cube (KCT), which involves patients replicating different tapping series on four black one-inch cubes placed in a row [12]. For this illustration, we used the WINSTEPS® *Exam1.txt* file, included 35 person ratings on 18 items. The set of items comprises different lengths of sequence, from 2 taps to 7 taps, and sequences with different orders. Thus, the KCT offers a suitable test in which expression (1) for entropy can be evaluated for a number of sequences, which not only have increasing numbers of taps but also some repeated taps of the same block (e.g., tapping sequence 1-4-2-3-4-1). The sequences with repeats can be expected to be easier to recall according to Brillouin's expression than a sequence of the same length but without repeats [11]. Tentatively, as no suchlike tapping sequence is available in KCT, a tapping sequence of 1-4-1 would be expected to be easier than 2-4-1

We found a high correlation (Pearson coefficient r = 0.96) in a linear regression of the measured task difficulty, δ, against the CSE estimates zR based on the three explanatory variables X = {*Entropy, Reversals, Average Distance*} (Figure 1). Likewise, the measured task difficulty, δ, is within the corridor of model uncertainties, i.e. $zR + UzR$ (Figure 2). The effects of dis-attenuation [13, 14] on the regression degree of correlation are quite small in the present case, reflecting that measurement uncertainty is small compared with other sources of scatter.

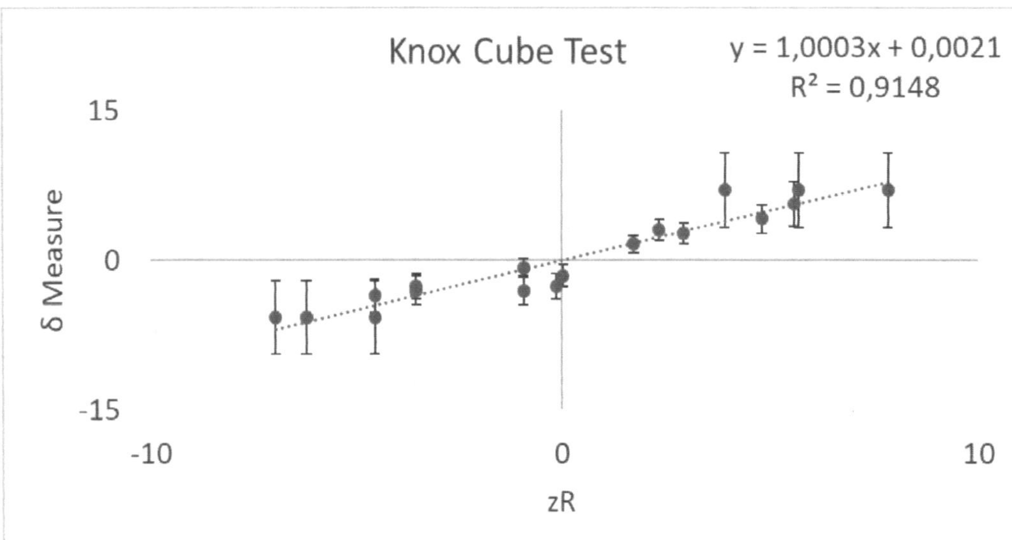

Figure 1. Linear regression of the measured task difficulty, δ, against the CSE estimates zR based on the three explanatory variables X = {*Entropy, Reversals, Average Distance*} for the series of KCT sequences.

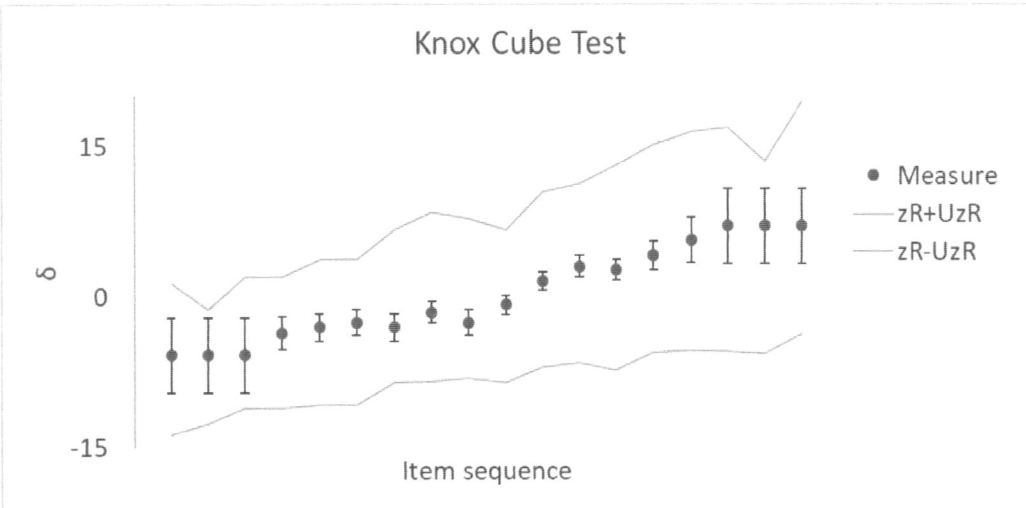

Figure 2. Predicted values of task difficulty, δ, for series of KCT sequences of increasing difficulty from a CSE, zR, based predominantly on entropy [eq. 1] (with minor additional contributions from the number of reversals and average distance of each sequence) compared with corresponding measurement values (dots with uncertainty intervals). The corridor of model uncertainties is shown as $zR \pm UzR$, uncertainty coverage factor $k = 2$.

4. A CSE for person ability – the aspiration for person centred cognition metrics

As commented by de la Torre and Minchen [15] in the context of educational assessment, students' scores are typically determined by identifying person locations along a single proficiency continuum using item response theory, do not in themselves naturally provide diagnostic information. To be able to provide diagnostic information to enhance classroom instruction and learning so-called cognitively diagnostic assessments are made where responses are modelled as a function of discrete latent variables.

Similarly, for patients with neurodegenerative diseases in health care, a CSE for person cognitive ability, $\theta = f[z_1,...z_m]$ can be formulated analogously to the task difficulty CSE. Instead of the explanatory variables, such as entropy, reversals and average distance, for task difficulty of KCT the CSE for person cognitive ability could be a function of diverse set of volumes of AD-related brain structures and neurometabolite concentrations.

A CSE for patient cognitive ability would be able to provide better metrological assurance of biomarker analyses compared to univariate studies. In the upcoming follow-up project, EMPIR NeuroMET2 18HLT09, this will further be examined to develop mathematical reference measurement procedures by combining the novel CSEs for patient ability and task difficulty which link metrological traceability in the cognitive and biomarker assessments thereby leading to improved measurement quality for better NDD early diagnosis and better monitoring of the effects of pharmaceutical intervention and disease progression.

5. Conclusion

The potential of synthesising novel certified reference 'materials' for metrological traceability for the separate probe and entity attributes based on ordinal or nominal responses becomes possible in terms of CSEs, i.e., experimental intervention/manipulation of either person attribute (e.g., ability) or entity attribute (e.g., difficulty) or both simultaneously, following logistic (Rasch) regression, and enabling prediction of the observed outcome (count correct). This work illustrates how CSEs and new metrological references for task difficulty for cognitive block tests, such as the Knox Cube Test, can be established in terms of, in particular, informational entropy as a dominant explanatory variable for cognitive task difficulty. Our analyses are in good agreement with earlier work such as the CSE proposed by earlier researchers, such as [16], using their LLTM model. We believe that the concept of entropy, as developed here, has extensive applications in a wide variety of contexts in explaining task difficulty and person ability when assuring the quality of categorical data on ordinal and nominal scales.

6. Acknowledgments

This work has been performed as part of the EMPIR 15HLT04 NeuroMET project which belongs to the European Metrology Programme for Innovation & Research (Horizon 2020), jointly funded by the EMRP participating countries within EURAMET (www.euramet.org) and the European Union.

References

[1] Cano S *et al.* 2016 *J. Phys.: Conf. Ser.* **772** 012025
[2] Cano S *et al.* 2018 *J. Phys.: Conf. Ser.* **1044** 012057
[3] Andrich D 2018 Advances in social measurement: A Rasch Measurement *Theory Perceived Health and Adaptation in Chronic Disease* ed F. Guillemin, *et al.* (Routledge: New York)
[4] Stenner A *et al.* 2013 *Front. Psychol.* **4** 1-14
[5] Cano S *et al.* 2018 *J. Phys.: Conf. Ser* .**1065** 072033
[6] EMPIR Project 15HLT04 NeuroMET 2016 *Innovative Measurements for Improved Diagnosis and Management Neurodegenerative Diseases 2016–2019* https://www.lgcgroup.com/our-science/national-measurement-laboratory/european-metrology-programme-for-innovation-and-re/neuromet/#.XBPL2lz7SUk
[7] Pendrill L 2018 *Meas. Sci Technol.* **29** 034003
[8] Stenner A and Smith M 1982 *Percept. Motor Skill.* **55** 415-426
[9] Stenner A, Smith M and Burdick D 1983 *J. Educ. Meas.* **20** 305-316
[10] Weaver W and Shannon C 1963 *The Mathematical Theory of Communication* (Univ. of Illinois Press)
[11] Brillouin L 1962 *Physics Today* **15**
[12] Knox H 1914 *JAMA* **LXII** 741-747
[13] Spearman C 1904 *Am. J. Psychol.* **15** 72-101
[14] Spearman C 1910 *Br. J. Psychol.* **3** 271-295
[15] de la Torre J and Minchen N 2014 *Psicologia Educativa* **20** 89-97
[16] Green K and Smith R 1987 *J. Educ. Stat.* **12** 369-381

Real-time spectral analysis using Prism Signal Processing

Manus Henry[1,2]

[1]Department of Engineering Science, University of Oxford.
[2]School of Electrical Engineering and Computer Science, South Ural State University, Chelyabinsk, Russia

E-mail: manus.henry@eng.ox.ac.uk

Abstract. The Prism is a new signal processing object which provides fully recursive FIR filtering i.e. the computational load is small and fixed irrespective of filter length. The Prism also has negligible design cost. These features are attractive to the next generation of Industry 4.0 sensors requiring low cost and flexible signal processing. Recent work has demonstrated how, using a few simple design rules, a chain of six Prisms can create narrowband filters with arbitrary central frequency and bandwidth, also with low design and computational cost. In this paper this technique is extended to provide spectral analysis of data, whereby a sequence of filters is applied to a data set to provide selective frequency analysis.

1. Prism signal processing

The Prism (precise, repeat integral signal monitor), is a recently developed signal processing object. It is based upon a Fourier-style double integration applied recursively via a sliding windows calculation. Either one or two outputs are generated. If there are two, these form a sine/cosine pair from which the properties of the input signal, such as amplitude, phase and/or frequency, can be calculated. Unlike conventional, convolution-based finite impulse response (FIR) filters, the Prism calculation is fully recursive, having a low computational cost which is independent of window length; its benefits include a linear phase response, and a low design cost, as the filter 'coefficients' are simply linearly spaced sine and cosine values. Although the frequency response design options for a single Prism are relatively limited, serial and/or parallel networks of Prisms can be created to provide a wider range of responses suited to a range of tasks including notch and bandpass filtering, or tracking multiple frequency components in a signal.

In [1], an introduction to Prism Signal Processing (PSP) is provided as with potential applications in the Internet of Things/Industry 4.0. Current applications include sensor validation [2] generally, pressure sensors [3], the condition monitoring of rotating machinery [4, 5], and the first application of Coriolis mass flow metering to fuel injection monitoring in internal combustion engines [6].

Consider a sinusoidal input signal (or indeed a signal component at a specified frequency)

$$s(t) = A\sin(2\pi ft + \phi), \tag{1}$$

where t (or τ) is time, A is the amplitude, f is the frequency, and ϕ is the initial phase offset at t = 0. The basic mathematical operation used in the Prism is to calculate double integrals of the form:

$$I^{h}_{[s|c][s|c]} = m^2 \int_{-\frac{1}{m}}^{0} [\sin|\cos](2\pi hmt) \left(\int_{t-\frac{1}{m}}^{t} [\sin|\cos](2\pi hm\tau) \cdot s(\tau) d\tau \right) dt \tag{2}$$

where [s | c] indicates choosing one of s (sin) or c (cosine), h is the harmonic number, a positive integer, and m is the characteristic frequency of the Prism. Equation (2) describes a family of Fourier–style double integrals whereby at each level the input signal is multiplied by a modulating sine or cosine function and integrated over the period of the characteristic frequency m. The harmonic number h determines how many whole cycles of the modulation function occur over the period of integration $1/m$. In the Prism itself, either one or two pairs of such integrals are evaluated each sample. The output of each integral pair is combined to provide a filtered output of the input signal:

$$G_s^h = I_{ss}^h + I_{cc}^h = A\,\text{sinc}^2(r)\frac{r^2}{r^2 - h^2}\sin(\phi - 2\pi r) \tag{3}$$

$$G_c^h = I_{cs}^h - I_{sc}^h = A\,\text{sinc}^2(r)\frac{h\,r}{r^2 - h^2}\cos(\phi - 2\pi r) \tag{4}$$

2. PSP narrowband filtering

As described in [7, 8], simple design rules have been developed for building narrowband FIR filters using a signal chain consisting of three pairs of Prisms. Figure 1 shows the frequency response of the generic Prism filter for an arbitrary central frequency c and bandwidth b.

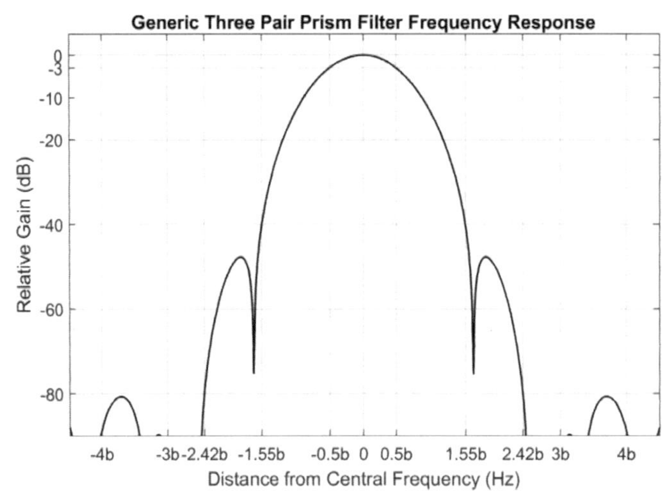

Figure 1. Frequency response of bandpass filter formed from three Prism pairs.

Prism narrowband filtering offers a significant speed-up compared with conventional FIR filtering. In [8], an FPGA implementation of real-time Prism narrowband filtering is described, in which a signal sampled at 2 MHz can be filtered around an arbitrary central frequency with a bandwidth as narrow as 0.1Hz. This requires a total Prism filter length of 192 million samples, but because the calculation is recursive, it is able to run in real time on a single FPGA, as long as sufficient memory is available to store the sample data. A conventional, convolution-based, non-recursive FIR filter would require 384 TMAC/s (i.e. 3.84 x 10^{14} multiply-and-accumulate operations per second i.e. supercomputer performance) to achieve this throughput, assuming such long filters could be designed.

The combination of instant filter design and fast filter throughout provides the opportunity for developing Prism spectral analysis, as described in the next section.

3. Prism spectral analysis

Prism spectral analysis is carried out by passing the data set to be analyzed through a series of frequency-specific detectors, one for each frequency at which amplitude information is required. Each detector consists of a narrowband filter followed by a Prism-based tracker ([1, 6]) to calculate the amplitude of the resulting filtered signal. Within the tracker, the two sine/cosine outputs of a Prism (given in equations (3) and (4)) are combined in a weighted root-sum square calculation to provide an estimate of the amplitude of the input signal on a sample-by-sample basis. Such amplitude estimates can only be provided once the narrowband filter and the tracker filter are filled with samples, and so one design consideration is the tradeoff between the narrowband filter length and the data set to be examined. To generate a single amplitude value to represent the corresponding frequency over the data set, the average value can be used, as discussed in the simulation example below.

The principle benefits of the Prism spectral analysis technique are flexibility and numerical accuracy, the latter arising from the suppression of frequency leakage via the narrowband filtering. Flexibility is provided because the range, spacing and filter bandwidths to be analyzed are entirely arbitrary. While it is certainly possible, as illustrated below, to perform an analysis across a large contiguous range of spectral frequencies, it is also possible to select only individual and/or sub-ranges of frequencies for detailed, narrow bandwidth analysis, thus avoiding the outlay of computational effort on spectral regions which are not of interest.

The computational requirements of Prism spectral analysis, based on recursive FIR techniques, are low enough to make the technique viable, but remain significantly higher than for the FFT itself. The computational load for an FFT is $o(N \log_2 N)$ for a data set of size N. By contrast, for Prism signal processing, the computational cost for *each frequency detector* is approximately 300 N. Thus only rarely would the technique be suitable for narrowband analysis across a wide frequency range. More typically, a mixture of coarse and narrow frequency detectors would be used, as discussed at the end of the paper.

4. Simulation Example

A comparison between conventional FFT and Prism-based spectral analysis techniques is given via a simulation example, based upon typical sensor signal characteristics from a Coriolis mass flow meter. Table 1 gives the frequencies and amplitudes of five components in the simulated sensor signal. Data sets were generated using 50 kHz sampling, either noise-free or with white noise (0.01 V standard deviation) added. Each simulation data set consisted of 2^{20} or approximately 1 million samples, corresponding to a time series of approximately 20 s duration.

Table 1. Frequencies and amplitudes of components in simulated sensor signal.

Frequency (Hz)	Amplitude (V)
80	0.2
160	0.02
190	0.01
240	0.015
320	0.001

Results for the FFT calculation were derived using the standard MATLAB call *fft()*. Note that a wide variety of FFT algorithms exist, with various advantages and disadvantages; an optimal choice will depend upon the characteristics of the data set and the analytic requirements. The intention here is simply to use a straightforward and widely accessible algorithm as a basis for comparison with the Prism technique.

The Prism spectral analysis was carried out over the range 12 Hz – 500 Hz, at intervals of 0.1 Hz. Note that all Prisms have a gain of zero at DC and so a non-zero lowest frequency must be selected. The filter bandwidth was also set to 0.1 Hz over the entire range, but these choices are arbitrary and independent for each detector. As discussed in [7], the sample length of the narrowband filter design shown in Figure 1 is approximately $1.92fs/b$ where fs is the sampling rate and b is the filter bandwidth. Each frequency detector also requires a tracker, but this is relatively short compared to the bandpass filter. For a detector at frequency $f \in [12, 500]$ Hz, the optimal tracker Prism length is fs/f samples; for the lowest frequency, 12 Hz, this amounts to approximately 4000 samples. Overall, therefore, the detector length is approximately 5×10^5 samples, compared with the data set length of approximately 10^6 samples. Accordingly, the detector 'warm-up' requires approximately half the data set, and a time series of approximately 5×10^5 amplitude values are generated for averaging. While a conventional convolutional FIR filter of order 5×10^5, applied to a data set of length N samples, would require 10^6 N arithmetic operations (i.e. 5×10^5 each of multiply and accumulate per sample), the Prism bandpass filter requirement is only 300 N. With different design choices, for example selecting a bandwidth of around 0.05 Hz, a longer bandpass filter would result in fewer amplitude values generated from the same data set.

Figures 2 - 5 show typical results obtained by applying FFT and Prism spectral analyses to noise-free and noisy data sets. In the noise free case the Prism analysis generates a significantly lower noise floor compared to the FFT, and in both cases the Prism narrow-band filtering results in a reduction in frequency leakage, resulting in sharp spectral peaks.

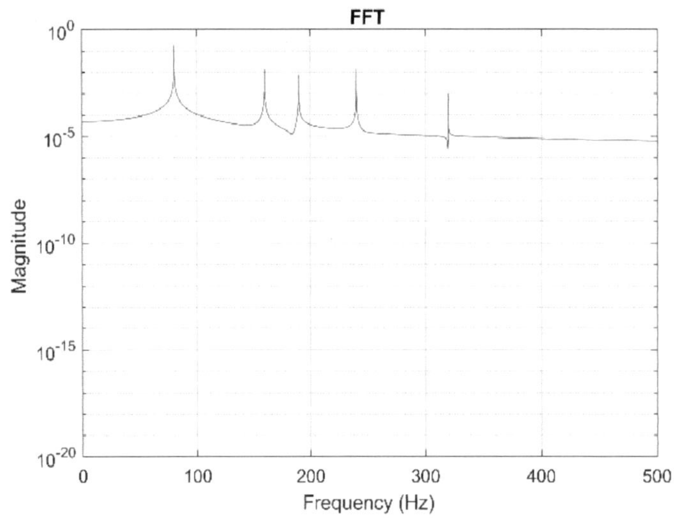

Figure 2. FFT analysis of simulated data set – noise free.

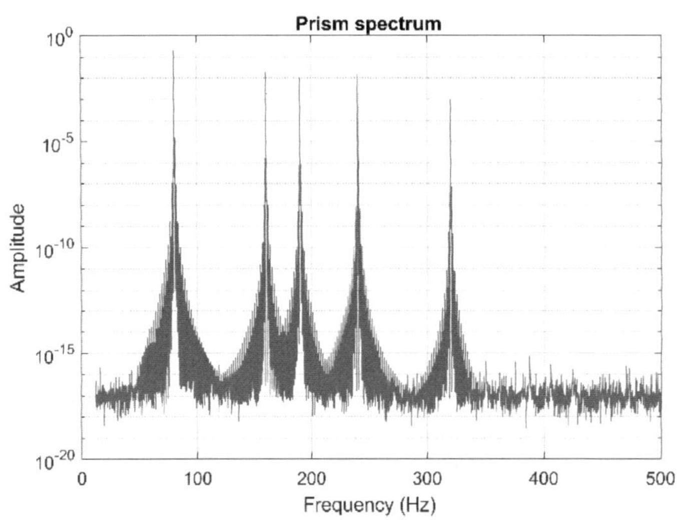

Figure 3. Prism analysis of simulated data set – noise free.

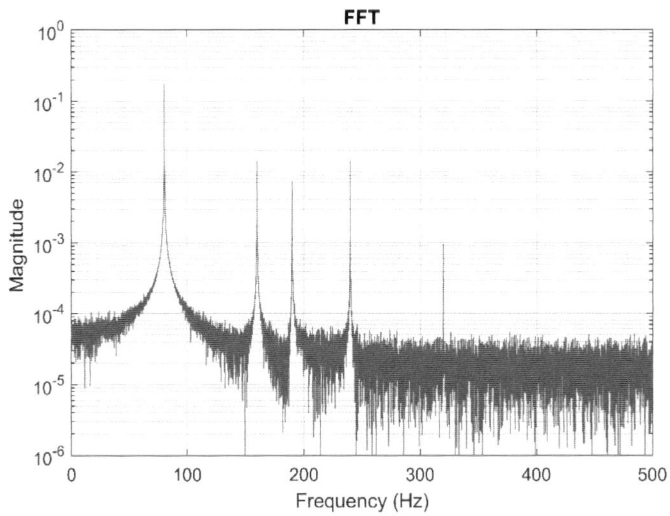

Figure 4. FFT analysis of simulated data set – with added noise.

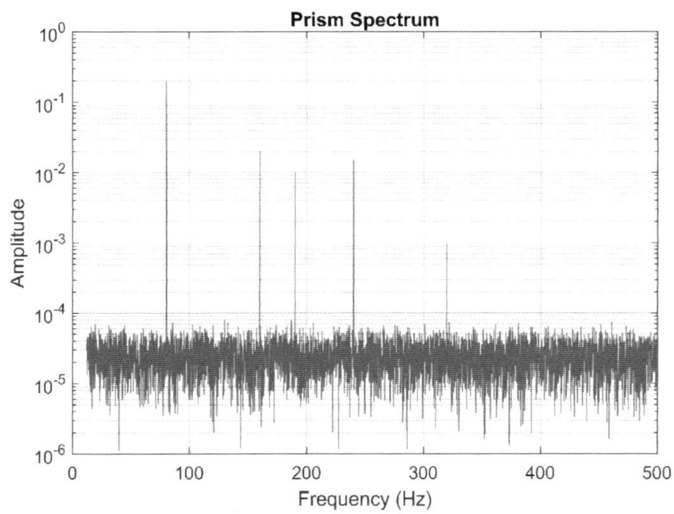

Figure 5. Prism analysis of simulated data set – with added noise.

A quantitative assessment of the Prism spectral analysis has been carried out as follows. For each case (noise-free and noisy), 100 data sets were generated. For both the noise-free and the noisy data sets, the initial phase of each frequency component (Table 1) was generated randomly, so that each of the 100 data sets is distinct. In addition, for the noisy data, different random noise sequences were added to each data set. Prism spectral analysis was applied to each of the two sets of 100 trials to obtain the calculated amplitude of each of the frequency components listed in table 1. Based on these results, the mean, mean error, and standard deviation were collated for the calculated amplitude for each component. The mean error and the standard deviation are scaled as a percentage of the true amplitude of the corresponding frequency component, so that the relative accuracy and variability can be assessed. The results are shown in tables 2 and 3.

In the noise-free case (Table 2), high amplitude accuracy is achieved, irrespective of the initial phase of the frequency components, so that the mean errors and standard deviations are small. In the noisy case (Table 3) the average errors remain small, while the standard deviation of the error increases as the amplitude of the signal component decreases relative to the noise level. For example, for the 80 Hz component, which has a true amplitude of 0.2 V, the calculated value had a standard deviation of approximately 0.1 % of the mean value, while for the 320 Hz component, with a true amplitude of 1 mV, the calculated value had a standard deviation of approximately 2 %.

Table 2. Results of Prism tracking of spectral component amplitudes in simulated sensor signal – no noise (100 trials).

Frequency (Hz)	True Amplitude (V)	Calculated Amplitude Mean (V)	Calculated Amplitude Mean Error (%)	Calculated Amplitude Standard Deviation (% of mean)
80	0.2	2.0000e-01	-1.1793e-08	1.4557e-10
160	0.02	2.0000e-02	-1.8900e-07	3.3356e-10
190	0.01	1.0000e-02	-3.3471e-07	1.0579e-09
240	0.015	1.5000e-02	-9.5902e-07	3.2710e-09
320	0.001	1.0000e-03	-3.0413e-06	4.9119e-09

Table 3. Results of Prism tracking of spectral component amplitudes in simulated sensor signal – with added white noise 0.01 V standard deviation (100 trials).

Frequency (Hz)	True Amplitude (V)	Calculated Amplitude Mean (V)	Calculated Amplitude Mean Error (%)	Calculated Amplitude Standard Deviation (% of mean)
80	0.2	2.0000e-01	-1.7747e-03	1.0151e-02
160	0.02	1.9999e-02	-3.1187e-03	1.0999e-01
190	0.01	9.9998e-03	-1.6958e-03	1.7003e-01
240	0.015	1.5001e-02	3.9012e-03	1.4967e-01
320	0.001	1.0006e-03	6.3456e-02	1.9849e+00

5. Conclusions

Prism signal processing is a recursive FIR technique offering low design and computational cost compared with conventional convolutional forms of FIR filtering. This paper has demonstrated its potential as an alternative approach to spectral analysis. Frequency leakage is minimized by the use of narrowband filtering, leading to accurate estimation of amplitude at each selected frequency component, and the technique affords complete flexibility with regard to the frequencies and bandwidths to be analysed. While its computational cost is unlikely to be competitive with the FFT over the full spectral range of a data set, Prism spectral analysis may prove useful for examining key regions of a data spectrum in detail (perhaps where these key regions are identified using conventional FFT or widely spaced, broader-band Prism spectral analysis), or for data sets where its high accuracy warrants the higher computational requirement.

References

[1] Henry M, Leach F, Davy M, Bushuev O, Tombs M S, Zhou F and Karout S 2017 *IEEE Ind. Electron. M.* **11**. URL 10.1109/MIE.2017.2760108
[2] Henry M P 2018 *Meas. Tech.* URL 10.1007/s11018-018-1345-1
[3] Henry M P, Bushuev O and Ibryaeva O 2017 *26th IEEE Int. Symp. on Industrial Electronics (19-21 June 2017, Edinburgh, Scotland, UK)* URL 10.1109/ISIE.2017.8001451
[4] Henry M and Sinitsin V V 2018 *Proc. IEEE Ind Cyber-Physical Syst ICPS (May 15-18, 2018, St.Petersburg, Russia)* pp 452-457. URL10.1109/ICPHYS.2018.8390747
[5] Henry M and Sinitsin V V 2018 *Proc. IEEE Ind Cyber-Physical Syst ICPS (May 15-18, 2018, St.Petersburg, Russia)* pp 458-463 URL 10.1109/ICPHYS.2018.8390748

[6] Leach F, Karout S, Zhou F, Tombs M, Davy M and Henry M 2017 *Flow Meas Instrum*. URL 10.1016/j.flowmeasinst.2017.09.009
[7] Henry M P 2018 *44th Annual Conf. of the IEEE Industrial Electronics Society IECON 2018 (October 21-23, 2018, Washington D.C., USA)*
[8] Owen J and Henry M 2018 *44th Annual Conf. of the IEEE Industrial Electronics Society IECON 2018 (October 21-23, 2018, Washington D.C., USA)* URL 10.1109/iecon.2018.8591619

Journal of Physics: Conference Series **1379** (2019) 012031 doi:10.1088/1742-6596/1379/1/012031

Comparing a compact hematology analyzer based on direct optical measurements using blue LED with the VCS reference technology for neutrophil differential count

VM Lorenzon[1] **and E Costa Monteiro**[1]

[1]Pontifical Catholic University of Rio de Janeiro - PUC-Rio, Postgraduate Programme in Metrology, Rio de Janeiro - RJ, Brazil

E-mail: beth@puc-rio.br

Abstract. Technologies for the automated blood cell count have been evolving over the last decades. Hematology analyzers enable identifying blood elements, including leukocytes differential. In the bloodstream, the most abundant leukocyte, the neutrophil, presents multi-lobed nuclei, requiring, therefore, combined measurement principles for their detection and classification. Due to the lack of white blood cells' certified reference materials or reference measurement method, it has been customary to evaluate the performance of new hematology analyzers by comparing them with traditionally available and well-consolidated systems. In 2015, a technology based on axial light loss using a blue LED, the Beckman Coulter DxH500, was launched. This hematology analyzer presents reduced cost, fewer chemical reagents, and a compact design. This work compares the performance of the DxH500 with an analyzer frequently employed in large-scale laboratory routine, the Beckman Coulter LH750, based on the well-established combination of volume (impedance), conductivity and light scatter principles. The study examined 310 paired samples taken out randomly from the routine of a general hospital laboratory. The analyses evaluated the correlation coefficient, p-value, and the linear regression equation. The equivalence of results provided by both analyzers for the characterization of neutrophils in the leukocyte population points to the performance adequacy of the new low-cost and portable technology, suitable for small-sized clinical laboratory's routines.

1. Introduction

White blood cells (WBC) perform an essential role in immune systems, defending the body against foreign particles, infectious diseases, and cancerous cells. [1]. Bone marrow produces all blood cells, including the different types of WBC. There are mainly five major types of white blood cells circulating in the bloodstream: basophil, eosinophil, neutrophil, lymphocyte, and monocyte. The detection, classification, and determination of both the absolute value and the percentage of the overall WBC count consist of essential strategies for the diagnosis of a wide range of diseases.

Since 1870, with the use of microscopes and dyes in a blood smear, it is possible to differentiate the types of white blood cells and to quantify the number of cells by blood count [2]. Commonly based on a count of 100 cells, this method can introduce statistical errors, distribution errors caused by the unequal distribution of cells in the smear, and subjective interpretation errors [3]. The results of a WBC cell count and differentiation are interpreted by comparison. The comparator can be a population reference interval, a decision point (from clinical studies), or a previous result on the same patient. For validating the comparison, the knowledge of methods' limitations and errors is required to provide a reliable clinical diagnosis [4, 5].

By mid-XXth century, the emergence of automated blood cell counters significantly reduced the time spent on the procedure of blood analysis. Hematology analyzers evaluate thousands instead of 100 cells per sample, being, therefore, considered more reliable [6]. These methods contribute to avoiding the statistical and distributional errors associated with the white blood cell count in a blood smear [2].

Over the ensuing decades, flow-based cytometers using light, impedance, conductivity by radio frequency measurements, or the combination of these principles, were developed. These devices provided a more significant number of parameters enabling identifying blood elements, including leukocytes differential [2].

The first principle applied in hematology analyzers was the impedance, also called the Coulter Principle [7-9]. It is an electronic method for counting and sizing particles. Blood cells act as electrical insulators while suspended in a conductive liquid (dilution reagent), guided by a gentle vacuum; it passes through an aperture. Each blood cell, when crossing into the sensitive zone, increases the resistance for the electric current flow established between two submerged electrodes. The conductivity principle, in turn, is the measurement of intracellular content using a high-frequency electromagnetic probe. The cell´s constituents and their chemical composition result in measurable changes in conductivity [9, 10]. The third principle, light scattering, characterizes the cell´s surface, distinguishing cell structure, shape, and reflectivity [9, 11].

The combination of the three principles (impedance, conductivity by radio frequency and light scatter) brought, in 1970, a hematology device that could differentiate WBC population in five parts: basophil, eosinophil, neutrophil, lymphocyte, and monocyte.

Currently, the well-consolidated combination of these three principles, called VCS technology (volume, conductivity and scatter), is the most frequently employed by automated hematology analyzers. An example of these VCS systems in use is the Beckman Coulter industry, LH750 model, launched in 2001, using a total of five types of reagents [12].

In 2011, an automated blood cell counter combining light scatter and fluorescence, the Sysmex XN, was launched. These instruments need fluorescence dyes polymethine based [13].

More recently, in 2015, the same manufacturer of LH750, Beckman Coulter, launched a more compact and less sophisticated hematology analyzer, DxH500, which combines the Coulter principle with the direct optical measurements of axial light loss (AxLL) with a blue led source incorporated [7]. This system uses three types of reagents [14]. Traditionally used in flow cytometry for abnormal cell's characterization, the axial light loss consists of the interruption of the optical path when a cell passes in front of the electromagnetic emission [15]. The amount of light falling on a sensor can be measured and varies depending on cell structure [7]. At the same aperture that measures impedance, an optical assembly projects the blue LED light onto an AxLL sensor [7].

Sensors development is continually evolving for improving detection sensitivity, signal to noise ratio and interferences, without the need for sophisticated and expensive instruments for the blood cell's count [16-18]. In the clinical laboratory practice, however, for quality control, background counts (limit of blank) need to be performed daily for identifying and avoiding any possible interference [19].

Combinations of principles introduced by manufacturers may result in distinct performances and interferences. Each type of instrument has warnings signalizing interference effects and emits suspect messages in the presence of abnormal cells, identified as the flagging system. The literature shows that automated hematology analyzers are superior for white blood cell (WBC) differential counts when evaluating well characterized (mature) cell types, whereas optical microscopy is better for differentiating cells based on nuances of cytological features, especially for immature cells [20]. The flags, or a combination of them, show the difficulty of the automated method in characterizing the cell, indicating the need for a smear review [16]. Strategies for dealing with flags are available in the International Consensus Group for Hematology Review (ICGHR) published in 2005 [20, 21].

The appropriateness of performance evaluation of the various technologies coming from the WBC faces, however, a challenging metrological demand [5]. Despite the advances on the worldwide equivalence of measurement results in clinical laboratories with the creation, in 2002, of the Joint Committee for Traceability in Laboratory Medicine (JCTLM) [4], up to the moment, there are no certified reference material or reference methods for WBC analysis available. With the lack of WBC reference materials or reference measurement procedure, while commutability of reference material is not established [8], it has been customary to evaluate the performance of new hematology analyzers by comparing them with traditionally available systems [22].

The literature shows comparisons between the compact hematology analyzer recently launched (DxH 500), which uses AxLL and incorporates a blue LED source, with the Sysmex XN-3000, a technology based on Light scatter and Fluorescence [23]. Another comparison available is performed between DxH500 and Beckman Coulter HmX, which is one of the first VCS systems launched and still in use [24], presenting a more limited flagging system than LH750. All comparisons were performed with selected samples without morphological WBC flags or suspect messages.

Among the leukocytes in the bloodstream, neutrophils are usually the most abundant and have multi-lobed nuclei, requiring the use of combined principles for their detection and classification [1]. Therefore, neutrophils without morphological changes are useful parameters for comparing automated techniques for differential WBC.

This work aims at comparing the most recently launched Beckman Coulter DxH500 hematology analyzer, a compact technology, with reduced operational complexity, based on AxLL with a blue LED source, with the Beckman Coulter LH 750, which is a VCS system that presents a more robust flagging system than the HmX. This preliminary study focuses on the neutrophils' differential, whose intracellular characteristics require combinations of measurement principles for automated detection and quantification.

2. Materials and Methods

Comparisons were performed during a one-week period in which the hematology analyzer DxH 500 (Beckman Coulter, Miami, USA) was placed in a hospital laboratory employing the LH750 (Beckman Coulter, Miami, USA). The laboratory is located at a general hospital that offers a diversity of medical specialties, such as surgery (including neurosurgery and transplants), oncology, orthopedics, pediatrics, and geriatrics.

Specimens were obtained from anonymized whole blood samples and submitted for routine blood count testing. Samples were stored at room temperature, and all analyses were completed within eight hours after sample entry into the laboratory. A total of 310 unselected samples were taken randomly from routine analyses. The average interval between the paired exams was one hour, with 46 min of standard deviation.

Background counts performed daily presented results within the acceptable values for both measuring instruments. The samples with positive flagging for the WBC count, or any suspect message that implies in the necessity of a smear review, were excluded from the analyses. One hundred six samples were removed from the comparison, representing 34.19 % of the total of 310 samples. As a result, 204 samples were used for paired comparison.

For data analysis and statistics, a two-tailed t-test was used on the paired results of neutrophils' percentage ($NE_{\%}$) in leucocytes differential to evaluate the p-value to ascertain whether or not there was any significant discrepancy between the $NE_{\%}$ mean values obtained by each one of the hematology analyzers. The correlation coefficient and linear regression equation, with the plot of the linear regression, were evaluated for the $NE_{\%}$ provided by DxH500 and LH750 for each of the 204 samples.

3. Comparative Analysis

The automated evaluation of the 310 whole blood samples, covering a wide variety of underlying medical conditions, using both DxH500 and LH750 hematology analyzers, resulted in a total number of flag signals that required the removal of 106 samples (34.19 % of entire WBS evaluated).

Both hematology analyzers presented indications for smear review by flagging of abnormal white blood cells, but not always for the same set of samples (table1). Forty-two samples were flagged exclusively by the DxH500 analyzer (13.55 % of the 310 WBS), and 43 were flagged only by LH750 VCS technology (13.87 %). Twenty-one samples were flagged by both analyzers (6.77 % of the total WBS).

Table 1. Number of samples flagged by each automated hematology analyzer (DxH500 and LH750) and their percentage regarding the total number of analyses (310 samples).

	Flagged Samples	**Percentage of the 310 samples**
DxH500	42	13.55 %
LH750	43	13.87 %
DxH500 and LH750	21	6.77 %
Total	106	34.19 %

After excluding the 106 flagged samples, the mean values for the percentage of neutrophils in leucocyte differential of the remaining 204 analysis using DxH500 and LH750 were, respectively, 61.07 % of leucocytes and 61.93 %.

The difference between the mean values for the percentage of neutrophils obtained by the two technologies was -0.86 ± 2.80 %, for a 95 % confidence interval. A paired, two-tailed *t*-test showed that there was no significant difference (*p*-value = 0.4148) between the $NE_{\%}$ means obtained from LH750 and DxH500 measurements. Figure 1 shows the plot of the $NE_{\%}$ values provided by both hematology analyzers for each of the 204 samples.

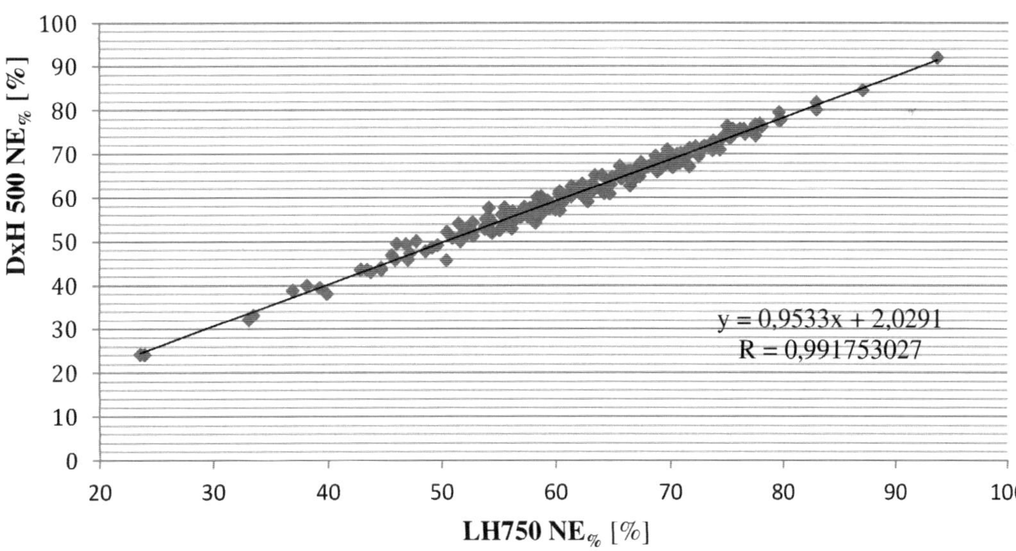

Figure 1. Linear regression, correlation coefficient R and linear regression equation *y* for neutrophils' percentage ($NE_{\%}$) in leucocytes differential for the paired test between DxH500 and LH750.

The linear regression analysis demonstrates a high correlation (R = 0.99), with a slope of 0.95, between both results of neutrophils' percentage measured by DxH500 and LH750, as the reference VCS technology.

Journal of Physics: Conference Series **1379** (2019) 012031 doi:10.1088/1742-6596/1379/1/012031

4. Discussion

Since its launch, in 2015, the literature presents two other comparisons for the results of neutrophils' percentage in leucocyte differential provided by the new DxH500 technology [23, 24]. One of these comparisons was performed with one of the first VCS technologies launched and still in use [24], model HMX, from de same industry LH750, Beckman Coulter, but with a reduced flagging system. The other study compared DxH500 with another recently launched hematology analyzer, Sysmex XN-3000, which combines Light scatter and fluorescence. All these studies removed flags from the set of whole blood samples' results being compared, and do not report the correspondence of signalized flagged samples between the compared technologies. Table 2 presents the correlation coefficients obtained by each study comparing DxH500 results for neutrophils' percentage in leucocyte differential obtained with each one of the three alternative technologies: HMX [24], XN-3000 [23], and, in the present study, LH750.

Table 2. Correlation coefficients between DxH500 and each of three alternative hematology analyzers, for neutrophils' percentage in WBC differential ($NE_{\%}$).

	Hematology Analyzer (*Physical Principle*)		
	HMX (VCS) [24]	**XN-3000** (*light scatter*) (*fluorescence*) [23]	**LH750** (VCS) [present study]
Correlation coefficient comparing with **DxH500** (*Coulter*)(AxLL *with blue* LED)	0.990	0.9894	0.9917

The high correlation coefficient evidenced in all studies shown in table 2 points toward the appropriateness of results for leucocyte differential using the compact and simple technology based on the impedance and AxLL with Blue LED, recently launched.

5. Conclusions

While certified reference materials or reference methods are not available in the JCTLM database to enable metrological traceability for leucocyte differential, the comparison of measurement methods combining diversified chemical and physical principles remains essential for in vitro diagnostics (IVD) assessment [25].

The present study compares the performance of a new technique, DxH500, designed for small routines, needing fewer chemical reagents, which combines a simple principle by impedance measurements with axial light loss using blue led, with one of the conventionally employed technologies, the LH750. This analyzer applies the well-consolidated measurement principles, the VCS (impedance, conductivity, and light scatter).

Although using different measurement principles, the present study showed a high correlation between the results of DxH500 and LH750 for the detection and quantification of neutrophils' percentage in leucocytes' differential. These results confirm the outcome equivalence for neutrophils' characterization, described in the literature, between the new hematology analyzer, DxH500, and the HxM, which is the most ancient VCS device still in use [24]; as well as with the also recently emerged device, XN-3000, based on the light scattering and fluorescence principles [23].

The present comparison completes the evaluation of the performance equivalence of results for neutrophil characterization provided by the compact new technique, with low-cost and few chemicals consume, appropriate for small clinical laboratory routines.

As future work, it is worth mentioning the extending of these studies for encompassing the whole set of white blood cells' analyses. The discrepancies observed between the two hematology analyzers regarding their review warning flags associated with abnormal cells consist of additional demand for investigations.

References

[1] Prinyakupt J and Pluempitiwiriyawej C 2015 Segmentation of white blood cells and comparison of cell morphology by linear and naïve Bayes classifiers *BioMed. Eng. OnLine* **14** 63
[2] Green R and Wachsmann-Hogiu S 2015 Development, history, and future of automated cell *CountersClin. Lab. Med.* **35** 1–10
[3] Rumke C L 1985 Statistical reflections on finding atypical cells *Blood Cells* **11** 141-144
[4] Graham R D J and Craig J 2016 The Joint Committee for Traceability in Laboratory Medicine (JCTLM) – its history and operation. *Clin. Chim. Acta* **453** 86–94
[5] Costa Monteiro E and Leon L F 2015 Metrological reliability of medical devices *J. Phys.: Conf. Ser.* **588** 012032
[6] Buttarello M and Plebani M 2008 Automated blood cell counts: state of the art. *American Society for Clinical Pathology Hematopathology* **130** 104-116
[7] Beckman Coulter Inc 2018 Technology and case studies: DxH 500 Hematology series https://www.beckmancoulter.com/ucm/idc/groups/public/@wsr/@literature/documents/document/glb_bci_157095.pdf
[8] Armbruster D 2017 Metrological traceability of assays and comparability of patient test results *Clin. Lab. Med.* **37** 119–135
[9] Robinson J P and Wallace H 2013 Coulter: decades of invention and discovery. *Cytometry Part A* **83A** 424438
[10] Coulter W H and Hogg WR 1970 Signal modulated apparatus for generating and detecting resistive and reactive changes in a modulated current path for particle classification and analysis. Patent# 3502974
[11] Auer RE and Tucker HE 1977 Method and apparatus for simultaneous optical measurement of particle characteristics. Patent# 4,038,556
[12] Beckman Coulter Inc 2018 Coulter LH Series Pak Reagent Kit (Ref 8547195)
[13] Briggs C, Longair I, Kumar P, Singh D and Machin S J 2012 Performance evaluation of the Sysmex haematology XN modular system *J. Clin. Pathol.* **65** 1024–30
[14] Beckman Coulter Inc 2018 DxH500 Series Lyse (Ref B36846)
[15] Steinkamp J A 1983 A Differential amplifier circuit for reducing noise in axial light loss measurements *Cytometry* **4** 83-87
[16] Zhou T, Song Y, Yuan Y and Li D 2019 A novel microfluidic resistive pulse sensor with multiple voltage input channels and a side sensing gate for particle and cell detection *Anal. Chim. Acta* **1052** 113 – 123
[17] Song Y, Yang J, Pan X and Li D 2015 High-throughput and sensitive particle counting by a novel microfluidic differential resistive pulse sensor with multidetecting channels and a common reference channel *Electrophoresis* **36** 495–501
[18] Song Y, Zhang H, Chon C, Pan X and Li D 2011 Nanoparticle detection by microfluidic Resistive Pulse Sensor with a submicron sensing gate and dual detecting channels-two stage differential amplifier *Sens. Actuators B Chem.* **155**(2) 930 –6
[19] Vis J Y and Huisman A 2016 Verification and quality control of routine hematology analyzers *Int. J. Lab. Hematol.* **37** 63–70
[20] Wei C *et al.* 2010 Development of the personalized criteria for microscopic review following four different series of hematology analyzer in a Chinese large scale hospital. *Chinese Med. J.-Peking* **123**(22) 3231-7
[21] Depoorter M, Goletti S, Latinne D and Defour Jp 2015 Optimal flagging combinations for best performance of five blood cell analyzers. *Int. J. Lab. Hematol.* **37** 63–70

[22] Fernandez T, Domack L B, Montes D, Pineiro R, Landrum E and Vital E 2001 Performance evaluation of the Coulter LH 750 Hematology Analyzer *Carden Jennings Publishing Co Laboratory Hematology* **7** 217–228

[23] Papa F, DiVeroli L, Belli M, Masi L and Rongioletti M 2016 Performance evaluation of the 5-part differential Haematology Analyzer Beckman Coulter Dxh500. *ISLH 2016 Abstracts* **38**(2) 83–84

[24] Pizarro M, Cepeda C and Laguer A 2016 Performance evaluation of the Beckman Coulter Dxh 500* when compared to Coulter Hmx for the five-part differential. *ISLH 2016 Abstracts* **38**(2) 85–86

[25] Armbruster D 2009 Measurement traceability and US IVD manufacturers: the impact of metrology *Accredit. Qual. Assur.* **14** 393–398

Abbe error compensation for a micro/nano CMM with a coplanar stage

Y X He[1], R J Li[1], K C Fan[1,2], Q X Huang[1], P H Hu[1], Y H Wang[1], W W Liu[1] and Y Hu[1]

[1] School of Instrument Science and Opto-electronic Engineering, Hefei University of Technology, No. 193, Tunxi Road, Hefei 230009, China
[2] School of Mechanical Engineering, Dalian University of Technology, No. 2, Linggong Road, Dalian 116024, China

E-mail: rj-li@hfut.edu.cn

Abstract: A micro/nano Coordinate Measuring Machine (CMM) with a stroke of 20 mm × 20 mm × 10 mm has been developed by our group. The "Pagoda structure" with Chinese style is adopted, which has fine mechanical stability. The base of the micro/nano CMM is made of granite to achieve high thermal stability. The stages are driven by piezoelectric motors with a resolution of 1 nm in three directions. The displacements of the stages in horizontal directions are detected by homemade Michelson interferometers. The optical axes of the two interferometers are consistent with their slide way of the coplanar stages respectively, which can reduce the Abbe errors in X and Y axes as much as possible. Semiconductor laser diodes are used as the light source of the interferometers due to low-cost and small size. A linear diffraction grating interferometer (LDGI) has also been developed to detect the displacement in Z direction. After subdivision, the measuring resolution can reach nanometer scale. The micro/nano CMM is compatible with different types of probes. A model of Abbe error and its compensation method are proposed in this paper. Experimental results show that the positioning errors were reduced from 435 nm to 91 nm and from 426 nm to 151 nm via Abbe error compensation method. The flatness of a 0-grade gauge block was tested by the micro/nano CMM, the repeatability of 44 nm (K=2) was achieved.

1. Introduction

With the rapid development of ultra-precision manufacturing technology, Micro-Electro-Mechanical Systems (MEMS), various miniature components have been manufactured. In order to measure these small components with complex geometrical features, ultra-precision micro/nano-CMMs are required. The resolution of the micro/nano CMM needs to reach 1 nm, the accuracy should also reach nanometer scales. Several micro/nano CMMs have been developed by other institutions or universities around the world, such as: (a) The micro-CMM with a stroke of 50 mm×50 mm×50 mm developed by NPL (National Physical Lab.), its resolution can reach 0.31 nm in three directions and the uncertainty is 50~100 nm. This CMM has been used in industrial fields [1-2]; (b) The CMM with a stroke of 25 mm×40 mm×25 mm developed by PTB (Physikalisch-Technische Bundesanstalt), its repeatability can reach 10 nm and the uncertainty is 100 nm [3-5]; (c) The X/Y/Z nano stage developed by SIOS corporation, whose displacement is detected by fiber laser interferometer, it has a linear resolution of 0.1 nm while the stroke is 25 mm×25 mm×5 mm [6].

Both Taiwan University and our University have developed a micro/nano CMM with a coplanar X/Y stage. The stage is driven by two piezoelectric motors and the displacements in X/Y directions are detected by homemade Michelson interferometers. The design of coplanar X/Y stage conforms to the Abbe free principle. However, Abbe error in Z direction still exists. The performance of our micro/nano CMM can be improved by compensating the Abbe error in Z direction.

Abbe error compensation method is proposed and some experiments are conducted in this study. The effectiveness of the method has been verified. The measurement performance of the micro/nano CMM has also verified initially through testing a flatness of a gauge block.

2. Structure and sensors of the micro/nano CMM

2.1. Mechanical structure and Z-spindle

Figure 1 shows the structure of the micro/nano CMM. The base of the CMM and the moving stage is made of granite and invar respectively. The structure adopts the pagoda type with Chinese style, which has very good rigidity and symmetry. The Z-spindle with the probe is suspended from the top of the machine and only have the freedom in vertical. The two-dimensional X/Y stage is placed under the Z-spindle. A linear diffraction grating interferometer (LDGI) is used to measure the displacement in Z direction [7], while the horizontal displacements are detected by two homemade miniaturized Michelson interferometers.

(a) (b)

Figure 1. Micro/nano CMM.
(a) schematic diagram; (b) photography.

2.2. Coplanar stage and miniaturized Michelson interferometer

The two slide ways of the X/Y motion stage are coplanar in horizontal plane. This design conforms to the Abbe principle which can improve measurement accuracy. Figure 2 shows the structure of the coplanar stage. The base of the stage is made of invar which has good thermal stability. Two piezoelectric motors with a resolution of 1 nm are used to drive the stage.

The displacements in X/Y directions are detected by two homemade miniaturized Michelson interferometers with low-cost and good integration. The Michelson interferometer using a semiconductor laser diode has a resolution of 1 nm by electronic and software subdivision method.

2.3. Angle measurement module based on autocollimator

The X/Y stage usually generate pitch and yaw angle errors when doing a linear motion. Since an Abbe offset still exists in Z direction, Abbe error need to be detected and compensated. An angle measurement module based on autocollimator is used to measure the pitch and yaw errors in real time during the stage movement. Figure 3 shows the principle of angle sensor [8]. A laser beam passes through the polarizing beam splitter, then reflected by the measuring mirror. The reflected beam will pass through a focusing lens and project on the surface of the quadrant photo detector (QPD). The pitch and yaw angles can be detected by applying appropriate resistances to QPD and signal processing circuit.

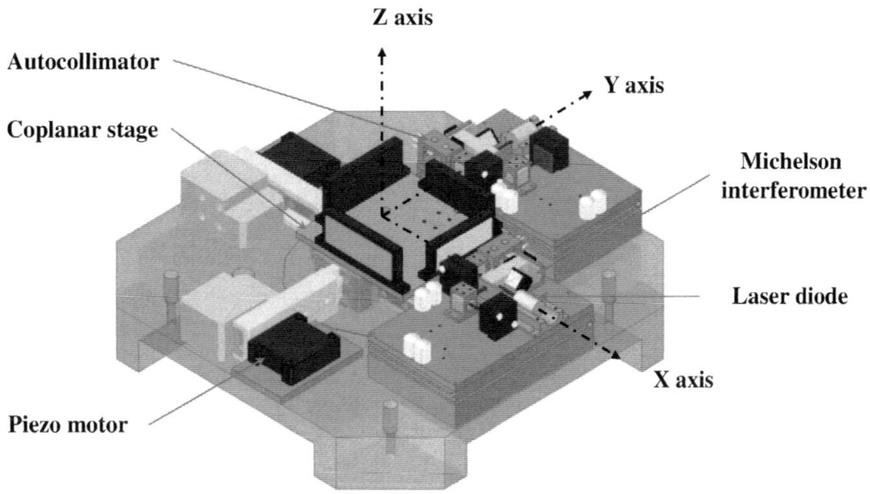

Figure 2. Co-planar X/Y stage.

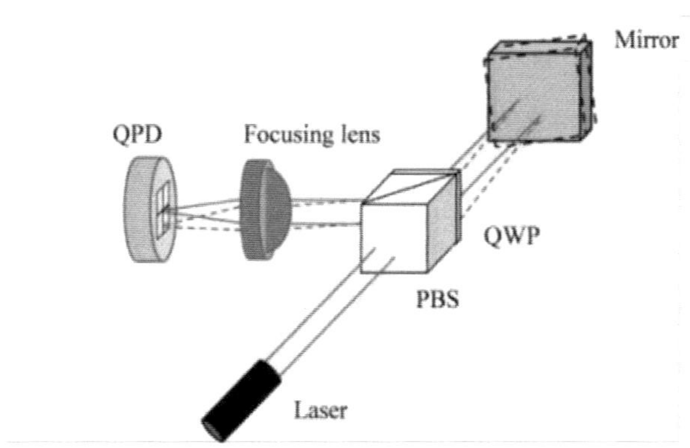

Figure 3. Principle of the 2D angle sensor.

2.4. Contact scanning probe

The probe can greatly affect the measuring accuracy of the micro/nano CMM. A contact scanning probe with nano-scale resolution has been developed by our group [9]. Figure 4 shows the photo of the probe.

The 3D displacements of the probe tip are transferred into a displacement in Z direction or two-dimensional angle of a plane mirror adhering to the upper surface of an elastic mechanism. A Michelson interferometer is used to detect the mirror's displacement in Z direction. An autocollimator is used to detect the angles of the mirror and the probe has a resolution of 1 nm, a stroke of 40 μm×40 μm×20 μm (X×Y×Z) and a repeatability of 30 nm in each direction.

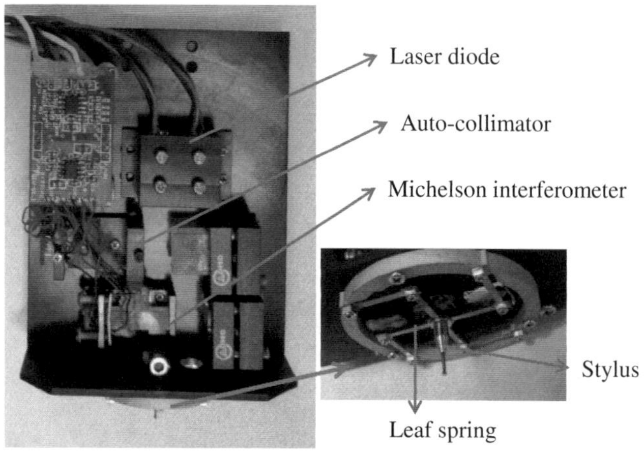

Figure 4. Photo the contact scanning probe.

3. Abbe error compensating model

According to the Abbe principle, the positioning errors of the X/Y stage can be calculated by the Abbe offset (L) and the angle error (θ). It can be expressed by the following equation [10]:

$$\delta = L \times \theta \tag{1}$$

The contact point and the Abbe offset are shown in figure 5.

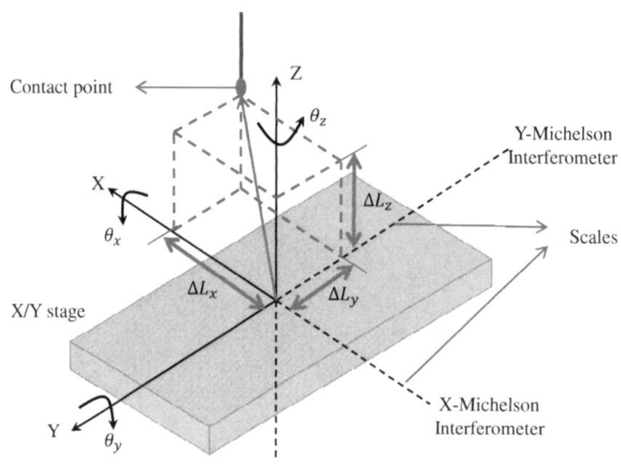

Figure 5. Abbe offset of the micro/nano CMM.

The Abbe error in three directions can be expressed by:

$$\begin{bmatrix} \delta_x \\ \delta_y \\ \delta_z \end{bmatrix} = \begin{bmatrix} 0 & \theta_z & \theta_y \\ \theta_z & 0 & \theta_x \\ \theta_y & \theta_x & 0 \end{bmatrix} \bullet \begin{bmatrix} \Delta L_x \\ \Delta L_y \\ \Delta L_z \end{bmatrix} \tag{2}$$

where δ_x, δ_y, δ_z are the Abbe error in each direction, θ_x, θ_y, θ_z are the angles in each axis and ΔL_x, ΔL_y, ΔL_z are the Abbe offset in each direction. In this equation, θ can be precisely obtained by the 2D angle sensors. However, it is quite difficult to measure ΔL accurately, a calculation method using "Least Square Method" has been proposed to solve this problem. The displacements of the stage in each direction can be measured by the Michelson interferometers integrated in the CMM and reference interferometers at the same time, then the positioning errors (δ_x, δ_y, δ_z) of the CMM can be obtained. At the same time, θ_x, θ_y, θ_z can be detected by an angle sensor. Therefore, the Abbe offsets (ΔL_x, ΔL_y, ΔL_z) can be calculated by Eq. 2 using "Least Square Method".

4. Experiments

4.1. Abbe error compensation

Each of the three linear stages of a CMM has three angle degrees of freedom. Angle error and Abbe offset always exist. In order to obtain the Abbe errors, both the angle errors and the Abbe offset need to be measured. However, no proper angle sensor can be used to measure the angle errors of the Z stage currently due to the limited space. Therefore, we have only finished the Abbe error compensation in X and Y axes.

Two SIOS laser interferometers (SP2000) were used as references. The X/Y stage moved at an interval of 1 mm from 0 to 10 mm. The displacement motions of the stage measured by the integrated interferometers and SIOS interferometers and the angles of the stage were recorded. Then, both the Abbe error and Abbe offset (ΔL_x=1.520 mm, ΔL_y=0.017 mm) can be obtained. Figures 6 and 7 show the positioning errors of the X/Y stage with and without compensating the Abbe error. From figures 6 and 7, we can see that the positioning errors in X and Y directions have been reduced from 435 nm to 91 nm and from 426 nm to 151 nm respectively after compensating the Abbe error.

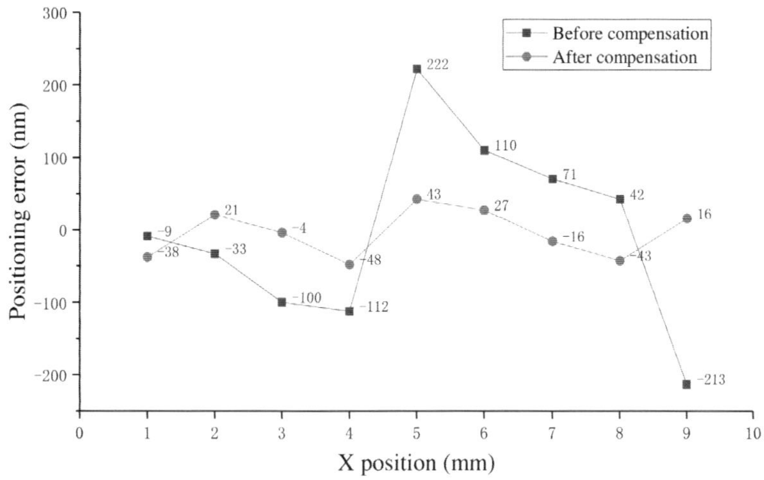

Figure 6. Comparison of results before and after Abbe error compensation (X position).

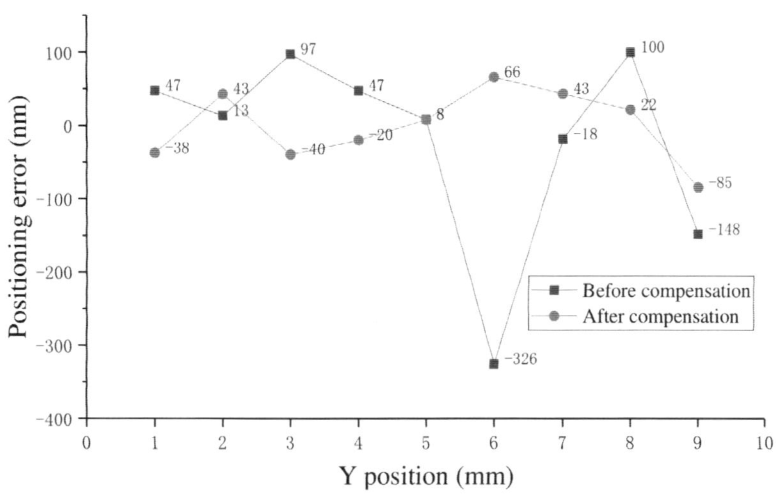

Figure 7. Comparison of results before and after Abbe error compensation (Y position).

4.2. Flatness test

In order to verify the performance of the micro/nano CMM, the flatness of a 0-grade gauge block was tested. The coordinate data of the six points shown in figure 8 were measured by the CMM, a plane equation was fitted in accordance with the principle of "Least Square Method". The measurement precision of flatness mainly depends on the measurement precision of Z coordinate. The Abbe offset of Z-axis stage is 120 mm and the angular motion is within ± 15 arcsec by calibration. The Abbe error in Z direction is less than 2.5 nm. Therefore, it is so small that can be ignored. Therefore, the flatness of the gauge block can be obtained. The results are shown in table 1, the repeatability is 44 nm (K=2).

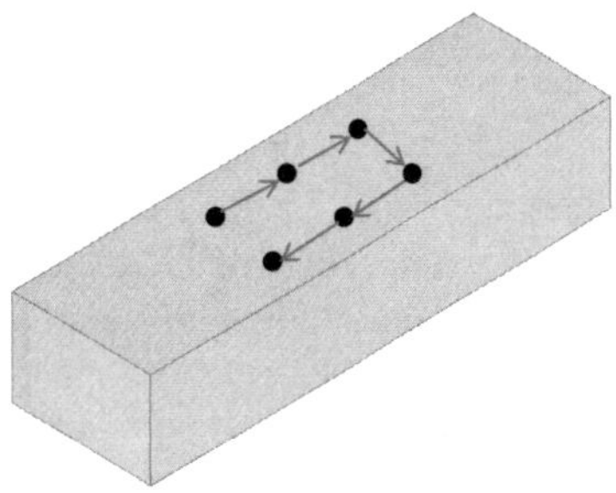

Figure 8. The gauge block and the flatness measurement path.

Table 1. The measurement results of flatness.

Times	Flatness (μm)
1	0.387
2	0.370
3	0.430
4	0.380
5	0.371
Average	0.388
Standard deviation	0.022

5. Conclusions

On the basis of a micro/nano CMM developed by our group, Abbe error compensation method is introduced and verified. In comparison with a SIOS interferometer (SP2000), experimental results validated the effectiveness of the method. The repeatability of the flatness measurement can reach 44 nm (K=2). Much more works to promote the performance of the CMM still need to be carried out in future.

References:

[1] Leach R K and Murphy J 2004 The design of coordinate measuring probe for characterising truly three-dimensional micro-structures *4th EUSPEN Int. Conf. (May 30- June 3, 2004, Glasgow, UK)* pp 230-1

[2] Peggs G N, Lewis A J and Oldfield S 1999 Design for a compact high-accuracy CMM *Annals of the CIRP* **48** pp 417-20

[3] Brand U, Kleine-Besten T and Schwenke H 2000 Development of a special CMM for dimensional metrology on microsystem components *ASPE 15th Annual Meeting in Scottsdale (Arizona)* pp 22-7

[4] Hughes E B, Wilson A and Peggs G N 2000 Design of high accuracy CMM based on multilateration techniques *Annals of the CIRP* **49** pp 391-4

[5] Brand U 1998 *Metrology for Microsystems* (Braunschweig: Physikalisch-Technische Bundesanstalt)

[6] SIOS Corporation 2014 *Nanopositioning and Nanomeasuring Machine* https://sios-de.com/wp-content/uploads/2016/02/NMM-1_e_2014.pdf

[7] Cheng F and Fan K C 2011 Linear diffraction grating interferometer with high alignment tolerance and high accuracy *Appl. Optics* **50** pp 4550-6

[8] Fan K C, Hsu S H and Zhou H 2014 Self-volumetric error compensation of a micro-CMM *58th Ilmenau Scientific Colloquium*

[9] Li R J, Fan K C, Huang Q X, Zhou H, Gong ER and Xiang M 2016 A long-stroke 3D contact scanning probe for micro /nano coordinate measuring machine *Precis. Eng.* **43** pp 220-9
[10] Fan K C, Yen H M and Li K Y 2014 A new concept of volumetric error analysis of machine tools based on Abbe principle *The 3rd Int. Conf. on Design Engineering and Science ICDES 2014 (August 31 – September 3, 2014, Pilsen, Czech Republic)*

Method selection to evaluate measurement uncertainty in microflow applications

J A Sousa[1], E Batista[1], O Pellegrino[1], A S Ribeiro[2], L L Martins[2]

[1]IPQ – Portuguese Institute for Quality, Caparica, Portugal
[2]LNEC – National Laboratory for Civil Engineering, Lisbon, Portugal

E-mail: jasousa@ipq.pt

Abstract. Microflow applications are gaining increasing importance in medicine, and are indeed crucial in areas such as neonatology and oncology. However, in terms of uncertainty evaluation the very small levels of flow entail problems characterized by a "close to the physical limit" situation. The same happens in nanovolume applications. In these cases the Law of Propagation of Uncertainty framework (GUM) may not to be adequate, since the interval of coverage probability will, in some circumstances, contain negative values, which is physically difficult to explain. Monte Carlo methods may be a good alternative when there are not many negative readings in the application (limit-of-detection problem), but may also prove inadequate otherwise, since the required functional relationship of the model will produce a peak of values centred at the zero value, which is also not very credible to describe satisfactorily the real situation. In the latter, the application of a Bayesian method will be the better approach to deal with the evaluation of measurement uncertainties. The whole process of the revision of the international metrological standardization lies in this principle, of selecting the best method to tackle the particular problem in hand, and not to apply blind recipes to the evaluation of uncertainty in measurement, and there is a new EMPIR project focused on providing examples to BIPM documents to strengthen this concept.

1. Introduction

The overall message conveyed by the new EMPIR (European Metrology Program for Innovation and Research) project 17NMR05 EMUE (Examples of measurement uncertainty evaluation) is that the choice of method to the evaluation of measurement uncertainty must be based on the particular features of the problem in hand, and should not be based on blind recipes [1]. It will contribute to the set of Joint Committee for Guides in Metrology (JCGM) documents related to measurement uncertainties (e.g., JCGM 100, JCGM 101, JCGM 102, JCGM 103, JCGM 106), particularly to JCGM 110 which will collect a number of examples to support this view and help users to select the best method for their particular problem. To provide a flavour of that approach, the present study will interrelate with another EMPIR project, MeDD II (Metrology for drug delivery) [2]. It is aimed at studying the development of new calibration methods requiring traceable measurements of volume, flow and pressure of existing drug delivery devices and inline sensors operating at very low flow rates, which increasingly take place in hospitals and health centres, expanding the existing metrological infrastructure, which has been a matter of applied research funded by other EMPIR projects [3].

The most commonly used form of therapy in health care is infusion therapy [4], which implies that drug delivery is an important topic in this sector. Due to the widespread applications in critical health care, infusion errors are often made, with reported dramatic effects in different applications in the health sector, especially in neonatology. There are various examples where adverse incidents, morbidity and mortality, can be traced back to poor inaccurate dosing [5, 6].

For instance, the accuracy of the flow rate set point adjustments based on the patient vital signs is insufficient to ensure safe delivery of drugs. Also the increasing implementations of novel microfluidic solutions in medical applications, such as organ-on-a-chip technology, urge the development of metrological infrastructure for validating quality and reproducibility. And a vital part of this infrastructure lies in the ability of producing quality measured values, which can only be achieved if adequate measurement uncertainty values can be obtained.

2. Objectives

This MeDD II health project has the overall goal of enabling traceable measurements of volume, flow and pressure of existing drug delivery devices and inline sensors that work at a flow rate lower than 100 nL/min, in order to minimize inaccurate measurement results, thus improving dosing accuracy in each infusion line.

To that purpose, one specific objective of MeDD II is to develop and validate novel calibration procedures for existing drug delivery devices (e.g. infusion pumps, pain controllers and infusion pump analyzers) with measurement traceability to a primary standard and with a target uncertainty value of 2 % in a 5 nL/min up to 600 mL/min range and in addition to develop a proof-of-concept of an on-chip microfluidic pump used as transfer standard in drug discovery and organ-on-a-chip applications in 5 nL/min to 100 nL/min flows. Another specific objective is to develop new nanoflow metrology techniques based on optical technology to measure flow below 100 nL/min for both steady and fast changing flow. Measurements will be performed for flow sensors and flow generators for Newtonian liquids. The target uncertainty is 1 % (k=2) for steady and 2 % (k=2) for fast changing flow rates.

To exemplify the mathematical treatment of the uncertainty evaluation, experimental values arising from an insulin pump calibration, will be used to prove the clear advantage of using a Bayesian method to quantify measurement uncertainties, when compared to other methods.

In terms of impact, it is known that drug delivery errors account for a significant percentage of medical errors, and depending on the drug type, the patient characteristics and the applied therapy, dosing errors can have severe consequences [7], including a substantial number of fatalities. Important examples can be found in chemotherapy, in oncology, in anaesthesia, in the operating theatre and in nursery wards, especially for the neonates. Therefore, any attempt to prevent adverse events by improving the knowledge on actual doses can already make an enormous difference for the individual patient, especially new born babies, and has a significant impact on the health sector as a whole. And, of course, measurement uncertainty plays a vital role in this equation, since a reliable uncertainty evaluation will lead to an improved confidence in the actual dosing prescribed and provided.

3. Uncertainty budget and possible approaches

3.1. Most readings are positive

A typical uncertainty budget of a flow measurement is illustrated in table 1, resulting from a real experiment with an insulin pump. In this particular test, most of the readings gave positive values as shown in figure 1, where the dispersion of values can be perceived. In this situation, the use of the GUM uncertainty framework (GUF) can first be attempted especially when the repeatability is assumed to have a Gaussian behaviour (which is the case in Table 1). The main uncertainty sources were considered in the mathematical modelling of the flow measurement (Q) expressed by

$$Q = \frac{1}{T_F - T_I}\left[\left((M_F - M_I) \times \left(1 - \frac{D_{tube}{}^2}{D_{tank}}\right)\right) \times \frac{1}{\rho_W - \rho_A} \times \left(1 - \frac{\rho_A}{\rho_B}\right) \times [1 - \gamma(t - 20)]\right] + M_{evap} \quad (1)$$

with mass measurements (M), densities values of water (ρ_W), air (ρ_A) and mass pieces (ρ_B), evaporation rate (M_{evap}), water temperature (t), time(T), expansion coefficient (γ), diameters (D), and measurements repeatability.

Table 1. GUM "uncertainty budget" for the insulin pump.

Quantity	*PDF*	Best estimate	Standard uncertainty $u(x_i)$	c_i	u_i
T_I (s)	Gaussian	2.65×10^{2}	7.00×10^{-4}	4.62×10^{-11}	3.23×10^{-14}
M_I (g)	Combined	2.244425	3.78×10^{-5}	3.27×10^{-4}	1.23×10^{-8}
ρ_W (g/ml)	Combined	0.998604	5.57×10^{-4}	-1.42×10^{-7}	-7.88×10^{-11}
ρ_A (g/ml)	Rectangular	0.001208	2.89×10^{-6}	1.24×10^{-7}	3.58×10^{-13}
ρ_B (g/ml)	Normal	8.00	2.50×10^{-3}	2.66×10^{-12}	6.66×10^{-15}
t (°C)	Combined	1.80×10^{1}	7.22×10^{-1}	-1.41×10^{-12}	-1.02×10^{-12}
γ (°C^{-1})	Rectangular	1.00×10^{-5}	2.89×10^{-7}	2.82×10^{-7}	8.16×10^{-14}
M_{evap} (ml/s)	Rectangular	1.04×10^{-7}	1.47×10^{-8}	1.00	1.47×10^{-8}
D_{tube} (cm)	Normal	0.09	0.001	-1.38×10^{-8}	-1.38×10^{-11}
D_{tank} (cm)	Normal	1.36	0.001	9.13×10^{-10}	9.13×10^{-13}
δQ_{rep} (ml/s)	Normal	0.00	1.99×10^{-8}	1.00	1.99×10^{-8}
M_F (g)	Combined	2.255857	3.78×10^{-5}	-3.27×10^{-4}	-1.23×10^{-8}
T_F (s)	Gaussian	3.32×10^{3}	7.00×10^{-4}	-4.62×10^{-11}	-3.23×10^{-14}
Flow rate (ml/s)		$\mathbf{2.45\times10^{-7}}$			$\mathbf{3.03\times10^{-8}}$

In this first case study, it is apparent from figure 1 that almost the entire set of readings is non negative, despite having very small magnitude. There are other examples, as will be seen in section 3.2, where this is not the case and the range of values is much larger, including also negative values.

The example of figure 1, though, represents a good response of the insulin pump with a relatively small repeatability, and two possible approaches will be discussed to quantify the corresponding measurement uncertainty, namely the GUF and the Monte Carlo method (MCM).

In the first approach, and for majority of situations, repeatability is treated as having a Gaussian distribution, assuming that data can be well represented by the corresponding mean value and standard deviation [8]. The associated uncertainty is taken as the standard deviation of the mean, which is the standard deviation divided by the square root of the number of readings. The result from this approach can be seen in figure 3, from where it is reasonable to conclude that the GUF approach is acceptable, which was expected since there is only a mild non linearity in the model (1), and there is not a dominant non Gaussian source of uncertainty in table 1. The result of the MCM is displayed in figure 3, with a difference between approaches of about 10 % in terms of the coverage interval for a confidence level of 95 %, thus the GUF approach would be acceptable in most applications.

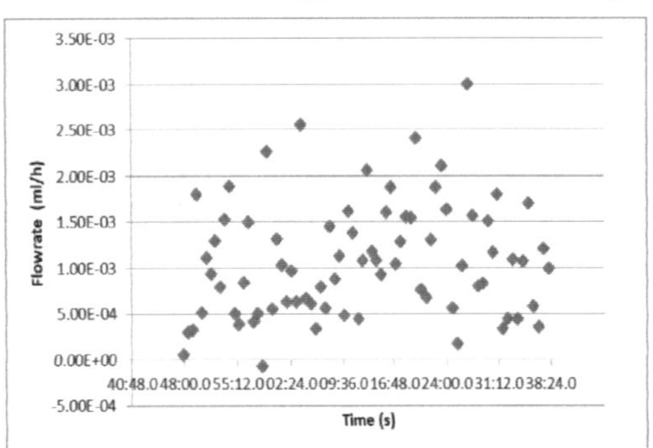

Figure 1. Flowrate readings for the case study 1 when most readings are positive.

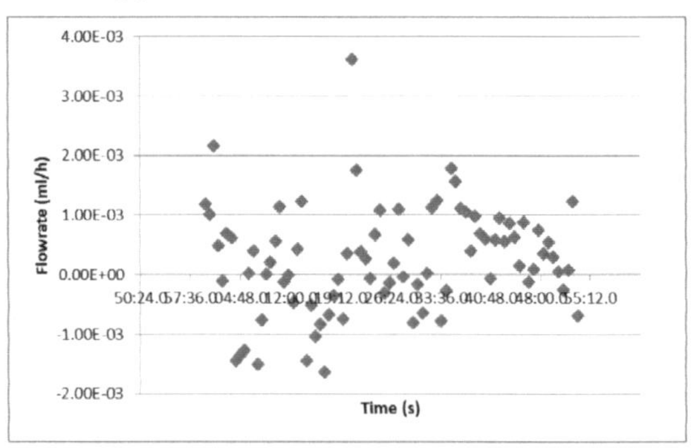

Figure 2. Flowrate readings for the case study 2 with a significant number of negative readings.

However, if it is assumed that data in figure 1 can be represented by a rectangular distribution, based on the fact the scatter of the data only allows using their limits as a valid information [8], then the comparison between the two approaches reveals a striking difference as illustrated in figure 4. The difference between the GUF and MCM is quite apparent in the graph and is accounted as being more

Journal of Physics: Conference Series **1379** (2019) 012033 doi:10.1088/1742-6596/1379/1/012033

than 200 % in the coverage interval for a level of confidence of 95 %. This allows an obvious conclusion that the use of the GUF, under these assumptions, would clearly underestimate the measurement uncertainty associated with the insulin pump.

An interesting feature of the MCM in this situation is the concentration of values at zero shown in figure 4, where a spike near the origin is apparent. In this case study 1 where the concentration of values is not too large, one can reasonably conclude that a greater number of negative values would dramatically increase this effect.

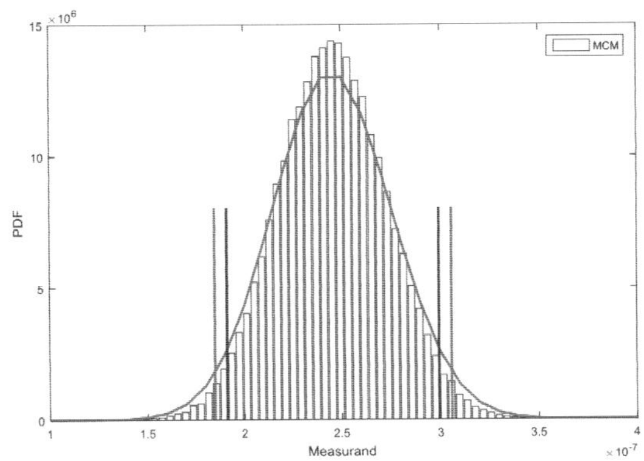

Figure 3. Simulations using the GUM and MCM methods for the case study 1.

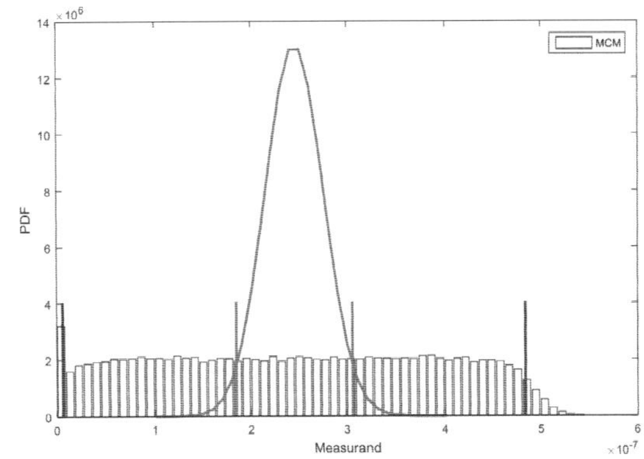

Figure 4. Simulations for the case study 1 when repeatability is taken as rectangular.

3.2. Significant number of negative readings

A different situation arises when one is faced with data as depicted in figure 2. In this case study 2 the number of negative values is significant and the spike of values found in figure 4 would be much enlarged. Therefore neither the GUF nor the MCM method will be adequate to describe this latter situation, as these methods concern the probability density of the measurand estimate. The Bayesian method rather concerns the probability density of the measurand, considering the data produced by the measurement and possible prior information [9]. Indeed, this method should be capable of dealing well with this situation, where the distribution of values is truncated at zero (prior knowledge), and thus redistributes the values over the rest of the integral domain.

The average flow rate Q can be estimated from measurements of mass and time according to a relationship of the form

$$Q = \left(\frac{M_F - M_I}{T_F - T_I}\right) f(\boldsymbol{a}) \tag{2}$$

Where $f(\boldsymbol{a})$ is a correction factor depending on parameters $\boldsymbol{b}$ of the specific instrument and environmental effects. Given distributions associated with each quantity they can be propagated through to a distribution for Q using a Monte Carlo method, as a measurement equation. An observation equation approach to the analysis of the measurement [10] uses a relationship of the form

$$M_F = \left(M_I - M_{evap}\right) + (T_F - T_I)Qg(\boldsymbol{a}) + \epsilon_F \tag{3}$$

Setting $\boldsymbol{b} = (M_I, T_I, T_F, \boldsymbol{a}^{\mathrm{T}})^{\mathrm{T}}$ with non-informative priors, we can write (3) as

$$M_F = \phi(Q, \boldsymbol{b}) + \epsilon_F = h(\boldsymbol{b}) + k(\boldsymbol{b})Q + \epsilon_F \tag{4}$$

The prior for Q should include the constraint that $Q \geq 0$ since there is no physical mechanism for a negative flow. The posterior distribution for the parameters Q and $\boldsymbol{b}$ is such that

$$p(Q, \boldsymbol{b}|M_F) \propto p(M_F|Q, \boldsymbol{b})p(\boldsymbol{b})p(Q) \tag{5}$$

where the prior $p(\boldsymbol{b})$ has been assigned [11]. The form of the observation equation (4) is precisely that of the instrument response model considered in [10] which allows for a particularly simple Markov chain Monte Carlo (MCMC) algorithm to be implemented in the basis of a GUM supplement [8] sample generated using the measurement model. The posterior distribution $p(Q|M_F)$ will differ from the distribution derived using the GUMS1 approach since $\partial\phi/\partial Q = k(\boldsymbol{b})$ is not constant and the prior distribution $p(Q)$ must impose the constraint that $Q \geq 0$.

The result of the simulation using the MCMC with the constraint indicated above proves the advantage of the Bayesian approach in limit-of-detection problems, as illustrated in figures 5 and 6. In fact, the change of mass is similar to the uncertainty in the mass measurement. The Bayesian posterior is very much like a truncated Gaussian, as was expected. A simple MCM simulation would result in all the negative values allocated at zero creating an unfeasible representation of the flow rate distribution.

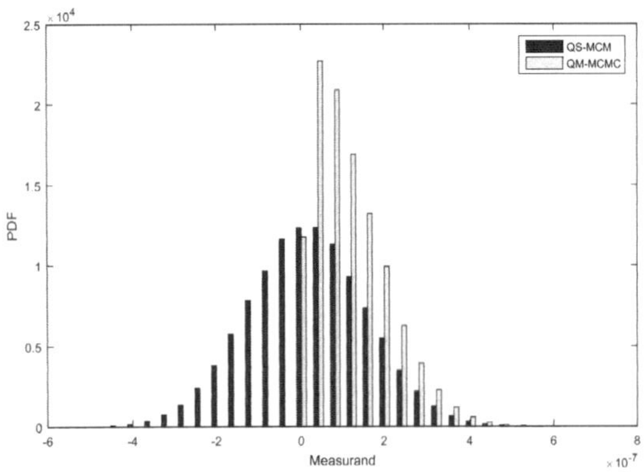

Figure 5. MCMC simulations truncated at zero compared with MCM simulations without any constraint for the case study 2.

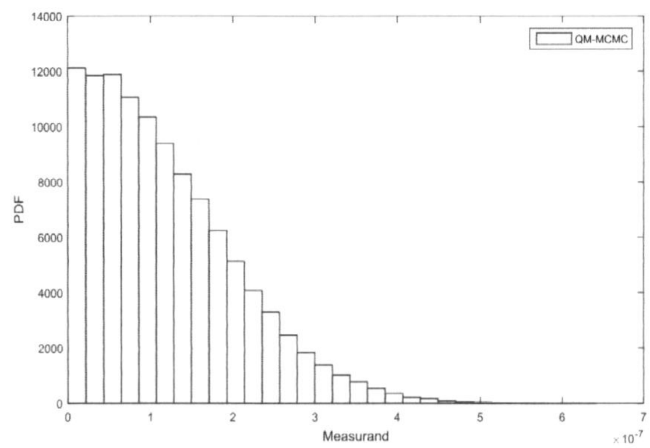

Figure 6. MCMC simulations for the case study 2 with constraint $Q \geq 0$.

4. Conclusions

The aim of this work is to provide guidance on method selection with respect to the evaluation of measurement uncertainty. Clearly the choice of method depends on the particular problem under evaluation and no blind recipe should be used. Depending on the conditions of the problem, the GUM uncertainty evaluation may be sufficiently adequate, whereas in other circumstances alternative approaches should be applied instead, e.g., the Monte Carlo method or the Bayesian method. What this study indicates is that the MCM approach is a convenient alternative to the GUM uncertainty evaluation approach, with the advantage of having often a simpler application, despite requiring a software implementation. However, this robust alternative method may also prove inadequate in certain circumstances such as in limit-of-detection type of problems. In the latter class of problems, the constraints imposed to the measurand by the use of a prior fits well to the Bayesian approach and is clearly a better method to evaluate measurement uncertainty in a number of problems that have application not only in microflow but also in nanovolume, surface characterization (roughness) or chemical analytical measurements in substances with a high degree of purity.

Acknowledgments

Thanks are due to Professor Alistair Forbes of NPL for a thorough revision of the paper and invaluable suggestions for improvement. The EMPIR project "17NRM05-EMUE" is carried out with funding of the European Union under the EMPIR. The EMPIR is jointly funded by the EMPIR participating countries within EURAMET and the European Union.

References

[1] EMPIR project 17NRM05 EMUE – Examples of measurement uncertainty evaluation. Program available on www.euramet.org
[2] EMPIR project MeDD II – Metrology for drug delivery. Program available on www.euramet.org
[3] EMPIR project MeDD I – Metrology for drug delivery. Program available on www.euramet.org

[4] Lee P T, Thompson F and Thimbleby H 2012 Analysis of infusion pump error logs and their significance for health care *British Journal of Nursing* Apr 26-May 9; **21**(8) S12, S14, S16-20
[5] Heinemann L, Fleming GA, Petrie JR, Holl RW, Bergenstal RM and Peters AL 2015 Insulin pump risks and benefits: a clinical appraisal of pump safety standards, adverse event reporting, and research needs: a joint statement of the European Association for the Study of Diabetes and the American Diabetes Association Diabetes Technology Working Group *Diabetes Care* **38**(4) 716-22
[6] Ma H, Lovich M A and Peterfreund R A 2011 Quantitative analysis of continuous intravenous infusions in pediatric anesthesia: Safety implications of dead volume, flow rates, and fluid delivery *Paediatr. Anaesth.* **21** 78–86
[7] Husch M, Sullivan C, Rooney D, Barnard C, Fotis M, Clarke J and Noskin G 2005 Insights from the sharp end of intravenous medication errors: implications for infusion pump technology *Qual. Saf. Health Care* Apr; **14**(2) 80-6
[8] JCGM 2008 *JCGM 101:2008 Evaluation of Measurement Data – Supplement 1 to the "Guide to the Expression of Uncertainty in Measurement" – Propagation of Distributions Using a Monte Carlo Method* (Joint Committee for Guides in Metrology)
[9] Calonico D, Levi F, Lorini L and Mana G 2009 Bayesian inference of a negative quantity from positive measurement results *Metrologia* **46** 267-71
[10] Forbes A B and Sousa J A 2011 The GUM, Bayesian inference and the observation and measurement equations *Measurement* **44** 1422-35
[11] Gelman A, Carlin J B, Stern H S, Dunson D B, Vehari A and Rubin D B 2014 *Bayesian Data Analysis* third edition (Boca Raton, Florida: Chapman & Hall/CRC)

Journal of Physics: Conference Series **1379** (2019) 012034 doi:10.1088/1742-6596/1379/1/012034

A dual measurement and stimulating system to monitor and promote NNS

J M Pereira[1,2], V Viegas[1,3], O Postolache[1,4] and P S Girão[1,5]

[1] Instituto de Telecomunicações, Lisboa, Portugal
[2] ESTSetúbal, Instituto Politécnico de Setúbal, Setúbal, Portugal
[3] Centro de Investigação Naval (CINAV), Escola Naval, Base Naval de Lisboa – Alfeite, Almada, Portugal
[4] ISCTE – Instituto Universitário de Lisboa, Lisboa, Portugal
[5] Instituto Superior Técnico, Universidade de Lisboa, Lisboa, Portugal

Email: dias.pereira@estsetubal.ips.pt

Abstract. The present paper presents a dual system that can be used to measure and to generate non-nutritive sucking (NNS) signals. The NNS measurement is based on a pressure sensor that is pneumatically connected to a silicone pacifier. By its turn, the NNS stimulator is based on a miniature pump and a miniature electro-valve that are operated by opposite phased signals to produce the pulsed pressure signal that is pneumatically applied to the silicone pacifier. The paper highlights some specific issues related with the calibration of the pneumatic signal measuring chain. Processing and wireless transmission capabilities are provided by a low-cost solution based on a microcontroller. Calibration and experimental results are also included in the paper.

1. Introduction

Measurement of non-nutritive sucking signals in neonatology services, namely in neonatal intensive care units (NICU), is a very important topic. This topic has a strong clinical appeal and a different NNS related subjects had been studied in the last 20 years highlighting the relationship between NNS behavior and baby's physiological variables. The effectiveness of NICU affects directly the improving of the survival rate of premature babies. However, it is important to refer that NICU fixed and variable costs are extremely high [1]. As a reference of the direct costs in NICU, it can be cited an annual cost over 4 billion USD in the United States for low-birth-weight infants through their first year of life [2]. This paper proposes a low-cost and dual measurement and stimulating system that can be used to measure non-nutritive sucking signals of premature babies and to stimulate an autonomous nutritive sucking capability.

Considering the importance of this topic, this paper presents a low-cost and safe solution that can be used to measure and to generate very low amplitude pressure signals that are required to analyze and to stimulate, nutritive and non-nutritive sucking activities, respectively. It is important to underline that one of the main challenges associated with the development of the prototype that is presented in this paper is associated with the very low pressure measuring and generating amplitude ranges of the NNS pressure signals with the required accuracy. The paper also focusses some issues related with calibration [3-7] that is an essential concern whenever low amplitude measurements are at stage. It is

Journal of Physics: Conference Series **1379** (2019) 012034 doi:10.1088/1742-6596/1379/1/012034

also important to refer that proposed system enables an easy implementation of auto-calibration routines since it integrates a dual measurement and stimulating system that have an inherent feedback topology.

Finally, it is important to underline that although there exists at least one commercial solution available that can be used for some kind of NNS measurements, its cost is very high, its flexibility is low [8-9] and from the specification data available it is not clear its dual capability to measure and stimulate NNS signals. Moreover, these commercial solutions are as usual closed solutions in terms of hardware and software.

2. Measurement system

The block diagram of the dual measurement and stimulating NNS system is represented in figure 1. The system includes a common silicone pacifier that is connected through two pneumatic tubes to the input of the sucking pressure measurement (SPM) unit and to the output of the sucking pressure stimulator (SPS) unit. The microcontroller-based unit [10] includes 8 A/D conversion channels, 22 digital I/O ports, 6 of which are PWM, and a Flash Memory of 32 KB, of which 2 KB are used by the boot loader. Wireless transmission capabilities are easily supported by connecting the Arduino unit to a Bluetooth Module HC-05 [11].

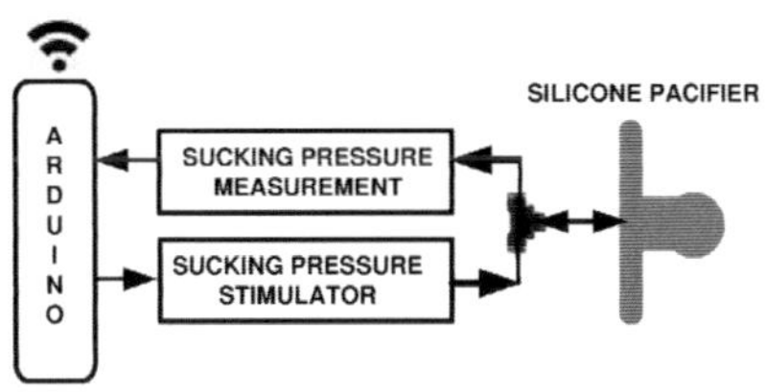

Figure 1. Measurement system block diagram.

The SPM unit performs the signal conditioning of the electrical signal delivered by the pneumatic pressure sensor and the SPS unit generates the sucking signal patterns that are applied to the silicone pacifier. Figures 2 and 3 represent the schematic diagram of the SPM and SPS units, respectively. The circuits include several COTS devices, the pneumatic pressure sensor [12], a low-cost miniature pneumatic pump [13] and a miniature electro-valve [14] similar to the ones that are used in commercial blood pressure measurement devices [15]. Regarding the main specifications of the sensors and actuators, it is important to refer that the pneumatic pump delivers a maximum output pressure equal to 15 p.s.i. and the electro-valve has a maximum flow rate amplitude equal to 1.2 l/min. The pressure transducer has a relative pressure measuring range equal to 15 p.s.i., a linearity error lower than 0.25 % of its span and a repeatability plus hysteresis error lower than 0.2 mV for a nominal supply voltage of 10 V.

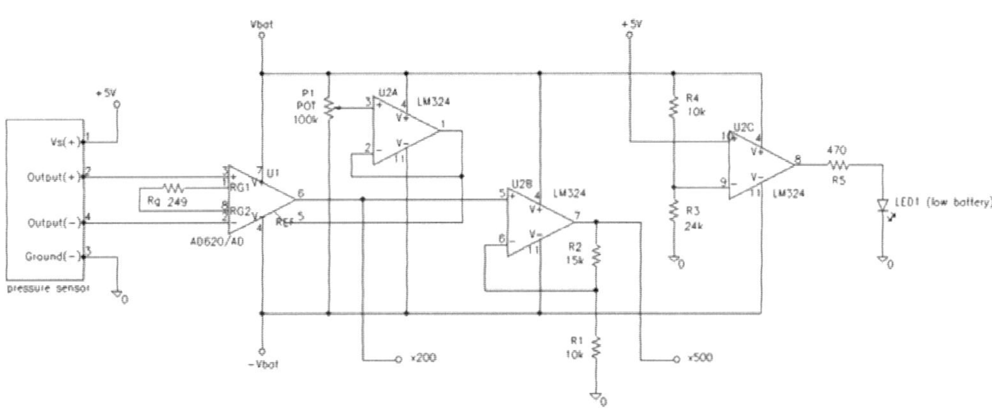

Figure 2. Schematic diagram of the SPM unit.

Journal of Physics: Conference Series **1379** (2019) 012034 doi:10.1088/1742-6596/1379/1/012034

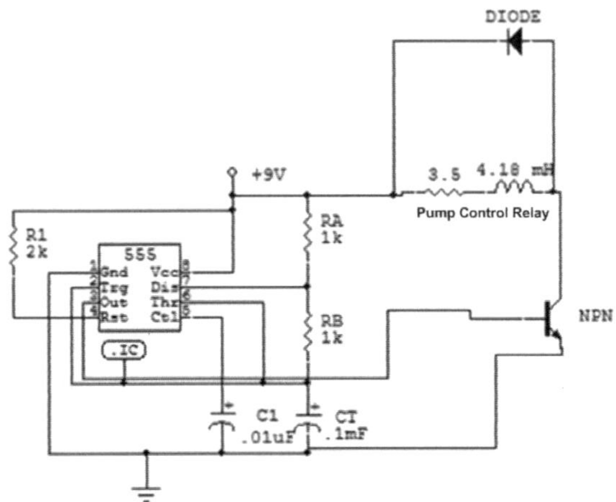

Figure 3. Schematic diagram of the SPS unit.

The software that runs in the Arduino device performs several simple tasks related with HW and SW configurations, data processing and packing, data transmission and some self-test functions to monitor the proper working conditions of the system. More complex functions that require higher data processing capabilities are performed remotely in a Smartphone or Laptop of the NICU supervision center that can communicate, by its turn, with several NNS measurement and stimulating units.

3. System calibration

The required accuracy of the measurement system implies the calibration of the pneumatic pressure sensor measurement chain. Figure 4 depicts the devices that were used for calibration purposes that include a pneumatic pump [16], an inclined U-tube manometer [17], a pressure indicator and pressure loop calibrator [18], and the pressure sensor unit under calibration.

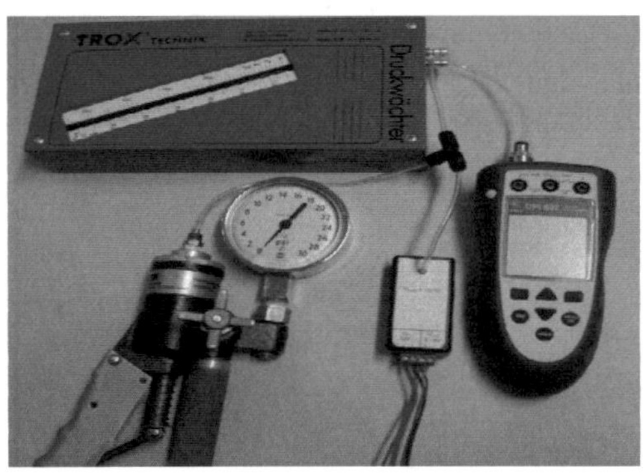

Figure 4. Calibration apparatus set-up.

Figure 5 represents the calibration results that were obtained after three calibration cycles. Using a linear curve fitting of the full set of the calibration points, the linear relationship between pressure and voltage values is given by:

$$V_0=0.368 \cdot p+8.6 \quad (1)$$

where V_0 represents the output voltage of the conditioning circuit in mV, p represents the input pressure in Pa, and the constant, 8.61 mV, is an offset deviation caused by the non-linearity error of the sensor plus conditioning circuit. The linearity of the measuring chain is experimentally confirmed, and the correlation coefficient is almost equal to one (cc=0.99996).

Regarding hysteresis evaluation, the upward and downward relationship between pressure and voltage values are given by:

$$V_0=0.360 \cdot p-0.2 \quad \text{(upward)}$$
$$V_0=0.364 \cdot p-0.5 \quad \text{(downward)} \qquad (2)$$

where, as previously, V_0 represents the output voltage of the conditioning circuit in mV and p represents the input pressure in Pa. From the previous relationships, it is possible to confirm that the hysteresis of the input pneumatic measuring chain is lower that 0.91 % of the measuring range span.

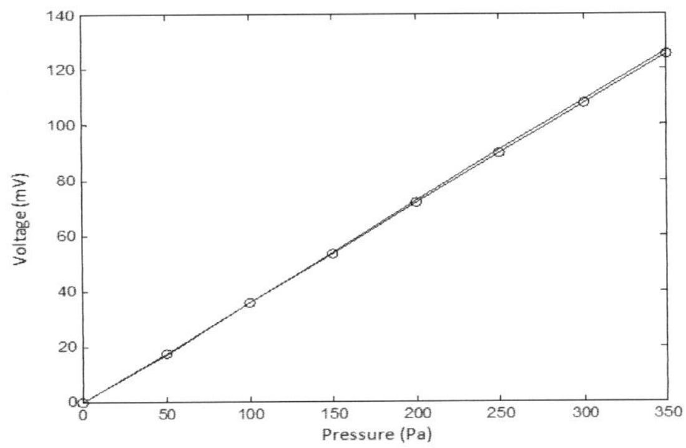

Figure 5. Calibration results.

It is important to underline that the expected uncertainty of the instrument after calibration is mainly related with the uncertainty of the pressure measurements since the voltage measurements obtained from a 5 1/2 digit multimeter [19] is about 0.13 mV in the selected DC voltage measurement range. The error that is associated with the pressure measurements is 0.1 % of the FS (200 mbar) that corresponds to 20 Pa. Using the calibration curve relationship, it is possible to obtain an overall voltage measurement accuracy better than 7.5 mV, and an accuracy better than 25 Pa in terms of pressure measurements. This accuracy level is acceptable for the instrument purposes since the main pressure parameter related with NSS measurements is the maximum pressure amplitude that is typically higher than 500 Pa, which means that a maximum relative error lower than 5 % is achieved.

4. Experimental results

A low-cost data acquisition board (NI USB-6008) and a simple LabVIEW routine was developed to obtain some tests of the proposed NNS dual measurement and stimulating system. Figure 6 represents NNS measurement results when the proposed system was tested with a preterm infant. In this case, it was acquired a NNS signal with a burst duration almost equal to 7 s, the suction frequency is approximately equal to 1,3 Hz and the maximum pressure suction amplitude (PSA) in equal to 1200 Pa.

Concerning stimulating NNS signal stimulation, figure 7 represents the experimental results that were obtained with a data acquisition rate of 1 kS/s for a stimulating sucking pattern with a frequency of 4.8 Hz, 12 pulses per sucking burst and an inter-burst interval equal to 1 s. The signal with a square waveform profile corresponds to the pneumatic pump control signal and the signal with a ramp waveform profile corresponds to the associated pressure sensing signal.

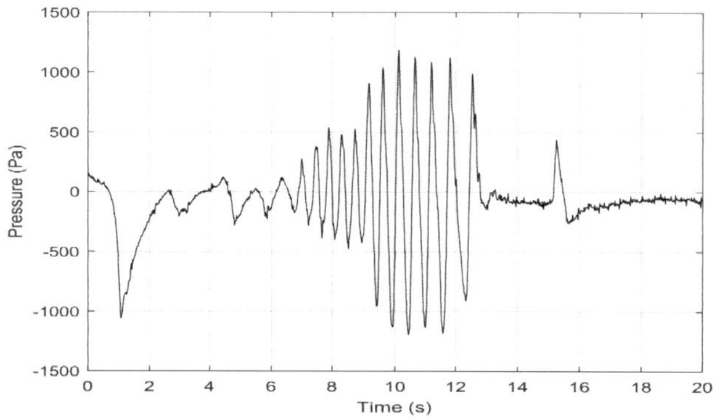

Figure 6. Measurement of a NNS burst ($f_{suction}$=1,3 Hz; $burst_{duration}$=7 s; PSA_{pk}=1200 Pa).

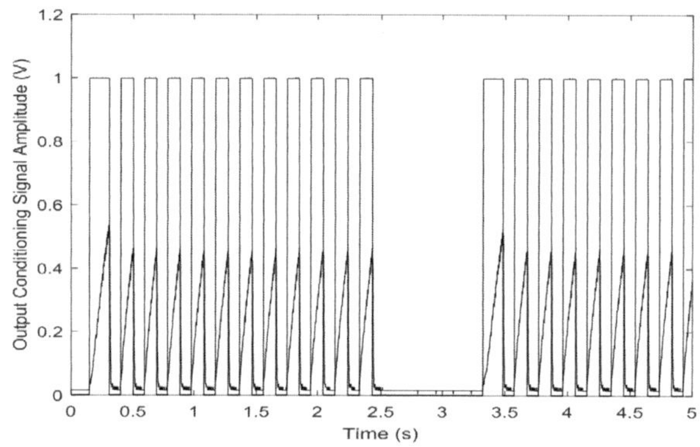

Figure 7. NNS signal stimulation ($f_{suction}$=4,8 Hz).

By its turn, figure 8 represents the experimental results that were obtained in similar testing conditions but for a stimulating sucking pattern with a frequency of 1,7 Hz. As it is clearly visible, the output conditioning signal amplitude of the pressure signal, in this case, is much higher than the one that was previously obtained, since the transfer function associated with these signals has a low-pass filter first order behavior, being the time constant dependent on the pneumatic characteristics of the pacifier, pneumatic tubes and, obviously, air compressibility.

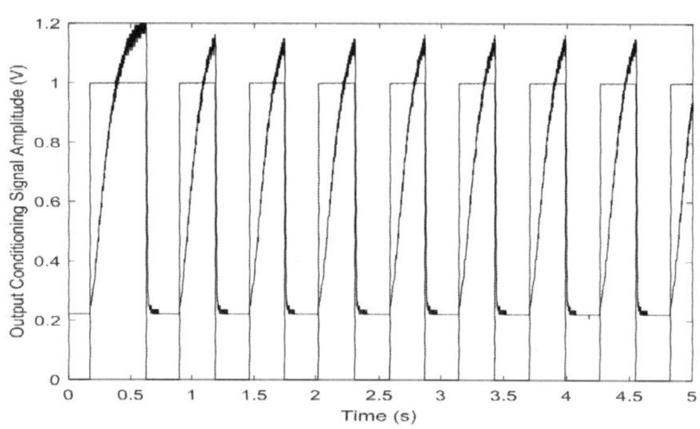

Figure 8. NNS signal stimulation ($f_{suction}$=1,7 Hz).

5. Conclusions

The experimental results that were obtained confirm that the proposed system can be successfully used to measure and stimulate sucking abilities of premature babies. The system integrates measurement and stimulating capabilities that can be used separately or together. The sensing unit includes mainly a pressure sensor, a conditioning circuit and a microcontroller-based device. The stimulation unit includes a miniature and low-cost pneumatic pump, a miniature electro-valve and a driver circuit. Besides its low-cost, the system exhibits a high flexibility in terms of configuration and operation. Concerning flexibility, it is important to underline that the proposed system can generate sucking patterns with different parameters, such as, amplitude, frequency, number of pulses, sucking burst per minute, and duty rate between stimulating and measuring periods, being possible to perform an on-line adjustment of any of these parameters according to the premature infant evolution.

Acknowledgements

The work was supported by Fundação para a Ciência e Tecnologia, Project TAILORPHY - PTDC/DTP-DES/6776/2014, and Instituto de Telecomunicações.

References

[1] Kornhauser M and Schneiderman R 2010 How plans can improve outcomes and cut costs for preterm infant care *Managed Care Review*
[2] Narang A, Kiran P S and Kumar P 2005 Cost of neonatal intensive care in a tertiary care center *Indian Pediatric* **42** 10 989-97
[3] Frank R 2013 *Understanding Smart Sensors* (Boston/London: Artech House)
[4] Ponmozhi J, Frias C, Marques T and Frazão O 2012 Smart sensors/actuators for biomedical applications: review *Measurement* **45** 7 1675-88
[5] Pertijs M, Aita A, Makinwa K and Huijsing J H 2010 Low-cost calibration techniques for smart temperature sensors *IEEE Sens. J.* **10** 61098-105
[6] Meijer G, Pertijs M and Makinwa K 2014 *Smart Sensor Systems: Emerging Technologies and Applications* (John Wiley & Sons)
[7] Pereira J M, Postolache O and Girão P S 2005 Adaptive self-calibration algorithm for smart sensors linearization *Proc. 22th IEEE Instrumentation and Measurement Tech. Conf. (Ottawa)* vol 1 pp 648-652
[8] NTrainer New device helps premature babies get off feeding tubes better http://www.science20.com
[9] Site of Care Systems, Neonatal healthcare software http://www.siteofcare.com/products/nursery
[10] Store.arduino.cc, Arduino Nano https://store.arduino.cc/arduino-nano
[11] Electrofun Módulo Bluetooth HC-05 Para Arduino https://www.electrofun.pt/comunicacao/modulo-bluetooth-arduino-hc05
[12] Honeywell Temperature compensated pressure sensor (Ref. 40PC001B*)* http://www.sensortechnics.com
[13] Shenzhen aliElectric Technology Co., Ltd, KPM14A mini monitor with solder point https://ru.aliexpress.com/item/32234902089.html
[14] Xiamen AJK Technology Co., Ltd, Pressure relief mini solenoid valve AJK-F051 https://www.ajkpump.com/pressure-relief-mini-solenoid-air-valve_p49.html
[15] Hartmann Tensoval Blood pressure monitor for reliable upper arm measurement https://hartmann.info/en-za/products/diagnostics/blood-pressure-monitors/upper-arm-products/tensoval%C2%AE-comfort-classic
[16] AMETEK Portable pneumatic tester – model T730 http://www.testequipmentdepot.com/ametek/pressurepumps/t730.htm
[17] Airflow Developments Type 504 manometer https://www.bsria.co.uk/instrument/sales/product/airflow-developments-type-504-manometer

[18] GE Sensing, GE Druck DPI 800/802 Pressure indicator and loop calibrator https://www.instrumart.com/products/18294/ge-druck-dpi-800-802-pressure-indicator-and-loop-calibrator
[19] Keithley 2110-220 - Bench digital multimeter 2110 series 1 kV 10 A 5.5 digit https://www.tek.com/digital-multimeter/broad-purpose-digital-multimeters-manual-3

Musical instruments for the measurement of autism sensory disorders

E Benoit[1], S Perrin[1], S Donnadieu[2], C Dascalu[2], G Mauris[1], J Favory[3], C Dautremer[3]

[1] Univ. Grenoble Alpes, Univ. Savoie Mont Blanc, LISTIC, Annecy, France
[2] Univ. Grenoble Alpes, Univ. Savoie Mont Blanc, CNRS, LPNC, 38000 Grenoble, France
[3] Conservatoire Rayonnement Regional, Annecy, France

E-mail: eric.benoit@univ-smb.fr

Abstract. The analysis of sensory disorders in children with autism is a long process based on the observation of children in a specific environment. It cannot be treated frequently and monitoring the evolution of sensory disorders is either impossible or limited to a period of more than one year. The objective of our study is to define and then use dedicated musical instruments to perceive the intensity of children's responses to different stimuli in order to measure their sensory disorders.

1. Introduction

According to the American Psychiatric Association, autism spectrum disorders (ASD) are neurodevelopmental disorders characterized by deficits in social communication and the presence of limited and repetitive behaviours. Abnormalities in the sensory treatment of ASDs are frequently reported at all ages and may involve all sensory modalities (e.g., apparent indifference to pain, avoidance of certain sounds or textures, perception of unusual odour of objects, search for visual experiences of light or movement [1]). Given the behavioural, neurophysiological and anecdotal evidence of the sensory characteristics of ASDs, the most recent edition of the Diagnostic and Statistical Manual of Mental Disorders [2] presents *hyper- or-hypo-reactivity to sensory inputs or special interests in the sensory aspects of the environment* as a limited and repeated type of behaviour.

Some authors have taken an interest on music and its impact under people with ASD. Indeed, audio stimuli have a greater impact on them than on people without ASD [3]. The response depends of the nature audio stimuli: song [4] and music [5] have a greater influence compared to typically developing people. And conversely, language stimulation cause a less important response compared to typically developing people [4, 6]. People with ASD seemed to have better musical skills than those associated with language [7-10]. Another benefit of using music for people with ASD is that it seems beneficial to themselves [11]. Therefore, a musical environment is preferred to stimulate people with ASD, but the audio stimuli is not the only one involved in musical activities. Vision, tactile and vestibular stimuli are also involved and cause various responses depending on the specific sensory sensitivity of people.

The idea that initiated our study is to take the opportunity of the existence of the musical activities dedicated to children with ASD to analyze the specific sensory sensitivity of these children in order to identify their sensory disorder. The basic approach of our study is based on the hypothesis that the sensitivity features related to the disorders of the children can be measured by a set of dedicated musical instruments. Such instrument will them play two roles. The main role involving the musical aspects of the instrument will be to stimulate the children by the use of multiple stimuli. According to the metrological aspects of the instruments, the second role will be to perform measurements of the sensory sensitivity.

The horizon of new possibilities related to the confirmation of this hypothesis is extremely large. First, and it's the most important, the periodicity of the measurement will be reduced to one week instead of one year. This will allow to consider to follow the evolution of sensory disorders.

From a measurement science viewpoint, the definition of the measuring instruments needs not only to define the main quantities under the global measurement, but also the link to physical quantities able to be measured by physical instruments.

We present the context of our study in the next section. Then we give the definition of the quantity under measurement. After, we present the possible measurands for the musical instruments and finally, we tackle sensitivity in order to calibration stage to be done.

2. Context of study and methodology

One of the final goals of our study is to design an augmented musical instrument capable of measuring the sensory characteristics of ASDs. The first step in doing this is to determine what are the important characteristics of people with ASDs to use. In this particular context, the traditional measurement based on questionnaires [12] cannot be done. We chose to use visual observation as a starting point to allow an autism specialist to make a diagnosis.

The Annecy conservatory (CRR) develops from several years specific musical activities for children with ASD. During these music workshops, children have the possibility to use various kind of musical instruments under the supervision of music teachers. To assess sensory sensitivity with reference methods, six observation sessions from the music workshop were filmed. The content of the sessions included two programs of 9 and 8 sequences established by the music teacher. The first program contains the following sequences: body percussion to music, playing with a phonotonic and djembe, playing with boomwhackers, instrument cards, playing the statue with hoops, bells, "African music", meditative music, violin. The second program contains other sequences: "melody of first names", electronic drums, conductor's playing, string instrument sequences. In order to best assess the sensory abilities of young people with ASD, two additional sessions were developed in collaboration with the music teacher. The proposed musical program included the following sequences: two opening songs played with sticks and boomwhackers, play with tissues under the strings (guitars and ukuleles), Xylophones with hard and soft red sticks and variations in blade colors, "The lion is dead tonight" with the boomwhackers, Ira Congo sung a capella to finish. A sensory game was also interspersed with the proposed musical sequences to evaluate visual, auditory, tactile, proprioceptive and vestibular modalities. The sensory objects used in this sensory game were chosen to test each modality: the double bass (hearing), pin balls (tactile), a jumping ball (vestibular), bright eggs (vision), sensory balls (proprioceptive).

We filmed the music workshop for several children. Then, we annotated the videos using ELAN software to characterize the symptomatic attitude of ASDs. These observations are being analysed to check their consistency with the conventional diagnosis. These results related to Psychology sciences are not presented in this paper that is concerned by the Measurement science; we focus on the ability of ASD to be measured by sensors. In the long term, these sensors will be integrated into musical instruments made available during the workshops. The next step in this process is to design such a musical instrument that can measure the sensory profile of children with ASD. To do this, it is necessary to associate, to each symptomatic sensory attitude to be measured, the sensors that allow it. This article focuses on the proposal of this association to establish the sensory profile by measuring ASDs from sensors that will be placed in an object, particularly in a musical instrument made *sensitive* in this way.

3. Definition of the quantity under measurement

Sensitivity studies in children with ASDs indicate that sensory characterization is achieved by identifying the intensity of hypersensitivity and hyperreactivity to stimulation related to hyposensitivity and hypersensitivity respectively.

Actually, responsiveness is never considered as a one-dimensional quantity because a child's sensitivity can be both hypo and hyper. This appears especially on the result given by an analysis based on Dunn's sensory profile [13]. Bogdashina, who considers not only the sensory disorders but also any sensory experiences, proposed to split the global sensory sensitivity on 7 modalities that are: vision, hearing, taste, smell, tactile, proprioception, vestibular [14]. Even if the hyposensitivity and hypersensitivity can be detected simultaneously for a given modality, this situation occurs rarely. The sensitivity is then defined by its threshold. A low threshold indicates a hypersensitivity and a high threshold indicates a hyposensitivity.

In the field of Psychology, the measurement approach is based on the clinical observation of the behavior's manifestations of the hypo/hyper sensory sensitivity. For example, approaching a noisy object to the ear is a manifestation of a hyposensitivity on the hearing modality. This approach conforms with the basic principles of measurement. The involved process performs observations of the manifestations of a quantity. Indeed, this quantity is the sensory sensitivity threshold, and each manifestation observation contributes to the production of an observation statement that defines an intersubjective evaluation as defined by Mari [15].

Bogdashina proposed to evaluate separately the sensitivity for the 7 modalities with a questionnaire made of 232 questions. Each question is related to the occurrence of a behavior that expresses an hypo or hyper sensitivity. For each modality, aggregation is performed by counting the number of observations of hypersensitivity-related behaviours and the number of non-observations of hyposensitivity-related behaviours. The final measurement result is obtained by a normalisation with the number of questions for the given modality.

The measurand, i.e. the quantity under observation that we will name the *multimodal sensitivity*, is defined by a set of 7 measurands: formally the sensory sensitivity thresholds for each of the 7 modalities. The measurement of measurands related to each modality is then performed with an ordinal scale, i.e. a scale that preserves the ordinal relations.

In order to create references for a future calibration of the augmented musical instruments, our study was first focused on the analyse of the sensitivity of 4 children with ASD by the way of psychological approaches. The chosen approach is an analysis based on the Bogdashina questionnaire limited to the 5 modalities: vision, hearing, tactile, proprioception and vestibular

Figure 1 presents a synthesis of the analyse of the sensory sensitivity of the 4 children with the Bogdashina's questionnaire. For confidentiality reasons, the children names had been changed.

4. Definition of the possible measurands for the musical instruments

As the Conservatory organizes musical activities, musical instruments will be augmented to allow them to perceive their environment, and more specifically to perceive the interaction that children have with them. They may also be able to stimulate children through the production of sounds, lights or movements. As augmented musical instruments can only detect or act on physical quantities, they cannot directly measure children's *multimodal sensitivity*. If we consider each modality (measuring the taste sensitivity and the smell sensitivity is a quite complex process and these two modalities are not included in our study) the *multimodal sensitivity* is restricted in our study to 5 modalities: vision, hearing, tactile, proprioception, vestibular.

Journal of Physics: Conference Series **1379** (2019) 012035 doi:10.1088/1742-6596/1379/1/012035

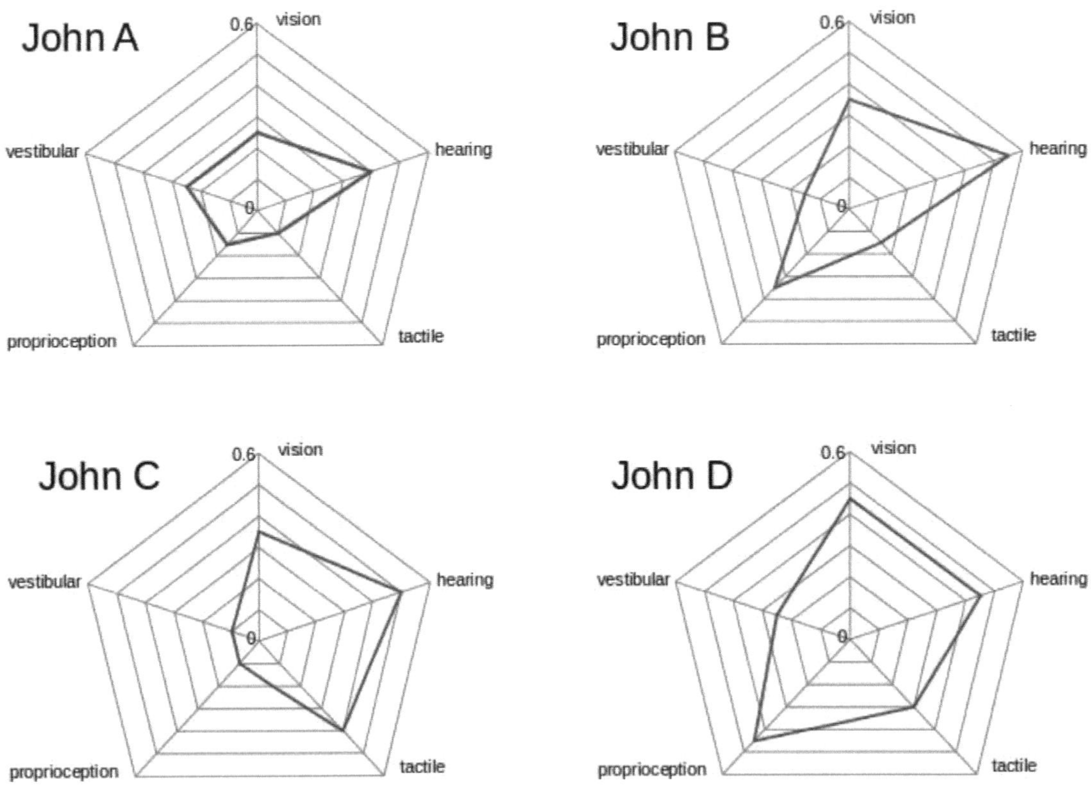

Figure 1. Bogdashina synthesized result for 5 sensory modalities. On the scale, 0 leads to a sever hyposensitivity and 0.6 leads to an important hypersensitivity, and 1 to a sever hypersensitivity

In our approach, each augmented musical instrument (AMI) is formally a set of sensors and actuators, and is able to communicate with the other instruments. Actually, an AMI holds all the properties of an actor of the Internet of Things (IoT). Furthermore a group of AMIs is defined as a measuring experiment around the children, each measuring experiment being defined relatively to the observation of a children behavior and each behavior being a manifestation of a hypo/hypersensitivity of a given sensory modality. The goal is to let the AMIs observe and identify the hypo/hypersensitivity manifestations.

The proposed tables 1 to 4 list the behaviors under observation for each modality. These lists, defined by a psychologist, are similar to the Bogdashina's questionnaire but are dedicated to the observation by sensors. The column 2 of each table associates with each behavior one or more quantities that must be measured by the AMIs in order to identify the occurrence of the given behavior. The column *Devices* gives the pertinent sensors, actuators and processes the AMIs must hold to participate to the measuring experiment (detailed in table 5).

According to a group' ability to perform multiple experiments to detect the occurrence of the listed behaviours, the link between the intensity of the behaviour and the *sensitivity* remains to be established. Two main options are available: apply existing methods based on knowledge of the relationship of each behaviour with hyposensitivity/hypersensitivity, or establish this relationship through a learning method. Especially, the Rash method, already successfully used in the field [16], is under study.

Journal of Physics: Conference Series **1379** (2019) 012035 doi:10.1088/1742-6596/1379/1/012035

Table 1. Hearing sensitivity.

Behavior	Quantity under observation	Devices
Draws near an AMI producing noise	distance (child/AMI)	s_1 s_2 s_3
Fends off an AMI producing noise	AMI motion	s_4 p_1
Expresses interest for an AMI producing noise	touch frequency	s_5 s_6
Turns his head to his talking peers	AMI motion if used as a pointer	s_4 p_2
Turns his head to a noise	AMI motion if used as a pointer	s_4 p_2
Reacts to a singing voice	AMI motion if used as a pointer	s_4 p_2 p_3
Reacts to a speaking voice	AMI motion if used as a pointer	s_4 p_2 p_3
Sings spontaneously	sound (singing character of song)	s_7 p_1
Reproduces a short melodic sequence	distance between sequences	a_1 s_7
Requests a louder sound	pressure, frequency (on a button)	a_1 s_8
Requests a tone change	pressure, frequency (on a button)	a_1 s_8
Requests a tempo change	pressure, frequency (on a button)	a_1 s_8
Requests a timbre change	pressure, frequency (on a button)	a_1 s_8

Table 2. Vision sensitivity.

Behavior	Quantity under observation Devices	
Looks the AMI	head orientation	s_1 s_2 s_3
Get closer to the AMI when flashing	distance (head/AMI)	s_4 p_1
Caresses the AMI when flashing	vibration, acceleration, frequency	s_5 s_6
Exchange of glances towards the same direction	eye direction	s_3 s_9
Eye-to-eye contact with peers	eye direction	s_3 s_9
Eye-to-eye contact with the teacher	eye direction	s_3 s_9
Is oriented towards the teacher	AMI motion and location	s_3 s_4
Requests a hue change	pressure, frequency (on a button)	s_8 a_3
Requests a brightness change	pressure, frequency (on a button)	s_8 a_3
Requests a saturation change	pressure, frequency (on a button)	s_8 a_3

Table 3. Tactile sensitivity.

Behavior	Quantity under observation	Devices
Manipulates the AMI	AMI motion	s_4
Touches the AMI	distance (fingers/AMI), frequency	s_5 s_6 s_8
Caresses the AMI	vibration, acceleration, frequency	s_5 s_6 p_1
Fends off an AMI with a given texture	AMI motion	s_3 s_4 s_5 s_6 a_4
Requests a texture change	pressure, frequency (on a button)	s_5 s_6 s_8 a_4
Uses AMIs with similar textures	frequency, duration	s_8 a_4
Hits on the AMI	AMI motion	s_4 s_5 s_6 s_8
Moves away from the AMI	distance (user/AMI)	s_1 s_2
Manipulates or touches the AMI when it vibrates	AMI motion, distance (fingers/AMI)	s_4 a_5

Table 4. Vestibular and proprioception sensitivity.

Behavior	Quantity under observation	Devices
Moves the AMI in rhythm	AMI motion	$s_4\ p_1$
Sways with the AMI	AMI motion	$s_4\ p_1$
Swings with the AMI when it makes noise	AMI motion	$s_4\ p_1\ a_2$
Turns with the AMI	AMI motion	$s_4\ p_1$
Manipulates the AMI abruptly	AMI motion	$s_4\ p_1$
Coordinates his movements to play the AMI	distance (fingers/AMI)	$s_5\ s_6\ s_8$
Appropriately touches the AMI to produce a sound, a rhythm	distance (fingers/AMI)	$s_5\ s_6\ s_8$
Contact with his own body: he touches his mouth, his hands	skeleton motion	$s_{10}\ s_{11}\ p_4$

Table 5. Sensors, actuators and processes involved.

s_1 : telemeter	s_2 : proximity sensor	s_3 : geolocation device
s_4 : 9Dof gyro/compass/accelerometer	s_5 : capacitive sensor	s_6 : inductive sensor
s_7 : microphone	s_8 : push button	s_9 : eye-tracker
s_{10} : camera	s_{11} : depth camera	
a_1 : music synthesizer	a_2 : voice synthesizer	a_3 : LED
a_4 : adaptable texture	a_5 : vibration motor	
p_1 : signal processing	p_2 : acoustic localization	p_3 sound recognition
p_4 : gesture recognition		

5. Conclusion and perspectives

The ultimate objective of our work is to automate the production of the sensory profile of children with sensory autism disorders (ASD). We take the opportunity of music workshops to convert musical instruments into augmented musical instruments by making them sensitive to the children who manipulate them. This paper presents the early phase of this work. Its shows the possibility to link the most recent approaches in psychology related to the identification of the sensory sensitivity with the measurement science. As a reference, we produced sensory profiles using the Dunn/Bogdashina observation grid. We proposed an alternative observation grip dedicated to the observations performed by the augmented musical instruments.

The next steps in this work is the finalization of the firsts augmented musical instruments able to follow the proposed observation grid and with respect to the safety of the children. After the firsts observation statements given by the instruments, a calibration phase based on the use of the reference sensory profiles is planned for the next year. Finally, a toolbox including a set of connected augmented musical instrument will be proposed. The automation of the measurement of sensory profiles is a great interdisciplinary challenge. It needed to interpret a psychological analysis with the concepts of measurement science in order to maximise the common data sources between the 2 approaches. The results are very important for children with ASD. Compared to manual analysis, the automation allows to drastically reduce the analysis period and to reduce the bias introduced by the subjectivity of the observers. It opens up new possibilities not only in monitoring sensory disorders but also in their early detection. Even if using musical instrument came from an opportunity, they have the merit of reducing the impact of the bias due to children's fatigue in relation to a simple object or toy.

Acknowledgement

The authors acknowledge the *Fondation Savoie Mont Blanc* for the important financial support to this study

Journal of Physics: Conference Series **1379** (2019) 012035 doi:10.1088/1742-6596/1379/1/012035

References

[1] Thye M D, Bednarz H M, Herringshaw A J, Sartin E B and Kana R K 2018 *Dev. Cogn. Neuros-Neth.* **29** 151–167 ISSN 1878-9293
[2] American Psychiatric Association and American Psychiatric Association DSM-5 Task Force 2013 *Diagnostic and Statistical Manual of Mental Disorders* DSM-5. (American Psychiatric Association) ISBN 0890425558
[3] Thaut M H 1987 *J. Autism Dev. Disord.* **17** 425–432 ISSN 01623257
[4] Lai G, Pantazatos S P, Schneider H and Hirsch J 2012 *Brain* **135** 961–75 ISSN 14602156
[5] Blackstock E G 1978 *J. Autism Child. Schiz.* **8** 339–53 ISSN 00219185
[6] Boddaert N, Belin P, Chabane N, Poline J B, Barth´el´emy C, Mouren-Simeoni M C, Brunelle F, Samson Y and Zilbovicius M 2003 *American Journal of Psychiatry* **160** 2057–2060 ISSN 0002953X
[7] Bonnel A, McAdams S, Smith B, Berthiaume C, Bertone A, Ciocca V, Burack J A and Mottron L 2010 *Neuropsychologia* **48** 2465–2475 ISSN 00283932
[8] Heaton P 2009 *Philos. T. R. Soc. B* **364** 1443–17 ISSN 14712970
[9] Bouvet L, Simard-Meilleur A A, Adeline P, Mottron L and Donnadieu S 2014 *Cognition* **131** 367–372
[10] Bouvet L, Mottron L, Valdois S and Donnadieu S 2016 *J. Autism Dev. Disord.* **46** 1553–1561 ISSN 15733432
[11] Simpson K and Keen D 2011 *J. Autism Dev. Disord.* **41** 1507–1514 ISSN 01623257
[12] Baranek G T, David F J, Poe M D, Stone W L and Watson L R 2006 *J. Child Psychol. Psyc.* **47** 591–601 ISSN 00219630
[13] Dunn W 2007 *Infants & Young Child.* **20** 84–101 ISSN 0896-3746
[14] Bogdashina O 2003 *Sensory Perceptual Issues in Autism and Asperger Syndrome: Different Sensory Experiences- Different Perceptual Worlds* (Jessica Kingley Publishers) ISBN 1843101661
[15] Mari L 2003 *Measurement* **34** 17–30
[16] Lundqvist L O and Lindner H 2017 *J. Autism Dev. Disord.* **47** 2080–2091 ISSN 15733432

Towards biological quantity theory for nominal property metrology in polyenzymatic devices with living cells

P I Belobrov[1], A A Evstrapov[2], E N Esimbekova[1,3], I A Denisov[1,4], K A Lukyanenko[1,4], E D Osipova[5], A S Yakimov[5]

[1] Laboratory of Bioluminescent Biotechnology, Siberian Federal University, Krasnoyarsk, 660041, Russia
[2] Institute for Analytical Instrumentation RAS, St. Petersburg, 190103, Russia
[3] Laboratory of photobiology, Institute of Biophysics SB RAS, Krasnoyarsk, 660036, Russia
[4] Laboratory for Digital Controlled Drugs and Theranostics, Federal Research Center "Krasnoyarsk Science Center SB RAS", Krasnoyarsk, 660036, Russia
[5] Research Institute of Molecular Medicine and Pathobiochemistry, Krasnoyarsk State Medical University named after Prof. V.F. Voino-Yasenetsky, Krasnoyarsk 660022, Russia

E-mail: peter.belobrov@gmail.com

Abstract. Here we discuss the concepts of "biological quantity" and "nominal property" within the framework of the problem of biological measurements based on new specific results of biological analysis using a microfluidic platform and chips developed by our team earlier. It was shown that based on different microfluidic platforms it is possible to develop chips with a polyenzymatic bioluminescent system NAD(P)H:FMN-oxidoreductase-luciferase (Red + Luc), which can be used in various areas of biological analysis. Thus, disposable microfluidic chips with Red + Luc system suitable for field and indoor use were developed using continuous flow microfluidic platform. The use of droplet-based microfluidic platform allowed to develop microfluidic chips with Red + Luc system for long-term continuous measurements of water samples, for example, in places of waste disposal by industrial enterprises. The reference for comparing different biological quantities with each other in the proposed system was a photodetector, which converted non-numeric values and nominal properties recorded by a biological module Red + Luc into numerical variables. Such a reference was implemented as a portable luminometer based on silicon photomultiplier. The results allow to perform other biological measurements and to start the discussion of modern biological concepts in the language of biological measures.

1. Introduction

Galilei created fundamental physics by formalization of physical phenomena four centuries ago. The great final is the system of quantum units of Nature was adopted at 16/11/2018 at 26th meeting of the CGPM without any artefacts. A matter-of-course the question arises: "what Nature?" with exact answer: "of course, physical!" The chemical and, especially, biological description of Nature contain much more non-quantitative values having the qualitative character. The problem is reflected in metrological concept "nominal property", which is written "propriété qualitative" in French and in Russian has many meanings: "formal, nominal, and specified or indicator property". Therefore, the decade of joint discussion the “biological quantity” and "nominal property" allow formulating the problem of biological measurement [1-3].

The traditional approach to measuring biological variables is to apply physical measures to biochemical processes. More complex than “depletion interaction” was introduced “biophysicochemical interaction” [4]. Bio-numbers [5] and quantitative biochemistry in living cells [6] were born. Metrological results of this direction reflected in the new edition of the metrological dictionaries [7, 8].

We think that any cell measures the other cells (without microscopes and micrometers!) in the cell-to-cell communication by membrane-derived vesicles and exosomes with different kinds of molecules (chemical signals). Theoretical description of those processes is “cell-to-cell measurement”. The qualitative variables can be quantified in the space of infinite dimension by recurrently perhaps. Therefore, we must here note that the movement to this limit will always be through more and more ingenious formalizations using new physical, chemical and biological measures [9, 10].

As all cells, tissues, organs etc. are differ from each other, as we cannot directly use any cell etc. as reference. Luca Mari suggested good idea for novel treatment of nominal properties in biological metrology when we apply some measurement procedure for a comparison [11], i.e. quantify of biological objects and processes. Our approach is to develop biological measures and digitize those using microfluidic technologies and devices.

Microfluidic platform is basis of microphysiological system development [12-14]. "Organs on chips" and "tissue-engineered organ constructs" have direct application to the rapidly developing field of systems biology attempts to span scales with integrative, bottoms-up modeling coupled with massively parallel biological measurements [12].

Here we briefly describe our results in application of polyenzymatic chains with bacterial luciferase with different microfluidic technologies for biological measurements in the scope of biological metrology [15-19].

2. The choice of the converter of biological signals into physical

In the description of the physical phenomena of nature it is common to use the fundamental physical constants as reference parameters for comparison and interpretation of experimental data. These constants, for example, include the Avogadro constant N_A, the Boltzmann constant k, the Planck constant h, and others. The interaction of biological objects (organisms, tissues, cells) among themselves is much more complicated than the interaction of objects of a purely physical nature. When describing this interaction, it is also necessary to take into account new properties and biological variables (including non-numerical) that occur in interacting objects under the influence of self-organization and the desire for biodiversity. For example, there are no two absolutely identical cells in nature. In this connection there is the problem of choosing the biological converter of non-numeric variables into numerical variables which can be measured by physical methods.

Enzymes can be one of the possible biological converters. Unlike cells, enzymes after synthesis always have the same structure and physicochemical properties. Errors in the structure of the enzyme during its synthesis do not allow it to adopt the correct conformation, which excludes it from strictly aligned biological chains of interaction of organisms. At the same time, enzymes, being biological objects, can be used as indicator systems when describing the interaction of biological objects with each other or with the environment.

Luciferase enzyme is an example of such a biological converter. This enzyme is bioluminescent, that is, during a chemical reaction with its participation, light quanta (hν) are emitted, as shown in the reaction scheme below. Here we used polyenzymatic bioluminescent system NAD(P)H:FMN-oxidoreductase-luciferase (Red + Luc):

$$FMNH_2 + RCOH + O_2 \xrightarrow{\text{Luciferase}} RCOOH + H_2O + h\nu \qquad (1)$$

NAD(P)H:FMN-oxidoreductase

$$NADH + FMN + H^+ \longrightarrow FMNH_2 + NAD^+ \quad (2)$$

Luciferase catalyzes the oxidation of aldehydes with the emission of light in the blue-green spectral region (1). The luciferase reaction is coupled with reaction (2) catalyzed by NAD(P)H:FMN-oxidoreductase in order to provide luciferase with reduced flavin mononucleotide ($FMNH_2$).

Light quanta emitted during a chemical reaction can be accurately counted using photo detectors. In this work, we used silicon photomultiplier (SiPM), which allow us to register individual photons of light. Moreover, SiPM possesses a compact size that allows the development of portable measuring systems on its basis.

3. Biological measurements in microfluidic chips with Red + Luc system

On the basis of Red + Luc bioluminescent system, photodetectors and various microfluidic platforms, portable biosensors were developed for various biological measurements. Here we briefly describe these biosensors and discuss their application as a reference measuring systems for biological parameters.

3.1. Disposable polymer microfluidic chips

Previously we demonstrated [17] the integration of Red + Luc system with microfluidic platform of continuous flow. For that the surface of PMMA (Poly(methyl methacrylate)) substrate was channelized followed by immobilization of Red + Luc system in the reaction chamber and then sealed using solvent bonding. The water sample added through the inlet channel dissolved the components of Red + Luc system immobilized in reaction chamber of the chip and, thus, bioluminescent reaction started. These chips were tested for application in environmental bioassay. The proposed indication system demonstrated limit of detection for copper (II) sulfate at the level of 3 μM which was comparable with copper (II) sulfate LOD of traditional lux- biosensors based on bacterial cells (1 μM). Moreover, analysis time with disposable microfluidic chips was reduced from 3 to 5 h with cell-based bioassays to 1–3 min.

Then the integration of luciferase-based microfluidic chips with a portable Luminometer was performed in order to develop handheld enzymatic luminescent biosensor (Figure 1) [19]. For that the composition of Red + Luc system was optimized to achieve higher luminescence intensity and sensitivity for toxic compounds. The detection system of the portable luminometer was based on a SiPM PM6650-EB from KETEK.

This biosensor was suitable for field and indoor use. For example it could be used as an early warning field-deployable system for rapid detection of heavy metals salts and other toxic chemicals, which affect bioluminescent signal of enzymatic reaction.

3.2. Microfluidic chips for continuous biological measurements

Another biosensor was made using droplet-based microfluidic platform. For that a polydimethylsiloxane (PDMS) chip was designed and manufactured using soft lithography techniques. Photomultiplier tube H7828 from Hamamatsu was used as a photodetector. The Red + Luc system was split into two solutions which mixed and formed oil-in-water emulsion with droplet microreactors (see description of figure 2).

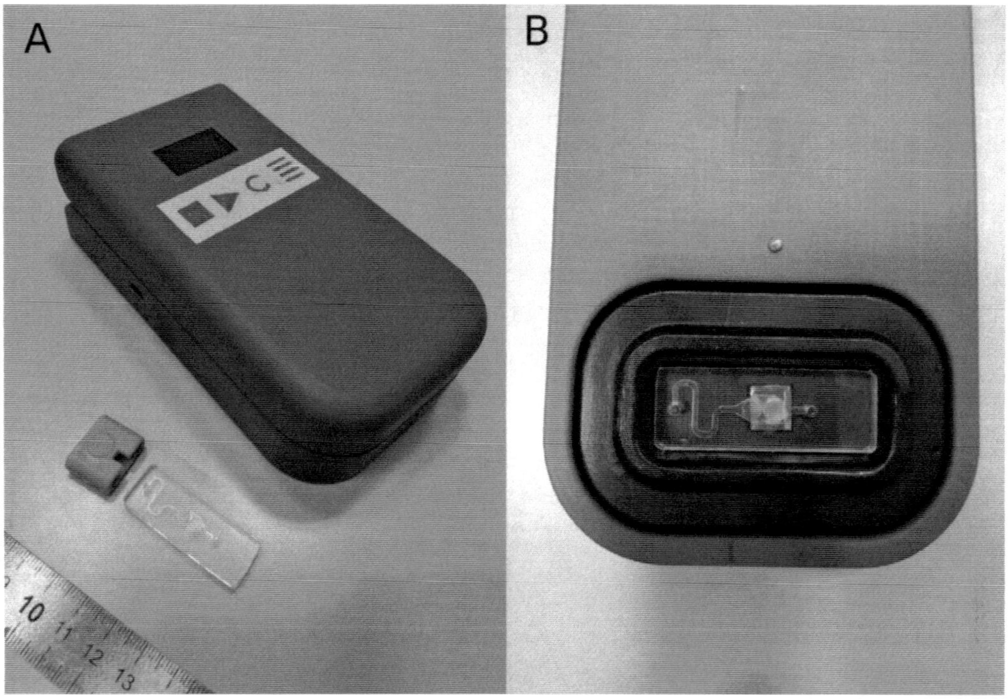

Figure 1. The handheld biosensor consisted of a portable SiPM-based luminometer, disposable microfluidic chips with immobilized bioluminescent enzyme system and a sampler adapter for chips [19].

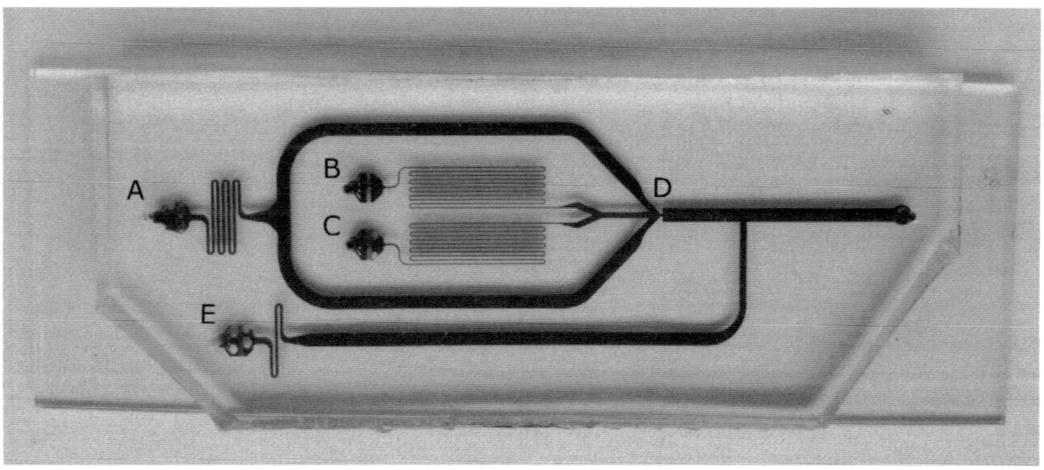

Figure 2. Overview of PDMS chip for Red + Luc system based on droplet microfluidic platform.

Red + Luc system was fed into the chip in the form of two solutions. One solution (B) contained enzymes and NADH. The other solution (C) contained aldehyde, FMN, and the analyzed sample. After merging of B and C, the flow is compressed by the oil phase (A) at the point of hydrodynamic focusing (D) forming water-in-oil emulsion. Water droplets contained equal volume fractions of B and C. The volume of oil phase (E) was additionally introduced into the flow between the droplets to increase the distance between them and mix reagents. After passing the additional channel (E), the content of the droplets was mixed and the bioluminescent reaction started. Droplets passed through the capillary of the chip opposite the aperture of the photodetector, where the signal was recorded.

Mathematical modeling of the formation of droplets with a volume of 0.1–10 μl showed that the distance between adjacent droplets is much smaller than their diameter; therefore, an additional oil channel was introduced into the design of the microfluidic chip, allowing to adjust the distance between the droplets. Also, computer simulation showed that the deformation of droplets near this channel led to effective mixing of the reactants, which was confirmed experimentally.

The developed chip was tested for application in environmental bioassay as the disposable one. Results indicate that long-term continuous measurements in droplet microreactors with Red + Luc system reached 1 mg/l LOD for copper (II) sulfate, which was at the MPC level. Also at the level of the MPC (0.2 mg / l), it was possible to detect benzoquinone in water.

The application of droplet-based microfluidic platform allowed to develop microfluidic chips with Red + Luc system for long-term continuous measurements of water samples, for example, in places of waste disposal by industrial enterprises or water treatment plants.

4. Conclusions

In this paper, a mechanism for conversion of non-numerical biological variables into measurable numerical variables on the example of environmental bioassay was proposed. Enzymes, in particular luciferase, were proposed to use as a biological converter. This approach allows to take a fresh look at the problem of searching for a biological reference to perform biological measurements and to discuss modern biological concepts in the language of biological spaces, topologies, and measures that are useful for novel version of metrological dictionary (VIM4).

References

[1] Mari L 2009 On (kinds of) quantities *Metrologia* **46**(3) L11

[2] Dybkaer R 2010 ISO terminological analysis of the VIM3 concepts 'quantity'and 'kind-of-quantity' *Metrologia* **47**, 127

[3] Sapozhnikova K, Chunovkina A and Taymanov R 2014 "Measurement" and related concepts. Their interpretation in the VIM *Measurement* **50** 390-396

[4] Nel A E, Mädler L, Velegol D, Xia T, Hoek E M V, Somasundaran P, Klaessig F, Castranova V and Thompson M 2009 Understanding biophysicochemical interactions at the nano-bio interface *Nat. Mater.* **8** 543

[5] Milo R and Phillips R. 2015 *Cell Biology by the Numbers* (Garland Science) 400

[6] Honigmann A and Nadler A 2018 The next frontier: Quantitative Biochemistry in living cells *Biochemistry-US* **57** 47-55

[7] Nordin G, Dybkaer R, Forsum U, Fuentes-Arderiu X and Pontet F 2018 Vocabulary on nominal property, examination, and related concepts for clinical laboratory sciences (IFCC-IUPAC Recommendations 2017) *Pure Appl. Chem.* **90** 913-35

[8] Petersen U M, Padró-Miquel A, Taylor G, Hertz J M, Ceder R, Fuentes-Arderiu X and Dunnen J T 2018 Properties and units in the clinical laboratory sciences part XXIV. Properties and units in clinical molecular genetics (technical report 2017) *Clin. Chim. Acta* **484** 122-31

[9] Taymanov R E and Sapozhnikova K V 2017 Measurements of multiparameter non-physical quantities *Pribory (Devices)* **10** 40-45 (in Russian)

[10] Kasum O, Perovic A and Jovanovich A 2018 Measures and metrics of biological signals *Front. Physiol.* **9** 1707

[11] Mari L 2017 Toward a harmonized treatment of nominal properties in metrology *Metrologia* **54** 784

[12] Wikswo J P 2014 The relevance and potential roles of microphysiological systems in biology and medicine *Experimental biology and medicine* **239** 1061-72

[13] Edington C D, Chen W L K, Geishecker E, Kassis T, Soenksen L R, Bhushan B M, Freake D 2018 Interconnected Microphysiological Systems for Quantitative Biology and Pharmacology Studies *Sci. Rep.- UK* **8** 4530

[14] Cooper M, Charest J L and Coppeta J 2019 Design principles for dynamic microphysiological

systems *Microfluidic Cell Culture Systems* (Amsterdam, Netherlands: Elsevier)1-29

[15] Lukyanenko K A, Belousov K I, Denisov I A, Yakimov A S, Esimbekova E N, Bukatin A S, Evstrapov A A and Belobrov P I 2017 Active mixing of immobilised enzymatic system in microfluidic chip *Micro Nano Lett.* **12** 377-81

[16] Lukyanenko K A, Denisov I A, Yakimov A S, Esimbekova E N, Belousov K I, Bukatin A S, Kukhtevich I V, Sorokin V V, Evstrapov A A and Belobrov P I 2017 Analytical Enzymatic Reactions in Microfluidic Chips *Appl. Biochem. Micro.* **53** 775-80

[17] Denisov I, Lukyanenko K, Yakimov A, Kukhtevich I, Esimbekova E and Belobrov P 2018 Disposable luciferase-based microfluidic chip for rapid assay of water pollution *Luminescence* **33** 1054-61

[18] Evstrapov A A *et al.* 2017 *Microfluidic Technology of Biological Analysis for Natural and Industrial Water* (Novosibirsk: SB RAS) p 176 (in Russian)

[19] Lukyanenko K A, Denisov I A, Sorokin V V, Yakimov A S, Esimbekova E N and Belobrov P I 2019 Handheld enzymatic luminescent biosensor for rapid detection of heavy metals in water samples *Chemosensors* 7 16

Information measurement systems in the digital society

G F Malykhina[1,2], A N Vasilyev[1] and D A Tarkhov[1]

[1]Peter the Great St. Petersburg Polytechnic University, Russia
[2]Russian State Scientific Center for Robotics and Technical Cybernetics, Russia

E-mail: g_f_malychina@mail.ru

Abstract. The transition to the next technological order requires robotics and automation in the industry. Managing such process of production is impossible without scenario modeling, forecasting and diagnostic of possible emergencies based on measurement information received from various sensors. The models used must be adequate to the real state of the object, and not to its project, on the basis of which it was designed and manufactured. Such models are called twin models of physical objects. The objective of the research is to obtain the approaches to the creation of information-measuring systems capable of processing large volumes of measurement information for implementing dynamically changing twin models of physical objects, and to carry out continuous training and adaptation to changing conditions. The proposed method allows continuous extraction of knowledge from the measurement information, accumulation knowledge about the object during its life. It was shown that hybrid models with physical and mathematical substantiation and capability of learning based on measurement data are the possible solution.

1. Introduction

Globalization processes require a transition to a new technological structure associated with the creation of advanced information and communication technologies. These technologies will become the basis of the new digital infrastructure of the research and industrial complex [1]. Digital transformation in the technical field requires the development of information measuring systems and technologies in several directions: the creation of wireless networks of sensors (with the possibility of their movement), carrying out continuous acquisition of information from physical devices and the environment. The accumulation and storage of measurement information is associated with the implementation of a platform for intelligent big data analysis, creating internal cloud infrastructure of the enterprise or application of collective cloud services.

Information-measuring systems in the industry and transport are designed for monitoring, technical diagnosisand predicting the behavior of technical objects and therefore must interact with their digital model, which allows them to be classified as a class of cyber-physical systems. When wireless data transmission pays significant partin maintenance efficiency, fault tolerance, adaptability and the possibility of self-organization of the sensor network. In our research, the IEEE 802.15.4 standard is of interest. It regulates the organization of energy-efficient sensor networks, such as ZigBee [2].

The up-to-date approach is the use of a dynamically updated object model, which, in fact, is its digital twin [3]. Dynamically updated models have the opportunity to constantly receive new measuring information from the sensors of a technical object, from the sensors of control devices and the external environment, to summarize the newly obtained and archive data to change their parameters. The digital twin is constantly being trained on the basis of new and historical data, updates its parameters in real time, adapts to external conditions and predetermines the optimal behavior of a technical object in the current situations [4].

The necessity for continuous learning, adjusting the characteristics of the model when updating information has led to the fact that artificial intelligence, neural networks and machine learning are the methodical basis for its creation. However, the approach based on the continuous learning, adjusting the characteristics of the model when updating information led to the fact that the artificial intelligence, neural networks and machine learning are the methodological basis for digital twin creation. Though, the machine learning approach cannot be directly applied, it requires further development. The machine learning model is based only on data that forms flexible-controlled non-uniform flows coming from sensors and from other sources of information. The model improves with time as data accumulates. However, the machine learning model does not take into account knowledge of the causal relationships of a physical object; it represents an object as a blackbox; the approximations obtained by this model are not fully mathematically justified. The greatest difficulty is the prediction of the rapid deterioration of the object in the absence of such situations during training, so based only on machine learning models it is difficult to predict critical situations.

Physical and mathematical models require deep knowledge of the physics of objects, the physics of very complex processes of changing of the object properties; use a large amount of computation, making it difficult to work in real time, which makes it impossible or very difficult for the model to reflect on the changes occurring in the object.

Sophisticated technologies for digital processing of measurement information should be carried out in a common computing environment, using cloud resources with convenient network access to shared computing resources [5]. The Peter the Great St. Petersburg Polytechnic University has a supercomputer center organized, the computing environment of which supports the cloud system.

The objective of the survey is to develop modern, more complex methods of creating digital twins, which take into account both the physical and mathematical foundations of causal-investigatory relationships, and the constantly obtained measurement data specifying the model by means of machine learning. Such a model will have all the advantages of two classes of models.

2. Information processing technique

The objective of measurement informationprocessing is to create and maintain a neural network model of a physical object or process. If a physical object or process is characterized by a differential, integro-differential or differential-algebraic equation (ordinary or partial differential equation) with initial or boundary conditions of the following form:

$$A(u) = g,\ u = u(\vec{x}),\ \vec{x} \in \Omega \subset R^p,\ B(u)|_\Gamma = h, \tag{1}$$

here, $u = u(\vec{x})$ is the function characterizing the state of the object, $A(u)$ is a differential, integro-differential or differential-algebraic operator, i.e. an algebraic expression containing derivatives, integrals or algebraic equations of an unknown function $\boldsymbol{u}$, $B(u)$ is an operator defined by boundary conditions, Γ is a boundary of a region Ω.

We are looking for an approximate solution of problem (1) in the form of an output of an artificial neural network (ANN) of a given architecture:

$$u(\vec{x}, \vec{w}) = \sum_{i=1}^{N} c_i v(\vec{x}, \vec{a}_i), \tag{2}$$

where $\vec{w} = [w_1, w_2, .., w_N]^T$ is the weight vector $\vec{w}_i = (c_i, \vec{a}_i)$, which aggregates the linearly input parameters c_i, and the nonlinear input parameters $\vec{a}_i$.

Basic neuroelement is the function $v(\vec{x}, \vec{a}_i)$ defined by the choice of the type of neural network and the activation function.

Journal of Physics: Conference Series **1379** (2019) 012037 doi:10.1088/1742-6596/1379/1/012037

Sometimes it is advisable to use ANN with different basic elements (heterogeneous ANN) [6]. Usually, such networks are used when the desired solution behaves fundamentally different in different subdomains, for example, in problems with phase perturbation, jumps, etc. In such a situation, it often makes sense to build two or more neural networks of a smaller size, each of which corresponds to its own subdomain.

The ANS weights vector $\vec{w}$ is in the process of step-by-step network training, built in the general case by minimizing the error functional $J(\vec{w})$. The error functional for problem (1), is $J(\vec{w}) = J_1 + \delta_2 J_2 + \delta_3 J_3$ where J_1 is the assessment of the error of a differential equation, J_2 is the assessment of the error depending on the initial conditions, J_3 contain terms corresponding to experimental data, and δ_2, δ_3 are penalty factors. The third term is introduced to ensure that the digital twin takes into account the measurement data of a physical object. The impact of measurement results on the neural network model is characterized by value δ_3.

The calculation of integrals $J_1 = \int_\Omega (A(u) - g)^2 d\Omega$, and $J_2 = \int_\Gamma (B(u) - f)^2 d\Gamma$ in an analytical form, is possible only in exceptional cases, therefore we use a discrete form for functional when digital dataprocessing:

$$J_1 = \sum_{j=1}^{M} \left(A(u(\xi_j)) - g(\xi_j)\right)^2; \tag{3}$$

$$J_2 = \sum_{k=1}^{K} \left(B(u(\xi'_k)) - h(\xi'_k)\right)^2; \tag{4}$$

$$J_3 = \sum_{l=1}^{L} \left(u(\xi''_l) - u''_l\right)^2; \tag{5}$$

here u''_l are the results of observations of the object corresponding to the points ξ''_l. Due to the fact that the initial equations (1) correspond to the simulated object inaccurately, we are not looking for the exact minimum of the functional $J(\vec{w})$, but a point $\vec{w}_\eta$ in the space of weights, $J(\vec{w}_\eta) < \eta$ which defines the so-called η – solution $u_\eta = u(\vec{x}, \vec{w}_\eta)$. The number $\eta > 0$ is chosen so as to consider the constructed model sufficiently accurate. The smaller the number η, the more accurate the model (1).

Computational experiments have shown that the use of a fixed set of test (trial) points $\xi_j, j = 1,..,M$ is impractical, since in this case small errors in these test points may be accompanied by large errors at other points in the region Ω. An increase in the number of test points in order to cover the area Ω sufficiently dense allows us to avoid a similar situation in the case of a relatively small number of neurons (terms in expression (2)). At the same time, the computation time increases dramatically, which is significant in a situation when we want to model an object in real time. The solution to this problem was the use of periodically regenerated sets of test points $\{\xi_j\}$ in the region Ω and, if necessary, regenerated sets of test points $\{\xi'_k\}$ on its boundary Γ. Re-generation of test points after a certain number of steps of the network learning process makes it more stable [7].

In this case, we organize calculations as the process of minimizing a set of functional, each of which is obtained by a specific choice of test points and is not completely minimized (between regenerations of the test set, only a few steps of the chosen minimization method are performed). Such an approach, in particular, makes it possible to circumvent the problem of falling into a local extremum, which is characteristic of most global nonlinear optimization methods. In the process of optimizing the error functional, we are obliged to include new observations as additional components in the sum J_3. If it is

Journal of Physics: Conference Series **1379** (2019) 012037 doi:10.1088/1742-6596/1379/1/012037

necessary to take into account the obsolescence of information over time, then you should throw out obsolete observations or use a formula $J_3 = \sum_{l=1}^{L} \delta_l \left(u(\xi_l'') - u_l''\right)^2$; with variables δ_l

In most situations, the random distribution of test points, generated over a certain number of training epochs using a uniform probability density function (PDF) or any other PDF, which ensures a more stable course of training. In some cases, it is advisable to use non uniform PDF for test points — thickening them near features (discontinuity boundaries, angles, etc.) or in areas with large errors. It is also possible to regenerate only a subset of the set of test points. The number of test points is a compromise between the requirement of accuracy and stability of the computational process (the more points, the better these indicators) and the limited time to build and adapt the model (for a digital twin, these processes must go in real time during its operation).

3. Heating simulation

Consider the task of controlling the heating of the ship's premises. According to the [8], the temperature field in the presence of a heating source satisfies the equation:

$$\rho C \frac{\partial T}{\partial t} = \operatorname{div}\left(k \operatorname{grad}(T)\right) - h\left(T - T_\infty\right) + f(t, \mathbf{x}) \qquad (6)$$

with the initial conditions $T(\mathbf{x},0) = T_0(\mathbf{x})$, with boundary condition $grad(T)\mathbf{n} + h(T - T_\infty)_\Gamma \big| = 0$, where ρ is the density, C is the heat capacity, k is the heat conductivity of the substance, h is the coefficient, T_∞ is the ambient temperature, f is the heating source, div is the divergence, grad is the gradient, Γ is area boundary, $\mathbf{n}$ is outer normal line.

The problem of the analysis is to determine the influence of the current measurements results obtained by sensors under the conditions of gradual changes in the object using two approaches: traditional modeling based on the Finite Element Method (FEM) for solving equations, and the proposed neural network method.

To solve this problem was applied FEM method of the PDF solution on the grid obtained by the Delaunay triangulation with the number of nodes from 25 to 1089. The standard deviation (std) measure of inaccuracy was found in the range from 0.001 to 0.0004. These results can be made as referencemeasure of inaccuracy. In the neural network method, an architecture of perception with one or two hidden layers, having dimensions (2-1) and (2-2-1) were used. The hyperbolic tangent was used as the activation functions of the hidden layers, the output layer was linear. The perception was trained using 25 randomly selected data. In the table 1 the estimation results shows the errors of the neural network model testing at the nodes of the Delaunay mesh. As was expected, std-error increases as compared with the exact solution.

Table 1. Errors of the FEM and neural network model in the Delaunay mesh nodes.

Number of nodes	Root-mean-square errors in the exact calculation	Root-mean-square errors of perceptron (2-1), training with 25 samples	Root-mean-square errors of perceptron (2-1) generalization	Root-mean-square errors of perceptron (2-1-1) training with 25 samples	Root-mean-square errors of perceptron (2-1-1) generalization
25	0,01	0.04	0.044	0.02	0.040
81	0.004		0.066		0.047
289	0.001		0.059		0.045
1089	0.0004		0.057		0.042

The temperature graphin a cabin with a point source of heat is shown in figure 1.

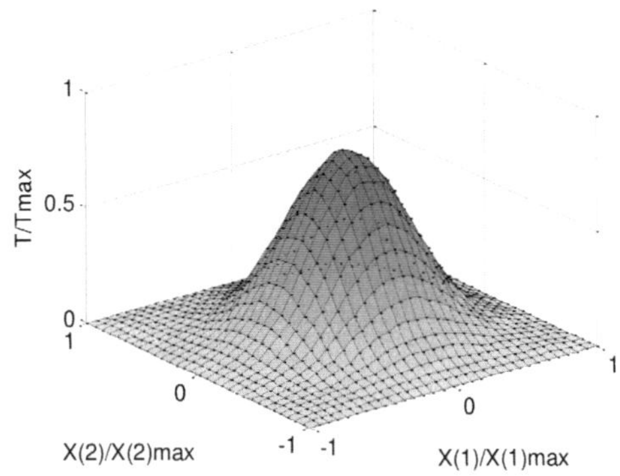

Figure 1.A temperature distribution $T/T_{\max}$ in the range $\left\{ \frac{x_1 - \bar{x}_1}{x_{1\max}}, \frac{x_2 - \bar{x}_2}{x_{2\max}} \right\}$ having a point source of heat.

Table 2. Errors of the traditional FEM model in the Delaunay mesh nodes during model degradation

Number of nodes	Root-mean-square errors in the exact calculation	Root-mean-square errors in the exact calculation provided model become degraded
25	0,01	0.22
81	0.004	0.33
289	0.001	0.40
1089	0.0004	0.43

In the case of a gradual change in parameters of physical object or in environmental, the network model, originally built on equation (6), takes into account the measurements obtained and performs continuous additional adaptation. Let the model receive additional measurements from 25 sensors placed arbitrarily. Then root-mean-square errorsof the neural network generalization in different nodes of space practically do not increase. The simulation results are shown in table. 3.

Table 3. The errors of the neural network generalization provided model become degraded

Num-ber of nodes	Root-mean-square errors of the perceptron (2-1) training with 25 samples	Root-mean-square errors of the perceptron (2-1) generalization	Root-mean-square errors of the perceptron (2-1-1) training with 25 samples	Root-mean-square errors of the perceptron (2-1-1) generalization
25	0.04	0.043	0.03	0.040
81		0.063		0.050
289		0.055		0.053
1089		0.057		0.052

4. Problem of controlling the heating of the ship's premises

The neural network model is used in the intelligent system for early detection of fire on the ship. Sensors located inside the ship's premises continuously transmit fire risk measurements to the center via a wireless interface, which is based on the measurement results and the Navier-Stokes mathematical model for heating and heat transfer. Mathematical model has high computational complexity therefore is implementing at the SPbPU supercomputer center [9-10].

Additionally, the proposed approach made it possible to determine the optimal location of the sensors in the space of the room using a specially developed genetic algorithm. The fitness function of the algorithm is obtained on the basis of the recurrent neural network object model.

5. Conclusion

Information-measuring systems in a digital society are designed not only to measure a variety of parameters, but also to generalize measurements in order to gain knowledge about the physical object. Therefore, the modeling of measurement objects is an integral function of the information-measuring system.

Modern information-measuring diagnostic and forecasting systems should constantly update their models in real time. Information-measuring system designed to control the current state of a physical object, to predict its optimal behavior in the next period of time.

The proposed hybrid models take into account the causal relationships characteristic of a physical object or its block, expressed as a partial differential equation and update measurements.

In contrast to the traditional FEM method of solving PDEs, it was proposed to use solutions using neural networks, which, without complete retraining, allow the model to be adapted to newly acquired data.

The proposed method has been used in an intelligent multisensory early warning system for fire detection on a ship.

References

[1] Benghozi P, Krob D, Lonjon A and Panetto H 2014 Digital enterprise design & management *Proc. of the 2nd Int.Conf. on Digital Enterprise Design and Management DED&M,* ed Benghozi P and Krob D (Paris: Springer) p 151

[2] Malykhina G F and Tarkhov D A 2018 Digital twin technology as a basis of the industry in future *18-th PCSF Professional Culture of the Specialist of the Future. The European Proc. on Social &Behavioral Sciences (Future Academy EPSBS),* vol **LI45** pp 416-428

[3] Uhlemann T H-J and Steinhilper C L 2017 The digital twin: realizing the cyber-physical production system for Industry 4.0 *The 24th CIRP Conf. on Life Cycle Engineering* vol **61** ed S Takata, Y Umedaand S Kondoh, (London: Procedia CIRP) pp 335–340

[4] Uhlemann TH-J, Schock C, Lehmann C, Freiberger S and Steinhilper R 2017 The digital twin: demonstrating the potential of real time data acquisition in production systems *Proc. Manufacturing* **9** pp.113-120

[5] Kang D-K 2015 A Study of resource management for fault-tolerant and energy efficient cloud datacenter *Cloud Computing 6th Int. Conf., Cloud Comp.* ed Zhang, Yin *et al.* (Daejeon, South Korea) p 44

[6] Lazovskaya T N, Tarkhov D A and Vasilyev A N 2017 Parametric neural network modeling in engineering *Recent Patents on Engineering* **11** pp 10–15

[7] Lazovskaya T and Tarkhov D 2016 Multilayer neural network models based on grid methods *IOP Conf. Ser.- Mat. Sci.* pp 158-62

[8] Cook RD, David S M and Plesha M E 1989 *Concepts and Applications of Finite Element Analysis 3rd ed* (New York: John Wiley & Sons)

[9] Guseva A I, Malykhina G F and Militsyn A V 2017 Early fire prevention in the plant *Int. Conf. on Industrial Engineering, Applications and Manufacturing (ICIEAM),* (Saint Petersburg: IEEE Explore) pp 1-4

[10] Guseva A I, Malykhina G F and Nevelskiy A S 2019 Neural network based algorithm for the measurements of fire factors processing *Advances in Neural Computation, Machine Learning, and Cognitive Research II. Selected Papers from the XX Int. Conf. on Neuroinformatics (October 8-12, 2018, Moscow, Russia)* ed Kryzhanovsky B, Dunin-Barkowski W *et al.* (Springer, Cham) **799** pp 160-166

[11] Malykhina G F, Guseva A I, Militsin A V and Nevelskii A S 2017 *Proc. of the II Int. Scientific Conf. "Convergent cognitive information technologies"* (Moscow: MSU) pp 289-296

Journal of Physics: Conference Series **1379** (2019) 012038 doi:10.1088/1742-6596/1379/1/012038

Metrology of nerve-similar sensing systems in anthropomorphic robots and living organisms

G Y Buymistriuk

Optical fibre measuring systems laboratory of the Moscow Plant "Fizpribor", Saint-Petersburg, 194044, Russia

E-mail: bgy812@mail.ru

Abstract. The features of metrological support of nerve similar sensor measuring systems with the use of neural networks are considered. It is established that the metrological verification of such measuring systems, by analogy with living organisms, is possible by comparing their responses, invariant to the external unstable environment. It is shown that from the standpoint of classical metrology, instrumental, methodical, static, dynamic, systematic and random errors of nerve-like sensory measurement systems with neural network results are reduced to two types – methodical and systematic errors of neural network training. It is proved that without using of reference calibration tools for embedded sensors, it is possible to experimentally evaluate systematic and random errors in the presence of several similar measurement systems by multiple pairwise comparisons of their responses, and the obtained systematic error can be introduced in the form of a metrological correction in the results of responses.

1. Introduction

Humans and other animals demonstrate a remarkable ability to map sensory information from their skin onto internal notions of hardness, texture, and temperature in order to reason about their physical condition and environment. For example, human hands serve not only prehensile functions but also as powerful sensory organs. Recent advances in stretchable fiber optics and electronics, soft robotics, and non-convex optimization methods for deep neural networks (NN) now offer us building blocks on which we can start to replicate this tactile perception synthetically with different sensing mechanisms [1].

Similarly to an effective biological nervous system of living organisms, the possibility of creating small-sized distributed optic fibre sensors embedded in robot structures using additive technologies opens up the prospect of implementing a sensor network (artificial nerves) for monitoring dynamics of temperature, load and defect by continuously measuring parameters (pressure, movement, curvature, temperature, humidity and acoustic emission) during operation of people with artificial limbs and of anthropomorphic robots. Because any embedded sensors are not extracted during all term of operation of objects, it is impossible to calibrate (verify) by means of traditionally in metrological laboratories. However, due to the nonlinearity of the response characteristics of embedded sensors and their functioning in a complex internal and external environment, there is a problem of our uncertainty in the reliability of sensory information. This problem can be solved with the help of NN-approach and statistical methods of error estimation. Embedded measuring optical fiber sensor networks on the basis of small-factor sensing structures of micron and submicron scale form are a class of cyber-physical intelligent measuring systems with NN processing and fusion of sensory information [2-4].

The main metrological requirements for such distributed sensor systems are an invariance of the measurement result from the influencing quantities (factors) and a statistical correction error of the measurement result.

2. Neural network approach to metrological verification of nerve-similar sensors

Currently, there is no complete and universal system engineering and metrological description of the NN. Measurement systems based on NN are very specific and differ from classical formalized measurement systems. The choice of NN structure is carried out in accordance with the features and complexity of the task. The operation of the NN depends on the value of synaptic connections, therefore, given a certain structure of the NN, which meets any problem, the network developer needs to find the optimal values of all variables, weighting factors, that is, to make the training of the NN. The learning process requires a correct assessment of the quality, reliability of the result, evaluation of possible errors of the NN. One way to get such an estimate is to test the examples, and the learning quality indicator — determining the predictive ability of the NN— is to calculate the percentage of correctly recognized examples. Determine the confidence factor Q reliable recognition of fuzzy test examples:

$$Q = \frac{M1 - M2}{R * 100\%}$$

where $M1$ is the maximum response of the NN; $M2$ is the response of the NN close to the maximum; R is the level of reliability.

A neural network works with empirical examples with a fuzzy set of solutions. Therefore, the NS, in addition to the decision on the class of the input example, calculates the confidence factor Q in this decision, which depends on the given level of reliability R, the specific value of which is selected in the range of 0.9 -0.99, for example - 0.98 and determines the degree of "training" of the NN. The formula shows that the confidence of the NN depends on the difference between the highest response and the closest to it level. If Q is greater than 100 %, that takes a single value of the signal (the decision). The formula R is in the denominator correctly. Indeed, with a large value of reliability R and the magnitude of the difference in the numerator will also be the greatest (the response of one neuron $M1$ will be much greater than the response $M2$ of another neuron). In addition, Q expresses the confidence of a particular NN trained on specific examples. That is, if the input example differs from the example in the knowledge base, a better trained NN will doubt more than a NN with less knowledge (experience).

Therefore, the arithmetic mean of the percentages of the confidence values obtained by testing each example, with a known result, gives us the necessary confidence percentage. From the point of view of classical metrology errors in one of the classifications are divided into methodological and instrumental. The division of errors into methodological and instrumental ones is related to the level and volume of a priori information used in the description of the measuring procedure and measurement results.

3. Estimation of accuracy of neural network models

Artificial NN is a very specific measurement tool. Inaccuracy of classification is determined by: incorrect choice of structure and poor-quality training of NN. Error of architecture selection and training of the NN relate to methodological errors. Learning errors are also dynamic and systematic. The application of NN models is attracted by the fact that they provide invariance of transformation in relation to noise and instability of input measuring signals.

In fact, artificial NN are universal approximators, which makes it possible to effectively use them as models of various nonlinear signal converters, including measuring sensors, devices and systems. Errors of NN models are caused by inaccuracy of estimates of synaptic weight coefficients of NN obtained as a result of model training, as well as inaccuracy of technical implementation of artificial NN based on digital computing devices with limited bit capacity. For the above reasons, the synaptic weights of neural network W_{ij} and V_j may deviate from their nominal values. In the general case, these deviations are random in nature and can be expressed in the form of a root mean square deviation (RMS) σ_{Wij}, σ_{Vj} or in the form of limit values errors $\pm\Delta_{Wij}$, $\pm\Delta_{Vj}$, where $i = 1,..., m; j=1,..., n,$ so it is necessary to estimate the RMS output of the model from its nominal value σ_y and the error limit of this signal Δ_y.

Obtaining these estimates for the NN model of nerve-similar sensing measurement system, we use an approach that has been used in Metrology to evaluate results of indirect measurements [5].

In indirect measurements, the value of the desired value Y is found by direct measurements of other quantities $X_1, X_2,...,X_N$ as $Y=F(X_1, X_2,...,X_N)$. In our case, the output signal of the NN model depends on its parameters and the input signal $Y=G(W_{ij}, V_j, x)$.

We assume that the values of synaptic weights of the NN model W_{ij}, V_j are statistically independent, then the estimation of the output signal dispersion is:

$$\sigma_y^2 = \sum_{j=1}^{n}\left(\frac{dG}{dV_j}\right)^2 \sigma_{V_j}^2 + \sum_{i=1}^{m}\sum_{j=1}^{n}\left(\frac{dG}{dW_{ij}}\right)^2 \sigma_{W_{ij}}^2 \quad (1)$$

and the interval estimation of the output signal error of the network model is:

$$\Delta_y = \pm K_p \sqrt{\sum_{j=1}^{n}\left(\frac{dG}{dV_j}\right)^2 \Delta_{V_j}^2 + \sum_{i=1}^{m}\sum_{j=1}^{n}\left(\frac{dG}{dW_{ij}}\right)^2 \Delta_{W_{ij}}^2} \quad (2)$$

Experimental estimation of errors of nerve-similar sensing systems of the same level of accuracy without application of standard verification tools is made by multiple pairwise comparison of their responses. The number of sensor measurement systems to be compared must be at least three. As a result of statistical processing of the results of multiple n comparisons of X_z and X_k responses, the arithmetic means are obtained:

$$\Delta_{zk} = \left[\sum_{j=1}^{n}(X_{zj} - X_{kj})\right]/n \quad (3)$$

The deviation of the arithmetic mean of its errors from the middle of dispersion of systematic errors of all measuring instruments, which is determined by the middle of the measuring instrument with a minimum deviation of the arithmetic mean, in comparison with others, is taken as a systematic error of a particular means of change. The obtained estimation of systematic error maybe introduced as an amendment to the output signal of the measuring system [6].

4. Results of modeling and applications

Simulation modeling of nonlinear neural network (three-layer perceptron, Hopfield neural network with back propagation of error, etc.) of a nerve-like sensor system allowed us to carry out a comparative analysis of NN errors (universal approximator) and polynomial approximation model of the 3rd order. During the simulation, the parameters of both models were changed by 1 % of the nominal values and the conversion functions were calculated according to the formula (1) and interval error estimates according to the formula (2)

It is obvious that in the given range of input signals the errors of the NN model do not exceed 2 %, while the errors of the polynomial model are in the range from 4 to 12 %.

Metrological characteristics of NN models of nerve-similar measuring transducers (sensors with neural network approximator) are better than those of classical transducers with polynomial approximation. Most robots today have sensors on the outside of their bodies that detect things from the surface. Optical fiber sensors are built into the body so that they can actually detect forces transmitted through the thickness of the prosthesis or robot, as we and all organisms do when we feel pain, heat, solid and liquid objects. New artificial nerves and nerves-like sensors can radically change the limb prosthetics of living organisms and improve the functionality of anthropomorphic robots. Researchers have created a new type of artificial nerve that can feel touch, process information and communicate with other nerves, as nerves do in our own body, and now managed to connect optical fiber artificial neuron to a biological counterpart [1].

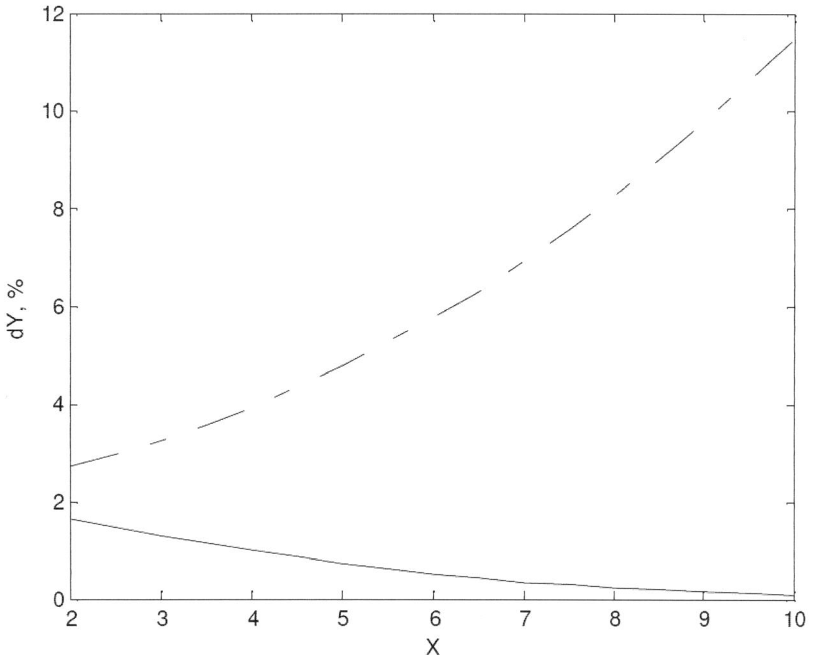

Figure 1 The boundary values of the errors of the models neural network model, polynomial model.

5. Conclusions
Embedded nerve-similar sensor systems can be metrologically controlled by their intellectualization using fiber optics and neural networks.

Neural network models of nerve-like measuring transducer are resistant and invariant to noise and internal defects and are the most suitable candidate for use in the sensorization of cyber-physical systems - anthropomorphic robots and artificial limbs of living organisms.

References
[1] Zhao H, O'Brien K, Li S and Shepherd R F 2016 Optoelectronically innervated soft prosthetic hand via stretchable optical waveguides. *Sci. Robot.* 1, p eaai7529

[2] Buymistriuk G Y 2017 Metrology challenges of nerve-similar optical fiber and distributed wireless sensor systems and networks *Proc. Int. Scientific and Practical Conf. "175 years of the D.I. Mendeleyev Institute forMetrology (VNIIM) and National Measurement System"*, (*St.Petersburg*, June14-15, 2017) p 222 (in Russian))

[3] Taymanov R, Sapozhnikova K and Ionov A 2017 topical metrology problems in the era of cyber-physical systems and Internet of Things. *18th Int. Congress of Metrology (September 19-21, 2017, Paris, France)* p 6

[4] Buymistriuk G Y 2014 Technologies of fusion of sensors information for control in emergencies. *Control Engineering Russia,* 5(53) p 47 (in Russian)

[5] Granovski V F and Siraya T N 1990 Methods of experimental data processing in measurements (Leningrad: Energoatomizdat) p 173 (in Russian)

[6] Kudryashova J F 1991 Determination of metrological characteristics of the measurement means by comparison of the same level of accuracy. *Meas. Tech.* 3 p 221

Journal of Physics: Conference Series **1379** (2019) 012039 doi:10.1088/1742-6596/1379/1/012039

Measurement and compensation of electrostatic forces between conducting surfaces at distances in the micrometer range by means of PTB's nanonewton force facility

Vladimir Nesterov

Physikalisch-Technische Bundesanstalt (PTB), Germany

E-mail: vladimir.nesterov@ptb.de

Abstract. The measurement and compensation of electrostatic forces between conducting surfaces play a major role in various experiments, including measurements of the Newtonian constant of gravitation, measurements of the Casimir force, precision tests of general relativity in space, searches for hypothetical forces (fifth forces), gravity on elementary particles, heating in ion traps, and the physics of Rydberg atoms [1, 2]. Although the actual value of the contact potential was not of direct interest, it was necessary to eliminate it accurately in these experiments in order to avoid residual electrostatic forces that may limit the accuracy and precision of such experiments. We demonstrate how PTB's nanonewton force facility [3, 4] can be used to accurately determine and compensate for the contact potential difference. To measure the contact potential difference as well as the distance between a force sensor and a measured object, a new three-voltage method was developed. A new method of adjusting the parallelism of the plates is also presented. The methods and the results of measuring the contact potential difference and the distance in vacuum are presented for a so-called plane-plane geometry and a ball-plane geometry. Measurements of the temporal and spatial variations of the contact potential difference were performed. Measurements were carried out at distances in the micrometer range between the metallic surfaces. A significant dependence of the contact potential difference on the distance between the conducting surfaces and on the time was found.

1. Introduction

Compared to other surface forces, electrostatic force has a very long range. The contact potential difference is obtained from the difference in the work functions between the conducting surfaces. The work function may vary over a polycrystalline metal surface due to inequivalent crystal facets or due to non-uniform contamination, producing an electrostatically patchy surface (patch effect). It is well known that the work function of a metal surface depends on the crystallographic plane it lies on; as an example, for gold, the work functions are 5.47 eV, 5.37 eV, and 5.31 eV for surfaces in the <100>, <110>, and <111> directions, respectively. The electrostatic potential along a chemically clean metal surface varies along the length of the typical size of surface crystallites, which can vary from the submicron scale to the millimeter scale or larger. Variations on the order of 0.16 V are expected for atomically clean surfaces; however, surface contamination generally seems to reduce these variations to about 10 mV. The baking surfaces in an ultra-high vacuum increased the patch variations, while exposure to air decreased them. It would appear that preferential adsorption of background contaminants is the reason for this, possibly in addition to the migration of existing adsorbates along the surface. For example, water molecules present on the surface, even in an ultra-high vacuum, have a substantial intrinsic dipole moment. Depending on the coverage and experimental conditions, water on a surface forms different low-dimensional structures ranging from isolated monomers and clusters to one-dimensional (1D) chains and two-dimensional (2D) ordered overlayers. The presence of intrinsic or applied gradient electric fields will lead to an increase in the adsorption (as well as the distortion and migration) of these water structures. Migration and distortion can also occur with surface crystallites. This leads to a change in the distribution of charges and potentials on the surface. A significant temporal dependence of the contact potential difference between the metallic surfaces at distances in the micrometer range has been reported in [7]. Surface electric noise, i.e., the non-uniform and non-stationary distribution of charges and potentials on a surface, poses a great experimental challenge in modern precision force measurements. In this paper, we present an improved version of PTB's nanonewton force facility, which was first published in [3, 4]. The facility is based on a disc pendulum with an electrostatic control system that reduces the stiffness of the pendulum as well as its deflection from a selected position. The improvement of this facility includes a new electrostatic control system in order to generate an electrostatic compensation force and to reduce the angular oscillation of the disc pendulum in a high vacuum. For example, the angular movements of the disc pendulum have been reduced from 10^{-2} rad to $2{\cdot}10^{-7}$ rad in a vacuum of 10^{-6} mbar. To control the mutual arrangement of the upper edges of the plates, a method analogous to confocal laser scanning microscopy (CLSM) was used. The facility is able to measure horizontal forces in the range below 1 μN with a resolution below 5 pN and an uncertainty below 2.7 % for a measured force of 1 nN at a measurement duration of about 20 s [4]. We demonstrate the possibility of using the nanonewton force facility to accurately determine the contact potential difference. To measure the contact potential difference as well as the distance between a force sensor and a measured object, a new three-voltage method was developed. This method makes it possible to compensate for the contact potential difference in real time. For plane-plane geometry and plane-ball geometry, which will be described later, the method used to measure the contact potential difference in a vacuum and the measurement results are presented below. A new method of adjusting the parallelism of the plates (so-called plane-plane geometry) is presented. Measurements of the temporal and spatial variations of the contact potential difference were performed. A significant dependence of the contact potential difference on the distance between the conducting surfaces was found.

2. New design of the nanonewton force facility

The facility consists of a measuring part and an identical reference part in order to reduce thermal drift and seismic noise [3, 4]. The new functional diagram of the measuring part is presented in figure 1. The facility is based on a disc pendulum with electrostatic reduction of its deflection and stiffness in the X-direction. The facility uses a disc pendulum (1) between two electrodes (2) with an electrostatic control system for force measurement. The disc pendulum is made of gold-coated silicon plates (size 30 mm x 30 mm, thickness $T_p = 0.425$ mm) that are suspended on two gold-coated tungsten wires (1d) ($d_w = 10 \pm 1$ μm in diameter and approx. 424 mm in length). The deflection of the pendulum due to a force is measured by means of an interferometer (8) (attocube FPS3010). The facility can measure horizontal forces in the range below 1 μN, with a resolution below 5 pN and an uncertainty of 2.7 % down to a measured force of 1 nN at a measuring duration of about 20 s [4]. One of the main objectives of the new upgrades was the modification of the disc pendulum design to achieve greater flexibility. A new measuring disc-pendulum was developed wherein a rod (1a) is mounted horizontally through a hole in the center of the pendulum; a gold-coated silicon plate – the "sensor plate" (1b) (size 20 mm x 20 mm, thickness 0.425 mm) – is attached to one side of the rod and a counterbalance (1c) to the other side. The rod (1a) is made of Zerodur with a coefficient of thermal expansion of less than 10^{-8} K^{-1}. To ensure electrical conductivity and to eliminate static charges, all elements are coated with a 100 nm gold layer. The main source of measurement uncertainty and instability in a high vacuum is the free angular oscillations of the disc pendulum around the pendulum's main axes. Such oscillations have a high quality factor and can achieve amplitudes of 10^{-2} rad. To detect these oscillations, a laser beam (9) is reflected from one end of the rod (1a) onto a quadrant photodiode (10). Each of the electrodes (2) consists of four segments that can be controlled individually by using the signals from the quadrant photodiode (10). The angular oscillations are compensated by means of two additional independent electrostatic control systems, the first of which compensates for angular oscillations relative to the Y-axis, and the second of which compensates for the angular oscillations relative to the Z-axis.

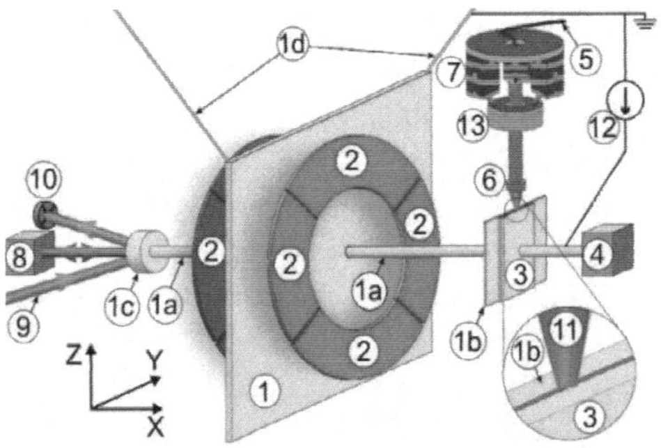

Figure 1. Functional diagram of the measuring part of the nanonewton force facility: (1) gold-coated disc pendulum; (1a) gold-coated Zerodur rod; (1b) sensor plate; (1c) counterbalance; (1d) disc pendulum suspension (thin gold-coated tungsten wires); (2) outer conductive electrodes (each consists of four segments); (3) probe plate; (4) 6D positioner; (5) single mode fiber laser system; (6) focusing optical system; (7) 6D positioner (Hexapod); (8) interferometer; (9) laser beam; (10) quadrant photodiode; (11) focused laser beam; (12) voltage source; (13) rotator.

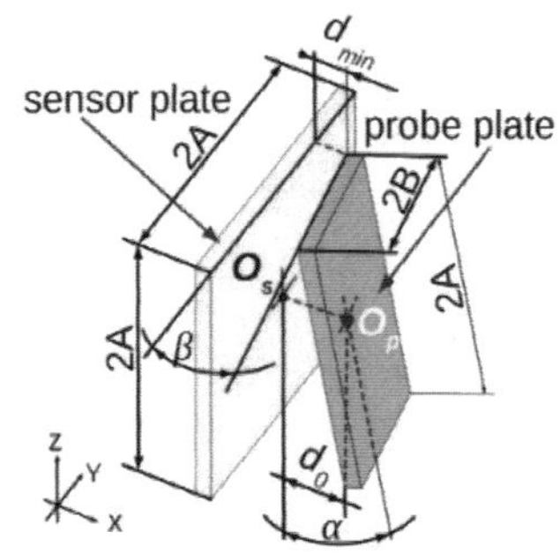

Figure 2. The location of the sensor (1b) and of the probe plates (3) ($2A$ = 20 mm; $2B$ = 4 mm).

The angular oscillations can be reduced by several orders of magnitude. For example, in a vacuum of 10^{-6} mbar, the angular oscillations of the disc pendulum have been reduced from 10^{-2} rad to $2 \cdot 10^{-7}$ rad. The quadrant photodiode with the two electrostatic control systems also makes it possible to measure the torque that acts on the disc pendulum (sensor plate). The nanonewton force facility contains three independent electrostatic control systems. Each of these systems is a PID (proportional-integral-derivative) real-time control system with a sampling period of about 1 μs. Control voltages are applied independently to each of the eight segments from the FPGA card. The FPGA card operates under the control of a real-time system and provides voltages from 0 V to 10 V. For a so-called plane-plane geometry, a gold-coated silicon plate (size 20 mm x 4 mm, thickness 0.425 mm) was used as a probe plate (3). This plate is attached and moved using a 6D positioner (4). A slit whose width is in the sub-micrometer to millimeter range can be created between the probe plate (3) and the sensor plate (1b). The electrical voltages may be applied between the plates (3, 1b) by means of the voltage source (12). The electrostatic force arises between the sensor plate (1b) and the probe plate (3). The location of the sensor plate and of the probe plate is presented in figure 1.

This force is measured by means of the nanonewton force facility. To control the arrangement of the upper edges of the plates (3, 1b), a method analogous to confocal laser scanning microscopy (CLSM) is used. The light is provided by a single mode fiber laser system with a wavelength of λ = 1550 nm (5) and an output power P (0 W< P < 2 W). The laser beam is focused by means of an optical system (6) until it forms a spot radius of about ω = 4 μm (11) on the upper edges of the plates. The whole optical system is mounted on a compact 6D positioner called a hexapod (7). The hexapod is used as a scanning system. The power of the radiation reflected from the upper edges of the plates and returned to the single mode fiber is measured. For the so-called plane-ball geometry, the probe plate (3) is replaced with a ball (hemisphere).

Journal of Physics: Conference Series **1379** (2019) 012039 doi:10.1088/1742-6596/1379/1/012039

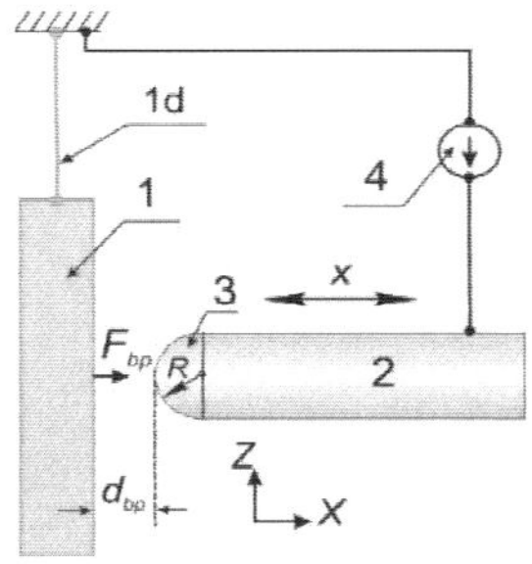

Figure 3. Ball-plane geometry: (1) disc pendulum (sensor plate); (1d) disc pendulum suspension (thin gold-coated tungsten wires); (2) movable mounting rod; (3) probe ball; (4) voltage source.

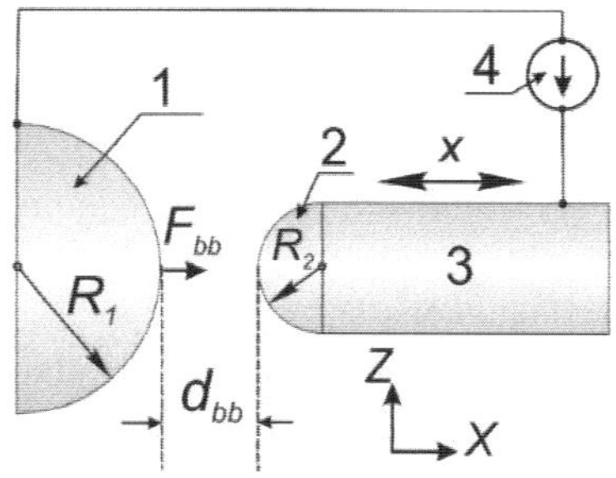

Figure 4. Ball-ball geometry: (1) sensor ball; (2) probe ball; (3) movable mounting rod; (4) voltage source.

3. New three-voltage method

In this section, we first consider a so-called plane-plane geometry (Figure 1 and 2), which is a most complicated case. To measure the contact potential difference δu as well as the distance d_0 between the center of the sensor plate (1b) and the center of the probe plate (3) (Figure 1 and 2), a new three-voltage method was developed (first presented in [5]). Let us assume that an electrical voltage u exists between the sensor plate and the probe plate. The beginning of the Cartesian coordinate system is located in the center of the sensor plate at the point O_s (Figure 2). Let us assume that the centers of the plates (points O_s and O_p) are located on the X-axis. The required position of the plates is controlled by means of confocal laser scanning microscopy (CLSM). The angles of rotation between the plates are α (relative Y-axis) and β (relative Z-axis). We will assume that the dimensions of the $2A$ and $2B$ plates considerably exceed the distance d_0 between them ($d_0 << 2A$, $d_0 << 2B$). In this case, the edge effects can be neglected. Assuming that $\alpha << 1$ and $\beta << 1$, the electrostatic force $F(u)$ between the plates can be described by the following expressions:

$$F(u) = \frac{1}{2} \cdot \frac{\partial C_{pp}}{\partial d_0} \cdot u^2 = \frac{\varepsilon_0 \cdot u^2}{2} \cdot \int_{-A}^{A} dz \int_{-B}^{B} \frac{dy}{(d_0 - \alpha z - \beta y)^2} = \frac{T_{pp} \cdot u^2}{D(d_0)^2} \,, \tag{1}$$

$$T_{pp} = 2 \cdot \varepsilon_0 \cdot A \cdot B \,, \tag{2}$$

$$D(d_0)^2 = {d_0}^2 - (\alpha \cdot A)^2 - (\beta \cdot B)^2 \,, \tag{3}$$

where C_{pp} is the capacity between the plates, ε_0 is the vacuum permittivity.

The voltages u_0, $u_1$and u_2 are applied between the plates by means of the voltage source (12). At the same time, the forces $F(u_0)$, $F(u_1)$, and $F(u_2)$ are measured. Taking the contact potential difference δu into account, these measured forces can be written as:

$$F(u_0) = \frac{T_{pp} \cdot (u_0 + \delta u)^2}{D(d_0)^2} + \delta F \,, \tag{4}$$

$$F(u_1) = \frac{T_{pp} \cdot (u_1 + \delta u)^2}{D(d_0)^2} + \delta F \,, \tag{5}$$

$$F(u_2) = \frac{T_{pp} \cdot (u_2 + \delta u)^2}{D(d_0)^2} + \delta F \,, \tag{6}$$

where δF is the offset force (any force independent of the applied voltage).

Solving the system of equations (4-6), the contact potential difference δu and the parameter $D(d_0)^2 = ({d_0}^2 - (\alpha \cdot A)^2 - (\beta \cdot B)^2)$ are represented by means of the following expressions:

$$\delta u = -\frac{F(u_0)\cdot(u_2^2-u_1^2)+F(u_1)\cdot(u_0^2-u_2^2)+F(u_2)\cdot(u_1^2-u_0^2)}{2\cdot(F(u_0)\cdot(u_2-u_1)+F(u_1)\cdot(u_0-u_2)+F(u_2)\cdot(u_1-u_0))}, \tag{7}$$

$$D(d_0)^2 = \frac{T_{pp}\cdot((u_1+\delta u)^2-(u_0+\delta u)^2)}{F(u_1)-F(u_0)}, \tag{8}$$

$$D(d_0)^2 = d_0{}^2 - L^2, \quad L^2 = (\alpha\cdot A)^2 + (\beta\cdot B)^2 . \tag{9}$$

To determine the distance d_0, the probe plate (3) (Figures 1 and 2) is displaced in the X-direction by a known distance Δ. Using the three-voltage method and the equations (7, 8) at a distance d_0 (unknown) and at a distance (d_0 + Δ), the parameters $D(d_0)^2$ and $D(d_0+\Delta)^2$ are determined. Using equation (9), the distance d_0 is calculated by means of the following expression:

$$d_0 = \frac{D(d_0+\Delta)^2 - D(d_0)^2-\Delta^2}{2\cdot\Delta} \tag{10}$$

When the distance d_0 is known and equation (9) is used, the parameter L and the minimum distance between the plates d_{min} (Figure 2) are given by the following equations:

$$L^2 = (\alpha\cdot A)^2 + (\beta\cdot B)^2 = d_0{}^2 - D(d_0)^2 , \tag{11}$$

$$d_{min} = d_0 - L . \tag{12}$$

Note that the method can be easily expanded to include a ball-plane geometry (Figure 3) and a ball-ball geometry (Figure 4). Using of the so-called proximity force theorem (PFT) [6], the electrostatic force F_{bp} of the ball-plane geometry is given by means of the equations:

$$F_{bp} = T_{bp}\cdot\frac{u^2}{d_{bp}}, \qquad T_{bp} = \pi\cdot\varepsilon_0\cdot R , \tag{13}$$

where R is the radius of the ball, d_{bp} is the distance between the ball and plane (Figure 3). Replacing T_{pp} with T_{bp} and $D(d_0)^2$ with d_{bp} in equations (4-6) and using equations (8,13) the distance d_{bp} is calculated by means of the expression:

$$d_{bp} = \frac{T_{bp}\cdot((u_1+\delta u)^2-(u_0+\delta u)^2)}{F(u_1)-F(u_0)} . \tag{14}$$

Using the PFT, the electrostatic force F_{bb} of the ball-ball geometry is given by means of the equations:

$$F_{bb} = T_{bb}\cdot\frac{u^2}{d_{bb}}, \qquad T_{bb} = \pi\cdot\varepsilon_0\cdot\frac{R_1\cdot R_2}{(R_1+R_2)}, \tag{15}$$

Where R_1 and R_2 are the radiuses of the balls, d_{bb} is the distance between the balls (Figure 4). Replacing T_{pp} with T_{bb} and $D(d_0)^2$ by d_{bb} in equations (4-6) and using equations (8,15), the distance d_{bb} is calculated by means of the expression:

$$d_{bb} = \frac{T_{bb}\cdot((u_1+\delta u)^2-(u_0+\delta u)^2)}{F(u_1)-F(u_0)} . \tag{16}$$

The contact potential differences δu of the ball-plane geometry and the ball-ball geometry, as well as those of the plane-plane geometry, are given by equation (7). Note that the contact potential difference δu determined from equation (7) does not depend on the calibration of the force meter used. Note also that equation (7) can be used for any given geometry and relative position of the conductive surfaces.

4. Adjustment of the parallelism of the plates

The parameter L characterizes the degree of the parallelism of the probe plate (3) relative to the sensor plate (1b). The parallelism of the plates corresponds to the zero value of the parameter L. Another method of setting the parallelism of the plates is to minimize the torque applied to the sensor plate (1b) by the probe plate (3) (Figures 1 and 2). Using CLSM (Section 2), an angle of $\beta < 10^{-4}$ rad between the plates is achieved in practice. To simplify the calculations, we assume that the angle β between the plates is zero. In this case, the torque $M_Y(\alpha)$ applied to the sensor plate (1b) by the probe plate (3) can be written as

$$M(\alpha)_Y = \int_{-A}^{A} \frac{\varepsilon_0 \cdot u^2 \cdot B \cdot z \cdot dz}{(d_0 - \alpha \cdot z)^2} = \frac{4 \cdot \varepsilon_0 \cdot u^2 \cdot A^3 \cdot B}{3 \cdot d_0^3} \cdot \alpha \cdot \left(1 + \frac{6}{5} \cdot \left(\frac{\alpha \cdot A}{d_0}\right)^2 + O\left(\left(\frac{\alpha \cdot A}{d_0}\right)^4\right)\right). \quad (17)$$

If the angle α between the plates is small ($\alpha \cdot A \ll d_0$), it follows from equation (17) that the dependence of the torque $M_Y(\alpha)$ on the α is linear. The parallelism of the plates ($\alpha = 0$) corresponds to the zero value of the torque $M_Y(\alpha)$. After parallelism of the plates has been achieved, the desired angles of inclination α and β can be set by turning and displacing the probe plate (3) by means of the 6D positioner (4).

5. Measurements

Figures 5 and 6 show the measurements of the contact potential difference δu and the distance d_0 between the plates (1b, 3) for plane-plane geometry by means of the new three-voltage method. The voltages $u(t)$ are applied between the plates (1b, 3) by means of the voltage source (12). Figures 7 and 8 show the measurements of the contact potential difference δu (blue solid line) and the distance d_0 (brown solid line) between the plates (1b, 3) for plane-plane geometry by means of the new three-voltage method, when distance changes cyclically. To change the distance between the plates, the probe plate (3) is moved stepwise relative to the sensor plate (1b) with a step size of 20 nm and a time step of 90 s. During each step of the probe plate with a time step of $\tau_p = 90$ s, the voltage changes three times with a basic time interval of $\tau_u = 30$ s. The mean values of the force are calculated from the signal measured after it has passed a Bessel low-pass filter (0.05 Hz, 12th order) in the interval from 21 s to 30 s after the start of the voltage interval. During each step, the contact potential difference (compensation voltage) δu and the distance d_0 are calculated. Thus, the contact potential difference δu using the three-voltage method is measured and compensated to zero for each step of the probe plate. The measurements were carried out in a vacuum of about $2 \cdot 10^{-7}$ mbar.

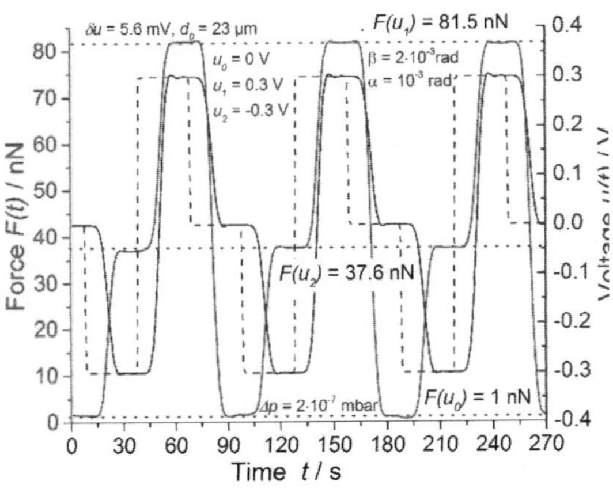

Figure 5. Demonstration of the new three-voltage method without compensation of the contact potential difference: applied voltage = blue dashed line; applied voltage after the low-pass filter = blue solid line; measured force after the low-pass filter = red line.

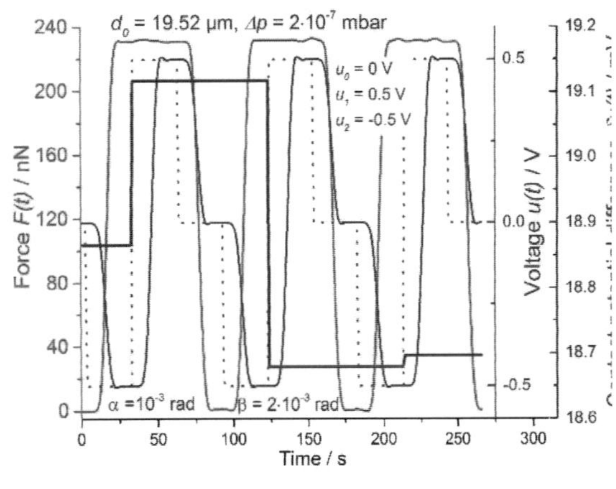

Figure 6. Demonstration of the new three-voltage method with the contact potential difference compensation: applied voltage = brown dashed line; applied voltage after the low-pass filter = brown solid line; measured force = red line; measured and compensated contact potential difference = blue solid line.

The probe plate (3) was tilted around the vertical Z-axis (β= $2\cdot10^{-3}$ rad) and the horizontal Y-axis (α = 10^{-3} rad) relative to the sensor plate (1b). The measured contact potential difference $\delta u(t)$ versus time and its power spectral density are presented in figure 9 (vacuum $\Delta p = 2\cdot10^{-7}$ mbar). In this experiment, the distance between the plates $d_0 = 23.4$ µm ($d_{min} = 1$ µm) was constant.

For ball-plan geometry, the probe ball (3) (hemisphere with a radius $R = 4.2$ mm) is moved stepwise relative to the sensor plate (1) (Figure 3) with a step size of 5 nm and a time step of 60 s. The measured contact potential difference δu and temperature variations of the facility versus the distance d_{bp} in a vacuum $\Delta p = 0.1$ mbar are presented in figure 10.

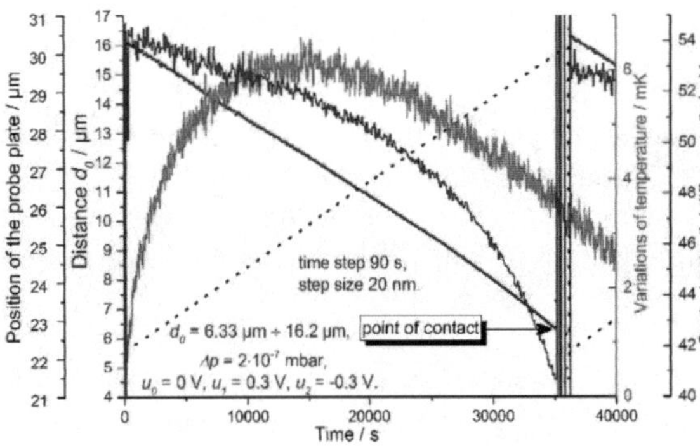

Figure 7. Measurements of the contact potential difference δu (blue solid line) and the distance d_0 (brown solid line) between the plates (1b, 3) for plane-plane geometry when distance changes cyclically: position of the probe plate = black dashed line; temperature variations of the facility = red line.

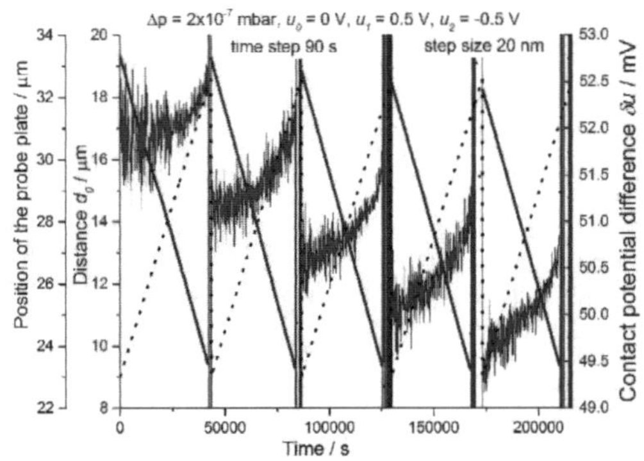

Figure 8. Measurements of the contact potential difference δu (blue solid line) and the distance d_0 (brown solid line) between the plates (1b, 3) for plane-plane geometry when distance changes cyclically (5 cycles): position of the probe plate = black dashed line.

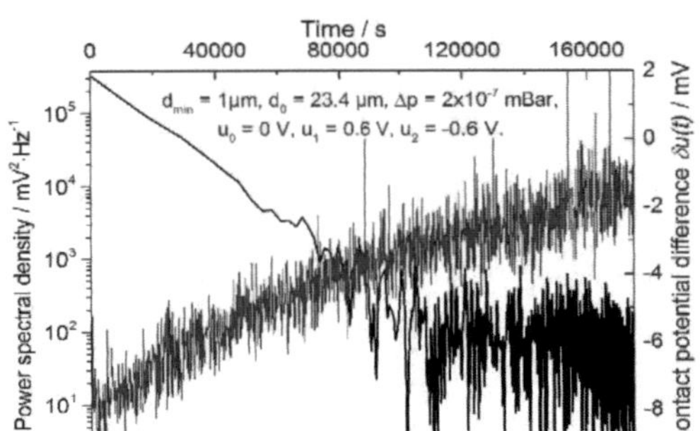

Figure 9. Measured contact potential difference $\delta u(t)$ versus time (blue line) and its power spectral density (black line) in a vacuum $\Delta p = 2\cdot10^{-7}$ mbar for plane-plane geometry.

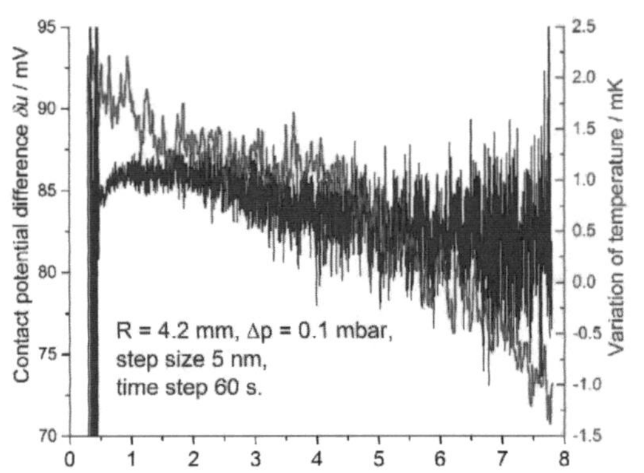

Figure 10. Measured contact potential difference δu (blue line) and temperature variations of the facility (red line) versus the distance d_{bp} in a vacuum $\Delta p = 0.1$ mbar for ball-plane geometry.

It should be noted that there is only a weak correlation of the contact potential difference δu with the temperature variations of the facility. It should also be noted that there is a strong variability in the dependences of the contact potential difference δu versus the distance d_0 (Figures 7 and 8). This behavior can be explained by structural surface changes as a result of the action of electric fields arising between the surfaces. These changes may include the structure of gold clusters that cover the surfaces. Temporary changes in the contact potential difference in some experiments exceeded 100 mV per day. Changes in the contact potential difference δu should be compensated during precision measurements.

Acknowledgements

This work was supported by the Deutsche Forschungsgemeinschaft (DFG NE 1550/2-1 and DFG NE 1550/2-2). I want to express my gratitude to my grandson Roman for the constant interest in my work.

References

[1] Behunin R O, Dalvit D A R, Decca R S and Speake C C 2014 *Phys. Rev. D* **89**, 051301(R)
[2] Sushkov A O, Kim W J, Dalvit D A R and Lamoreaux S K 2011 *Phys. Rev. Lett.* **107** 171101
[3] Nesterov V 2007 *Meas. Sci. Technol.* **18**, 360
[4] Nesterov V, Belai O, Nies D, Buetefisch S, Mueller M, Ahbe T, Naparty D, Popadic R and Wolff°H 2016 *Metrologia* **53** 1031
[5] Nesterov V, Nies D, Belai O, Buetefisch S, Kirchhoff J, Mueller M and Brand U 2018 *Proc. Int. Conf. CPEM* (Paris) p 950
[6] Blocki J, Randrup J, Swiatecki W J and Tsang C F 1977 *Annals of Physics* **105** 427-462
[7] Robertson N A, Blackwood J R, Buchman S, Byer R L, Camp J, Gill D, Hanson J, Williams S and Zhou P 2006 *Classical Quant. Grav.* **23** 2665-80

Metrologically coherent assessment for learning: what, why and how

Emily Pey-Tee Oon[1], U Hoi-Ka[1], W P Fisher Jr[2]

[1] Faculty of Education, University of Macau, Taipa, Macau SAR, China
[2] BEAR Center, Graduate School of Education, University of California, Berkeley, Berkeley California 94702 USA

E-mail: peyteeoon@umac.mo

Abstract. Assessment for learning aims to improve learning by feeding back information on the attainment of intended learning objectives to students and teachers. In this form of assessment, the student acts as a measuring instrument and learning attainment is the quantity to be measured. Current practices in classroom assessment, unfortunately, often are not grounded and have been focused on total correct scores without reference to the evidence hidden within them concerning the attainment of learning objectives. Our goal in this study is to clarify what assessment for learning is, why it is important, and how it works in a metrologically coherent context of equated assessments administered from calibrated item banks. The main focus lies in how item content can be framed according to intended learning outcomes that are meaningfully interpreted using a construct map. An example involves empirical datasets of 283 secondary three students from two formative tests designed by two science teachers. Coherent formative assessment of this kind applies a scientific model that belongs to the same class metrologists consider as defining measurement. The results illuminate the potential for establishing item banks that enable teachers to more efficiently and effectively assess and improve the attainment of learning objectives across different cohorts of students in a common metric.

1. Introduction

1.1. What assessment for learning is

Teaching practices in exam-oriented classrooms continue to be counterproductively driven by scores decades after improved methods have been made available [1]. Higher scores seem desirable, since a higher performance rewards both teachers and students. A lower score, in contrast, is seen as a sign of failure for both teachers and students. This kind of assessment, termed as assessment *of* learning focuses not on the attainment and promotion of learning outcomes but on students' abilities to perform on tests after instruction is completed.

Assessment *for* learning (formative assessment) differs, then, in that it occurs in classrooms during the course of instruction, and sets out to foster learning and improved outcomes. A large and significant body of published research studies have decisively established that formative assessments can move learning forward in highly productive ways not otherwise obtainable [2]. The promising potentials of formative practices, however, have not been fruitfully realized. Not only has formative assessment not been as widely implemented as one would expect given its proven value, it is often not understood as a source of information relevant to move learning forward, and even worse, it is often seen as a means of punitively evaluating teachers and students [3], especially in cultures where formative assessments are often conducted in a more formal manner [4].

1.2. Why assessment for learning is useful, and its problems

The attainment of learning objectives needs to be regularly assessed so that instruction can be planned to extend students' growth and to identify current levels of learning [3]. Assessment for learning may be structured formally or informally [4]. Informal formative assessments [5] are mainly verbal, as when teachers pose questions and students answer, allowing teachers to obtain multiple rapid indications about students' current states of understanding. In some cultures, education systems rely heavily on exam-based assessment of learning, feedback is sought in writing to obtain formal evidence about learning attainment. But instead of looking into the responses of students for evidence of what the student is prepared to learn next, total correct scores tend to focus attention only on an overall level of understanding. Summed total correct scores do not provide teachers with information useful to improve their ongoing instructional planning, and neither does it provide students with information useful to inform their learning processes relative to their objectives. It becomes instead a tool for ranking students, distorting the educational process. As a consequence, formative assessments implemented in this context, where a needed shift in values and practices has not been put into effect, have tended to become an additional interim assessment *of* learning. There is still much to be done with assessment of this kind are to be acted as agent *for* learning [1, 6].

2. How to plan and evaluate assessment for learning

2.1. Framing item content according to intended learning outcomes

Bloom's cognitive taxonomy [7] is often used as a guide in the framing of intended learning outcomes for knowledge and understanding development in education where construction of knowledge is conceptually interpreted in a progressive scale: (a) remember, (b) understand, (c) apply, (d) analyse, (e) evaluate, and (f) create.

Theoretically, test item ought to be written at a given level of cognitive challenge tapping an intended learning outcome. Taking an intended learning outcome (LO) on net forces for an example:

(a) A lower cognitive LO: students are able to *recall* that net force of a stationary object is zero
(b) A higher cognitive LO: students are able to *explain* that an object is stationary because either the net force acting on it is zero or no force is acting on it.

These contrasting assessment focal points are both borne out from an intended LO in which students are expected to understand that an object is not moving either because there are no forces acting on it or because the net force acting on it is zero. A correct response to an item with lower difficulty level implies that student has attained understanding that there is either no force or a net force of zero for a non-motion object. A correct response on a second item assessing higher cognitive understanding implies that net force applied to an object is actually proportional to its resulting acceleration [8]. Students who can answer the higher cognitive LO correctly most likely can answer the lower cognitive LO correctly. However, students who can answer the lower cognitive LO correctly may not be able to answer the higher cognitive LO correctly. Content difficulty for the higher LO is theoretically higher than the lower LO. Not basing expectations on personal or subjective judgements requires mapping content difficulty against cognitive complexities of intended learning outcomes relative to both theoretical and measurement analysis justifications.

2.2. Justification of theoretical and measurement analyses of intended learning outcomes

Items framed to assess higher and more complex cognitive LOs are likely to be estimated as being more difficult as compared to other items framed to assess lower and less complex cognitive LOs. Measurement models capable of framing and testing this kind of hypothesis are useful in quantifying learning outcomes explicitly as repetitions of an invariant unit. These model definitions and experimental tests are recognized by metrologists [9-10] as scientific approaches to measurement:

$$\ln[P_{nij} / (P_{nij1})] = B_n \, D_{ij}$$

which says that the log-odds of observed success for student n on item i at partial credit score j is equal to the difference between the estimate B of person n's ability and the difficulty estimate D of item i at rating j relative to the next lower rating of $j1$ [11-13]. Estimates expressed in these log-odds units (logits) are mathematically and experimentally proven to exhibit linear, interval, and additive properties [9-11].

Interpretation of these units is informed by construct theory, an explanatory model predicting variation in terms of causal relations [14-17]. For example, an item framed to assess content of middle difficulty ought to arrive at calibration lower than another item framed to assess content of high difficulty. If the measurement analysis results are not aligned with what has been expected theoretically, reasons to explain "why" and revisions of item content are in order. Rounds of iterations through this process to achieve stable relationships between theory and item calibrations are inevitable.

2.3. Formation of item strands based on theoretical construct map

Repeated patterns of calibrations emerging spontaneously across samples of students and across assessments indicate when construct definitions have achieved a stable balance between theoretical and measurement expectations. Theoretical terms can be communicated as strands in a construct map [14-15] outlining meaningfully varying amounts of the trait measured. Construct maps can be used to guide the writing of assessment items and to track learning progressions [15-17] from novice to expert understanding over time for each individual student.

3. An example

Three items were written from theory to assess different difficulty levels. Item 1 examines students' understandings at the lowest difficulty level, and item 3, the highest level (Table 1). The empirical scaling results, however, do not always align with theory. Two estimates (and uncertainties) of the difficulty of item 1, at 0.02 (0.07) and 0.03 (0.07) logits respectively, from two separate calibrations, are higher than estimates from the same samples for item 2, at 0.53 (0.1) and 0.46 (0.01) logits, even though the opposite was expected.

Item 2 was designed to examine if students are able to apply what they have learned about the food pyramid in general to a problem of actual daily relevance, whereas item 1 should elicit only the easier recall task. The measurement result has promoted the teacher who designed item 2 to think that this item is not at the level of applying existing understanding but instead involves simply recalling food categories according to different pyramid levels.

Theoretical and measurement results comparative alignments of this kind aim at defining a coherent assessment objective that serves to move student learning forward by identifying what is known, what is not known yet, where the student stands relative to the LO. Once stable relationships are established through multiple repeated measures, a provisionally definitive construct map can be built to inform interpretations of the trait being measured (Table 1). For an example, level 1 in the construct map outlines the minimum amount of knowledge students have achieved, and level 3 outlines the highest capacity of knowledge achieved by students with regards to the learning objectives. From here, teachers can feed back the information to students on what they have known and, most importantly, what they do not know yet. A response at a partial credit level may possibly indicate the student is hindered by misconceptions that must be addressed to move learning forward. Conversely, information on student performance can be fed back to teachers, communicating on the effectiveness of their instruction. Learning activities can be modified if most students fail to achieve the learning objectives. Having results measured in a unit of measurement comparable across classrooms provides additional opportunities for teachers to learn from each other and from their own historical data.

Journal of Physics: Conference Series **1379** (2019) 012040 doi:10.1088/1742-6596/1379/1/012040

Table 1. Construct map pertains to learning objectives with their test items.

Learning Objective	Bloom's Taxonomy of Learning			Construct Map
	Remember (Item 1)	Apply (Item 2)	Analysis (Item 3)	
Students are able to tell the categories of food ingredients in each layer of food pyramid. Students are able to design their own menus by using the energy values of food and the principles of food pyramid.	The _____ is a guideline for us to choose our food in right proportions. **Answer:** food pyramid	From the following categories, which layer in food pyramid does the food belong to? Mango, broccoli, butter, biscuits, fresh milk, cheese, granola, chocolate 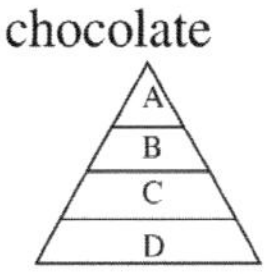 **Answer:** A: butter, chocolate; B: fresh milk, cheese; C: Mango, Broccoli; D: biscuits, wheat husks	William studied the food pyramid shown on the left. He then interpreted it as shown on the right, as below. 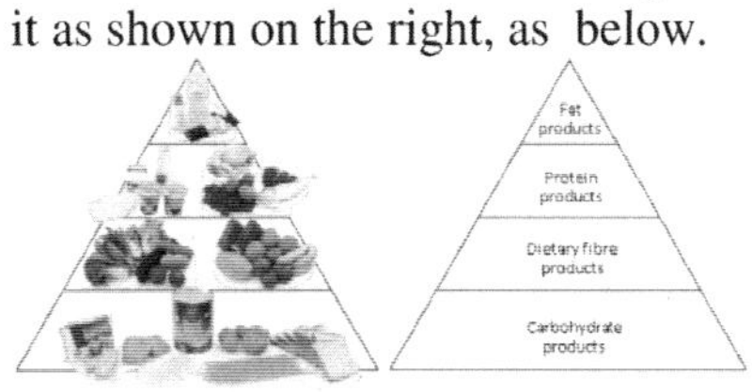 Explain why William's idea about each level is not completely correct. (a) Fat products (b) Protein products (c) Dietary fibre products (d) Carbohydrate products **Answer:** (a) This level actually includes salt and sugars, which are not fats. (b) The foods in this level also contain fats. (c) The foods in this level also contain carbohydrates, such as sugars and starch. (d) The foods in this level also contain dietary fibre.	0 - Cannot correctly describe the classification of each layer in the food pyramid 1 - Can clearly understand the classification of the layers in the food pyramid but cannot put all kinds of food into the corresponding categories 2 - Can clearly understand the classification of the layers in the food pyramid and put various foods into each corresponding categories 3 - Students can choose the right food from the food pyramid according to the needs of individuals.

4. Discussion

Student-focused assessment for learning reveals misconceptions useful for facilitating student progress toward learning objectives, where teachers are able to provide meaningful feedback to students in closing learning gaps. Teachers are similarly provided feedback on their own performance, so they can also evaluate and improve their instruction.

The example given demonstrates some of the student- and instruction-focused fundamental requirements for meaningful assessment for learning. This framework sets the stage for establishing item banks measuring the attainment of related learning objectives. Items in a bank can be created to provide reliable assessments across difficulty levels associated with specific learning objectives, and that are psychometrically validated as repeatedly and reproducibly measuring a construct. With tools like that available, teachers can select, apply, construct, and calibrate items for use in classroom assessments, adding to and working from the banks of questions.

5. Conclusion

Establishment of item banking of this kind will set stage to combine assessments for learning (formative assessment) and of learning (summative assessment) in a developmental coherence scale, as below:

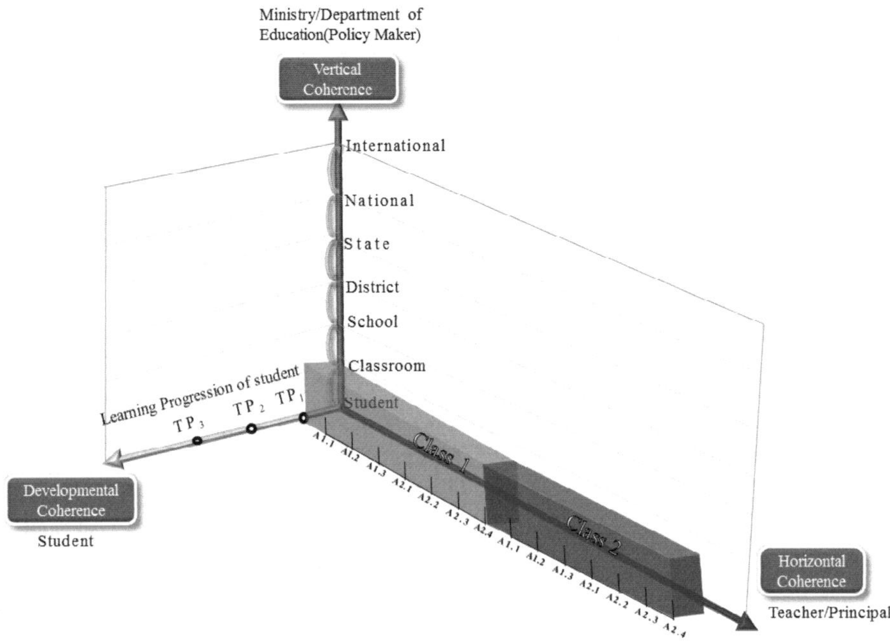

Figure. 1. Developmental, horizontal, and vertical coherence (TP=Time Point; A1.1-A2.4 are content domains) [18].

Summative evidence can be innovatively produced from accumulated formative assessments over time, instantiating the developmental coherence shown in figure 1. To achieve this purpose, formative evidence has to be collected in a reliable and valid form, meaning measures must be estimated and reported. This requires formative assessments and feedback to be structured in formal ways documenting progress [4]. But in this context, students' growth in understanding can be charted over time, allowing meaningful interpretation to move learning forward and to inform instruction. Capacities for taking missing data into account, adaptively administering items, and resiliently modifying LOs as educational needs and circumstances change suggest a productive future for this approach.

References

[1] Black P and Wiliam D 2018 *Assessment in Education: Principles, Policy & Practice* **25** 1-25
[2] Hattie J 2008 *Visible Learning* (New York: Routledge)
[3] Heritage M 2007 *Phi Delta Kappan* 140-5
[4] Dolin J, Black P, Harlen W and Tiberghien A 2018 *Transforming Assessment* eds J Dolin, R Evans (Switzerland: Springer Nature) pp 53-80
[5] Duckor B 2014 *Educ. Leadership* **71** 28-32
[6] Yan C 2015 *Improving Schools* **18** 5-19
[7] Anderson L W and Krathwohl D R 2001 *A Taxonomy for Learning, Teaching, and Assessing (*New York: Longman*)*

Journal of Physics: Conference Series **1379** (2019) 012040 doi:10.1088/1742-6596/1379/1/012040

[8] Alonzo A and Steedle J 2009 *Sci. Educ.* **93** 389–421
[9] Mari L and Wilson M 2014 *Measurement* **51** 315-27
[10] Pendrill L and Fisher W P Jr 2015 *Measurement* **71** 46-55
[11] Rasch G 1980 *Probabilistic Models* (Chicago: University of Chicago Press)
[12] Andrich D 1978 *Psychometrika* **43** 561-73
[13] Masters G N and Wright B D 1997 In W J van der Linden and R K Hambleton Eds, *Handbook of Modern Item Response Theory* pp 101-21 (New York: Springer-Verlag)
[14] Roberts L, Wilson M and Draney, K 1997, June *The SEPUP Assessment System: An Overview.* BEAR Report Series, SA-97-1 (Berkeley: University of California)
[15] Wilson M 2005 *Constructing Measures* (Mahwah, New Jersey: Lawrence Erlbaum)
[16] Wilson M R 2009 *J. Res. Sci. Teach.* **46** 716-30
[17] Wilson M R and Sloane K 2000 *Appl. Meas. Educ.* **13** 181-208
[18] Fisher W P Jr, Oon E P-T and Benson S 2019 *Measurement* in review

The BEAR Assessment System Software as a platform for developing and applying UN SDG metrics

W Fisher[1] **and M Wilson**[1]

[1] BEAR Center, Graduate School of Education, University of California, Berkeley, CA (USA)

E-mail: wfisher@berkeley.edu

Abstract. There are avoidable obstacles delaying realization of the United Nations' Sustainability Development Goals (https://unstats.un.org/sdgs/) which have been introduced by the methods used in measuring sustainability impacts and in monitoring progress toward the UN goals. Scientific measurement offers practical advantages based in rigorously defined and meaningful quantitative units that facilitate close management, trust, and efficient communication. Were improved metrics to be incorporated in systems of metrological traceability, legally binding conformity assessments could eventually provide a basis for new sustainability accounting standards, economic models, and the means for tracking returns on investments in sustainability impacts. Toward these ends, there is a need for making access to improved measurement principles, methods, and results much more widely available. That purpose could be served by the BEAR Assessment System Software (BASS), which provides an online environment offering practical tools for instrument development, administration, calibration, and quality improvement; data analysis and measurement; and reporting. Each variety of sustainability stakeholder could make use of the system for their diverse purposes, advancing their unique self-interests further by cooperatively competing with other stakeholder groups than they could working alone.

1. Introduction

The United Nations' 17 Sustainability Development Goals (SDG; https://unstats.un.org/sdgs/) present actionable targets for the changes needed to bring about safe, peaceful, and productive lives for all humanity, and for all life on earth. The call to action combines goals for prosperity with goals for protecting the planet. Many believe these are mutually contradictory goals, since improved environmental quality often leads to economic growth that in turn then reduces environmental quality.

But sustainability values can be measured objectively [1-4], opening new possibilities for aligning financial values and measurements. Commerce and science could be integrated in ways extending longstanding uses of measures of properties such as volume, mass, area, and kilowatts in trade [5]. Prosperity could be reconceived, so that authentic wealth and genuine productivity are no longer expressed in idiosyncratic and incommensurable terms. Institutional rules, roles, and responsibilities could be changed so that destroying authentic human, social, and environmental wealth can be seen as unprofitable, following the established pattern by which objective measures are conducive to the development of markets [6-10].

Most SDG metrics developed to date, however, do not support developments in this direction. Existing work plans (https://unstats.un.org/sdgs/tierIII-indicators/) treat each of the 232 different SDG indicators as a separate universe (of the total 244 indicators, nine overlap in multiple targets). Preliminary study of the SDG precursor, the UN's Millennium Development Goals, shows that advanced measurement models [1-4, 8, 9, 11-17] can be productively applied to data involving country-level counts, ratings, and indicators [18]. These results suggest it is unlikely that the 232 SDG indicators measure separate constructs. More likely, many of the indicators overlap in their focus, and present opportunities for combining observations of larger issues using composite measures.

The reasons why the SDGs have not yet been situated in this larger context stem from longstanding cultural assumptions about what is measurable and what is not. An extensive array of governmental, economic, educational, and other institutions have developed widely used measures that are neither scientifically rigorous, meaningful, nor practical. Decades of research across a wide array of fields has shown how those misconceptions came about, how they can be corrected, and how we can productively rethink and reshape our relationships with each other and the world by improving the quality of the information we use to represent those relationships [19]. We have within our grasp the liberating power of distributed cognitive models enacted via multilevel common measurement languages as adaptable to local situations as they are traceable to universal standards. Communicating and implementing that power requires a complex coordination of efforts from a wide range of interested stakeholders [8-9, 19]. Our aim is provide some initial indications of what is possible and how it might be accomplished using available tools, such as the BEAR Assessment System Software [19-22].

Though virtually all existing SDG metrics were developed with no reference to these metrological potentials, there are examples of widely used, high quality measures that could be adopted as SDG management tools [21-28]. After describing some of these examples, the results of analyses of several SDG datasets obtained online will be reported. Finally, another way to approach making SDG efforts measurable will then be taken up, using the ten targets for SDG 4 on education as an example.

2. The BEAR Assessment System Software

The BEAR Assessment System was developed at the Berkeley Evaluation and Assessment Research (BEAR) Center at the University of California, Berkeley Graduate School of Education. It is an implementation of four principles:

1. A developmental perspective that specifies the measurand to be assessed, its influence properties, and how it is identified, from initial conceptions to higher order proficiencies.
2. A match between the focus of the assessment and the object of management, with the aim of ensuring that measures and associated uncertainty estimates are meaningfully and usefully actionable, putting measurement and management in a mutually supportive, active dialogue.
3. Appropriate feedback, feed forward, and follow-up that engage users in the assessment process and that brings higher order thinking and creativity fully into play to improve outcomes.
4. Generation of evidence of measurement quality, so that end users and other stakeholders have confidence in the reported outcomes, and can readily interpret and use assessment findings to improve outcomes.

The Berkeley Assessment System *Software* (BASS) guides users in the design, development, and delivery of assessments, surveys, and other tools, following the four principles of the BEAR Assessment System. Progress toward goals can be monitored and reported in a context structured with rich, informative interpretive aids. BASS is intended for use in close dialogue with goal-oriented processes and content. Depending on the focus of the assessment area involved, professionals in the relevant substantive areas, such as teachers, health care clinicians, social workers, environmental managers, and public service providers, are supported in understanding results in practical, actionable terms. In education, for instance, learning progressions defined by curricula and assessment scales are used to diagnose student comprehension and learning needs in ways that feedback the information needed to advance toward outcome goals. Similar formative dimensions for developmental sequences, organizational learning, healing processes, and other change variables have been defined and scaled in thousands of research studies published over the last 50 years and more.

BASS builds on more than a decade of research supporting evidence-based assessment in a wide range of fields. This system allows easy management of all kinds of assessments involving surveys, multiple choice tests, performance or portfolio assessments, on-line examinations, game-based tasks, checklists, rankings, etc. In education, for instance, BASS might be used to design, administer, score, analyze, and report in-class activities, homework assignments, end-of-unit tests, and more. In order to make the system usable to instructors in classrooms, teachers are placed at the center of the design process, just as in other areas end users in other professions are at the center.

The development of the user tools in BASS (see figure 1) emphasizes (1) flexibility in designing, choosing, and assigning assessments, (2) simplifying scoring, and (3) reports written in ways that make them accessible and useful to a wide range of stakeholders, such as clients, patients, advocates, citizens, voters, students, parents, managers, line workers, etc. Assessments could include computerized versions of paper-and-pencil tasks, or electronically created assessment tasks (such as interactive graph or card-sort problems).

Reports are customizable for a wide range of measurement applications at the levels of individuals, small or large groups, as well as for theory-oriented instrument and construct studies. Reports are furthermore annotated with qualitative interpretive guidelines and various statistics indicating when and where unintended influences may have affected responses.

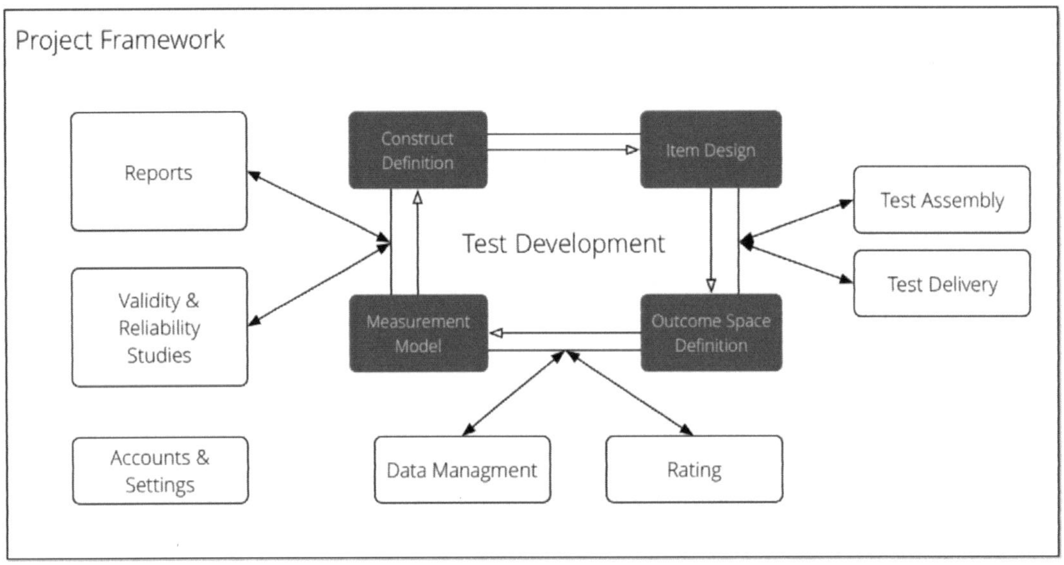

Figure 1. Structure of the BEAR Assessment System Software.

3. Developing SDG metrics in BASS

3.1. An inventory of measurands, influence properties, and sources of uncertainty

We manage what we measure. Attention is focused and collective efforts are coordinated via qualitative and quantitative methods of communicating purposes, intentions, processes, and results. Instruments metrologically traceable to unit standards are well established as a primary factor in the organization of institutions supporting efficient markets [5-7, 10].

Broad scale success in realizing the SDGs demands close attention to the quantification of the objects of interest. Because the goals and targets specified in the SDGs were not formulated with respect to the theories and methods of measurement science, they must be re-examined from the perspective of rigorous, meaningful quantification. The purpose of this reconsideration will be to identify and inventory the measurands (quantifiable properties) of sustainable development, along with the relevant influence properties and sources of uncertainty, such as test length and instrument targeting.

The results of this review could be publicly released for coordinated efforts by governments, academic institutions, research organizations, philanthropists, and corporate foundations interested in contributing to the fulfillment of the SDGS. BASS provides guidance needed for measurand definition, indicator development and administration, data gathering, analysis, and the reporting of comparable measurements with their associated uncertainties for quality improvement and research applications.

3.2. Unidimensional constructs, and multi-unidimensional combinations

Many of the SDG targets involve measurands representing different facets of sustainable development relevant to the same measured entities, such as individual persons, or firms, governments, ecosystems, etc. Though measured quantities vary along single dimensions, when multiple constructs are measured simultaneously, otherwise unavailable information takes shape in the multivariate relationships between and among the factors involved. Height and weight, for instance, are usefully combined in the Body Mass Index, which enables a more informed perspective on individual health than can be afforded by the two separate measures. Established methods [16, 17] for estimating and reporting these kinds of multidimensional results should be applied.

3.3. Examples

Examples of sustainable development measurands fall into three categories, those that are: (1) rigorously defined and quantified in the terms of measurement science, (2) less well defined but with available data for analysis, and (3) undefined except as SDG targets.

SDG 4 on education includes a target concerning the universal achievement of literacy, which of course is a longstanding area of interest in education and educational measurement. Literacy, in the domain of reading comprehension, is measured in an interval metric for over 32 million students in the US per year [23, 24]. Over a hundred thousand books and millions of magazine articles have their reading difficulties available in this unit. About half of the US states' departments of education report annual end-of-year tests in this unit, and dozens of reading curricula integrate assessment and instruction using it. Publishers express the reading difficulty of their books and magazine articles in this unit, enabling teachers to target reading assignments at the ability levels of individual students. Aggregate statistics on reading comprehension help state departments of education determine which schools are meeting accountability standards, while also providing educators, curriculum developers, book and magazine publishers, administrators, and testing agencies information they need to work together in common cause.

Similar kinds of measures in wide use focus on mathematics skills [20], with many rigorously designed and constructed scales used in managing outcomes in effective learning environments [25], developmental psychology [26], health care [27], and physical education [28].

Examples of the second level of measurand definition include data from a number of available sources focused on promoting progress toward the attainment of SDG 13 on Climate Action, and SDG 16 on corruption, reducing armed violence, freedom of the press, and eliminating violence against children. Data on these constructs have been obtained and analyzed. Results show multiple opportunities for improvement, but also positive implications for creating SDG communications and management systems incorporating more meaningful units of comparison.

Finally, an example of the third, undefined measurands includes SDG 4 education targets that could be used as raw material in an example of how to form the basis for measuring manageable outcomes. The first target states:

> By 2030, ensure that all girls and boys complete free, equitable and quality primary and secondary education leading to relevant and Goal-4 effective learning outcomes.

This target as written speaks to too many different targets to be measurable. A basic rule of thumb in psychological and social measurement is that, to be scalable, interpretable, and actionable, the question being answered has to be identifiable and reconstructable from the response. This first SDG 4 target breaks out into 16 separate statements, though these can immediately be reduced to eight by changing "all girls and boys" to "all children." But evaluating ratings of agreement and disagreement with the remaining statement still requires the assumption that each response will aggregate the attainment of the target equally across all eight different areas of focus. There will be no way to tell from the response if only one of the eight statements was rated, or which of over 40,000 (8!) alternative combinations of the eight statements was addressed.

What could be done, however, is to separate each implied statement into its own target, in the manner shown for SDG 4's first target in the table 1 below. The content from each target could be incorporated into BASS in this form as starting points in the development of construct maps, item writing, and scoring design processes, and hence, lead to sound measurements on this target.

4. Future directions

The formative assessment application BASS is built on the idea of measurement scales that stand for generalized structures exhibited across a range of psychological and social constructs relevant to the UN SDGs. The stability of these structures has been established for decades [23-26], suggesting that various

Table 1. Examples restating the first SDG 4 Education target as assessment criteria.

In this country, all children complete:	
	free primary education.
	equitable primary education.
	quality primary education.
	free secondary education.
	equitable secondary education.
	quality secondary education.
.	educations leading to relevant learning outcomes.
	educations leading to Goal-4 effective learning outcomes.

assessments could be linked together in common systems [23, 27], not unlike the metrological systems of weights and measures we take for granted in commerce and the natural sciences [1-10]. Common measurement languages like these will eventually impact human, institutional, and information resources in ways that will make educators, clinicians, managers, researchers, citizens, students, patients, advocates, employers, and others better able to work together to identify and meet needs for meaningful, precise, and actionable sustainability measures.

Acknowledgments

This work was supported in part by a grant from the US Institute for Education Sciences (award R305A120217). Technical staff contributing to the software portion of the project include David Torres Irribarra, Rebecca Freund, Ana Maria Albornoz Reitze, Shazi Khan, Daniel Stanfield, Sevan Tutunciyan, Richard Vorp and Mike Kendall (all with the University of California, Berkeley). The CDP data were obtained by Angelica Lips da Cruz of the ALDCPartnership.com in Stockholm, Sweden.

References

[1] Mari L and Wilson M 2014 *Measurement* **51** 315-27
[2] Pendrill L and Fisher W P Jr 2015 *Measurement* **71** 46-55
[3] Wilson M 2013 *Psychometrika* **78** 211-36
[4] Pendrill L R 2018 *Meas. Sci. Technol.* **29** 034003
[5] Ashworth W J 2004 *Science* **306** 1314-17
[6] Barzel Y 1982 *J. Law Econ.* **25** 27-48
[7] Poposki N, Majcen N and Taylor P 2009 *Accredit. Qual. Assur.* **14** 359-68
[8] Fisher W P Jr 2009 *Measurement* **42** 1278-87
[9] Fisher W P Jr 2012 *Standards Engineering* **64** 1. 3-5
[10] Miller P and O'Leary T 2007 *Account. Org. Soc.* **32** 701-34
[11] Rasch G 1980 *Probabilistic models* (Chicago: University of Chicago Press)

[12] Wright B D and Masters G N 1982 *Rating scale analysis* (Chicago: MESA Press)
[13] Andrich D 1978 *Psychometrika* **43** 561-73
[14] Wilson M 2013 *Measurement* **46** 3766-74
[15] Stenner A J, Fisher W P Jr, Stone M H and Burdick D S 2013 *Front. Psychol.* **4** 1-14
[16] Briggs D and Wilson M 2003 *J. Appl. Meas.* **4** 87-100
[17] Gochyyev P and Wilson M 2018 *BEAR Seminar Series* University of California, Berkeley
[18] Fisher W P Jr 2019 *J. Appl. Meas.* in review
[19] Fisher W P Jr and Wilson M 2015 *Pensamiento Educativo* **52** 55-78
[20] Torres Irribarra D, Freund R, Fisher W P Jr and Wilson M 2015 *J. Phys.: Conf. Ser.* **588** 012042
[21] Wilson M and Sloane K 2000 *Appl. Meas. Educ.* **13** 181-208
[22] Wilson M 2009 *Journal of Research in Science Teaching* **46** 716-30
[23] Fisher W P Jr and Stenner A J 2016 *Measurement* **92** 489-96
[24] He W and Kingsbury G G 2016 *J. Phys.: Conf. Ser.* **772** 012022
[25] Cavanagh R F 2015 *Learning Environments Research* **18** 349-61
[26] Dawson T L 2004 *J. Adult Dev.t* **11** 71-85
[27] Cano S, Melin J, Fisher W P Jr, Stenner A J, Pendrill L and EMPIR NeuroMet 15HLT04 Consortium 2018 *J. Phys.: Conf. Ser.* **1065** 072033
[28] Mok M M C *et al.* 2015 *J. Appl. Meas.* **16**(4) 379-400

Journal of Physics: Conference Series **1379** (2019) 012042 doi:10.1088/1742-6596/1379/1/012042

Measuring for management in science, technology, engineering, and mathematics learning ecosystems

J Morrison[1], **W P Fisher Jr**[2],

[1] Teaching Institute for Excellence in STEM, Cleveland Heights, OH, USA
[2] BEAR Center, Graduate School of Education, University of California, Berkeley, CA USA

E-mail: janmorrison@tiesteach.org

Abstract. Rapid global expansion of multi-stakeholder ecosystems for learning in Science, Technology, Engineering, and Mathematics (STEM) demand close attention to the information infrastructure needed for sharing best practices and improving outcomes. The existing conceptual model of ecosystem development and elements needs to be translated into a metrological measurement model tailored to ecosystem management. An evaluation tool not designed from measurement principles included 38 questions applied across three years. An initial review of the instrument conducted before data were made available found multiple ambiguous items and rating categories requiring respondents to summarize diverse aspects of their experience in a single rating, with no capacity to reconstruct which aspects were included by any given respondent. Response data from 36 ecosystems in 2016, 38 in 2017, and 110 in 2018 produced more uncertainty than expected given the numbers of items, rating scale categories, and responses. No common factor structures across items could be identified. Stated expectations already on the record concerning STEM ecosystem development characteristics and focal partners, attributes, and goal areas provide a basis for an instrument redesign likely to result in meaningful measures advancing a theory of action fit for the purpose of STEM learning ecosystem management.

1. Introduction

Meaningfulness in language requires a capacity for expression that represents things in the world in ways that make them repeatably identifiable no matter who is speaking or which particular words or phrasings are used. As Mundy [1] put it,

> The hallmark of a meaningless proposition is that its truth-value depends on what scale or coordinate system is employed, whereas meaningful propositions have truth-value independent of the choice of representation, within certain limits. The formal analysis of this distinction leads, in all three areas [measurement theory, geometry, and relativity], to a rather involved technical apparatus focusing upon invariance under changes of scale or changes of coordinate system.

Measurement models requiring sufficient statistics, separable parameters, and invariant comparisons substantiate the independence of measures from the questions asked by evaluating the fit of data and the explanatory capacities of predictive theory [2, 3]. Models of this kind specify requirements for measurement, and support the calibration of instruments, in accord with the definition of quantity

[1] To whom any correspondence should be addressed.

Journal of Physics: Conference Series **1379** (2019) 012042 doi:10.1088/1742-6596/1379/1/012042

accepted by metrologists [4-6]. When, as is commonly the case, this definition is ignored, measurement is assumed to be achieved by assigning numbers to observations according to a rule, and meaningfulness is not taken into consideration in the design of assessment instruments or in the analysis of data from them. This meaninglessness is plainly evident in the wide acceptance of the fact that summed scores mean something different depending on what questions were asked. What is less widely understood is just how meaningless those scores might be. The fact is, however, that reliability and significance test results may appear reasonable even when instrument content includes nonsense words or items written in an uninterpretable language, or respondents evaluate blank entries associated with a rating scale [7].

The practical value of meaningfulness extends beyond technical concerns with empirical evidence of invariance and explanatory models predicting it, to matters of manageability. Clear and actionable connections between what needs managing and what we measure are created when we begin with the end in mind. How will we know when a goal has been achieved? How will we know where we are at in relation to the desired goal? How will we know if we are moving in the right direction? How will we find out what needs to be done next? How can we learn from our own and others' experiences in pursuing shared goals?

These questions point to the relevance of the principles of formative assessment [8, 9] and coherent frameworks integrating assessments across learning environments and accountability demands [10,-11]. When the content of goals are separated into steps in a process, and the sequence of those steps has a clear logical order, where progress depends on the accomplishment of prerequisite elementary objectives, then observations can be evaluated for how sensibly they conform to expectation, and expectations can be revised in light of data. The end result is a customizable tool each stakeholder can use to plot their individual course in terms communicable to others.

Basic instrument design principles suggest that items should be written to focus on variation from less to more in an expected range, and to provide the needed level of precision required to support the intended decision process [12]. To these ends,

- items should be written so each one addresses a single issue, and only one single issue (no if-then phrases, no ands, buts, or ors), because it is impossible to reconstruct from the response which question was answered when items are multivalent;
- the items should define a coherent narrative of a developmental sequence or learning progression from the one thing that will be present or agreeable if only one thing is, all the way up to the last thing that will not be present or agreeable, if only one thing is not;
- enough questions should be asked to drive uncertainty down, relative to the expected variation, to the point needed for reliability and precision [13]; and
- concerning the rating scale:
    - Not Applicable and No Opinion response options are to be avoided,
    - even numbers of response options should be provided, to prevent overuse of middle categories allowing a response not associated with an actual decision, and,
    - six or eight response options should be used in the rating scale, to aid in avoiding floor and ceiling effects, and to augment reliability when possible via added consistent distinctions and score groups.

2. STEM Learning Ecosystem evaluations

The STEM Learning Ecosystems Initiative was officially launched by the STEM Funders Network at the June, 2015 meeting of the Clinton Global Initiative America. Multiple robust discussions among STEM educators, policymakers, funders, and other key stakeholders were informed by growing evidence of the need for cross-sector collaborations. Experience shows that those who are in the best position to transform STEM education are collaborating communities of active STEM learning educators. In this context, the STEM Learning Ecosystem Initiative empowers local communities to deliver stronger STEM learning results for more students, more powerful professional development for educators, and more meaningful partnerships for business and education leaders. That said, community partners working across sectors must not only coordinate their efforts, they must work at new, deeper levels to leverage in-, after-, and out-of-school opportunities to provide more students with quality learning.

Metrological common languages have the potential of enabling work at deeper levels in important ways. Measuring ecosystem development (and other relevant variables, such as student learning outcomes) in common metrics calibrated from assessments tailored to each local ecosystem's needs could streamline communications within and across stakeholder groups, and enhance capacities for sharing what works. The capacity of stakeholders to work at deeper levels depends on having maps of where we want to go, maps that represent information at the functionally discontinuous levels of individual (micro), group (meso), and population (macro) complexity [10].

Educators need to be able to learn from one another as they explore this new terrain of interdependent relationships across previously disconnected areas of STEM learning. Initial attempts at STEM learning ecosystem data collection were intended to take advantage of local evaluations already in use, with the expectation that improvements would be implemented over time. Following common practices, the evaluation items were not written with the intention of achieving meaningful measurement in the sense of items and respondents that would together define a relation of conjoint additivity and conditional independence.

Contrary to recommendations based on measurement principles, the STEM Learning Ecosystems evaluation items include many conjunctions. Even rating scale choices are multivalent, as with one rating of 4 labelled "Infrastructure is robust, placing many educators, and growing to meet demand, and is iterating in response to in-field observations." Separating out each subject-verb-object combination in any given item may result in as many as eight or more different statements. When the rating options also present multiple possible single rating criteria, and even more combinations of rating option pairs and triples, the number of possible items represented in any single rating can be 20 or more.

The items, furthermore, were not written from a construct map with an intention of defining a range of variation from less to more. The number of items was determined by considerations of content coverage and respondent burden, not by reverse engineering from the precision needed to support a decision process. Alternatively, a definitive bank of all possible relevant items could be developed, calibrated, and adaptively administered according to the needs of individual ecosystems and the precision demands of the relevant management needs [14].

Finally, though Not Applicable and No Opinion response options are not used, all items are associated with five categories instead of an even number that forces respondents to make a decision.

Given the predominance of items asking multiple questions, rating categories specifying multiple responses, no deliberate intention of posing questions varying from least to most, and no concern for measurement uncertainty relative to variation, it is not likely that data from this tool will fit a model requiring separable parameters.

3. Measurement modelling principles

In the absence of an intention to calibrate an instrument providing meaningful measures independent of the questions asked, empirical data evaluations may take the form of a dialogue between model fit statistics and the Principal Components Analysis (PCA) of the standardized residuals [15-17]. This PCA sets aside the primary dimension measured, which is constituted by the construct-relevant variance shared by all or most of the items. When assessment items are deliberately written to address a single construct and the responses indicate they generally succeed in this, items provoking construct-irrelevant variance are picked up by the model fit statistics. But when items represent multiple constructs in roughly equal proportions, fit statistics get muddled and cannot tell them apart. This is where PCA excels, however, as it is sensitive to groups of items sharing more variance with each other than with the other items.

Conversely, a very wide score distribution or a multimodal distribution will confuse the PCA, causing it to mistakenly identify multiple constructs even if the data stand for an established unidimensional construct like length. In this latter case, the PCA results can be checked to see if the items supposedly measuring different constructs actually produce uncorrelated ecosystem measurement estimates. If high disattenuated (i.e., estimated after accounting for measurement uncertainty) [18-20] correlations (over 0.85) between the pairs of measures produced from different subsets of items are produced, PCA results indicating separate constructs may be discounted, especially if theory supports the contention that the items work together to define the same dimension.

PCA results identifying unexplained variance contrasts with eigenvalues less than 1.4 mean that the amount of unexplained variance is at a level attributable to that associated with only a single item, or less. Factor loadings in this context will likely all be less than |0.40|, disattenuated correlations of the measures implied by the item clusters within a contrast will be over 0.85, and the ratio of the variance explained by the measures to the unexplained variance captured in the PCA contrasts will likely be 3-1 or higher. With a well-designed instrument and either a wide score distribution or multimodal data, unexplained variance contrast eigenvalues may be over 1.4, and loadings may be lower than 0.40 or higher than 0.40, but the ratio of explained to unexplained variance captured in the contrasts will still be higher than 3-1, and measures estimated from the separate groups of items will correlate highly, with the disattenuated correlations approaching 1.00. Instruments with separate groups of items measuring different constructs, however, will have high eigenvalues, loadings greater than |0.40|, low ratios of explained to unexplained variance, and disattenuated correlations lower than 0.85.

4. Instrument calibration results

The instrument was administered, with some modifications, over three years, with response data from 36 ecosystems in 2016, 38 in 2017, and 110 in 2018. Ratings from all 184 reporting by-year ecosystems on 43 total items (38 items common across all three years, and five used only in 2016) were fit to a probabilistic model of measurement requiring sufficient statistics, invariant conjoint additivity, and separable parameters [21-24]. In an initial analysis allowing all items to each define their own rating scale, even if the ratings shared content, of the 215 (43 times 5) possible categories, four were empty and unused. There were no extreme scores for either ecosystems or items.

As stated, the instrument was not designed with the intention of producing data likely to fit a measurement model of this kind. In addition, the data were not gathered with the intention of calibrating an instrument in this way (which would have necessitated a pilot study and larger sample size). The analyses applied to the data were, then, entirely motivated by empirical considerations, in the hope that something substantively significant could be learned, developed into theory, and applied in the future.

In this context, with the goal of increasing the likelihood of meaningful results, the rating categories were optimized so that higher ratings uniformly calibrate at higher levels on the scale [25]. This process involves rescoring the data to combine responses in adjacent categories when, among other possibilities, there are fewer than ten responses in a category, and/or the category transition thresholds are disordered. This rescoring makes results more clearly interpretable, enhances the fit of the data to the model, and increases the clarity of the relationship between scores and measures. If a significant amount of new data from this tool becomes available, the process would have to be started again from the beginning, due to the locally-dependent characteristics associated with small samples.

After optimization, there were 132 total rating categories across the 43 items, 167 responses on average per item, with a range of 36 to 184, and 39 responses per ecosystem on average, with a range of 38 to 43. The mean measure was 0.10 logits, less than one standard error above the middle of the item scale (0.00), indicating an on-target assessment. Model fit statistics appeared reasonable, with the ecosystem measures' mean square information-weighted fit (infit) average and standard deviation at 1.02 and 0.37, respectively; the mean square outlier-sensitive fit (outfit) statistics were similar. The item calibration infit mean and standard deviation was 0.99 and 0.16, with similar outfit results. For the ecosystems, mean square fit statistics ranged from about 0.4 to 2.2; and for the items, from about 0.7 to 1.5. These numbers indicate generally acceptable model fit, given the sample size, rating scale, and instrument length, with some expectation that there may be local disturbances in some ecosystem-item interactions.

Ecosystem measurement and item calibration reliabilities were lower than expected (0.93 and 0.86, respectively), given the uncertainty typically associated with around 40 well-targeted items with a five-point rating scale and moderate standard deviations [13]. Uncertainties (modeled standard errors) were in the expected ranges, but the standard deviations were low (0.9 for the measures, and 0.4 for the items).

Five PCA contrasts in the overall analysis had unexplained variance eigenvalues ranging from 2.4 to 4.4, with a 2.25-1 ratio of explained variance to the unexplained variance captured in those five contrasts. Figure 1 shows that of the log-scaled variance explained by the measures (M) is about twice as large as the unexplained variance captured in the first residual contrast (1 in the figure).

Measures implied by the three item clusters within each of the five unexplained variance contrasts were correlated. Five of the first six, and nine of the 15, disattenuated correlations of the implied measures were in the range of 0.56 to 0.83, with the remaining six of the 15 ranging from 0.88 to 0.96. In all of the five sets of contrasts, the correlations of clusters 2 and 3 were 0.83 or higher. Correlations tended to increase as the contrast eigenvalues decreased, as expected. The low ratio of explained to unexplained variance, and the low correlations, suggest the presence of multiple constructs.

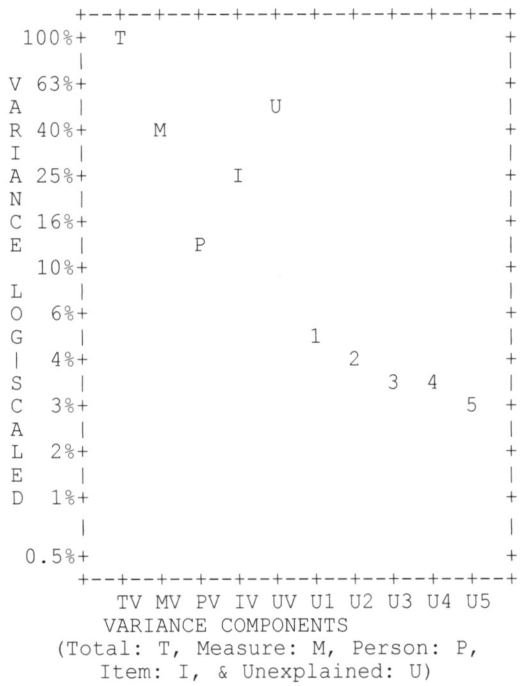

Figure 1. Standardized Residual Variance Scree Plot.

Analyses proceeded, then, by a series of experiments removing the first cluster of items in the first contrast with the highest eigenvalue and the lowest overall correlation of associated measures. The typical result of this process is that construct-irrelevant variance decreases in the remaining data, so that model fit and PCA results improve, though reliability may drop as uncertainty increases due to the removal of items.

That did not occur. The first re-analysis removed the ten items in the first contrast, in the expectation that the highly correlated measures associated with the two sets of items in the second and third contrasts would then work together to measure the same construct. But the PCA results for this restricted analysis were almost identical to the first analysis' results.

After eight analyses and the removal of 30 of the 43 items, there were still three PCA contrasts with eigenvalues over 1.4, four of 15 disattenuated correlations under 0.85, and, worst of all, a 1-1 ratio of explained variance to the unexplained variance captured in five contrasts. Further studies of various subgroups of items identified in these analyses repeated this kind of pattern for the overall data set including all three years as well as for each individual year.

With the removal of one further item, however, there was only one contrast with an eigenvalue over 1.4, the three correlations of the ecosystem measures implied by its item clusters were all 0.95 and higher, and the ratio of explained variance to the unexplained variance captured in that one contrast was 6-1. This 12-item scale may be unidimensional, though the ambiguities present in the item and rating scale contents suggest the model fit may be more a matter of random variation than meaningful variation.

5. Conclusions

Much can be learned from existing data in the context of pragmatic measurement principles, even when the invariance requirements of meaningful comparisons are not included in the design of the tool used.

Future efforts at measuring the development of STEM Learning Ecosystems will be able to build on expectations that have already been articulated concerning progressions in the quality of partnerships, collaborations, and outcomes. Details specified in, for instance, a 2016 slide presentation set the stage for a theory of ecosystem development that could be embodied in an assessment profile [26]. Should the results of the present analyses usefully correspond with expectations like these, formulated as they are from experience, they could provide a useful basis for proceeding toward a formatively useful ecosystem evaluation platform.

Acknowledgments

This study was funded by the STEM Funders Network, USA.

References

[1] Mundy B 1986 *Synthese* **67** 391-437
[2] Stenner A J, Fisher W P Jr, Stone M and Burdick D 2013 *Front. Psychol.* **4** 1-14
[3] Wilson M 2005 *Constructing measures* (Mahwah, New Jersey: Lawrence Erlbaum)
[4] Mari L and Wilson M 2014 *Measurement* **51** 315-27
[5] Pendrill L and Fisher W P Jr 2015 *Measurement* **71** 46-55
[6] Cano S, Pendrill L, Melin J, and Fisher W P Jr 2019 *Measurement* in press.
[7] Maul A 2017 *Measurement: Interdisciplinary Research & Perspectives* **15** 1-19
[8] Fisher W P Jr 2013 *Assessment and Learning* **2** 6-22
[9] Wilson M R 2009 *J. Res. Sci. Teach.* **46** 716-30
[10] Fisher W P Jr and Oon E P-T 2019 *Kybernetes* in review
[11] Wilson M 2004 *Towards coherence between classroom assessment and accountability* (Chicago: University of Chicago Press)
[12] Fisher W P Jr 2006 *Rasch Meas. Trans.* **20** 1072-4
[13] Linacre J M 1993 *Rasch Meas. Trans.* **7** 283-84
[14] Barney M and Fisher W P Jr 2016 *Annu. Rev. Organ. Psych.* **3** 469-90
[15] Smith R M 1996 *Struct. Equ. Modeling* **3**(1) 25-40
[16] Linacre J M 2018 *A user's guide to WINSTEPS v. 4.2.0* (Chicago, Illinois: Winsteps.com)
[17] Smith E V Jr 2002 *J. Appl. Meas.* **3** 205-31
[18] Muchinsky P M 1996 *Educ. Psychol. Meas.* **56** 63-75
[19] Schumacker R E 1996 *Rasch Meas. Trans.* **10** 479
[20] Wright B D 1991 *Rasch Meas. Trans.* **5** 147
[21] Rasch G 1980 *Probabilistic Models* (Chicago: University of Chicago Press)
[22] Andrich D 1978 *Psychometrika* **43** 561-73
[23] Fisher W P Jr and Wright B D 1994 *Int. J. Ed. R.* **21** 557-664
[24] Bond T and Fox C 2015 *Applying the Rasch model* (New York: Routledge)
[25] Linacre J M 1999 *J. Outcome Meas.* **3** 103-22
[26] Ottinger R and Solomon G 2016 *STEM Learning Ecosystems Initiative update* Presented at the APLU Science & Mathematics Teaching Initiative, May 19, Washington, DC. http://www.aplu.org/projects-and-initiatives/stem-education/science-and-mathematics-teaching-imperative/smti-conferences-meetings/SMTI%202016%20National%20Conference/Solomon-CS.pdf

Coherence in measuring student evaluation of teaching: a new paradigm

Fan Huang[1]**, Emily Pey-Tee Oon**[1] **and W P Fisher Jr**[2]

[1]Faculty of Education, University of Macau, Taipa, Macau SAR, China
[2] Research Associate, BEAR Center, Graduate School of Education, University of California, Berkeley, Berkeley California 94702 USA

[1]E-mail: hefu411@126.com

Abstract. Student Evaluations of Teaching (SET) have been adopted worldwide as a standard practice to enhance teaching and learning quality. However, improvement efforts are threatened by various problems involving ordinal scores and interpretive frameworks restricted to individual item sets and uninformed by a well-defined invariant construct. In this study, an approach to framing SET items according to theory rather than personal or subjective decisions is described. A sample of 920 students from the University of Macau participating in this study provided responses to 42 SET items. Applying a scientific model of measurement, the data collected were analysed and interpreted. Justification between theoretical and measurement analysis informed the creation of a theoretical SET construct map. That map and the associated item hierarchy were used as reference sources for configuring SET reports and recommendations. A new coherent paradigm for SET documentation is proposed for further study.

1. Introduction: framing SET items according to theory

Growing concerns with quality assurance in higher education focuses on students as the most reliable and often the only audience of information indispensable to the feedback needed to enhance teaching and learning effectiveness. The use of SET has increased rapidly since the 1960s, and in 1990, more than 90% of higher education institutions in the United States used the rating data [1]. Nowadays, SET have accordingly been adopted worldwide as a standard practice [2-4].

However, improper constructs used, questionnaire design, data analysis, reporting, information use in policy making decision and lack of a coherent system documenting SET measures lead to relationships between variables be over-interpreted or overlooked, and they could also provide unwarranted, mistaken information concerning course quality and instructors' effectiveness. The quality improvement efforts may be threatened by the following problems:

(a) On subjective selection of constructs and poorly framed item content that lack of theoretical support
(b) Lack of a coherent system for documenting and applying SET measures.

The problems mentioned above may set obstacles to provide instructors with information that can guide them to make changes in curriculum design or in teaching behaviours, and what's worse, may lead to erroneous personnel decisions made by administrators, which could influence the instructors' future careers. Therefore, poor SET practices and misuse of data may be harmful for course quality and instructors' development. Thus, framing SET items according to theory rather than personal or subjective decisions and a new coherent paradigm for SET documentation are proposed in this study.

2. Research objectives

Accordingly, we aim to explore the following three research objectives in this study:

1) To ascertain SET constructs (Course and Instructor) theoretically based on a body of SET pertinent studies.
2) To map measurement analysis on content against SET theories.
3) To project a coherent system in documenting SET measures capable to compare SET findings across contexts.

3. Theoretical framework

3.1. Justification between theoretical and measurement analysis

Less difficult items are reflective of most basic essential aspects of course quality, and so they carry less weight in the interpretation of the measures. More difficult items are reflective of aspects beyond essential expectations, and so they carry more weight in the measurement of course quality. Moderate course quality expectations are indicated by items with agreeabilities between these two distinctions (Figure 1). For instance, an instructor providing only responses to direct questions exhibits less quality compared to an instructor who also inspires students' interest in the subject. Items pertaining to the former (responsiveness during class) should be less difficult to agree with and items pertaining to the latter (interest inspiration), more difficult to agree with. In other words, it is theoretically less difficult to agree with responsiveness during class but more difficult to agree with interest inspiration. Instructors that garnered high ratings of agreement on interest inspiration measure higher than instructors whose that garnered most agreement on responsiveness during class.

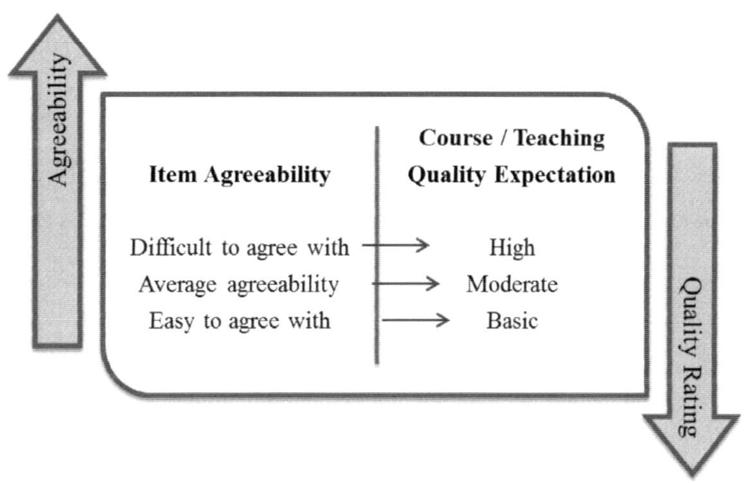

Figure 1. Mapping measurement analysis against theoretical justification.

3.2. Scientific measurement theory

Evidence of SET was evaluated by fitting the response data to a scientific measurement model explicitly hypothesizing an invariant relation between student agreeability and SET item difficulty:

$$\text{In}[P_{nij} / (P_{nij-1})] = B_n - D_i - K_j$$

which says that the log-odds of any student choosing any given category on any SET item is equal to the difference between the estimate B of student n's ability and the difficulty estimate D of item I, and the additional difficulty K of rating category j [5-7]. Assessment data must reasonably fit a model of this kind if we are able to evaluate the teaching effectiveness.

4. Methods

4.1. Sample
Students were invited to voluntarily participate in a research survey. A convenience sampling method was employed, whereby students were chosen from different educational levels at the University of Macau (UM). In pilot study, a total of 241 responses were collected from seven classes. In main study, 679 responses were collected from 28 classes.

4.2. Instrument
In essence, 36 items with six response categories (1 = Very Strongly Disagree, 2 = Strongly Disagree, 3 = Disagree, 4 = Agree, 5 = Strongly Agree, 6 = Very Strongly Agree) were constructed for the pilot study questionnaire, guided by the literature and research. 42 items were included in the main study.

4.3 Expected reliability
Given 36 items and a six-point rating scale, 180 (36×5) distinctions in total were made for 90 each of the two focal constructions (Course and Instructor). The expected uncertainty was equal to 0.2 logits. Measurement reliability is conservatively estimated at 0.96, for a true SD of 1.0 logits [8].

5. Results

5.1. Data quality assessment
Before conducting further investigations, data quality was first assessed. All the data were imported to Winsteps software, version 3.81.0 [9]. The SET data were evaluated for their capacity to support estimation of linear units of measurement before further investigations were undertaken.

5.1.1. Model fit and reliability In the pilot study, the person and item reliability values were 0.96 and 0.97. Compared to the expected reliability, the results show that person reliabilities in both pilot and main study are no less than expectation. The information-weighted (infit) and outlier sensitive (outfit) mean square fit statistics [7] for the 18 items were between 0.58 and 2.26, with the infit and outfit for item A11 (I spent a lot of time studying for this course) being at 2.29 and 2.64, respectively. The person and item reliability values from main study were 0.97 and 0.99. The infit and outfit statistics [7] for the items were between 0.63 and 2.64, with the infit and outfit for item A11 (time studying) being at 1.79 and 2.09, respectively.

5.1.2. Usefulness of response categories The shapes of category probability curves are presented in figure 2. Category 2 (strongly disagree) showed no recognizable peak (Figure 1). Depending on the status of the other indicators (number of responses, fit statistics, mean values of the number of students who answered in each category, etc.), future research on these data should collapse Categories 1, 2, and 3, or perhaps only 2 and 3, for a more optimal representation of response category usefulness.

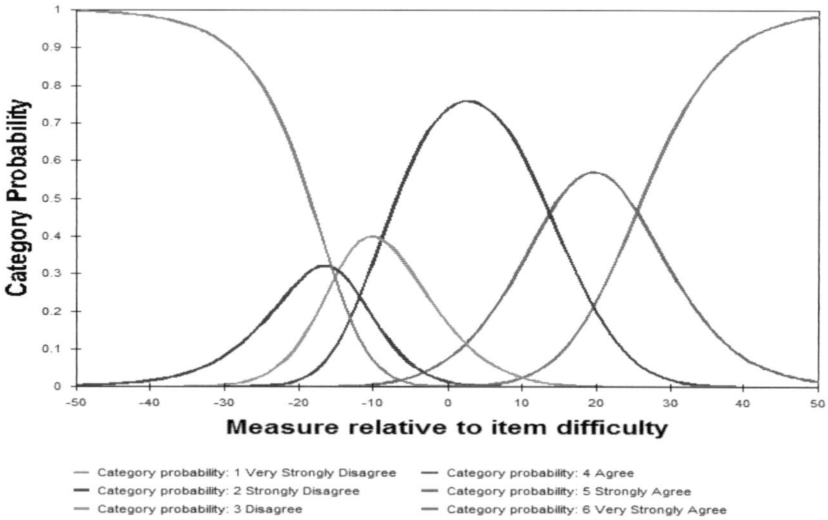

Figure 2. Probability curve of six response categories.

5.1.3. Invariant results between the pilot and the main study The item difficulties estimated from the pilot study correlated 0.96 (disattenuated, 0.98) with the main study (Figure 3), which indicated that the results from two studies remain invariant.

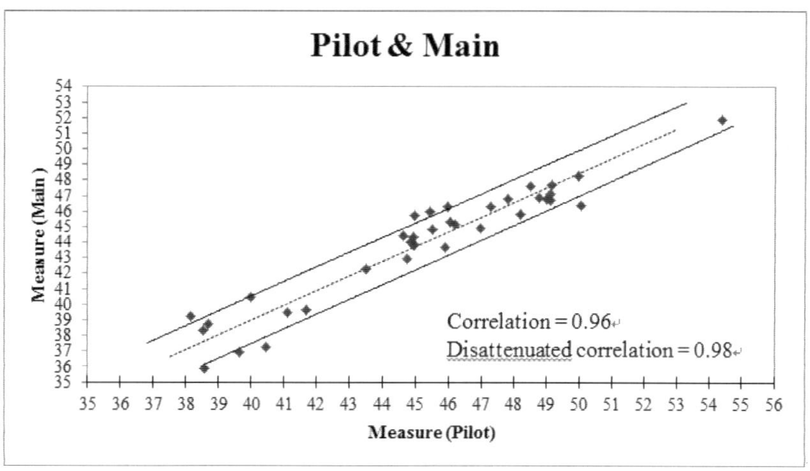

Figure 3. Pilot study item calibrations vs Main study estimates.

5.2. Formation of item hierarchy based on construct map

In order to achieve stability between theory and item calibrations, rounds of iterations are inevitable. Three theoretically intended groups (difficult to agree with, average agreeability and easy to agree with) were developed based on construct map (Figure 4). The construct map and the associated item hierarchy were used as a reference sources for configuring SET reports and recommendations.

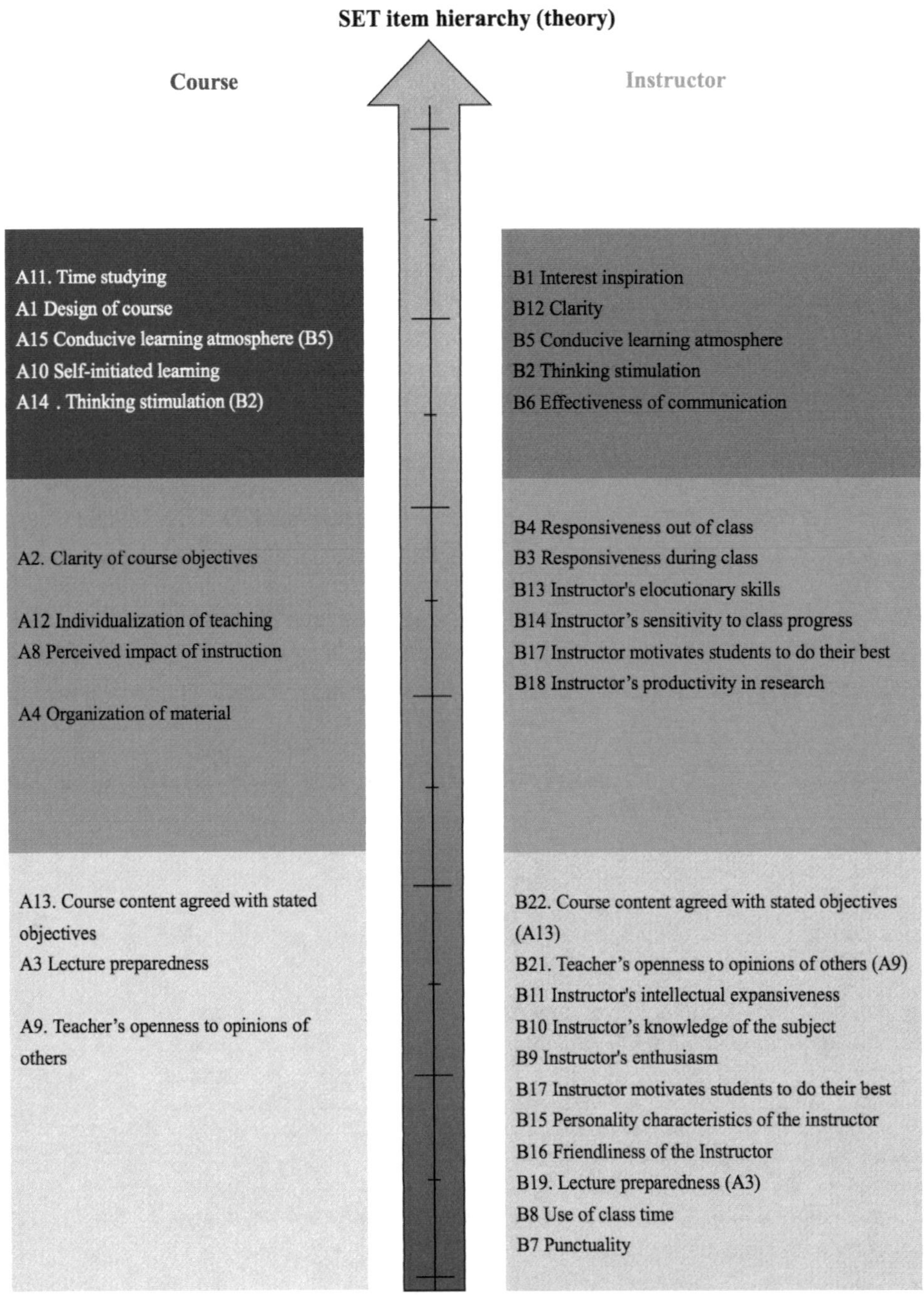

Figure 4. SET construct map.

6. Discussion: Propose a new coherent paradigm for SET documentation

In this study, a new paradigm was provided to overview teaching effectiveness of every instructor scientifically, which is a shift of the status quo in evaluating teaching effectiveness. In figure 5, a coherent system was developed which can be used to project SET according to horizontal coherence (across faculties), vertical coherence (as individuals, to classes, schools, universities, etc.) and developmental coherence (from different time points).

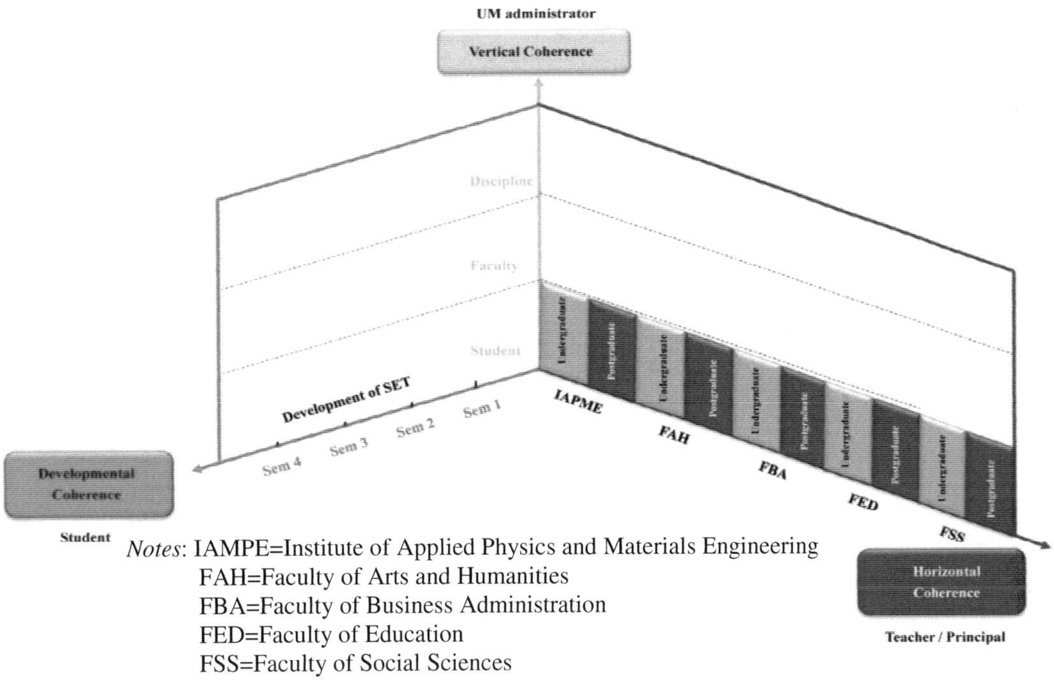

Figure 5. UM 3D figure.

7. Conclusion

A new paradigm was proposed for coherence in measuring SET, which set objective foundation for administrators and instructors making reasonable policies or adjusting meaningful teaching activities to enhance course quality, teaching and learning effectiveness.

References

[1] Seldin P 1999 *Changing practices in faculty evaluation* (San Francisco: Jossey Bass)

[2] Denson N, Loveday T and Dalton H 2010 Student evaluation of courses: what predicts satisfaction? *High. Educ. Res. Dev.* **29** 339-56

[3] Kember D, Leung D Y and Kwan K 2002 Does the use of student feedback questionnaires improve the overall quality of teaching? *Assess. Eval. High. Educ.* **27** 411-25

[4] Narasimhan K 2001 Improving the climate of teaching sessions: the use of evaluations by students and instructors *Qual. High. Educ.* **7** 179-90

[5] Andrich D 2010 Sufficiency and conditional estimation of person parameters in the polytomous Rasch model *Psychometrika* **75** 292-308

[6] Rasch G 1960 *Probabilistic models for some intelligence and attainment tests* (Copenhagen, Denmark: Paedagogiske Institut)

[7] Wright B D and Masters G N 1982 *Rating scale analysis: Rasch measurement* (Chicago, Illinois: MESA Press)

[8] Linacre J M 1993 Rasch-based generalizability theory: Reliability and Precision (S.E.) Nomogram *Rasch Measurement Transactions* **7** 283

[9] Linacre J M 2014 WINSTEPS (Version 3.81.0) [Computer Software]. Chicago: Winsteps.com.

Journal of Physics: Conference Series 1379 (2019) 012044 doi:10.1088/1742-6596/1379/1/012044

Advances in modelling visual symptoms and visual skills

M K Powers[1] and W P Fisher[2]

[1] Gemstone Foundation, Tarzana, CA, USA,
[2] University of California, Berkeley, CA, USA

E-mail: maureenpowers@gemstonefoundation.org

Abstract. Though it seems to be a natural process of physical development, not everyone learns on their own how to use their eyes effectively. Poor binocular vision skills negatively impact a wide range of daily life activities, including reading. This study extends previous work using existing data in multidimensional causal Rasch models of vision. Optometry, survey, and reading assessment data are evaluated in a common frame of reference. Thirty-three students performed vergence and tracking exercises in about 12 to 38 sessions each involving a sequence of five computerized training modules. The number of training modules experienced in any given session ranged from one to five. There were 4,064 total measured cases, with about 123 measures per student. Paired comparison t-tests of early and late measures show significant gains in vision quality. Two regression models predict reading scores on the basis of vision data. Results support further study of how training in Functional Binocular Vision might contribute to improved outcomes in reading education.

1. Introduction

Functional Binocular Vision (FBV) is a construct we have previously shown to be related to a set of quantifiable items on a symptom questionnaire (the Convergence Insufficiency Symptom Survey, CISS) [1, 2]. Our findings have established FBV as a separate, measurable component of the visual process that is not related to measures of acuity (the ability to see the eye chart [2]. Because many decisions about one's visual ability are currently based on measures of acuity alone, further investigation of the relationship between FBV and additional outcome measures becomes important.

Reading is a fundamental life skill that impacts nearly everyone but has been difficult to relate directly to binocular vision. Reading is not an automatic skill: Both the neuromuscular and cognitive systems must develop over time to provide optimal input to the brain for comprehension of text, and not all people achieve good reading performance. In particular, we know that many students with poor binocular skills also read poorly [3, 4], and at least one study shows that degrading binocular vision by inserting prisms between the eyes of good readers and their reading material (thus stressing the convergence system) can reduce reading rate [5]. One hypothesis that has developed from observations like these is that training visual skills can improve reading performance.

In this work, we take advantage of an existing data set where both vision and reading variables were measured in grade school students over time during a computerized visual skills training course [6]. By examining changes in vision variables over time (during training) and comparing their change to that in

[1] To whom any correspondence should be addressed.

reading outcomes (measured pre- and post- training), we hoped to provide new insight into the nature and strength of the relationship between binocular vision and reading.

2. Study Design

This study explores possibilities for productively modelling aspects of vision using a probabilistic approach prioritizing the estimation and comparison of linear measures [7-9] explained by causal theory [10], which may be combined together in a more complex model of multiple constructs [11]. Visual optometric measurements, survey ratings, and reading assessment data are evaluated in a common frame of reference, with the aim of determining the extent to which an overall model of FBV might be feasible. Even though the data available for study were not assembled with this explicit aim in mind, they appear to justify a preliminary exploration of the possibility that further study is warranted.

2.1. Sample and training modules

Thirty-three students with binocular vision issues were enrolled in a visual skills course to train their eyes in vergence and tracking exercises hypothesized to strengthen the neuromuscular coordination needed for FBV. Each session of the course involved a sequence of five computerized training modules (Table 1). Students participated in from 1 to 50 sessions. Setting aside two students participating in three or fewer sessions, and one participating in 50, the remaining 30 participated in 12 to 38 sessions, with means of about 25 and standard deviations of 6 across the five modules.

Table 1. Module descriptions.

Number	Module name	Module purpose
1	Fast Focusing	Improve accommodative facility
2	Smooth Tracking	Improve smooth pursuit speed and accuracy
3	Jump Tracking	Improve saccadic eye movement speed and accuracy
4	Cross-Eyed Fusion	Improve convergence range
5	Wall-Eyed Fusion	Improve divergence range

The number of training modules experienced in any given session ranged from one to five, with the sequence always adhering to the same order of administration. If only one module was administered, it was always the same one, and so on. Of 862 administrations of the first module, the second module was administered 827 times. Of those 827, the third was administered 810 times; of those, the fourth was administered 792 times; and the fifth, 773 times. Thus, for the 33 students across all sessions and all modules, there were 4,064 lines of data.

2.2. Reading measures

The ability to read words aloud for a sustained period of 1 minute was measured using a standardized grade-level-based battery of 1-page stories where each grade level had several stories, in order to prevent memory interference. Stories were sufficiently long for each grade that no student could finish in 1 minute. Students were told to read out loud as rapidly as possible but to try not to make mistakes. Use of fingers was not allowed. A stopwatch was used to start and stop reading, and the total number of words and the number of errors (words added, omitted, or misread) was recorded.

2.3. Original data structure and basis for ratings

Ten indicators were tracked by the computer during the administration of the training modules (Table 2). Data were stored as numeric counts of seconds, as percentage fractions to four decimal places, and as other kinds of computed scores occurring in various numeric ranges. Frequency charts and histograms were produced for each of these, with each distribution broken out into eight equipercentile ranges. The values in each range were then assigned ratings from 1 to 8, where 1 was associated with the value indicating the worst possible performance, and 8, the best. One item's top percentile range was confined in a range between 0.99990 and 1.00000; scoring went to four decimal places, so these are set at 7.

2.4. Form of the measurement model

The measurement model formalizes the expectation of a unit quantity, hypothesizing an invariant relation between visual function and challenges:

$$\ln[P_{nij} / (P_{nij-1})] = B_n - D_{ij} \quad (1)$$

which says the log-odds of any student responding within the range of seconds represented by any given category on any visual challenge is equal to the difference between the visual function B of student n and the visual challenge D of item i at the level of category j, relative to the level of category j-1 [8-9]. Data quality is evaluated relative to the model via information-weighted and outlier-sensitive mean square statistics, both with expected values of 1.00, and in terms of principal components analyses of the standardized residuals [12-13].

2.5. Expected reliability

With nine eight-category items, and one seven-category item, Rasch generalizability theory [14] predicts a measurement uncertainty of about 0.3. Reliability is, of course, a function of the ratio of true variation to uncertainty [15]. However, these data are not a calibration sample chosen with careful attention to variation in the construct of interest. The range of variation is therefore restricted by the study's focus on students with identified vision and reading problems. A low ratio of variation to uncertainty, and low reliability coefficients, can therefore be expected.

As shown in Table 2, the first three items are included in all five modules, and so they serve to connect the 4,064 by-module-by session-by student measures in a common frame of reference.

Table 2. Items by module.

Number	Code	Focus	M1	M2	M3	M4	M5
1	AHT2	Average Hit Time	X	X	X	X	X
2	BHT2	Best Hit Time	X	X	X	X	X
3	OHP2	Overall Hit Percent	X	X	X	X	X
4	THR2	Tracking Hit Rate		X	X		
5	RTN2	Red Transition Number	X				
6	BTN2	Blue Transition Number	X				
7	RTA2	Red Transition Average	X				
8	BTA2	Blue Transition Average	X				
9	MB2	Maximum Break				X	X
10	MR2	Maximum Recovery				X	X

3. Results

3.1. Measurement analysis

Three measures were at the minimum score extreme; there were none at the maximum. Model fit statistics appeared satisfactory, and the mean measurement uncertainty was 0.29, near the expected 0.3. The true standard deviation (0.41) and measurement separation reliability were low (0.67), as expected.

Principal Components Analysis (PCA) of the standardized residuals shows more unexplained variance captured in the first three contrasts (41 %) than is explained by the measures (38 %). Contrast eigenvalues and loadings are high, and the measures implied by the different subsets of contrasting items have low disattenuated correlations. The analysis separates the items into contrasting groups by the module they appear in, suggesting quite different constructs are being measured [12, 13].

Anm';lksdm'csdkmother analysis racking the data across modules within sessions was also completed, but given the by-module significance of the PCA results, these five groups of measures were used.

3.2. Regression analyses

The repeated measures from each of the five module administrations were aggregated into four sets of mean values by averaging each module's measures within ranges of 1-7, 8-12, 13-19, and 20 or more sessions. The number of measures summarized in the 20 cells of the five modules by the four session groups ranged from about 150 to about 260, with about 750 to 1,250 measures within a session group (Figure 1).

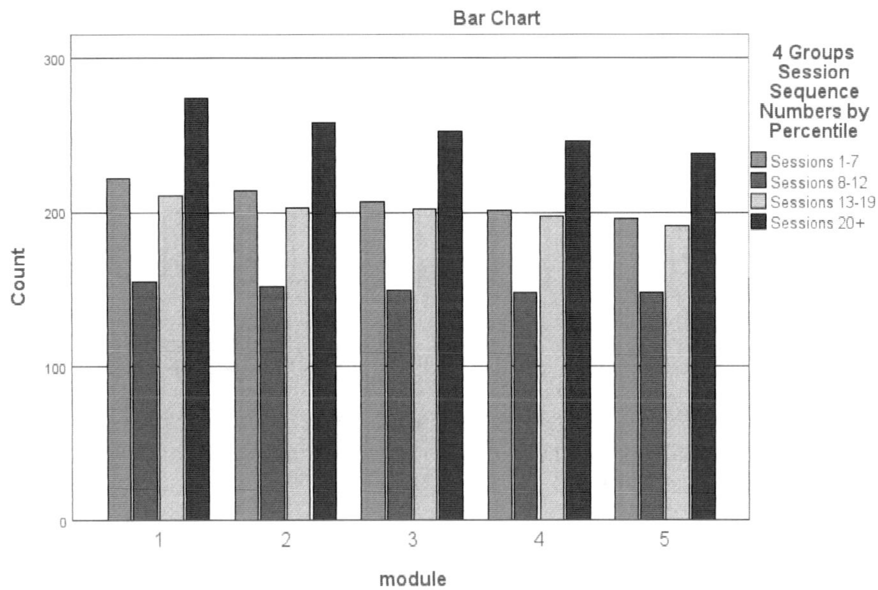

Figure 1. Numbers of measures by module by session group.

The 20 by-session, by-module means per student were then used in conjunction with previously reported [1, 2] pre- and post-intervention optometric measures of the same 33 students, and from the Convergence Insufficiency Symptom Survey, to predict oral reading fluency. The latter was measured via a log-odds transformations of counts of words read aloud correctly and incorrectly in one minute.

Of the 33 students, 28 had module 1, 2, and 3 mean measures in both the earliest and latest session groups (1 and 4). These were significantly different at a .001 level in a paired comparison t-test for module 1 (t=-6.05, 27 df, p=.000), but not for module 2 (t=-2.41, 27 df, p=.023), or module 3 (t=-0.30, 27 df, p=.76). Paired comparison t-tests also showed significant differences for module 4 (t= 5.46, 26 df, p=.000), and module 5 (t=-4.47, 25 df, p=.000). This shows that changes in mean optometric and symptom measures occurred during the training course.

Of the 33 students, 23 had reading measures. In the first regression model, measures from module 1 in session groups 1 and 2, and modules 4 and 5 in session group 1, predicted reading fluency (R=0.825, Rsq=0.68, adj. Rsq=0.61, SE=0.60, F change = 9.56, 4 df1, 18 df2, p=0.000). Figure 2 shows a scatterplot of reading fluency in logits and the regression standardized predicted value.

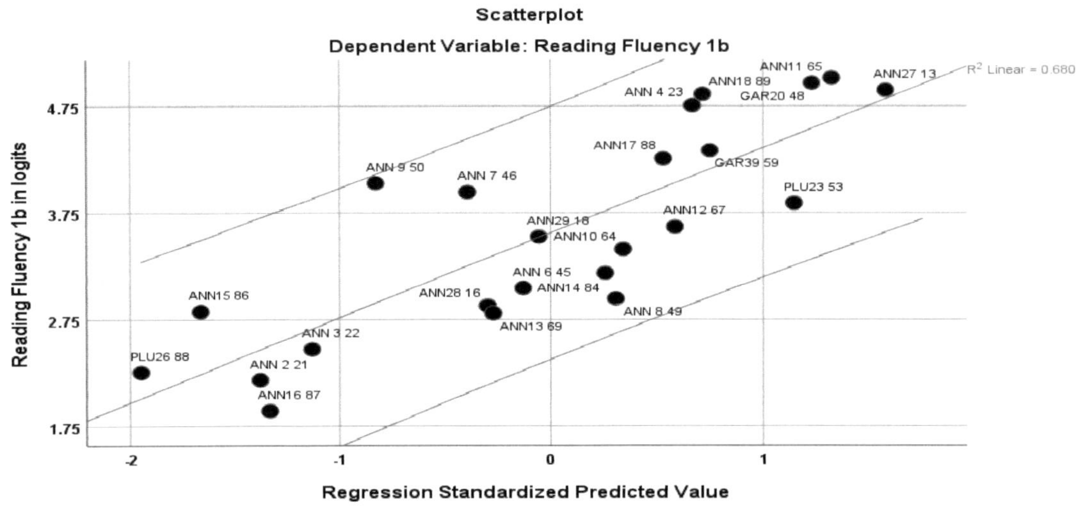

Figure 2. First model, 23 students' measures from module 1 in session groups 1 and 2, and modules 4 and 5 in session group 1 predicting reading fluency.

In the second model, for 16 students with visual acuity and reading measures, changes from the first to the fourth groups of sessions in modules 1, 4, and 5, along with changes in visual acuity, predicted reading fluency (R=0.767, Rsq=0.589, adj. Rsq=0.44, SE=0.74, F change = 3.94, 4 df1, 11 df2, p=0.032). Figure 3 shows a scatterplot of reading fluency in logits and the regression standardized predicted value.

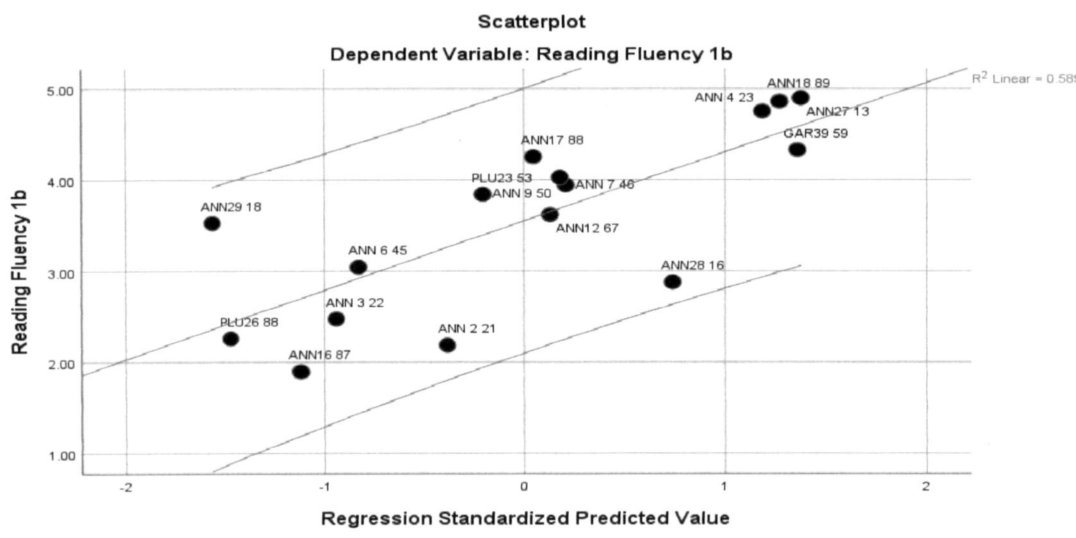

Figure 3. Second model, 16 students' changes from the first to the fourth groups of sessions in modules 1, 4, and 5, along with changes in visual acuity, predicting reading fluency.

4. Discussion

As expected from the results of the paired comparison t-test, the larger changes in visual capacity indicated in modules 1, 4, and 5 were associated with reading fluency, while the measures from modules 2 and 3 were not. Further study will be required to understand these differences; however they suggest that training accommodation and vergence skill may be more important for improving reading fluency than training tracking skills. To the extent that symptoms of vergence problems (on the CISS) involve blurring due to offset visual images when the ability to coordinate the eyes during reading, these results may indicate that poor vergence skill can have a larger impact on reading fluency than poor tracking skills. Whatever the precise mechanism, these results suggest that training in Functional Binocular Vision (FBV) can impact outcomes in reading education.

Broader interpretation suggests implications as well for current practices of vision screening in schools. Most states in the USA still use some form of the eye chart as their primary screening tool [16]. These instruments can be very good at identifying students with refractive error [17]. However, they are administered monocularly at a distance and thus cannot detect binocular problems. New instruments, perhaps similar to the CISS symptom survey, which are easy to administer and interpret, could be developed with more knowledge of their relationship to FBV.

Finally, a major benefit of the current approach is its quantitative nature. Vision measures taken in the clinic or field can vary widely due to different definitions and practices. When the measures are carefully defined and uniformly applied, not only are the characteristics of any particular variable easier to determine, but valid relationships among them can emerge.

5. Conclusion

The results overall indicate that continued investment of resources in the design and execution of a calibration study focused on better powered tests of an overall model of FBV will pay significant returns.

Acknowledgments

The data analyzed here were collected with support from National Institutes of Health research grant R01 EY07414 to MP.

References

[1] Powers M, Fisher W P Jr and Massof R W 2016 *J. Phys.: Conf. Ser.* **772** 012045
[2] Powers M and Fisher W P Jr 2018 *J. Phys.: Conf. Ser.* **1065** 132009
[3] Grisham D, Powers M and Riles P 2007 *Optometry* **78** 542-9
[4] Powers M, Grisham D and Riles P 2008 *Optometry* **79** 228-34
[5] Powers M, Miner G L, Morita Y, and Tyler C W 2010 *Invest. Ophth.Vis. Sci.* **51** 3068
[6] Powers M 2006 *J. Mod. Optic.* **53** 1313-23
[7] Rasch G 1980 *Probabilistic models* (Chicago: University of Chicago Press)
[8] Wright B D and Masters G N 1982 *Rating scale analysis* (Chicago: MESA Press)
[9] Andrich D 1978 *Psychometrika* **43** 561-73
[10] Stenner A J, Fisher W P Jr, Stone M H and Burdick D S 2013 *Front. Psychol.* **4** 1-14
[11] Briggs D and Wilson M 2003 *J. Appl. Meas.* **4** 87-100
[12] Smith R M 1996 *Struct. Equ. Modeling* **3**(1) 25-40
[13] Linacre J M 2008 *Rasch Meas. Trans.* **22** 1164
[14] Linacre J M 1993 *Rasch Meas. Trans.* **7** 283-4
[15] Fisher W P Jr 2008 *Rasch Meas. Trans.* **22** 1160-3
[16] https://nationalcenter.preventblindness.org/school-requirements-childrens-vision March 7 2019
[17] Tong L, Saw S-H, Tan D, Chia K-D *et al.* 2002 *Optometry Vision Sci.* **79** 650-7

Journal of Physics: Conference Series **1379** (2019) 012045 doi:10.1088/1742-6596/1379/1/012045

Actual metrological and legal issues of non-destructive testing

K V Gogolinskii and V A Syasko

Saint Petersburg Mining University, 2, 21st Line, St Petersburg 199106, Russia

E-mail: nanoscan@yandex.ru

Abstract. The article is devoted to analyzing problems of metrological support and legal regulation in the sphere of non-destructive testing (NDT). Differences between the measurement technique and NDT have been analyzed. The necessity to apply the principles of metrological traceability to the NDT technique, equipment and test pieces has been announced. It has been suggested to view NDT as multi-parameter measurements. An example of the reference measuring system for complex calibrations of metal coatings measures on metal bases to calibrate eddy-current thickness gages of coatings has been given.

1. Introduction

The modern development of NDT tools and methods is inseparably linked to the tendencies and challenges of the 4th industrial revolution taking place presently. It is underlain by the principles of digitization, i.e. a transfer to the digital methods of collection, transfer and processing of all data obtained, as well as of automation of processes in all spheres of industry, including design, development, manufacture, and use of products. In this connection the role of NDT considerably increases at all life cycle stages.

2. Specifics of legal regulation of NDT

Difficulties of metrological and legal support of NDT are related to the fact that NDT includes several mutually complementary concepts:

Testing is an experimental determination of quantitative and (or) qualitative characteristics of the properties of a test object as a result of it being affected by various physical fields (acoustic, electromagnet, irradiation, thermal radiation etc.).

Inspection is a verification of the object compliance with the established technical requirements. The requirements to material and products established in the manufacturing process critically depend on the specifics of the materials themselves, processes in the manufacture of the products and the conditions of their subsequent use. In this connection the material or end-product suitability criteria depends on many factors and can be identified only in the process of scientific and engineering research and trial operation. If NDT methods are applied for process inspection, only the manufacturer is responsible for their use. However, if NDT is applied for inspection of the technical state of potentially dangerous objects, such as: railway transport, nuclear power etc., the procedures and equipment applied must undergo attestation and certification in accordance with the legislation.

Evaluation is a quantitative assessment of the characteristics of an object closest to the concept of **measurement** in terms of metrology as a process of experimental obtainment of quantitative values of a quantity (hereinafter metrological terms are given in accordance with VIM [1]). In this case the mandatory requirement is to ensure metrological traceability of the result of the measurements to the reference through a documented unbroken chain of calibrations (VIM 2.41).

Thus, one and the same NDT procedure can be used for in-process testing, checking compliance with legal requirements and measurements of object parameters. A similar combination leads to significant problems both in methodological and legal aspects.

In accordance with the International Classification for Standards of the International Organization for Standardization (ICS ISO) [1], there are two different fields of standardization: 17 – Metrology and measurement. Physical phenomena and 19 – Testing, which includes the field 19.100 – Non-destructive testing. In the field of legal metrology, it is necessary to note the documents developed by the International Organization of Legal Metrology (OIML). The uses of measuring tools in Europe are regulated by international [3] and national [4] legislation. The requirements to the NDT systems are formulated in the document ENIQ (European Network for Inspection and Qualification) [5]. Even a shallow analysis of the basic terms and definitions in the field of measuring facilities and NDT demonstrates the essential similarity of approaches (table 1):

Table 1. Comparison of terms and definitions in metrology and NDT.

Terms Metrology/NDT	Definitions	
	Metrology (VIM)	NDT
Measuring instrument/ Inspection equipment	Device used for making measurements, alone or in conjunction with one or more supplementary devices	The means by which the inspection is implemented. For example, in the case of automated ultrasonics and eddy currents, the inspection equipment consists of cables, probes, pulser-receiver (only for ultrasonics), data acquisition and data processing tools and scanner[5]
Measurement procedure/ Inspection procedure	Detailed description of a measurement according to one or more measurement principles and to a given measurement method, based on a measurement model and including any calculation to obtain a measurement result	A definition of how an inspection is implemented for a specific inspection situation; a written description specifying all essential parameters and setting out the detailed steps and precautions to be taken when applying the specified inspection technique to the inspection situation [5]
Reference material/ Reference block	Material, sufficiently homogeneous and stable with reference to specified properties, which has been established to be fit for its intended use in measurement or in examination of nominal properties	Piece of material, with specified metallurgical, geometrical and dimensional characteristics, used for the calibration and assessment of equipment [6]

At the same time, NDT has a number of specific features making it different from measurements. The NDT specifics are reflected in the definition given in [5]: "**Inspection System -** All parts of the non-destructive inspection including equipment, inspection procedure and personnel which can influence the outcome and quality of the inspection". Unlike the measuring equipment subject to voluntary calibration or verification in accordance with the applicable legislation, the same document introduces the concept of **Inspection Qualification -** The systematic assessment, by all those methods that are needed to provide reliable confirmation, of an inspection system to ensure that it is capable of achieving the required performance under real inspection conditions.

Qualification of an inspection may require assessment of any NDT system, composed of any combination of NDT procedure, equipment and personnel. This qualification or assessment can be considered as the sum of the following items:

i) Technical justification, which involves assembling all evidence on the effectiveness of the inspection, including previous experience of its application, laboratory studies, mathematical modeling, physical reasoning and so on.

ii) Practical trials (blind or open) conducted on simplified or representative test pieces resembling the component to be inspected.

In the fields of industry which actively introduce new NDT technologies, guidance for application and validation of NDT systems, [7], as well as design and attestation of test specimens [8] are developed.

A consequence of NDT complexity and versatility is that for it, as compared to other types of measurement, testing and inspection, a multistage system for certification of NDT specialists has been created [9].

The above special features of legal regulation in the NDT field engender a series of problems related to the main tendencies of the 4th industrial revolution [10]. The necessity to automate NDT processes requires a reduction of human factor in measurements and inspection. As a consequence, the processes of initial setting at an object, measurement procedure, taking into account of the effects of parameters, processing of the measurement results should be automated. At the same time, there is a growing necessity to create means of objective quality control of the measurements being conducted and ensuring traceability of the obtained results.

In the authors' opinion, the above situation with the NDT procedures makes it difficult to use them in automated system of inspection and condition monitoring (CM) due to an individual approach to qualification of the inspection procedure and significant influence of human factor on the results of their application. To solve this problem, it is suggested to implement the concept of metrological traceability (VIM 2.41) in qualification of NDT procedure and test pieces.

3. Introduction of the principles of metrological traceability for NDT procedure

Non-destructive testing can be conditionally divided into flaw detection and measurements by NDT methods. In case of flaw detection, as a rule, certain objects are detected that are discontinuities or imperfections of materials (pores, caverns, disintegrations, cracks, foreign inclusions etc.), which are often problematic to characterize by any dimensional quantity. In case of measurements by NDT methods certain physical quantity (such as coating or tube wall thickness) can be determined objectively in units of measurement. However, there is no clear boundary between flaw detection and measurements. Thus, for instance, one and the same echo method in ultrasonic inspection and the respective equipment can be used to detect flaws in the volume of a part and to measure its linear dimensions. In the first case the indicator of a flaw is a reflected acoustic wave, in the second case the reflection time is measured, in which case it is necessary to know the ultrasonic velocity in the material in order to measure thickness. But in the cited example as well the problem of finding a flaw is easily transformed into a measuring task, namely, the determination of the flaw depth and its sizes. For flaw detection a key problem is the definition of the concept of a flaw itself. In terms of process inspection, not any detected flaw (discontinuity) is a defect (terms according to [6]), which is a cause of rejection or necessity to replace the part. Often a necessity arises to rank the detected defects by their influence on the suitability of the product or hazard of its use. It is in this that the NDT inspection and technical diagnostic critically differ from the measurement procedure. For the latter the basic indicators are their metrological characteristics.

The specifics of NDT procedures and measurements consist in a multitude of factors to be taken into account and controlled simultaneously during their performance. Apart from standard parameters taken into account during normal measurements such as metrological characteristics of measuring instruments and environmental conditions, they include: properties of the controlled object, material of which it is made, nature of possible defects, sounding signal parameters and many other things. A vivid example of effect of material properties on the NDT is the specifics of inspection of carbon-carbon composite materials (CCCM) actively replacing metal in aviation and space industry [11]. NDT ultrasonic methods are widely used for such products. Spatially reinforced carbon plastic materials fundamentally differ from the metal ones by a spatial anisotropy of the structure and mechanical properties and, as a consequence, ultrasonic velocity. Ultrasonic wave diffusion and attenuation pattern in CCCM also cardinally differs from metals and alloys, consequently, it is necessary to use transducers of another frequency and power. In no less degree the characteristic defects are different in these materials: the basic types of defects in metals are cracks; in carbon fiber-reinforced plastics, as a rule, disintegrations and irregularities of structure are observed. Thus, in qualification of a NDT procedure it is necessary to take into account specifics of the test object, the material of which it is made, defects characteristic for this material and product, and other factors able to have effects on the result of the inspection. At the same time, the solution of problems of qualification and validation requires a certain level of abstraction when describing the NDT procedures. As the overview of a NDT system generally the diagram of operation can be given from [7] (figure 1).

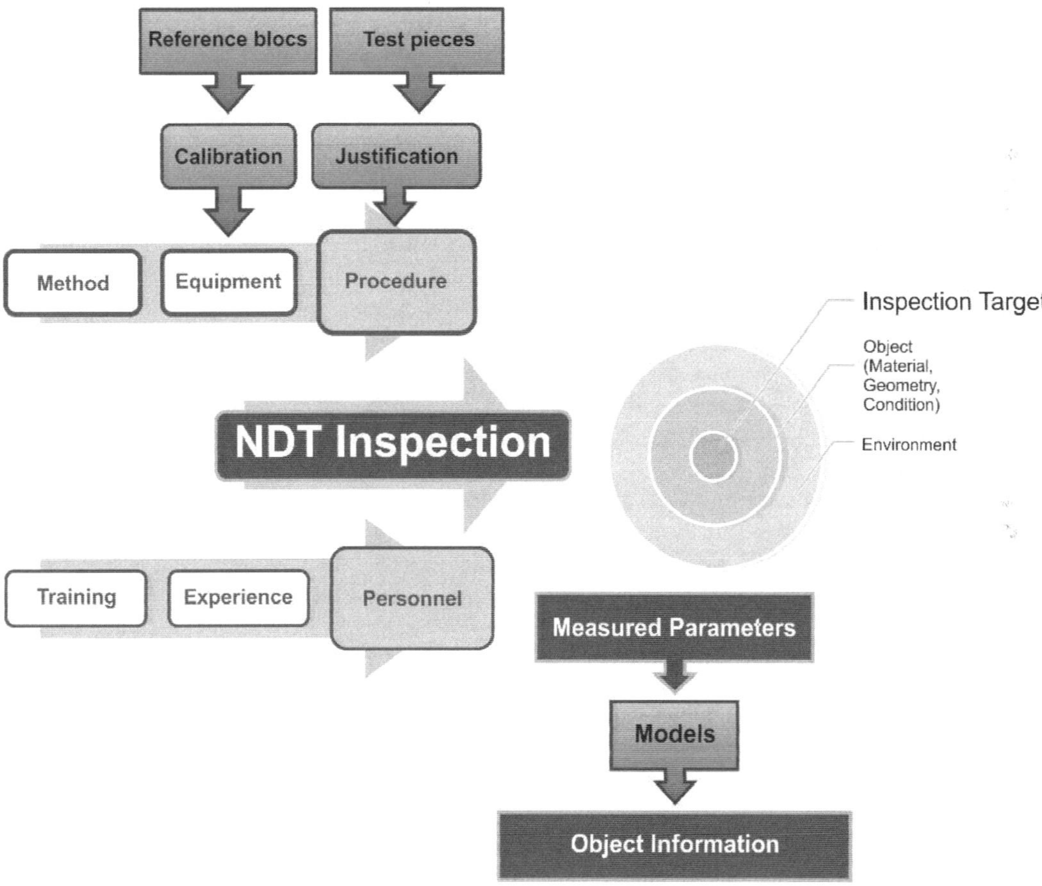

Figure 1. NDT as a provider of information regarding a defined object parameter.

In accordance with this diagram the inspection system consists of the components:

1. Procedure based on a specific method or several methods which are implemented in a set of respective equipment;
2. personnel having respective training and experience.

Information on the test object is calculated on the basis of the measured parameters taking into account the properties of the test object and the environmental parameters.

The authors deem it possible to add to the given diagram:

- reference blocks used to calibrate testing equipment and test pieces to determine the procedure characteristics in general;
- mathematical (physical) model used to calculate the results of inspection on the basis of the measured parameters as part of the inspection procedure. There is a modern tendency to use digital models describing the control object, the method and tool of inspection, and informative parameters of physical fields interacting with the controlled object, recorded by primary measuring transducers in the process of inspection.

In the system in question, to ensure unification and uniformity of NDT results, the equipment and procedure must be subjected to metrological attestation in order to ensure metrological traceability of the measurement results. Also in a number of cases reference blocks for NDT can be viewed as measurement standard (realized as material measure) in accordance with (VIM, 5.1) with the assigned values of the respective quantities with the known uncertainty and ensured traceability.

A similar approach is used in the Russian Federation. However, a reservation should be made here: the authors consider inefficient the practice to attest reference blocks of defects as reference standard of linear dimensions, for these specimens do not transmit linear dimensions to NDT tools (fault detectors). Such specimen can be deemed a «standard reference block» of defect taking into account the geometry of the entire specimen only, not only the defect itself, and the properties of the material of which it is made. In a similar way, it is not correct, by all appearances, to characterize the NDT procedure as the measurements procedure with typical metrological characteristics only. For many applications it is proposed to consider, as the basic characteristic of the inspection procedure, a probability of finding a defect in the specified range of geometric shapes and dimensions.

4. Non-destructive testing as multi-parameter measurements

As the basic principle of metrological confirmation of NDT equipment, procedures, and reference blocks it is proposed to consider NDT as multi-parameter measurements. The specifics of such measurements are that for the respective primary transducers it is impossible to divide informative and influence parameters unambiguously. These measurements are, as a rule, also indirect measurements. The measurements of the thicknesses of the walls or coatings with ultrasonic and electromagnetic methods can be an example. Particularly, eddy-current thickness gages implementing amplitude-difference, phase, and frequency-response methods of measurement based on an analysis of the electromagnetic field of eddy currents induced in the test object are widely used to measure the thickness of metal coatings [12]. The results of the measurements of these devices depend on several groups of parameters: electro-physical (conductivity of the materials of a coating σ_c and base σ_{base}, and specific permeability of the material of a base μ_{base}) and geometrical (coating thickness T_c, roughness, radius of surface curvature etc.). To ensure the traceability of thickness measurements by the described method, it is necessary to use reference standard assigned (certified) with over all specified parameters [13]. To solve this problem, a reference measuring system is being developed, ensuring measurement of conductivity of the base by Van der Pauw method, measurement of complex specific permeability of the metal on ring specimens using a permeameter, transmission of the value of this parameter to the base of a measure of an arbitrary shape with a flat surface, measurement of conductivity of the material of the cover of the measure using an eddy-current transducer with a wave excitation winding [14].

An approach to NDT as multi-parameter measurements will allow the self-control and metrological correctness maintenance methods applied in intelligent sensors to be used in it [15]. These methods are underlain by a measurement, in addition to the basic informative signal of additional parameters, by the changes in which one can judge whether the sensor readings correspond to the value being measured or they are related to a change in the measurement conditions, inherent characteristics of the sensor etc. This will enable automation of many NDT aspects and development of NDT self-adjustment procedures.

5. Conclusion

The existing specifics of qualification of NDT systems incorporating confirmation of equipment, procedures and personnel hinders their development in terms of automation and ensuring the uniformity of measurements. As a solution of this problem, it is suggested to introduce the concept of metrological traceability when attesting the NDT equipment, procedures, and reference blocks. It is necessary for solving this problem, in the authors' opinion, to view NDT as multi-parameter measurements. The creation of the respective metrological support tools should promote development of the NDT in accordance with today's requirements.

References

[1] JCGM 2012 *JCGM 200:2012 International Vocabulary of Metrology—Basic and General Concepts and Associated Terms (VIM)* 3rd ed (2008 version with minor corrections) (Sèvres: Joint Committee for Guides in Metrology)
[2] ISO 2015 *International Classification for Standards* (International Organisation for Standardization) https://www.iso.org/files/live/sites/isoorg/files/archive/pdf/en/international_classification_for_standards.pdf
[3] Directive 2014/32/EU of 26 February 2014 *OJ L* 96 29 March 2014 pp 149-250
[4] German Measures and Verification Act (MessEG) Bundesgesetzblatt – BGBl. I, p. 2722 (No. 43) 31.07.2013
[5] ENIQ 2007 *The European Methodology for Qualification of Non-destructive Testing - Third Issue* (Brussels-Luxembourg, European Commission)
[6] ISO 2005 *ISO/TS 18173:2005 Non-destructive Testing - General Terms and Definitions* (International Organisation for Standardization)
[7] Algernon D *et al.* 2018 *Proc. Int. Symp. on Non-Destructive Testing in Civil Engineering (NDT-CE)* (New Brunswick, NJ, USA: The American Society for Nondestructive Testing Inc.) pp 31-38 https://asnt.org/smt18papers
[8] Algernon D *et al.* 2018 *Proc. Int. Symp. on Non-Destructive Testing in Civil Engineering (NDT-CE)* (New Brunswick, NJ, USA: The American Society for Nondestructive Testing Inc.) pp 39-48 https://asnt.org/smt18papers
[9] ISO 2012 *ISO 9712:2012 Non-destructive Testing -- Qualification and Certification of NDT Personnel* (International Organisation for Standardization)
[10] Thayer P 2017 *Insight - Non-Destructive Testing and Condition Monitoring* **59**(9) pp 469-72
[11] Chung Deborah 2016 *Carbon Composites, 2nd Edition: Composites with Carbon Fibers, Nanofibers, and Nanotubes* (Butterworth-Heinemann) p 706
[12] Golubev S S, Syasko V A, Smirnova N I and Gogolinskiy K V 2016 *Proc. 55th Annual Conf. of the British Institute of Non-Destructive Testing (NDT 2016) (Nottingham, UK)* pp 166-174
[13] Golubev S S, Smirnova N I and Skladanovskaya M I 2017 *Meas. Tech.* **60**(6) pp 552-57
[14] Syasko V A, Golubev S S, Smorodinskii Ya G, Potapov A I, Solomenchuk P V and Smirnova N I 2018 *Russian J. of Nondestructive Testing* **54**(10) pp 698–710 (in Russian)
[15] Taymanov R, Sapozhnikova K, Danilova I and Druzhinin I 2015 Multy-channel intelligent measuring systems *Proc. XXI IMEKO World Congr. "Measurement in Research and Industry" (August 30 - September 4, 2015 Prague Czech Republic)* pp 961-6

Journal of Physics: Conference Series **1379** (2019) 012046 doi:10.1088/1742-6596/1379/1/012046

Accelerating the realization of the United Nations Sustainable Development Goals through metrological multi-stakeholder interoperability

A Lips da Cruz[1], **W P Fisher Jr**[2], **L Pendrill**[3], **A Felin**[3]

[1] ALDC Partnership, Stockholm, Sweden
[2] BEAR Center, Graduate School of Education, University of California, Berkeley, Berkeley California 94702 USA
[3] Research Institute of Sweden, Gothenburg, Sweden

E-mail: angelica@aldcpartnership.com

Abstract. The timeline for the fulfilment of the United Nations Sustainable Development Goals (SDG) diverse stakeholder interests can be accelerated many times over by creating incentives and rewards aligning individual and corporate self-interests with the greater good. The SDGs have been criticized for setting out ambitious goals with no accountability or enforcement mechanisms and little in the way of incentives or rewards past recognition for contributing to a better future. There is then an urgent and increasing need for quality assured sustainability and economic measurements capable of efficiently coordinating and aligning decision-making across investors, financial markets and governments, the private and public sectors, and institutions. Metrology is key to making the 17 UN SDGs measureable and manageable collectively by individuals globally, in accordance with the 169 targets. This massive global effort will require aligning metrological unit standards for the sustainability targets with financial values and economic key ratios, legally binding property (material and immaterial) rights, accounting standards, management practices incorporating new performance indicators and new standards in sustainable economic models. Just as the profit motive and individual and corporate priorities worked to create a new middle class over the last 200 years, so might they also be harnessed and channelled to create universally accessible opportunities for new ways of generating both financial and authentic wealth. The bottom line is that we liberate the power of common targets, language, and understanding with sustainable categorical standards to support adaptations to local situations in the context of traceability to universal standards.

1. Sustainability, complexity, and metrology

Today's development problems of humanly, socially, and environmentally just and sustainable relationships are enormously complex. Good intentions that prioritize a focus in a restricted range of domains, or heavy handed approaches autocratically imposing one perspective and excluding others, risk catastrophic imbalances worsening human suffering, social discontent, and environmental degradation. The difficulty of the challenges we face is multiplied by the seeming intractability of the need to simultaneously address:

[1] To whom any correspondence should be addressed.

Journal of Physics: Conference Series **1379** (2019) 012046 doi:10.1088/1742-6596/1379/1/012046

a. an array of loosely coupled horizontal relationships across social, economic, financial, legal, administrative, scientific, educational, health care, social service, environmental management, business, etc. ecosystems [1-2];
b. at least three vertical levels of complexity in the communications and information infrastructures of each of these ecosystems, including politics and international relationships [3-5]; and
c. equal compensation, wealth, and capital distribution among diverse groups safeguarding economic and civil society stability and evolution across the globe.

Fortunately, a model for the formation of these ecosystems and their interdependencies exists in a suggestive but problematic and incomplete way within current--but unsustainable--policies and practices [6-8]. Though this model met society's needs at its prior stage of development, today it must shift to accommodate the rapidly changing situation in the 21st century. Our aim is to show how this model can be expanded and developed to support broad scale entrepreneurial innovations advancing the realization of the United Nations Sustainable Development Goals (SDGs) (https://unstats.un.org/sdgs/).

1.1. Expanding the incomplete economic model in a fuller definition of sustainability

Today's unsustainable and incomplete economic practices have repeatedly been described as stemming in large part from a too-exclusive focus on manufactured capital and property, at the expense of the genuine wealth of fulfilled human potential, thriving communities, and environmental quality [9-10]. According to this point of view, capitalism can be seen as an incomplete work in progress. Profits are not associated with value creation. Profits are reaped via a linear process of resource usage and depletion. Income and expenditure reports are not accounting for all aspects and consequences of economic activities, production, and infrastructure. Because they are unmeasured and unmanaged, socially distributed returns on investment are used to justify pushing human, social, and environmental costs off the accounting spreadsheets as externalities. Sustainable alternatives likely to move capitalism toward a more complete expression of its positive potentials focus on including these externalities as accountable entries, also known as material issues, in economic models, finance, investment, and management [9-11].

The fulfilment of these potentials requires the creation of horizontally and vertically complex ecosystems incorporating multi-stakeholder information infrastructures analogous to those already in place for manufactured capital and property. Interoperable sustainability mechanisms hinge on the metrological traceability of scientific units of measurement as the means by which financial, legal, entrepreneurial, etc. self-interests are coordinated to provide returns on investments to individuals and disseminate creative advances for the greater good.

1.2. The roles of measurement science and metrological interoperability

The viability of metrological traceability and standardized units of measurement for human, social, and natural capital has been established in a number of recent works [12-15] rooted in previously validated, experimentally proven models, principles, and results [16-23]. The measurement models involved posit the form of interval units estimated on the basis of minimally sufficient statistics [17, 19, 23]. These models guide experimental tests of the hypothesized unit, the conditional independence of the model parameters, and the substantive meaning of the parameter estimates [18, 20, 22-23]. The results of these tests inform item banking, adaptively tailored instrument administrations, reporting and interpretation [18, 20, 22] throughout networks and ecosystems sharing the common language as the medium of coordinated decision making [12-13, 22, 24-27].

The key point at issue is that much of what we want and need to measure exists objectively in ways that can be usefully put to work in sustainability information systems. How to access and leverage this objectivity requires levels of technical expertise not widely available among those most concerned about sustainability. Researchers and managers in multiple fields often assemble groups of indicators carefully devised for the purpose of systematically observing pollution, or literacy, or health, or the prevalence of armed violence. But they then do not follow through to evaluating when, where, and how they have

Journal of Physics: Conference Series **1379** (2019) 012046 doi:10.1088/1742-6596/1379/1/012046

obtained objective evidence, and they cannot then use that evidence to achieve their ostensible goal of measuring the object of interest.

Instead, first, they report results for each indicator separately in a nonlinear ordinal unit, when they should have combined the observations across indicators in a model of the higher order construct that would meaningfully annotate a continuum of less to more scaled in a linear interval unit. Second, different groups assembling different indicators supposedly measuring the same thing do not work together to define a common language by equating their scales to a shared metric with known uncertainties. This project aims to lead the way toward fulfilling the intention to measure sustainable developments in way that inform effective sustainability management and communications.

2. The UN SDGs

The United Nations Resolution 70/1 "Transforming Our World: the 2030 Agenda for Sustainable Development" was unanimously adopted by the General Assembly in 2015. This 2030 Agenda, as it is known, sets out 17 Sustainability Development Goals (SDGs) in a comprehensive plan for sustainable prosperity, involving all people and the planet as a whole.

2.1. Shortcomings of typical approaches to sustainability metrics

With only very rare exceptions [26], efforts to date at quantifying progress toward the SDGs do not match their comprehensive scope with a similarly comprehensive measurement science. Numeric counts of events and sums of ratings are assumed to be, but actually are not, quantitative measures in the scientific sense of the term [12-27]. In addition, valid scientific units of measurement, such as tonnes of effluent, cubic meters of emissions, kilowatt hours, etc., do not, in and of themselves, stand as quantitative evidence of less and more sustainability. The status of these units for their intended use in quantifying physical amounts does not automatically translate into quantification of a separate domain of a new social construct not included in the conceptualization of the original purpose of measurement.

Finally, sustainability impact metric standards are commonly assumed to depend on consensus agreements regarding shared indicator or target content, where the same events are counted in the same numeric terms. These kinds of standards are insufficient for at least two reasons. First, the counting units literally do not add up. The size of the unit changes depending on the local specifics of what is counted. In the same way that sometimes three, and other times six, oranges will weight two kg, so, too, not all instances of literacy, health, or armed violence are created equal.

The second reason why consensus content standards are insufficient to sustainability measurement is that they narrowly restrict the definition of what counts as sustainability to the specific terms of the chosen indicators. A functional model of any measured construct, from literacy to health to poverty reduction, must feature an open architecture capable of encompassing any instance of the phenomenon of interest. It must be possible to add new indicators as they become relevant, and to drop old ones as they become irrelevant. Complaints about existing accountability systems often focus on how reporting impacts in the specific terms of a limited set of indicators fails to give credit where it is due, and moreover detracts from pursuit of the larger mission.

In short, sustainability needs to be modelled, estimated, calibrated, theoretically explained, and systematically measured at a given level of precision, with known uncertainties, in a standard unit. Until measurement of that quality is obtained, sustainability communications and efforts will be severely hampered by the lack of a common language capable of supporting a new economic system of sustainability incentives and rewards.

2.2. Reconceiving sustainability metrics in terms of measurement science

A science of experimental evaluations of evidence, calibrated instrumentation, and predictive theory is a technical undertaking that will require considerable investments of resources to create. Our purpose is to alert those interested as to the viability, feasibility, and desirability of a sustainability measurement science and associated metrology.

Journal of Physics: Conference Series **1379** (2019) 012046 doi:10.1088/1742-6596/1379/1/012046

Mapping impact measurement in support of the 17 SDGs then means abandoning the usual approach to sustainability metrics, where every indicator is treated as a separate universe, a silo, scaled in its own ordinal ranking "metric". In the new art and science of sustainability measurement, we take into account the interdependencies among the different SDGs (Figure 1). We know how all of our indicators work together to provide coherent, consolidated measures, and we know what data-based and theoretical consistencies make sense. Sustainable development occurs in broadly predictable sequences, in much the same way individual, organizational, social, and economic patterns of development unfold.

We identify capitalisation opportunities relevant to the SDGs, reallocating capital resources and mapping the world of sustainability to enable navigation towards the targets specified by the UN.

Figure 1. The 17 United Nations Sustainability Development Goals.

We focus here on describing how our approach to the SDGs aligns finance with environmental, social and governance considerations in ways analogous to existing alignments of finance with manufactured technology. Via these new alignments, individuals, companies, and countries will be able to increase their valuations and trustworthiness simultaneously, empowering themselves in market comparisons, with their brilliant solutions, their courage, and their empathic efforts. A new perspective, concept, and array of tools set the stage for transformations fulfilling the lofty mission of the 17 SDGs targets and beyond, financing sustainability and solutions for humanity, and indeed for all life on planet Earth.

Our methods are based on financial, economic, and sustainability data measurement metrics that (a) evaluate companies' and countries' data; (b) calibrate scientific instruments measuring in meaningful and objective interval units capable of serving as common measures for economic and sustainability value; and (c) enhance efficiency in essential ways by means of predictive, stochastic models. These mathematical models, unit definitions, and calibrated instruments form the basis of a new consolidated standard in the market.

We create a space enabling different stakeholders to work together, cooperatively competing to build on the best in the world, evaluating and rewarding comprehensive sustainability efforts. The knowledge technologies we deploy enable investors, credit granters, managers, customers, employees, and communities to systematically and collectively nudge companies and countries, leaders and steering boards, to work towards lasting realizations in sustainable growth.

Our methods facilitate the visualization of sustainability risks and opportunities by dramatically reducing data volume with no loss of information for each company or country in its comparisons in the markets. Results are expressed in a common interval metric across and within countries, companies, and sectors by region. All metrics are annotated with qualitative information, and every individual measure is always given along with its associated range of risk and uncertainty. The best part is our automated unique Finder tool, which identifies special risks and opportunities impacting investment decisions.

2.3. A three-stage plan for near term products
Over the course of the next two years, we will develop and deploy in three stages an online platform for complementary investment and management knowledge technologies supporting realization of the UN's 17 Sustainability Development Goals, 169 targets, and 233 official indicators (nine indicators are repeated in two or three different targets, making for 244 indicators in total).[2] We will initially focus on SDG 4 on Education, SDG 13 on Climate Action, SDG 16 on peaceful, inclusive, and just societies and institutions, and SDG 17 on sustainability partnerships. Our work will itself be an instance of SDG 9 on innovation and infrastructure where equality and diversity are key factors.

The three stages initially focus on investor and credit risk guidance, proceed from there to management quality improvement applications, and finally shift from company-level measures to individual-level measures. Each stage will begin with the identification and leveraging of existing high-quality scientific metrics and linear measures where these exist [24-26]. Existing ordinal data sources are being evaluated for their measurement potential. For instance, an extensive study has already been made of the Carbon Disclosure Project's (https://data.cdp.net/) data on regulatory, physical, and other risks and opportunities due to climate change. Other SDG data on armed violence reduction, freedom of the press, and other areas have also already been analyzed and evaluated. These efforts will be followed by working as needed to develop new linear measures that hew as closely as possible to the content of the SDG targets, making them freely available online for others to use in the national and regional efforts.

Our products will integrate data with calibrated instrumentation and predictive theory in an advanced approach to modelling bottom-up flows of self-organizing information. The principles of complex adaptive systems are put to practical use in creating useful tools communicating meaningful and actionable results to end users, with no need for time-consuming intermediate analyses and reporting. The mathematical models to be used have long-standing, established histories of successful application in a wide range of fields spanning the natural and social sciences. Existing software tools available to provide platforms for development are appropriately designed to support multiple stakeholder ecosystems [27]. End products will include scientifically calibrated metrics for each measurable SDG construct, investment and management interpretation guidelines for each metric, defined unit standards, APIs for third party online applications, etc.

3. Concluding comments
Currently in the market the usual approach to scoring sustainability indicators is separated from the financial aspects, resulting in numbers that do not add up and consolidate in a meaningful quantitative way. Ratings and scores are ordinal, meaning that they vary across different assessment schemes, and are locally useful for ranking companies and countries, but not for saying how much more or less sustainable one company's or country's policies, practices and solutions are than another's.

Decades of established research demonstrates alternative scaling methods supporting linear interval units of measurement that retain uniform meaning across different indicators, and that routinely include risk and uncertainty estimates. This quantitative information possesses a qualitative meaningfulness and scientific rigor as yet unknown to our competition. Real value lies in providing easy and universal access to meaningful uniform metrics and quantitative comparisons; this will be a primary focus going forward. But the qualitative results revealed by our strategic methods are where small local inefficiencies and a new class of sustainability investment risks and opportunities are identified. Sometimes companies and

[2] https://sustainabledevelopment.un.org/content/documents/11803Official-List-of-Proposed-SDG-Indicators.pdf

countries have special strengths or weaknesses that leaders, owners, managers, and investors could capitalize on, if only they knew about them.

The success of this project will focus attention on the need for a new institutional economics of human, social, and natural capital. This concerns the rules, roles, and responsibilities needed for establishing efficient markets [28-29]. Of particular importance will be the ways in which scientific metrology calibrates standardized instruments capable of mediating capital budgeting decisions via industry-wide consensus on common product definitions and legal title to shares of human, social, and natural capital stocks.

References

1. Galison P 1997 *Image and logic* (Chicago: University of Chicago Press)
2. Latour B 2005 *Reassembling the social* (Oxford, England: Oxford University Press)
3. Star S L and Ruhleder K 1996 *Info. Sys. Res.* **7** 111-34
4. Fisher W P Jr and Oon E P-T 2019 *Kybernetes* in review
5. Allen T F and Starr T B 2017 *Hierarchy: perspectives for ecological complexity* (Chicago: University of Chicago Press)
6. Beinhocker E D 2013 *J. Econ. Methodol.* **20** 330-42
7. de Soto H 2000 *The mystery of capital* (New York: Basic Books)
8. Teraji S 2018 *The cognitive basis of institutions* (New York: Academic Press)
9. Ekins P 1999 *Economic growth and environmental sustainability* (New York: Routledge)
10. Gleeson-White J 2015 *Six capitals, or can accountants save the planet? Rethinking capitalism for the 21st century* (New York: Norton)
11. Hawken P, Lovins A and Lovins H L 1999 *Natural capitalism* (New York: Little, Brown, and Co)
12. Cano S, Pendrill L, Barbic S and Fisher W P Jr 2018 *J. Phys.: Conf. Ser.* **1044** 012057
13. Cano S, Melin J, Fisher W P Jr, Stenner A J, Pendrill L and EMPIR NeuroMet 15HLT04 Consortium 2018 *J. Phys.: Conf. Ser.* **1065** 072033
14. Mari L and Wilson M 2014 *Measurement* **51** 315-27
15. Pendrill L and Fisher W P Jr 2015 *Measurement* **71** 46-55
16. Andrich D 1978 *Psychometrika* **43** 561-73
17. Andrich D 2010 *Psychometrika* **75** 292-308
18. Fisher W P Jr and Wright B D 1994 *Int. J. Ed. R.* **21** 557-664
19. Rasch G 1980 *Probabilistic Models* (Chicago: University of Chicago Press)
20. Wilson M 2005 *Constructing measures* (Mahwah, New Jersey: Lawrence Erlbaum)
21. Wright B D 1968 *Proc. of the 1967 invitational conference on testing problems* (Princeton, New Jersey: Educational Testing Service) pp 85-101
22. Wright B D 1977 *J. Educ. Meas.* **14** 97-116
23. Wright B D 1999 *The new rules of measurement* eds S E Embretson and S L Hershberger (Hillsdale, New Jersey: Lawrence Erlbaum) pp 65-104
24. Fisher W P Jr and Stenner A J 2016 *Measurement* **92** 489-96
25. He W and Kingsbury G G 2016 *J. Phys.: Conf. Ser.* **772** 012022
26. Mok M M C *et al.* 2015 *J. Appl. Meas.* **16**(4) 379-400
27. Fisher W P Jr and Wilson M 2015 *Pensamiento Educativo* **52** 55-78
28. Miller P and O'Leary T 2007 *Account. Org. Soc.* **32** 701-34
29. Fisher W P Jr and Stenner A J 2011 *IMEKO TC1-TC7-TC13 Joint Symposium* (Jena, Germany, 2011) ed G Linss http://www.db-thueringen.de/servlets/DerivateServlet/Derivate-24493/ilm1-2011imeko-018.pdf

Journal of Physics: Conference Series **1379** (2019) 012047 doi:10.1088/1742-6596/1379/1/012047

New tendencies in teaching measurement science in the higher education

A Maiurova[1], L Konopelko[2], Y Kustikov[2], M Kustikova[1], E Bykovskaia[1], and E Tyurikova[1]

[1] ITMO University, Saint-Petersburg, Russia
[2] D.I. Mendeleyev Institute for Metrology (VNIIM), Saint-Petersburg, Russia

E-mail: asmaiurova@gmail.com, marinakustikova@mail.ru

Abstract. This article dwells upon the organization of the educational process at ITMO University for specialization 27.04.01 Standardization and Metrology for masters according to the individual educational standard 3++ of ITMO University. The main professional standards and areas of activity for the implementation of this specialization and the process of creating competencies based on professional standards and areas of activity are described.

1. Introduction

In 2017, ITMO University as an advanced university of the Russian Federation developed and adopted new educational standards with the code 3++ (ES3++). Those standards allow educational program managers to create professional competencies(PC), which must be mastered by graduates, on the basis of professional standards, results of analysis of graduates' professional areas of activity, and requirements for professional competencies for graduates in the labor market, generalization of foreign experience and consultations with leading employers.

The previous educational standards with the code 3+ (ES3+) had clearly defined competencies, with all general cultural (GCC) and general professional competencies (GPC) being unified for all specializations. It offered a list of 28 professional competencies, divided by activities. In addition, organizations have the right to supplement the set of competencies of graduates, taking into account the focus of the master's program on specific areas of knowledge and activities. ES3+ of ITMO University for the Standardization and Metrology specialization contains 28 PCs [1]. Partners of the base organization helped to the PCs implementation, as this educational program is implemented in network form with D.I. Mendeleyev Institute for Metrology (VNIIM), Saint-Petersburg, Russia.

At ITMO University only the masters program is implemented for the Standardization and Metrology specialization. Therefore, this article will be devoted exclusively to the development of new level educational documents, and in particular to the educational standard 3++ for masters programs, and the experience of practical solutions to emerging problems.

2. Educational standards 3++ development

In December 2017, ITMO University created individual ES3++ for all specializations, and in the autumn of 2018 ITMO University began to implement those specializations with a recruited contingent.

ITMO University has created its own graduate competency model, which contains key competencies (KC), Soft Skills competencies (SS), general professional and professional competencies [2]. The first three are unified for all specialties, since their development allows the graduate to be successful in any field of his activity.

Formation of competence involves the organization and implementation of certain activities of students. Each competence contains several indicators of the competence achievement, describing the actions of graduates in the framework of the activity being mastered. For each indicator, separate knowledge, skills and abilities are specified.

For the development of KC, SS and GPC ITMO University introduced university-wide disciplines that are held for all masters of the university. For example, for the competences of the SS block these are disciplines such as Emotional Intelligence, Business and Scientific Ethics, Organization of Scientific Research, Internationalization of Scientific Research, Effective Team Management, Personal Efficiency and Time Management, and some others. Students independently choose one of them upon admission and attend classes simultaneously with students from other areas of specializations. This allows them to get acquainted with a large number of students, and accordingly, faster and easier to adapt to the life of the university.

If KC, SS and GPC were registered in ES3++ along with indicators, the PC needed to be installed by the educational program developers themselves based on existing professional standards and the list of graduate fields of activity that were previously proposed in ES3++. This system gives some freedom of action, since now you can create competencies that would fully comply with the developers' idea of the ideal graduate of the educational program.

According to the recommendation of the ITMO University Guidelines in ES3++, there should be 7-8 PCs, which is approximately four times less than in the previous ES3+ for this specialization. Professional competencies based on professional standards were created according to the scheme indicated in table 1.

Table 1. Algorithm of PC formation from generalized labor functions.

Stage	Level	Example
Stage 1	Generalized labor function	Organization of work to improve the quality of products in the organization
Stage 2	Professional competence	Able to organize work to improve the quality of products in the organization
Stage 3	Indicators of professional competence	• Organizes work on the development and implementation of new methods and means of technical control; • Monitors compliance with regulatory deadlines for product updates and preparing it for validation and certification; • Manages the employees of the technical control service.

To create PCs for the Standardization and Metrology specialization, the authors of the article used professional standards "Specialist in technical quality control of products" [3], "Metrology Specialist" [4], and "Product Certification Specialist" [5]. For the required level of qualification of the magistracy (level 7), generalized labor functions for creating competences were selected, labor functions for creating competence indicators. The results obtained are shown in table 2.

Table 2. Professional competencies based on professional standards.

Competence name	Name of the indicator of competence achievement	Professional standard
PC-2 Able to organize work to improve the quality of products in the organization	PC-2.1 Organizes work on the development and implementation of new methods and means of technical control PC-2.2 Monitors compliance with regulatory deadlines for product updates and preparing it for validation and certification PC-2.3 Manages the employees of the technical control service	40.010 Professional standard "Specialist in technical quality control of products"
PC-3. Able to organize work on the metrological support of the organization	PC-3.1 Analyzes the state of metrological support in the organization PC-3.2 Manages employees of the organization engaged in metrological support PC-3.3 Plans the activities of the organization's metrology service PC-3.4 Organizes work on the accreditation of the organization in the field of ensuring the uniformity of measurements	40.012 Professional standard "Metrology Specialist"
PC-4. Able to organize work to confirm the compliance of competitive products and services and quality management systems	PC-4.1 Carries out the organization to confirm the conformity of products and services in the organization PC-4.2 Implements the organization of confirmation of compliance with the organization's quality management system	40.060 Professional standard "Product Certification Specialist"

For competencies on the basis of professional activities, it was necessary to independently consider the very wording of competencies and indicators. The results are shown in the table 3.

Table 3. Professional competencies based on professional activities.
(a)

Competence name	Name of the indicator of competence achievement	Professional activities
PC-1 Able to conduct modeling of thermophysical processes in nature, technology and experiment	PC-1.1 Understands and analyzes thermophysical processes in nature, technology and experiment PC-1.2 Identifies the relationship of thermophysical processes in nature, technology and experiment PC-1.3 Creates models of thermophysical processes in nature, technology and experiment	The scope of numerical and full-scale modeling of thermophysical processes in nature, technology and experiment, the calculation and design of new heat engineering equipment

(b)

Competence name	Name of the indicator of competence achievement	Professional activities
PC-5. Able to conduct scientific research in the field of metrological provision of labor protection, fire prevention, environmental, biological, radiation and industrial safety and protection in emergency situations	PC-5.1 Develops the terms of reference for scientific research in the field of metrological support of labor protection, fire prevention, environmental, biological, radiation and industrial safety and protection in emergency situations PC-5.2 Performs the tasks of scientific research using empirical and theoretical research methods in the field of metrological support of labor protection, fire prevention, environmental, biological, radiation and industrial safety and protection in emergency situations	The scope of research in the areas of: labor protection, fire prevention, environmental, biological, radiation and industrial safety and protection in emergency situations
PC-6. Able to conduct research and development in the field of standardization, certification and quality control of products	PC-6.1 Develops the terms of reference for scientific research and development in the field of standardization, certification and quality control of products PC-6.2 Performs the tasks of scientific research and design development using empirical and theoretical research methods in the field of standardization, certification and quality control of products	The scope of research and development, standardization, certification and quality control
PC-7. Able to conduct an examination of methods and means of detecting dangerous objects	PC-7.1 Justifies the choice of methods and means of detection of dangerous objects PC-7.2 Gives recommendations on the accreditation of methods and means of detection of dangerous objects	Scope of expertise

Table 4. Variable-profile professional module (specialization).

"Specialization 1" Metrological support of devices and systems in medicine and optical engineering		"Specialization 2" Metrological support in the energy sector	
Reliability of devices and systems	4 crd	Reliability of technical systems	4 crd
Measuring systems and complexes	4 crd	Methods to ensure the unity and traceability of measurements	4 crd
Diagnostics of devices and systems	4 crd	Design principles	4 crd
Physico-chemical methods of analysis based on hybrid systems	3 crd	Physico-chemical methods of energy analysis	3 crd

As can be seen from tables 3 and 4, the developers of the educational program tried to illuminate as much as possible all types of tasks and objects of graduates' professional activity. On the basis of the obtained competencies a new curriculum was created.

At ITMO University educational program consists of several modules. PCs are formed only in the profile professional and variable-profile professional modules. Variable-profile professional module contains various specializations that a student may choose to master. This module takes at least 15 credits, each specialization must contain at least three disciplines. Table 4 shows the specializations that were developed by the authors for the direction of the specialization.

3. Conclusion
The educational program managers have created the new program with unique competence model in consequence of adaptation of the new educational standard. According to the results, the authors of the article can conclude that the new educational standard of ITMO University is more focused on students obtaining universal competencies as there are more than professional ones. These competencies will allow the graduate to better navigate general life situations, find their place in life and be successful in any situation. The number of PCs has been greatly reduced compared with the previous standard, however, now developers can create PCs and indicators for them themselves, thereby developing unique educational programs with authoring courses that would be updated according to professional standards and "foresight".

References
[1] ITMO University 2014 *Educational standard of higher education of the St. Petersburg National Research University of Information Technologies, Mechanics and Optics. Specialization 27.04.01 Standardization and Metrology. Qualification: Master* (St. Petersburg: ITMO University) (in Russian) http://edu.ifmo.ru/file/pages/104/27.04.01.pdf

[2] ITMO University 2018 *The educational standard of higher education of the St. Petersburg National Research University of Information Technologies, Mechanics and Optics - Master's Degree, which includes the specialization 27.04.01 Standardization and Metrology* (St. Petersburg: ITMO University) (in Russian) http://edu.ifmo.ru/file/pages/494/os_mag_biotehnologii_i_nizkotemperaturnye_sistemy.pdf

[3] *40.010 Professional Standard "Specialist in Technical Quality Control of Products"*, approved by order of the Ministry of Labor and Social Protection of the Russian Federation March 21, 2017 No. 292n (registered by the Ministry of Justice of the Russian Federation on April 6, 2017, registration number 46271) (in Russian)

[4] *40.012 Professional standard "Specialist in Metrology"*, approved by order of the Ministry of Labor and Social Protection of the Russian Federation dated July 24, 2017 No. 47507 (registered by the Ministry of Justice of the Russian Federation on July 24, 2017, registration number 47507) (in Russian)

[5] *40.060 Professional standard "Specialist in product certification"*, approved by order of the Ministry of Labor and Social Protection of the Russian Federation of October 31, 2014 No. 857n (registered by the Ministry of Justice of the Russian Federation on November 26, 2014, registration number 34921) (in Russian)

Analysis of the effect of lens components on the optical properties of polarization radiation in experimental studies of biomarkers

E Tyurikova[1], M Kustikova[1], A Stupnikov[1,2] and A Maiurova[1]

[1]ITMO University, 49 Kronverksky Pr, Saint-Petersburg, 197101, Russia
[2]D.I. Mendeleyev Institute for Metrology (VNIIM), 19 Moskovskiy Pr, Saint-Petersburg, 190005, Russia

E-mail: asmaiurova@gmail.com, ek.tyurikova@yandex.ru

Abstract. This article dwells upon the investigation of the radiation polarization structure in analysing biological tissues. The systematic investigation of the polarization parameters of lens systems with non-planar media boundaries was performed to increase the accuracy parameters of polarization instruments used to analyse biological tissues. A review of existing solutions, developed mathematical model and experimental setup were given, and series of experiments are described.

1. Introduction

Under the authority of anthropogenous impacts, biological systems change their properties. According to these changes it could detect the conversion in entire ecological system. The local plants, the disappearance or appearance of certain living organisms, variance in the composition of the soil or air can serve as markers. However, more significant information can be obtained by studying the morphologic composition of the analytical, physical and chemical methods at the molecular level of such biomarkers. One of the effective method of analysis such biomarkers is using optical polarization devices.

Polarization is a property of a light wave, the vector characteristic of which has a component that is orthogonal to the direction of light wave at a given point in space. However, when radiation passes through a mirror-lens system, the wave structure is changed. These changes lead to errors that are not taken into account when obtaining the results of studies using polarization methods.

The purpose of this article is a systematic investigation of the polarization parameters of lens systems with non-planar media boundaries, the result of which is an experimental setup scheme and a method for obtaining the results. To achieve that it is necessary to describe the optical properties of polarization radiation, which are changed with the passage of lens components, develop a block diagram of the experimental setup for studying the data of optical properties, and formulate a methodology for conducting experiments to study the optical properties of polarization radiation.

Impotent part of the work is an implementing for biomarkers analysis in ecological monitoring. Biomarker is a characteristic or mark which indicated about changing in organism or system. In ecological monitoring, it can be certain plants or bacteria. Worsening or improving the environment can be detected by shift or changing of polarization radiation.

Journal of Physics: Conference Series **1379** (2019) 012048 doi:10.1088/1742-6596/1379/1/012048

2. Methods and results

During the work, the following methods of describing polarized radiation during the passage of lens systems were identified: if fully polarized radiation is used, then using the method and Jones matrices; if partially polarized, using the method and Muller matrices. The method of three-dimensional tracing was also selected to reduce the calculation error.

The most suitable values for a quantitative assessment of the properties of fluctuating optical radiation are the Stokes parameters introduced in 1852. Through them it was possible to express the various characteristics of the radiation under study and their variations in real time.

This method is based on the analysis of the radiation polarization parameters when calculating the four Stokes parameters:

$$S_0 = I_0 = I_X + I_Y, \quad (1)$$

$$S_1 = I_X - I_{Y^\circ}, \quad (2)$$

$$S_2 = I_{45^\circ} - I_{-45^\circ}, \quad (3)$$

$$S_3 = I_r - I_l. \quad (4)$$

All Stokes parameters have intensity dimension. Based on these expressions, the first Stokes parameter is the total wave intensity; the second is the intensity difference of the orthogonally polarized components; the third parameter represents the intensity difference between the linearly polarized components measured in a linear basis, the position of which differs by π / 4 angle from the position of the linear basis determining the first two parameters; the fourth parameter is the intensity difference of the two opposite circularly polarized components [1, 2].

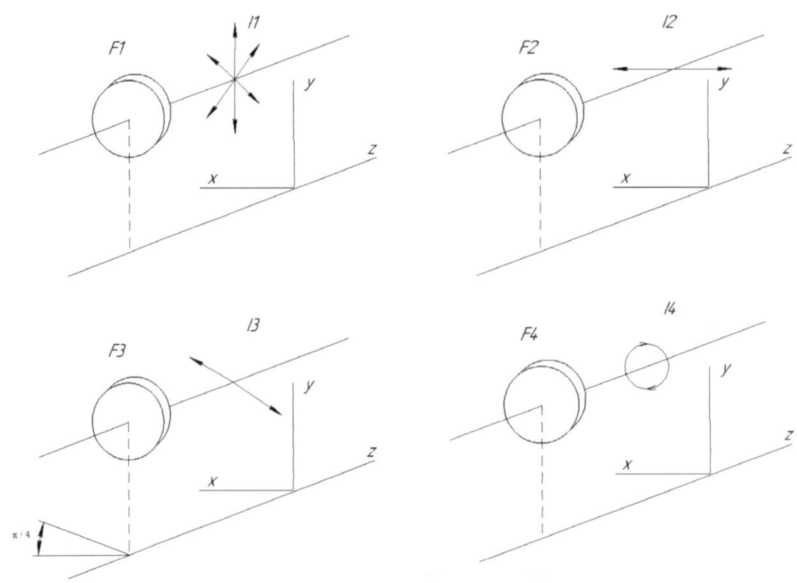

Figure 1. Method of experimental determination of Stokes parameters.

The method of experimental determination of parameters is based on the idea of four filters F1, F2, F3 and F4 (Figure 1), which transmitted 0.5 frequencies of incident nonpolarized radiation.

In this case:

- The filter F1 has anisotropic properties. Radiation with intensity I1 passes through F1;
- The filter F2 transmits a radiation with intensity I2. Radiation after F2 has principally horizontal direction of vector E oscillation;

- The filter F3 transmits a radiation with intensity I3 and vector E oscillations at an angle of π / 4 to the horizontal direction;
- The filter F4 transmits a right circulation polarized radiation with intensity I4.

According to these parameters, the equations of the basic parameters of polarized radiation can be used for detecting the changes in the radiation state: azimuth (5), ellipticity (6) and degree of polarization (7).

$$P = \frac{\sqrt{(S_1^2 + S_2^2 + S_3^2)}}{S_0}, \quad (5)$$

$$\alpha = \frac{1}{2} arctg(S_2 / S_1), \quad (6)$$

$$\beta = \frac{1}{2}\arcsin\left[\frac{S_3}{\sqrt{(S_1^2 + S_2^2 + S_3^2)}}\right], \quad (7)$$

Obviously, Stokes parameters are the most universal tool for analyzing the polarization state of quasi-monochromatic optical radiation in real time. This method is convenient when conducting a real experiment on the installation, since all parameters of polarized radiation can be set on the installation and read through the program.

It is necessary that the developed installation includes the following elements:

- Emission source (ES), with a small angle of divergence of the light beam to create monochromatic collimated radiation;
- Polarizer (P), to create a linearly polarized radiation;
- Analyzer (A), forming polarized radiation, described by the Malu law. Special requirements for the quality and homogeneity of creating polarized radiation are applied to P and A;
- Phase plate (PP), which creates circular radiation at the output of the laboratory setup. The ES should radiate at the working wavelength (or close to it) of the PP, therefore it is more expedient to use the LED as the ES;
- Emission receiver (ER), for accurate reception of the signal from the installation output and for imaging, readable by the Personal Computer (PC). It is necessary that the ER could break the image into small areas and read their intensity. For this purpose, it is recommended to use a receiver with a single matrix field. The matrix receiver also should be taken with a large number of pixels so that it is possible to track small increments of intensities when changing the orientation of the lens.

All elements must lie on the same optical axis. To study the polarization properties of the passage of radiation, it is necessary to enter a prototype - a lens between the P and A. For processing the results, the Stokes method was used.

The following block diagram was developed (Figure 2).

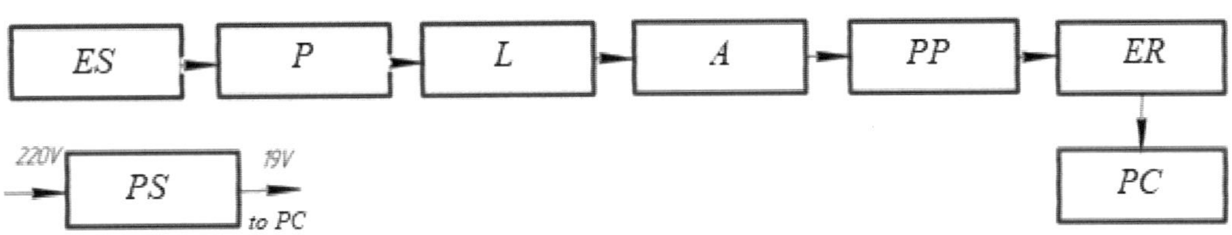

Figure 2. Structural combined experimental installation: ES – emission source, P – polarizer, L – lens, A – analyzer, PP - phase plate, ER – emission receiver, PC – personal computer, PS – power supply.

Journal of Physics: Conference Series **1379** (2019) 012048 doi:10.1088/1742-6596/1379/1/012048

An LED with a radiation maximum at a wavelength of λ = 523 nm was used as a source in the scheme, and an EVS VEC-545-USB television camera with a CMOS sensor was used as ER. The television camera used in the installation allows to work both in automatic mode and with adjustable parameters. In particular, one of the features is the disabling of the gamma correction during photometric measurements to ensure the linearity of the light-signal characteristics. In addition, there is the possibility of both automatic adjustments of exposure and amplification, and manual with the aim of matching the desired spot scattering with radiation conditions.

The setup used a quarter-wave PP operating at a wavelength of λ = 532 nm.

For each of the lenses at different angles of incidence of polarized radiation (0°, 5°, 10° and 15°), it is necessary to determine the Stokes parameters, on the basis of which the ellipticity, azimuth and degree of polarization of the transmitted radiation are calculated.

The method of measuring the polarization parameters of radiation using a laboratory setup consists of the following steps:

1. At the entrance of the P comes the radiation from the LED. With the help of P, the maximum transmission of polarized radiation through it is achieved (linearly polarized radiation is obtained).
2. Set A to 0° and take readings without entering the PP. We also take readings for position A 90°, + 45° and -45°.
3. We enter the PP after A and achieve circular radiation at the output of the experimental installation. We take readings at positions A -45° and + 45°.

A LabView program was formed to analyze the resulting scattering spot by the magnitude of the signal, which makes it possible to investigate test samples in real time. The program evaluates the position of the circle of dispersion and accordingly adjusts the size of the analyzed area of the matrix. This area is generally the size of $m \times n$, where m and n are the number of unit cells horizontally and vertically, respectively. In turn, each unit cell has the dimension $a \times a$, where a is the number of pixels. Thus, the Stokes parameters and, therefore, the polarization parameters of the transmitted radiation are calculated for each of the unit cells [3].

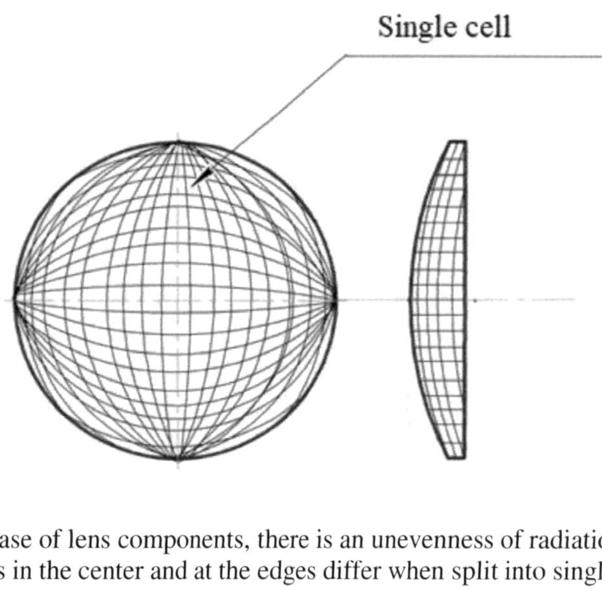

Figure 3. Splitting the surface of the lens on a single site.

In case of lens components, there is an unevenness of radiation distribution. The curvature values of the lens in the center and at the edges differ when split into single cells (Figure 3). The consideration of radiation using matrices was proposed.

3. Conclusion

Research of the polarization parameters of lens systems in experimental studies of biomarkers is made. In the course of the work the Stokes method was chosen to describe the optical properties of polarized radiation, and polarization parameters such as azimuth, ellipticity and degree of polarization were selected to record changes in the radiation structure. A block diagram of the experimental setup for studying the data of optical properties was developed and the main blocks were selected. The methodology for conducting experiments on the study of the optical properties of polarization radiation was formed. The installation can be used in environmental monitoring. It can be applied for changes detecting in biomarkers.

References

[1] Grishkanich A S, Bespalov V G, Vasiev S K, Gusarov A S, Kascheev S V, Elizarov V V and Zhevlakov A P 2015 Monitoring radioactive contamination by hyperspectral lidar, *Proc. SPIE Adv. Env., Chem. and Bio. Sens Technol.* XII 9486 p 94860X

[2] Qian X F and Eberly J H 2011 Entanglement and classical polarization states *Opt. Lett.* 36 4110–4112

[3] Tyurikova E P and Ryzhova V A 2014 Research of polarizing properties of lens components of optical system *Proc. Sensorica-2014* (St.Petersburg, Russia) 3 146-147

Journal of Physics: Conference Series **1379** (2019) 012049 doi:10.1088/1742-6596/1379/1/012049

Actual measuring technologies of Industry 4.0 and analysis of their realization experience

R Taymanov[1], **A Pronin**[1], **K Sapozhnikova**[1], **Iu Baksheyeva**[2] **and I Danilova**[1]

[1] D.I. Mendeleyev Institute for Metrology, 19 Moskovsky pr., St. Petersburg, Russia, 190005
[2] St. Petersburg State University of Aerospace Instrumentation, 67 Bolshaya Morskaya str., St. Petersburg, Russia, 190000

E-mail: k.v.s@vniim.ru, baksheyeva@rambler.ru

Abstract. Civilization development requires a multifold increase in the volume of measuring information. Such an increase, to a significant extent, is caused by the necessity to solve qualitatively new measuring tasks. To meet this need is possible by the transition to new measuring technologies. The experience of the D.I. Mendeleyev Institute for Metrology in developing actual technologies, in particular, metrological self-check and virtual tests, proves their efficiency and discloses realization features.

1. Introduction

For this century, the requirement is characteristic to provide a fast growth in the number of sensors as well as other measuring instruments (MIs) and measuring systems (MSs), their computerization and the increase in the MS channel number. Such a need is caused by the origination of cyber-physical systems, spread of "smart" transport, "smart" healthcare, "smart" home, "smart" cities, "smart" energetics, and "smart" manufacturing.

Another actual requirement of modern society is the increase in the variety of multidimensional quantities that should be measured.

This requirement is caused by globalization and raising the living standards of society, as well as the need to evaluate the quality of products, services, and human properties including the state of health, professional skills, capacities for certain abilities, etc. These evaluations, at least, should be characterized by metrological comparability and compatibility of results. Development of the models for measuring multidimensional quantities accompanied by the assessment of a significance coefficient of each input quantity in the measurement model, is of a particular demand with respect to the measurements of time-varying quantities, e.g., quantities related to the state of human health or forecast of disease development.

The society requirements bring forth attempts to meet them, which leads to the increase in the number of corresponding publications. As it is noticed in [1], tendencies in measurement science development reflected in these publications, play a forerunner role in the Industry 4.0 era.

Interdisciplinary projects become an important part of metrologists' activities [1]. It is not by chance that IMEKO TCs organize joint Symposia more and more often.

We observe a snowballing growth in the volume of measuring information required by society. However, the increase in the metrological reliability of MIs is significantly slower. This fact exacerbates

Journal of Physics: Conference Series **1379** (2019) 012049 doi:10.1088/1742-6596/1379/1/012049

the issue of providing the trustworthiness of information both at the stage of its receiving and transmitting through communication channels to automatic control systems.

Hereinafter, measuring information is called trustworthy if it is obtained from MIs that are "healthy" in the metrological sense and operate under environment conditions permissible for them.

A conventional system of metrological maintenance, which is based on periodic calibration of MIs at one- or two- year interval, cannot meet the increased requirements of society in trustworthy measuring information, in terms of economic and safety reasons. No wonder that the issues of measurement trustworthiness of cyber-physical systems, their reliability, resilience (survivability) as well as lifecycle including maintainability, are included in [2].

Among actual trends of measurement science development, the following tendencies should be mentioned (references below are of an illustrative but not exhaustive character):

- processing big data volumes and visualizing high-dimensional information [3];
- providing access, storage and transmission of measuring information, which is protected from noise as well as unintended and unauthorized access, change, damage, destruction, or use [2], including cases of measuring systems with hundreds and thousands of channels that are separated in space by long distances [4];
- organizing an automatic validation of MSs, automatic correction of measurement uncertainty and even calibration [1];
- testing MIs and evaluating measurement uncertainty by using simulation and virtual processes [3, 5].

To develop new measuring technologies is necessary, which will enable producing MIs with automatic functions, which provide required trustworthiness and accuracy, labour costs reducing in the course of MI operation.

Realization of two requirements of society mentioned above, can be efficient only in case of developing international regulations on the new measurement technologies and requirements for measurement models as well as for MIs and MSs based on these technologies. The standardization in this field should rely not only upon scientific hypotheses, but also upon the experience of engineering.

Measuring technologies enabling fulfilment of these requirements have been developed at the D.I. Mendeleyev Institute for Metrology (VNIIM), which is a leading scientific centre of the Russian Federation in the field of metrology. These technologies were developed along with specialists from other organizations. The experience has shown that these new technologies offer opportunities for solving other actual tasks. This can serve as an indicator that in the 21st century, the investigations in the measurement science open up wide horizons.

Some measurement technologies aimed at automation of providing measurement trustworthiness, are considered below.

2. Automatic checking the trustworthiness of measurement results

2.1. Metrological self-check

Metrological self-check (MSC) is an automatic check of the metrological "health" of a MI or MS in the course of its operation carried out using an accepted reference value generated with the help of an additional (redundant) embedded instrument (a measuring transducer or material measure) or an additional parameter of an output signal [6]. The term "accepted reference value" corresponds to the same term given in [7] and the term "reference value" according to [8]. If the MSC is accompanied by evaluations of uncertainty or error, it can be called "self-validation".

The organization of the MSC is aimed at monitoring changes of metrological characteristic from their values fixed in the course of previous calibration. The MSC provides application of supplement data as reference values. For obtaining information on a measurand, these supplement data are not necessary, and, accordingly, are redundant.

The technology that enables checking the state of metrological "health" of MIs in the course of their operation, was proposed in [9] and later justified theoretically and experimentally in [10-12] and other papers. In regulations, for the first time, it was reflected in Russian document [13], later, partly, in [14, 15] and in more expanded form in [6, 16]. Unfortunately, this complex of publications is not sufficiently known to specialists who deal with Industry 4.0 issues. For example, in [1] published in 2016, its authors write that such a field neither has yet been considered in measurement theory nor taken up in practice.

Two main types of the MSC can be used: a direct MSC or diagnostic one. The former provides that in a MI or MS, a material measure or measuring transducer that is characterized by a higher accuracy, is embedded. As a rule, their measurement range is smaller or response time is greater. The direct MSC reminds calibration. An embedded cell (capsule) containing some metal with a known value of the melting (hardening) temperature, i.e., the metal fixed point, can be a simple example [17, 18]. When the temperature of the environment changes, the metal is melting or hardening and a "plateau" in a diagram "temperature – time" is forming. When the temperature corresponds to the metal fixed point, the deviation of the measured temperature from the reference value, enables estimating the uncertainty.

However, to reduce significantly labour costs related to metrological maintenance due to the noticeable extension of a calibration interval is possible only if the metrological reliability of the embedded material measure or measuring transducer is much higher than that of the MI or multichannel MS. The direct MSC can be efficiently applied for measurements of electric quantities, but for non-electric ones it can be used only in rare situations.

The diagnostic MSC provides automatic monitoring and, if necessary, correcting the uncertainty, but these operations do not concern the uncertainty as a whole. They relate to its component that is critical, i.e., predominant or inclined to fast growth. (In some cases, it can be a group of components.)

This circumstance requires that at the stage of developing a MI or MS with the MSC of such a type, theoretical and experimental investigations should be carried out, which are aimed at determining the critical uncertainty component as well as factors significantly influencing the uncertainty. The efficiency of the development as a whole, depends on the difference between the chosen uncertainty component and others. After that, the task to reveal or artificially introduce space, time and/or informational redundancy in the MI or MS should be posed. In many instances, the decisions of such tasks can be of an inventive character. The redundancy should enable measuring some supplementary quantity, the value of which strictly depends on the critical uncertainty component, but weakly depends on the factors influencing the critical component. This value can be applied as the accepted reference value.

This additional quantity allows evaluating the time variation of the critical uncertainty component through the deviation of the above-mentioned quantity from its reference value determined at the stage of calibration. If the critical uncertainty component increases up to a dangerous level, it can be automatically corrected. Meanwhile, corresponding information comes to an automatic control system.

This measuring technology makes it possible to extend the calibration interval several times. The efficiency of such a technology was confirmed in experiments when developing MIs and MSs designed for measuring displacement, pressure, level and specific electrical conductivity of liquids, flow rate, etc. A number of regulations, including state standards [6, 16], support the technology. According to the terminology [6], sensors and MSs provided by the MSC function are called intelligent ones.

Spaceships and nuclear reactors more than other technical facilities require trustworthy information coming from embedded MIs and MSs, since for many years, they are practically not available for traditional metrological maintenance.

This circumstance became an incentive for VNIIM to develop an intelligent MI that determines the position of a control rod in the nuclear reactor of the VVER-1000 type. Within the range of 4 m, it applies a code scale. Sensors of such MIs are located inside the primary circuit of the reactor, their housing being mounted at a reactor cover. They operate at temperatures up to 325 °C, under high pressure and a strong radiation flux.

In the tube sensor housing with the length about 6 m, a set of inductance coils, i.e., sensing elements, is fixed. The sensor is located inside a moving tube rack of approximately the same length. A set of bushings made of magnetic and nonmagnetic steel is located inside the cavity of the drive rack rigidly connected with the control rod.

In the developed intelligent MI, any single sensor fault (e.g., any signal wire breaks or any coil fails) does not lead to noticeable metrological failure.

The result is achieved due to additional inductance coils that provide the redundancy of the MI code scale, application of a Ni-Cr wire in them as well as several special supply modes [11 and others].

Experimental investigations of such MIs were carried out at a nuclear plant during a multi-year period. They have shown that the lifetime of these sensors can be extended up to 60 years without metrological maintenance, the trustworthiness of measurement results being confirmed automatically in the course of operation.

However, such a lifetime is necessary for comparatively limited number of applications. In the Industry 4.0 era, for most MIs and MSs, to extend the lifetime without metrological maintenance up to 10-12 years is sufficient. This time interval corresponds to their period of obsolescence.

2.2. Removing the influence of part of critical uncertainty component

For a number of MIs, a factor that significantly influences measurement results, is a quantity changing with time, which characterizes the environment or a measurement object. Sometimes, the value of such a change is unpredictable. In similar cases, to measure the value of dangerous influencing quantity and make corrections is not always expedient.

For example, the humidity of the environment is a dangerous factor for a capacitance sensor that provides measuring the gap between a housing and shaft rotating inside it. The humidity determines a significant proportion of the critical uncertainty component. To provide the invariance of measurement results [19] regarding humidity values, is more efficient. Typically, for achieving such a technical result, the redundancy is also necessary.

In particular, the influence of possible humidity changes on the results of measuring the gap, is removed in the VNIIM capacitance sensor. To achieve this effect, in the sensor two measuring transducers are designed, electrodes being fixed in the housing. Each of them measures the gap through the capacitance between the electrode and the shaft. One of the electrodes is round, while the second electrode is performed in the form of a ring. It has the same area and surrounds the first electrode, being displaced along the direction of a shaft radius by a known distance.

The ratio of signal values obtained from displaced electrodes enables achieving the required effect.

The approach based on the redundancy can be applied for inductance sensors measuring the gap to the rotating shaft that is being magnetized, sensors measuring specific electrical conductivity of liquid, as well as in a number of the other cases.

2.3. Self-check in multi-channel measuring systems

The most efficient way for organizing MSC in multi-channel MSs consists in revealing interconnections between the signals obtained in different measuring channels that are metrologically "healthy". A characteristic example is an intelligent MS intended for monitoring the state of fastening assemblies of the turbine unit cover at a hydroelectric power plant that has been developed at VNIIM. 80 bolts are located at equidistant points along the cover circumference. Displacement sensors being installed at each eighth bolt measure the displacement (elongation) of the bolts relative to a turbine housing. In this MS, the MSC relies upon the property of such covers to be stiff. As a result, when the cover moves up, down, or at an angle, the values of the signals from metrologically "healthy" measuring channels should correlate with each other since the signals reflect the displacement of the same plane. The uncertainty of the displacement measurement is less than 5 μm, while measurement range is ± 3 mm. In the course of the turbine unite operation, according to the MS operation algorithm, the sensors are periodically queried and measurement results come to a data processing unit. The query frequency is constant (about 2 Hz). The algorithm includes preliminary and basic processing.

The preliminary processing enables fulfilling procedures aimed at increasing the efficiency of hardware application and initial statistic data processing. The former task is solved by "thinning" the original data set to decrease the frequency of their further processing. A data smoothing of "thinned" data in non-crossing time windows by calculation of medians, plays the role of the initial statistic processing. The median is applied since it allows removing outliers and provides the higher accuracy of calculations taking into account available hardware resources. The values of the "tinning" data frequency and time window durability were chosen by simulating the procedures based on signal records obtained from analogues turbine units. Results of simulation also showed that smoothing in a sliding window is not expedient.

Further, smoothing data (the median values obtained in each window) are subjected to the basic processing that includes the procedures revealing current sensor defects (e.g., sticking of a moving sensor part, its detachment, and so on) as well as those determining a slow drift of the measurement uncertainty with its possible correction. To develop each of these procedures, to analyse the signal records obtained from analogues turbine units was necessary. This enabled estimating the parameters of the proposed procedures. For example, to reveal the situation related to sticking of the moving part, requires that the standard deviation (or another measure of dispersion) calculated in each smoothing window should be accumulated for a certain number of windows or certain time. The accumulation duration was chosen based on mathematical simulation.

Time cycles of the procedure operation should be divisible by the smoothing window duration. The shortest cycles corresponding to revealing current defects, e.g., outliers and simultaneous two-directional variation of signals of different sensors, require application of only two smoothing windows that are neighbouring. These cycles use information concerning signal differences calculated for each sensors in the neighbouring windows. The longest cycles aimed at detecting slow uncertainty drift, apply all the data obtained in the course of the previous time of the turbine unit operation.

Current results of the MSC are recorded into a data logger, a time interval being devised to smoothing window duration.

These results include information on:

- significant events that were detected by the corresponding procedures,
- alerts and forecasts related to further operation of the MS,
- possible slope of the turbine unit cover,
- uncertainty drift and necessity to make self-correction.

The MSC provides confidence in the trustworthiness of measurement results and allows defect sensors to be detected. It gives grounds to carry out periodic calibrations or verifications of sensors, as a rule, in the course of planned shutdowns of turbine units. So, as a rule, metrological maintenance procedures can be carried out significantly less often than it would be necessary for MSs without MSC.

The efficiency of intelligent multi-channel MS based on the MSC, stimulated VNIIM to develop MSs with similar functions for other applications.

For example, the MSC was also applied in the MS controlling the ecological safety of the aquatic environment which is used in the system designed for supplying citizens with drinking water. It uses crayfish as "biosensors". Considering this feature, self-checking provides the identification of metrological faults at the technical part of the system, and, separately, anomalous deviations in the work of "biosensors" [20].

2.4. Virtual testing

Mathematical modelling and virtual experiments are applied in various measurement fields, in particular, in coordinate-measuring machines, for evaluating measurement uncertainty, which is reflected in ISO standard [21]. According to [3], in the era of Industry 4.0, to transfer the method used for virtual coordinate measuring machines to all relevant classes of MIs, is expedient.

In our opinion, this approach is especially actual for multi-channel MSs, the test program of which should take into account the influence of both measurand values covered all their permissible measurement range and various influencing factors. This feature can lead to a noticeable increase in the measurement uncertainty and even to erroneous estimates of the situation being analysed. Of course, such a situation is the most typical for sensors of measuring channels that are very far from each other, e.g., in a system measuring pure water flow rate, which is designed for water commercial accounting within a town or several its districts.

However, as practice shows, even in a multi-channel MS embedded in a unit, combinations of impacts on different measuring channels may appear to be unexpected for metrologists. To take into account all potentially possible versions of impact combinations in the test program is inappropriate for many reasons, including the economic ones. That is why to prepare the test program, it is appropriate to use information gained at the operation of previously used MSs.

At the VNIIM, a virtual test procedure was developed (with the participation of the co-author, a GUAP specialist) applied to the intelligent MS designed for monitoring the state of fastening assemblies of the turbine unit cover at hydroelectric power plants.

Virtual test using an identical MS, was performed in a laboratory. Samples from the signal records played the role of the source of test signals coming to the inputs of measuring channels with disconnected sensors. These records were obtained in the course of operation of similar MSs, which were not provided by the MSC, at the hydroelectric power plant for the two preceding years. They included both records related to transient modes of the turbine units and those simulating sensor signals in the course of operation modes causing special concerns of engineers.

The virtual test allowed checking the MS operation, including the MSC function, for variety of signals coming through different channels to the data processing unit.

2.5. Prospects for the automation of monitoring the measurement trustworthiness

Although works related to the improvement of the MSC theory as well as development of intelligent MIs and MSs mainly are still concentrated at the VNIIM, at present, MSC technology is also used by other Russian organizations, in particular, in Omsk, Moscow, Chelyabinsk, etc. Similar work related to particular measurements, is underway in the UK (Oxford University) and in some other countries.

However, not in all the cases, the self-check takes into account the experience gained in other countries and organizations, which creates obstacles to its widespread introducing into series produced MIs and MSs. On the other hand, not always a diagnostic MSC is focused on monitoring the critical uncertainty component. In some MIs and MSs, to reduce the cost of "intelligent" devices manufactured for sale, only a small part of the critical component is monitored. Application of such technical decisions actually leads to unfair competition of manufacturers and to consumer fraud.

At present, an unjustified delay of MSC widespread introduction is one of the most important problems that should be solved in order to accelerate moving to the Industry 4.0 era.

Crashes of Boeing 737 Max 8 aircrafts in Ethiopia and Indonesia testify that this problem is actual. A crash reason was associated with sensor defects in the Maneuvering Characteristics Augmentation System (MCAS). The Boeing corporation has a software providing alarm signals in case of sensor fault (it can be considered as a simplified version of the MSC). However, before the last crash, only special customers' requests made the company install this software for a particular fee.

After the analysis of the crashes caused the deaths of hundreds of passengers, the decision was made to install the corresponding software mandatory in all the Boeing 737 MAX aircrafts.

In our opinion, the necessity to use a MSC technology should be related to all the automatic control systems that incorporate sensors a fault of which can result in devastating consequences.

In this regard, there is an urgent need for preparation and publication of international standards both describing known methods of organizing the MSC and establishing requirements for the characteristics of automatic checking the reliability of measurement results as well as for justification of their fulfilment.

3. Conclusion

The development of new measurement technologies based on the fast processing of measurement data including the use of various types of redundancy and cloud computing, has become one of the most relevant trends of measurement science in the coming era. These technologies will contribute to the widespread adoption of intelligent MIs and multichannel MSs in almost all the areas of human life.

To accelerate introducing and improving the actual measurement technologies, a number of international standards should be prepared, which should contain the description of the most demanded technologies and requirements for the metrological characteristics of measuring instruments and systems that are based on them.

References

[1] Imkamp D, Berthold J, Heizmann M, Kniel K, Manske E, Peterek M, Schmitt R, Seidler J and Sommer K-D 2016 J. *Sens. Sens. Syst.* 5 pp 325–335
[2] Griffor E R, Greer C, Wollman D A and Burns M J 2017 NIST Special Publication 1500-Framework for Cyber-Physical Systems: Overview Version 1.0 **1** https://doi.org/10.6028/NIST.SP.1500-201 112
[3] Metrology for digitalization of the economy and society 2017 https://www.ptb.de/cms/fileadmin/internet/forschung_entwicklung/digitalisierung/preview_Metrology_for_the_Digitalization_of_Economy_and_Society.pdf
[4] Taymanov R, Sapozhnikova K and Ionov A 2017 *18th Int. Congr. of Metrology (September 19-21, Paris, France) EDP Sci. Web of Conf.* ed C Corleto 09006
[5] Schmitt R H and Voigtmann C 2017 *Tech. Mess.* **84** 5 pp 296-309
[6] GOST 2010 *GOST R 8.673-2009 Intelligent Sensors and Intelligent Measuring Systems. Basic Terms and Definitions* (Moscow: Standartinform) (in Russian)
[7] ISO 1994 *ISO 5725-1:1994 Accuracy (Trueness and Precision) of Measurement Methods and Results – Part 1: General Principles and Definitions* (International Organisation for Standardization)
[8] JCGM 2012 *JCGM 200:2012 International Vocabulary of Metrology—Basic and General Concepts and Associated Terms (VIM)* 3rd ed (2008 version with minor corrections) (Sèvres: Joint Committee for Guides in Metrology)
[9] Sapozhnikova K V, Taimanov R Ye and Kochugurov V V 1988 *Testing, Checking and Diagnostics of Flexible Production Systems* (from the materials of the seminar hold at the Blagonravov IMASH of the Academy of Science in 1985) (Moscow: Nauka) pp 269-273
[10] Taymanov R and .Sapozhnikova K 2003 *Proc. of the XVII IMEKO World Congr, (Dubrovnik)* pp 1094-1097.
[11] Sapozhnikova K and Taymanov R 2011 *Nuclear Power - Control, Reliability and Human Factors* ed P Tsvetkov pp 3-26
[12] Sapozhnikova K, Pronin A, Druzhinin I and Taymanov R 2016 *14th IMEKO TC10 Workshop on Technical Diagnostics (Milan, Italy, June 27-28, 2016)* pp 181-186
[13] MI 1991 *MI Recommendation 2021-89 GSI. Metrological Assurance of Flexible Manufacturing Systems. Fundamentals* (Moscow: Standartinform) (in Russian).
[14] BS 2005 *BS 7986:2005 Specification for Data Quality Metrics or Industrial Measurement and Control Systems* (London: British Standards Institute)
[15] VDI/VDE 2006 *VDI/VDE Guideline 2650 Requirements Regarding Self-monitoring and Diagnosis in Field Instrumentation. General Faults and Fault Conditions of Field Devices* (Berlin: VDI/VDE)
[16] GOST 2011 *GOST R 8.734–2011 Intelligent Sensors and Intelligent Measuring Systems. Methods of Metrological Self-checking* (Moscow: Standartinform) (in Russian)
[17] Froehlich T, Augustin S, Mammen H, Blumroeder G, Schalles M and Hilbrunner F 2011 *Proc. of the "Sensor+Test Conf. 2011" (Nurnberg, Germany, 07-09 June 2011)* pp 732-737
[18] Ongrai O, Pearce J V, MacHin G and Sweeney S J 2011 *Meas. Sci. Technol.* **22**(10) pp 105103-105107
[19] Petrov B N, Viktorov V A, Lunkin B V and Sovlukov A S 1976 *Principles of Independence in Measurements* (Moskow: Nauka Publ.) (in Russian)
[20] Lubimtsev V A, Kholodkevich S V and Druzhinin I I 2017 *Pribory* 10 pp 34-39 (in Russian)
[21] ISO 2008 *ISO 15530-4:2008 Geometrical Product Specifications (GPS) –Coordinate Measuring Machines (CMM): Technique for Determining the Uncertainty of Measurement– Part 4:*

Evaluating Task-specific Measurement Uncertainty Using Simulation (International Organisation for Standardization)

Journal of Physics: Conference Series **1379** (2019) 012050 doi:10.1088/1742-6596/1379/1/012050

Raman and CRDS isotopic resolution spectroscopy for biomedicine applications

Aleksandr Grishkanich[1,2], Yan Chubchenko[1,3], Marina Kustikova[1], Aleksandr Zhevlakov[1], Leonid Konopelko[1,3], Andrey Lunev[2], Irina Noskova[1], Nikita Paklinov[4]

[1] ITMO University, Saint-Petersburg, Russia
[2] St.Petersburg State Electrotechnical University (ETU), Saint-Petersburg, Russia
[3] D.I. Mendeleyev Institute for Metrology (VNIIM), Saint-Petersburg, Russia
[4] Tyumen Industrial University, Tyumen, Russia

E-mail: grishkanich@ya.ru

Abstract. SRS-sensor was developed for measuring absolute amount ratios of $^{13}C/^{12}C$ isotopes for medical application. Measuring absolute amount ratios of $^{13}C/^{12}C$ isotopes allows detection of Helicobacter pylori for medical diagnostics of human health. SRS method is used to measure the amount of carbon dioxide isotopes in exhaled air to detect Helicobacter pylori. This is used to determine influence of the bacterium on the occurrence of gastritis, gastric and duodenal ulcers. This paper includes methods and metrological characteristics for measuring isotopes ratios.

1. Introduction

Helicobacter pylori is a spiral-like gram-negative bacterium, one of the most frequent and persistent diseases in the world, such as gastric ulcer, lymphoma, stomach cancer, gastritis, and the World Health Organization has classified Helicobacter pylori as carcinogens in the first group [1].

Presence of Helicobacter pylori is not a disease of itself but is associated with several disorders of the upper gastrointestinal tract.

Several methods of detecting Helicobacter pylori exist.

The histological method has a sensitivity of 72-100 %, a specificity of 81-97 %. For carrying out such a method, qualified personnel and a morphological laboratory are required. The waiting time for results is a week, which is a significant drawback of this method.

The bacteriological method has 100 % specificity, but the sensitivity of this method requires a certain providing of a microbiological laboratory and qualified personnel.

Radionuclide methods for recording Helicobacter pylori based on urease activity. Urea labeled isotopes ^{13}C, ^{14}C is given to the patient. Urease Helicobacter pylori decomposes urea to carbon dioxide, while leaving the labeled carbon. Which diffuses into the blood, is exhaled via the lungs, and can be measured in the exhaled air.

^{14}C is a radioactive isotope that has limitations in use and is strictly controlled, so it is less common than ^{13}C isotope, which is not radioactive. When using urea ^{13}C, samples are collected before and 30 minutes after ingestion. The $^{13}C/^{12}C$ ratio is measured in both samples. The result is expressed as the difference between the two measurements. Requirement in the initial value is due to the diet of patients with diuretic syndrome. For accuracy, the volume of urea taken ^{13}C is 100ml. The amount of ^{13}C of the isotope can be determined using a gas mass spectrometer, infrared or laser equipment. The sensitivity of this method is 99 %, specificity is 98 %.

IR-gas analyzers have high, but often insufficient selectivity.

Mass spectrometers have a high price and, due to the need for a vacuum, it is difficult to use outside the laboratory.

Measuring of $\delta^{13}C_{VPDB}$ allows exploring the actual issues of the modern world, such as global climate change, detection of falsification of food products, doping control, the determination of origin of explosives, drugs, petroleum products and natural gas, medical diagnostics of human health.

The application of CRDS for $\delta^{13}C_{VPDB}$ measurements gave an opportunity for making in situ, continuous $\delta^{13}C_{VPDB}$ observation in ambient conditions. Fundamentally, it can achieve accuracy similar to mass spectrometry [2].

This paper reviews a method for the analysis of human infection with Helicobacter pylori by measuring the ratios between $^{13}C/^{12}C$ isotopes in exhaled air. Existing methods of gas analysis do not always provide great speed or accuracy of analysis. Developed method can provide a high accuracy analysis of par with mass

2. SRS-sensor

The monostatic optical system with Cassegrain configuration was used for the experiment. The laser beam was moved into the region of atmosphere to be analyzed by a system of rotating mirrors. The probing was done at λ = 261.7 nm by utilizing solid state Nd:YLF laser with diode pumping and frequency conversion of the harmonics. The spectrometric channel is a double polychromator with dispersion addition and three detectors, for recording in nitrogen, methane, and hydrogen sulfide atmospheres. The polychromator is based of two concave holographic gratings. Verification of obtained data was done in Mendeleev Institute of Metrology. Verification was done by utilizing CRDS gas analyzer. Three samples of gases were prepared, on with isotopically pure $^{12}CO_2$, second one with $^{13}CO_2$ and the last one with a mixture of those gases. All samples were analyzed with CRDS gas analyzer, then with SRS-sensor. Obtained data was then compared.

3. CRDS calibration methods

During the experiment the accuracy of measurements of gas analyzer based on the principle of radiation damping spectroscopy in the resonator, which provide lower detection threshold and better speed than alternative methods.

CRDS-method allows to register a small absorption of radiation which passes through the sample. This makes it possible to measure the percentages of impurities at the level of ppb and ppt. It also became possible to measure the ratio of carbon isotopes $^{13}C/^{12}C$ in carbon dioxide on a par with isotope mass spectrometers.

When studying the absorption spectra of a substance with a small absorption coefficient in ordinary spectroscopy, it is necessary to increase the optical path length by placing the test substance in a multi-path optical cell. Reflecting from its end mirrors, light repeatedly passes through it. In this case, we must take into account the loss of reflection of light from the mirrors. In the intracavity laser spectroscopy method, where the laser cavity plays the role of a multi-way cell, these losses are compensated by the amplification of light in its active medium. As a result, the sensitivity of the method increases sharply, which makes it possible to detect unprecedented low concentrations at levels of 104 atoms / cm^3.

The recommended by the World Meteorological Organization expanded uncertainty of $\delta\ ^{13}C_{VPDB}$ measurements in atmospheric air is 0,1 ‰. To ensure such quality of measurements it is necessary to determine the factors that affect measurements of $\delta\ ^{13}C_{VPDB}$ and develop a method for making corrections.

There are several methods for decreasing the expanded uncertainty of δ $^{13}C_{VPDB}$ measurements by CRDS: proper calibration, averaging time, correction for concentration dependence, and gas matrix effects.

Two different methods of calibration exist. Method used in conducted experiments is described by [3] and requires two or more CRM's with different molar fractions of $^{12}CO_2$ and $^{13}CO_2$. If two CRM's are used the calibration equations are

$$x_{at}^{12} = \frac{x_{2t}^{12} - x_{1t}^{12}}{x_{2m}^{12} - x_{1m}^{12}}(x_{am}^{12} - x_{1m}^{12}) + x_{1t}^{12}$$

$$x_{at}^{13} = \frac{x_{2t}^{13} - x_{1t}^{13}}{x_{2m}^{13} - x_{1m}^{13}}(x_{am}^{13} - x_{1m}^{13}) + x_{1t}^{13}$$

where

x_{am} – measured $^{12}CO_2$ or $^{13}CO_2$ mole fraction value in analyzed material; x_{1m} – measured $^{12}CO_2$ or $^{13}CO_2$ mole fraction value in CRM №1; x_{2m} –measured $^{12}CO_2$ or $^{13}CO_2$ mole fraction value in CRM №2; x_{1t} – reference $^{12}CO_2$ or $^{13}CO_2$ mole fraction value in CRM №1; x_{2t} – reference $^{12}CO_2$ or $^{13}CO_2$ mole fraction value in CRM №2.

The $\delta^{13}C_{VPDB}$ in analyzed material is calculated by formula

$$\delta^{13}C_{VPDB} = \left(\frac{x_{at}^{13} / x_{at}^{12}}{0,011237} - 1 \right) \cdot 1000$$

This method of calibration was chosen because it leads to lesser expanded uncertainty of δ $^{13}C_{VPDB}$ measurements.

Also influence of gas matrix composition and averaging time were studied.

The absorption line broadening and narrowing of $^{12}CO_2$, and $^{13}CO_2$ molecules can lead to growth of expanded uncertainty of δ $^{13}C_{VPDB}$. There are different causes of the absorption line broadening and narrowing such as random thermal motion and collisions. Different gas matrix compositions produce different absorption line broadening and narrowing which means different gas matrix compositions differ δ $^{13}C_{VPDB}$ measurements [4].

Using this method of calibration, calculations and measurements with nitrogen and argon were performed (Table 1).

Table 1. The experimental gas matrix offset factors and a prediction based on calculated Lorenz profile heights.

	δ $^{13}C_{VPDB(nitrogen)}$- δ $^{13}C_{VPDB(air)}$	δ $^{13}C_{VPDB(argon)}$- δ $^{13}C_{VPDB(air)}$
Result of calculation	8,4	-40,1
Result of measurement	5,3	-42,6

4. Results

During the experiment with Raman sensor the spectra of analyzed gases were obtained. It was found that for $^{12}CO_2$ Raman shift is 1388 cm^{-1}, for $^{13}CO_2$ – 1370 cm^{-1}. Stokes shifts for these isotopes lay close to each other (Figure 1). Thus, we can conclude that this method provides spectral selection at the level of $\lambda/\Delta\lambda >> 1000$. We were able to reach this level of spectral selectivity by utilizing high resolution polychromators and high sensitive photomultipliers. Experiments were conducted in UV range of spectra which provides more accurate results, but conducting analysis is IR range is better as Stokes shifts are not as close to each other as in UV range. Ratios of $^{13}C/^{12}C$ were measured using CRDS analyzer PICARRO G2131-I [5].

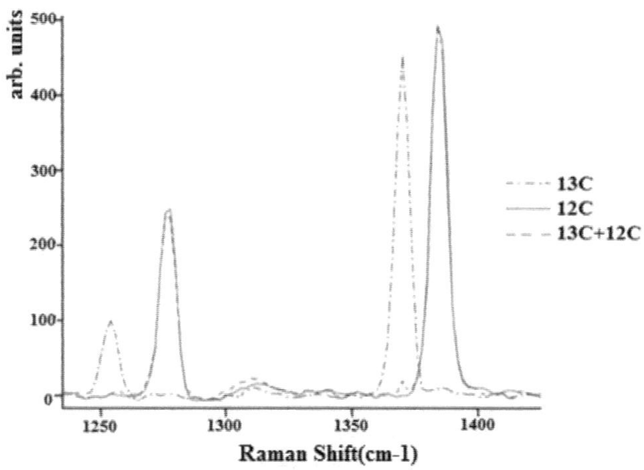

Figure 1. Raman shifts of CO_2 isotopes.

5. Conclusion

CO_2 isotopes have Stokes shifts located close to each other. The sensitivity of the Raman method is inversely proportional to the fourth power of the scattering wavelength, based on the difference in the Stokes shifts of the isotopes $^{12}CO_2$ and $^{13}CO_2$, equal to ($\Delta\nu = 2$ cm^{-1}), which in the wavelength range of less than 2.0 μm (ν>5000 cm^{-1}) The method allows spectral selection at a level $\nu/\Delta\nu = \lambda/\Delta\lambda >> 1000$. The method provides spectral resolution, noise immunity and sufficient selectivity for gas analysis. Using the methods of SCR and CARS, it is possible to resolve isotopic shifts of the indicator substances of the simulator in the ultraviolet and infrared ranges at levels not less than 5 cm^{-1}, which allows to judge the isotopic composition of methane and determine the source of methane generation from the $^{13}C/^{12}C$ concentration ratio. The CRDS method makes it possible to measure impurities at the ppb and ppt levels and the $^{13}C/^{12}C$ ratio in carbon dioxide at a level with mass spectrometers. CRDS was a method of verification of data.

Acknowledgements
The work is financially supported by RF Ministry of Education and Science, Grant RFMEFI57518X01800.

References

[1] Tonkic A, Tonkic M, Lehours P and Mégraud F 2012 Epidemiology and diagnosis of Helicobacter pylori infection *Helicobacter* **17** (Suppl 1) 1-8
[2] Friedrichs G, Bock J, Temps F, Fietzek P, Arne K and Wallace D W R 2010 Toward continuous monitoring of seawater $^{13}CO_2/^{12}CO_2$ isotope ratio and pCO_2: Performance of cavity ringdown spectroscopy and gas matrix effects *Limnol. Oceanogr.: Meth.* **8** 10 539-551
[3] Griffith D W T, Deutscher N M, Caldow C, Kettlewell G, Riggenbach M and Hammer S A 2012 Fourier transform infrared trace gas and isotope analyser for atmospheric applications *Atmos. Meas. Tech.* **5** 2481–98 doi:10.5194/amt-5-2481-2012
[4] Nakamichi S, Kawaguchi Y, Fukuda H, Enami S, Hashimoto S, Kawasaki M, Umekawa T, Morino I., Suto H and Inoue G 2006 Buffer-gas pressure broadening for the (30(0)1)(III) <-(000) band of CO_2 measured with continuous-wave cavity ring-down spectroscopy *Phys. Chem. Chem. Phys.* **8** 364-368
[5] Grishkanich A, Chubchenko Y, Elizarov V, Zhevlakov A and Konopelko L 2018 SRS-sensor $^{13}C/^{12}C$ isotopes measurements for detecting Helicobacter Pylori *Proc. SPIE* **10488** 104881A

Application of artificial neural networks for fusion of data from radar and depth sensors applied for persons' monitoring

Paweł Mazurek

Warsaw University of Technology, Faculty of Electronics and Information Technology, Institute of Radioelectronics and Multimedia Technology, Nowowiejska 15/19, 00-665 Warsaw, Poland

E-mail: p.mazurek@ire.pw.edu.pl

Abstract. The research reported in this paper is related to the fusion of measurement data from the impulse-radar sensors and infrared depth sensors applied in a system for unobtrusive monitoring of elderly persons. Three methods for data fusion, based on the artificial neural networks – one trained on real-world data, and two trained on synthetic data generated on the basis of two different models of the data – are compared with respect to their capacity of decreasing the uncertainty of position estimation in a series of experiments which involved the tracking of a moving person.

1. Introduction

The life expectancy has been growing in Europe for many years, while the healthy life expectancy has been slightly diminishing since the last decade of the XXth century (*cf.* http://www.healthy-life-years.eu/). Hence the growing importance of research on new technologies that could be employed in monitoring systems supporting care services for elderly persons. Those systems are expected to detect dangerous events, such as person's fall, but also to predict those events on the basis of acquired data [1]; it has been shown recently that the analysis of gait may contribute to the fall prevention [2].

There are two emerging categories of monitoring techniques that can be applied for non-intrusive monitoring of elderly persons, *viz.* radar-based and depth-sensor-based techniques [3]. Radar sensors and depth sensors operate according to different physical principles and, therefore, have specific complementary advantages and disadvantages: radar sensors offer a broad field of view and capacity of the through-the-wall monitoring, but have relatively low accuracy; on the other hand, the depth sensors have higher accuracy, but their coverage is limited, and they cannot detect occluded targets. One may expect that the combination of such sensors, followed by adequate data fusion, could be beneficial. It has been recently shown that the artificial neural networks can be effectively used for the fusion of measurement data acquired by means of several sensors tracking a moving object [4].

Before being used, an artificial neural network has to be trained, what in the case of tracking of an elderly person is particularly challenging task: involving the elderly persons in the experimentation aimed at the acquisition of the reference data could be problematic. One possible solution is the use of the synthetic data for training the neural network, but the effectiveness of this solution may depend on the accuracy of the model of the data. This paper is devoted to the assessment of the uncertainty of the position estimates obtained by means of the neural-network-based method for fusion of the data from the radar sensors and from the depth sensor, trained on the synthetic data.

2. Methodology of experimentation

In the experiments, three methods for data fusion based on the artificial neural networks – one trained on real-world data (hereinafter called NNR method) and two trained on synthetic data generated on the basis of different models of the data (hereinafter called NNS1 method and NNS2 method) – were compared with respect to their capacity of decreasing the uncertainty of estimation of a monitored person's position. The real-world data and the synthetic data, used for training, were representative of the movement according to the same trajectories; the real-world data used for testing the methods, however, were representative of the movement according to different trajectories. All the methods were based on the feedforward neural network with four input neurons (for two pairs of the x-y coordinates acquired by the radar sensors and the depth sensor), eight neurons in a single hidden layer, and two output neurons (for the fused x-y coordinates). In the experiments, the implementations of the neural networks, available in the MATLAB Deep Learning Toolbox, were used [5].

2.1. Acquisition of real-world data

The raw measurement data for experimentation were acquired by means of sensors configured as shown in figure 1. In the experiments, a person walked along ten straight-line trajectories which together covered all the reference points (see figure 2a), and along three square-shaped trajectories differing in length (see figure 2b); for each trajectory, 20 walks were performed.

2.2. Generation of synthetic data

The synthetic data for training the artificial neural networks were generated on the basis of the reference movement trajectories presented in figure 2a.

In the case of the first set of synthetic data (used for training the neural network of the NNS1 method), the x-y data representative of the measurement data acquired by means of a depth sensor were corrupted with zero-mean white noise with the standard deviation 0.05 m, while the synthetic data representative of the measurement data acquired by means of the impulse-radar sensors were corrupted with zero-mean red noise with the standard deviation 0.1 m. For each straight-line trajectory, $R = 20$ realizations of the data were generated for each type of sensor. The data acquisition rate for the depth sensor and for the impulse-radar sensors was set to 10 Hz.

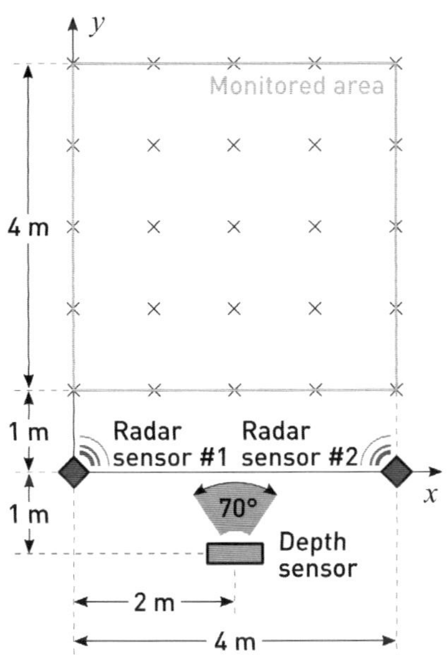

Figure 1. Experimental setup; the crosses indicate the reference points.

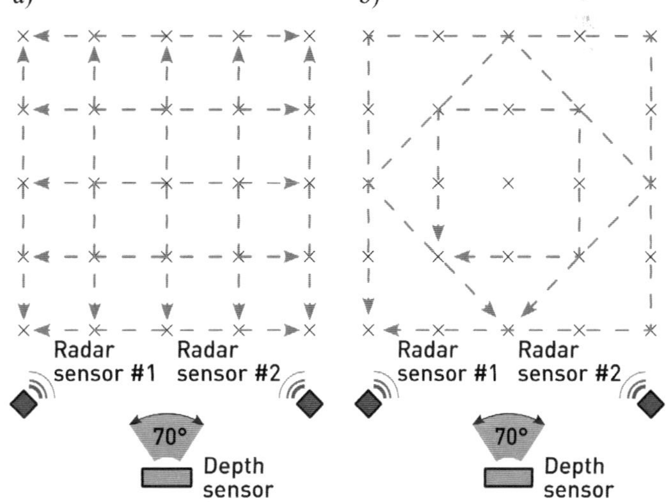

Figure 2. Movement scenarios considered in the experiments: a) straight-line trajectories, b) squares-shaped trajectories.

In the case of the second set of synthetic data (used for training the neural network of the NNS2 method), the silhouette of a moving person was modelled by means of an ellipse, with semi axes of 0.1 m and 0.25 m, whose center followed a sine-shaped trajectory oscillating around the reference trajectory: the amplitude of the sine-shaped trajectory, modelling the sway of a body during a single step, was 0.04 m, while the half of the period of that trajectory, modelling the length of a single step, was 0.5 m. The coordinates of the points lying on the circumference of the ellipse were corrupted with zero-mean white noise with the standard deviation 0.1 m. To take into account the fact that the depth sensors "can see" only one side of the body, the *x*-*y* data representative of the measurement data acquired by means of this sensor were calculated as the mean values of the coordinates of half of the points of the ellipse – the points being closest to the *x*-axis. The synthetic data representative of the measurement data acquired by means of the impulse-radar sensors were generated as in the case of the first set of synthetic data, but were smoothed using the moving-average filter.

All the trajectories used for training the networks, are presented in figure 3.

2.3. Criteria for performance evaluation

The experiments were aimed at assessing the uncertainty of the estimation of the person's position with respect to the assumed reference trajectory. The position estimates, obtained for each type of sensor and for each method for data fusion, were treated separately.

For each of the R sequences of the data representative of the person's position in the *x*- and *y*-dimension:

$$\left\{\hat{x}_n^{(r)} \mid n=1,\ldots,N^{(r)}\right\} \text{ and } \left\{\hat{y}_n^{(r)} \mid n=1,\ldots,N^{(r)}\right\} \text{ for } r=1,\ldots,R \tag{1}$$

the corresponding sequence of the absolute errors of the position estimation has been computed in the following way:

$$\left\{\Delta\hat{x}_n^{(r)} = \hat{x}_n^{(r)} - \dot{x}_n^{(r)} \mid n=1,\ldots,N^{(r)}\right\} \text{ and } \left\{\Delta\hat{y}_n^{(r)} = \hat{y}_n^{(r)} - \dot{y}_n^{(r)} \mid n=1,\ldots,N^{(r)}\right\} \text{ for } r=1,\ldots,R \tag{2}$$

where $\dot{x}_n^{(r)}$ and $\dot{y}_n^{(r)}$ denote the reference value for $\hat{x}_n^{(r)}$ and $\hat{y}_n^{(r)}$.

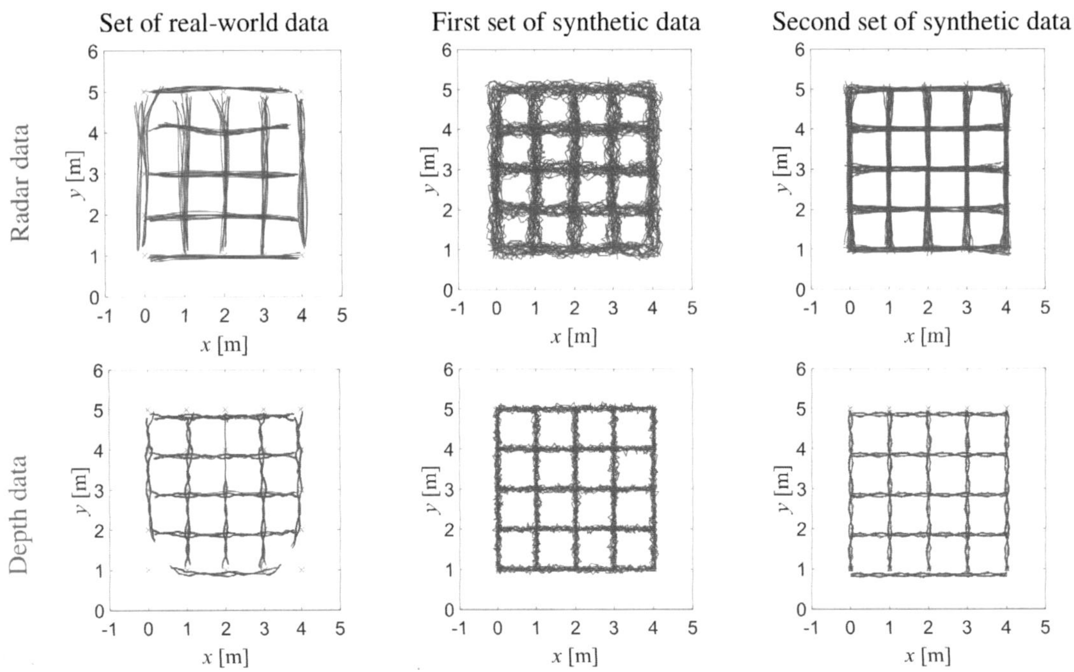

Figure 3. Data used for training the artificial neural networks: the set of real-world data (left column), the first set of synthetic data (middle column) and the second set of synthetic data (right column).

The sequences of the absolute errors in the x-dimension and in the y-dimension have been then used to determine R sequences of the position errors $\{\Delta\hat{p}_n^{(r)}\}$:

$$\left\{\Delta\hat{p}_n^{(r)} = \sqrt{\left(\Delta\hat{x}_n^{(r)}\right)^2 + \left(\Delta\hat{y}_n^{(r)}\right)^2} \,\middle|\, n = 1,\ldots,N^{(r)}\right\} \text{ for } r = 1,\ldots,R \tag{3}$$

The qualitative assessment of the performance of the methods for data fusion was based on the inspection of the empirical cumulative distribution functions $F(\xi)$ [6]. $F(\xi)$ is an estimate of the cumulative distribution function characterizing the position errors, defined as:

$$F(\xi) \equiv \frac{1}{M}\sum_{n=1}^{M} I(\Delta\hat{p}_m \leq \xi) \tag{4}$$

where $\{\Delta\hat{p}_m \mid m = 1,\ldots,M\}$ is a sequence composed of all the sequences of the position errors $\{\Delta\hat{p}_n^{(r)}\}$, and $I(\bullet)$ is the indicator function which takes the value 1 if the condition inside the brackets is met, and 0 otherwise.

The quantitative assessment of the performance of all the methods for data fusion was based on an indicator called A_{ECDF} and defined as the area under the empirical cumulative distribution functions $F(\xi)$ in the interval $\xi \in [0,1]$ m, taking the values from the interval $[0,1]$. The higher the value of the A_{ECDF}, the better the performance of the data fusion. Furthermore, this type of assessment was based on the following statistics of the distribution of position errors: the mean position error (abbreviated to MEAE), the median position error (abbreviated to MEDE), the maximum position error (abbreviated to MAXE) and the standard deviation of the position errors (abbreviated to STDE).

3. Results and discussion

The real-world trajectories used for testing are presented in figure 4, while in figure 5 – the trajectories obtained by means of the fusion of radar data and depth data. The empirical cumulative distribution functions characterizing the position errors are shown in figure 6, and the values of the uncertainty indicators, calculated on the basis of the position errors, are provided in table 1.

The analysis of the presented results is leading to the following conclusions:

- It can be seen that the data fusion has alleviated the problems occurring if only single-type sensors are used for monitoring, *viz.* the dispersion of the estimates have decreased, and the blind spots have disappeared.
- The estimates of the trajectories, obtained by means of the NNS1 method, are subject to bias introduced by the depth sensor – the depth sensor "can see" only one side of the body; this phenomenon was not taken into account in the generation of the first set of synthetic data, and therefore the bias has not decreased.
- The estimates of the trajectories, obtained by means of the NNS2 method, are more accurate than the estimates obtained by means of the NNS1 method – the bias introduced by the depth sensor has decreased since during the generation of the second set of synthetic data more realistic model of the person's movement was used. However, it can be noticed that the trajectories are slightly distorted around the point (4,1) – in this area the person was on the edge of the field of view of the depth sensor, and the resulting deformation of the silhouette of a person caused the corruption of the estimates of the person's position.
- The estimates of the movement trajectories, obtained by means of the NNR method, seem to be the most accurate – the bias and the distortion of the trajectories, introduced by the depth sensor, have decreased.
- The values of the uncertainty indicators, calculated on the basis of the fused data, are better than their values calculated on the basis of the radar data only or of the depth data only. Moreover, the values obtained for the NNR method and for the NNS2 method are very similar, and notably better than the values obtained for the NNS1 method.

Journal of Physics: Conference Series **1379** (2019) 012051 doi:10.1088/1742-6596/1379/1/012051

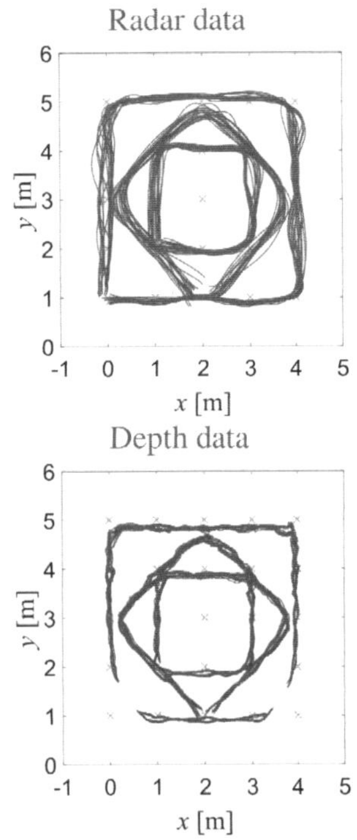

Figure 4. Real-world data used for testing the methods for data fusion.

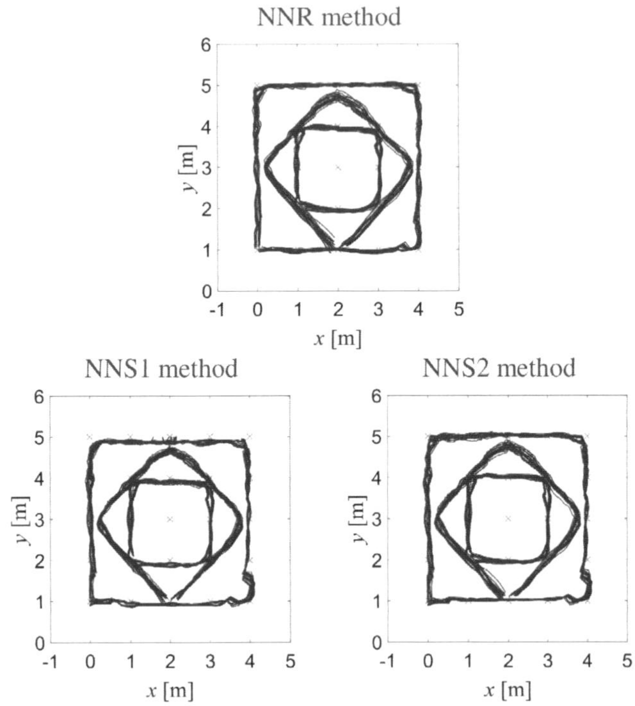

Figure 5. Estimates of the square-shaped trajectories obtained by means of the methods for data fusion.

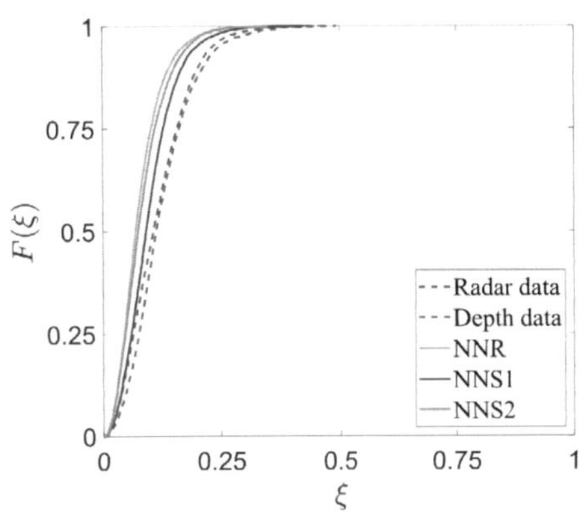

Figure 6. Empirical cumulative distribution functions characterizing the position errors.

Table 1. Values of the uncertainty indicators, calculated on the basis of the position errors.

Uncertainty indicator	Radar data	Depth data	NNR	NNS1	NNS2
MEAE [m]	0.12	0.12	0.08	0.10	0.08
MEDE [m]	0.11	0.11	0.07	0.09	0.07
MAXE [m]	0.49	0.46	0.37	0.43	0.35
STDE [m]	0.07	0.07	0.05	0.05	0.05
A_{ECDF}	0.88	0.88	0.92	0.90	0.92

During the experiments, various configurations of the artificial neural network were considered, and the following observations have been made:

- As long as the networks are trained sufficiently long, the number of neurons in the hidden layer varying from 6 to 10 does not influence the results of the data fusion.
- The best results are obtained for the networks trained with the Levenberg-Marquardt algorithm.
- All types of sigmoid transfer function – such as logistic function, hyperbolic tangent function or Elliot sigmoid function – may be effectively used, and they lead to similar results.

4. Conclusions

The novelty of the research results, presented in this paper, consists in identification of the effectiveness of neural-network-based methods for fusion of data from the impulse-radar sensors and from the depth sensors, when applied for unobtrusive monitoring of elderly persons in their home environment.

In the experiments, three methods for data fusion, based on the artificial neural networks, were compared with respect to their capacity of decreasing the uncertainty of estimation of a monitored person's position: one method (NNR) was based on the network trained on the real-world data, and two methods (NNS1 and NNS2) were based on the network trained on the synthetic data, but the sets of synthetic data were generated according to two different models of the data.

The experiments have demonstrated that the artificial neural networks trained on the synthetic data may be effectively used for fusing the data from the radar sensors and from the depth sensors – and yield as good results as the networks trained on the real-world data – but the model of the data, used for generation of the synthetic data, has to take into account all the relevant characteristics of the person's movement, and characteristics of the sensors used for monitoring.

Acknowledgments

This work has been accomplished partially within the project PL12-0001 financially supported by EEA Grants – Norway Grants, and partially within the project 2017/25/N/ST7/00411 financially supported by the National Science Centre, Poland.

References

[1] Chaccour K, Darazi R, Hassani A H E and Andres E 2017 *IEEE Sens. J.* **17** pp 812–22
[2] Stone E, Skubic M, Rantz M, Abbott C and Miller S 2015 *Gait Posture* **41** pp 57–62
[3] Wagner J, Mazurek P, Miekina A, Morawski R Z, Jacobsen F F, Sudmann T T, Borsheim I T, Ovsthus K and Ciamulski T 2017 *Measurement* **111** pp 420–31
[4] Sun B, Jiang C and Li M 2016 *Sensors* **16** p 1823
[5] Mathworks, *Matlab Documentation: Deep Learning Toolbox*. Available: https://www.mathworks.com/help/deeplearning/index.html (2018-12-14)
[6] van der Vaart A W 1998 *Asymptotic Statistics* (Cambridge: Cambridge University Press)

Combinatorial characterization of inrankings as weak orders induced by intervals

S V Muravyov and E Y Emelyanova

National Research Tomsk Polytechnic University
Pr. Lenina, 30, Tomsk, 634050, Russian Federation

E-mail: muravyov@tpu.ru

Abstract. The problem of reliable processing of heteroscedastic interval data occupies an important niche among urgent topics of measurement science. The paper is devoted to a combinatorial characterization of so called 'inrankings' which are weak orders induced by input intervals of the interval fusion with preference aggregation (IF&PA) procedure. The procedure transforms the given m initial real line intervals into inrankings, which are a specific case of weak order relations (or rankings) over a set of n discrete values belonging to these intervals. The new notation of inranking appears as a result of restrictions imposed on the ordinary rankings by interval character of the initial data. In the paper, the inranking spaces properties are investigated from the combinatorial theory point of view. It is shown that the inranking space is a subset of the set of all weak orders with a single symbol of strict order. The cardinality of inranking space is defined by the triangle number for the given number n of the discrete elements. Cardinalities of other adjacent spaces are considered.

1. Introduction

The role of interval data integration and interpretation both in theory and practice of measurement data fusion can hardly be overestimated [1]. Suffice it to recall an importance of the confidence interval parameters estimation in a traditional statistical framework with various applications both in physical and social sciences [2]. In this paper we deal with the problem of heteroscedastic interval data processing. Heteroscedasticity means observations heterogeneity manifested as changeable variance of a regression model random error [3]. The random errors heteroscedasticity results in inefficiency of estimates obtained using the least square method. In turn, this can lead to inadequate statistical conclusions on estimated quantity values and their quality.

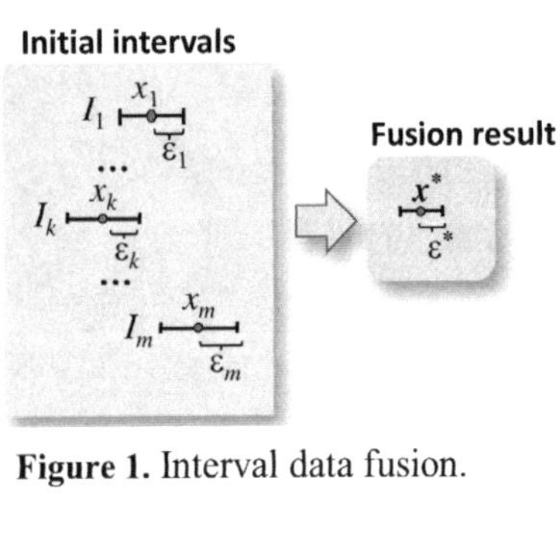

Figure 1. Interval data fusion.

Generally, the *data fusion* refers to a joint processing of data on some object obtained from multiple sources aiming to acquire fuller, more objective and accurate knowledge of a characteristic under investigation than knowledge derived from a single source [1]. The *interval data fusion* (figure 1) can be deemed as a procedure intended to find an interval to be consistent with maximal number of the initial intervals and with maximal likelihood including a representing them value x^*. Existing methods of the heteroscedastic interval data fusion, such as combination of weighted mean and χ^2-test [3], different versions of approval voting [4], etc. do not provide simultaneously a uniqueness of the fusion outcome and its *robustness* in terms of an independence on particular law of the interval data probability distribution. In [5], the paper authors have been proposed the *interval fusion with preference aggregation* (*IF&PA*) and have demonstrated the method outcomes' guaranteed improved robustness and accuracy by Monte Carlo experimentation. The IF&PA procedure has been successfully applied in inter-laboratory comparison data processing, enhancement of multisensor readings accuracy and providing energy-accuracy trade-off in presence of outliers in wireless sensor network [6, 7, 8].

The IF&PA transforms the given initial real line intervals into inrankings, which are a specific case of weak order relations (or rankings) over a set of discrete values belonging to these intervals. The inrankings appears as a result of restrictions imposed on the ordinary rankings by interval character of the initial data. Then, for the preference profile consisting of the inrankings, the IF&PA determines a single consensus ranking by a Kemeny rule [6, 9]; the highest ranked value in the consensus ranking is accepted as the fusion result.

The new notation of inranking possesses a series of interesting properties. In this paper, we investigate these properties from the combinatorial theory point of view. The properties are particularly important for a development of new solutions when building search techniques of the consensus ranking for a given preference profile consisting of the inrankings.

2. Terms and notations

Preference aggregation supposes a use of the notation of *ranking* (or *weak order*) of n elements of a set $A = \{a_1, a_2, \ldots, a_n\}$ in the form $\lambda = (a_1 \phi\ a_2 \ldots \sim a_s \sim a_t \phi \ldots \sim a_n)$. The ranking λ can be described as a union, $\lambda = \rho \cup \tau$, of the two relations: *strict order* ρ, i.e. $a_i \phi\ a_j$, and *tolerance* τ (deemed as indifference, or tie), i.e. $a_i \sim a_j$. Then the relation λ specifies a binary relation of *weak order* on the set A [10]. It is easy to show that $\rho = \lambda^{-1}$ and $\tau = \lambda \cap \lambda^{-1} = \rho \cup \rho^{-1}$. Set $\Lambda(n, m) = \{\lambda_1, \lambda_2, \ldots, \lambda_m\}$ of m rankings of n elements a_i, $i = 1, \ldots, n$, is called a *preference profile* for m and n given.

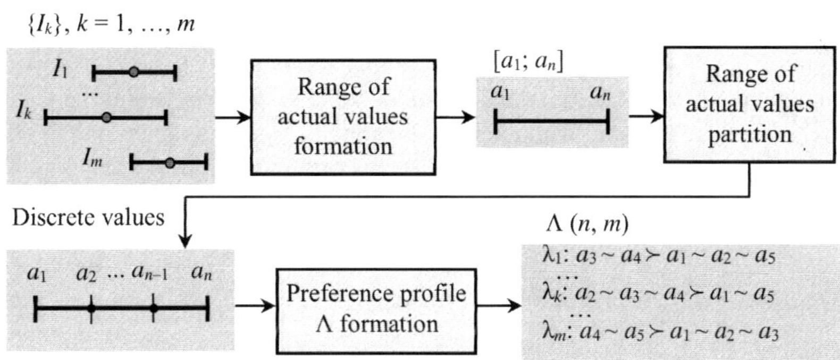

Figure 2. Stages of conversion of a set of intervals to a preference profile.

Consider a collection of m closed intervals $\{I_k\}_{k=1}^m$ in the real line, where each interval has a *middle point* x_k, a *lower bound* $x_k - \varepsilon_k$ and *upper bound* $x_k + \varepsilon_k$, so that $I_k = [x_k - \varepsilon_k, x_k + \varepsilon_k]$; $x_k, \varepsilon_k \in \Upsilon$. Introduce a *range of actual values* (RAV) $A = \{a_1, a_2, \ldots, a_n\}$, over which the strict order (inherited from the real line) $a_1 < a_2 < \ldots < a_n$ exists. The discrete set A can be produced starting from the given continuous intervals $\{I_k\}_{k=1}^m$ in three steps shown in figure 2. First, RAV's lower and upper bounds are chosen, i.e. $a_1 = \min\{x_k - \varepsilon_k \mid k = 1, \ldots, m\}$ and $a_n = \max\{x_k + \varepsilon_k \mid k = 1, \ldots, m\}$ correspondingly. Then, to generate elements $a_2, a_3, \ldots, a_{n-1}$, the obtained interval $[a_1, a_n]$ is partitioned into $n - 1$ equal subintervals of length $h = (a_n - a_1)/(n - 1)$, and the i-th RAV's element becomes $a_i = a_{i-1} + h$, $i = 2, \ldots, n$. Finally, the set A

= $\{a_1 < a_2 < \ldots < a_n\}$ of strictly ordered discrete values a_i, $i = 1, \ldots, n$, is used to shape the preference profile $\Lambda(n, m)$ representing the initial intervals $\{I_k\}_{k=1}^{m}$ [5]. The conditions for formation of the profile are discussed at the beginning of next section.

3. Inrankings and inranking spaces

For any interval I_k, $k = 1, \ldots, m$, we have $A = A_k \cup \bar{A}_k$, $A_k \cap \bar{A}_k = \varnothing$, where the subset A_k includes all the elements of A belonging to the interval I_k, and its complement $\bar{A}_k$ includes all the rest elements of A. The ranking induced by interval I_k will be called the *inranking* λ_k, if it satisfies the following four conditions for $i, j = 1, \ldots, n$:

$$a_i \in A_k \wedge a_j \notin A_k \Rightarrow a_i \succ a_j; \tag{1}$$

$$a_i, a_j \in A_k \vee a_i, a_j \notin A_k \Rightarrow a_i \sim a_j; \tag{2}$$

$$a_i \notin A_k \wedge a_j \in A_k \Rightarrow a_i < a_j; \tag{3}$$

$$a_i, a_j \in A_k \text{ are neighbors} \Rightarrow j \equiv i + 1. \tag{4}$$

Notice that the k-th ranking consists of two equivalence classes composed of elements of the sets A_k and $\bar{A}_k$. Elements of the class A_k are strictly preferred over elements of the class $\bar{A}_k$, i.e. always $\lambda_k = A_k \succ \bar{A}_k$. Hence, the ranking contains a single symbol of strict order $\succ$ and $n - 2$ symbols of tolerance $\sim$. ***Example 1.*** One of possible ranking for $n = 5$ could be $\lambda_k = \{a_2 \sim a_3 \sim a_4 \succ a_1 \sim a_5\}$, where $A_k = \{a_2 \sim a_3 \sim a_4\}$, $\bar{A}_k = \{a_1 \sim a_5\}$ and $A_k \succ \bar{A}_k = \{a_2 \sim a_3 \sim a_4\} \succ \{a_1 \sim a_5\}$.

For notational convenience, ranking elements will be denoted by their indexes, and subsets A_k and $\bar{A}_k$ will be shown by parenthesizing appropriate sets of indexes. Then the symbols $\succ$ and $\sim$ are omitted.

Example 2. The inranking from example 1, $\{a_2 \sim a_3 \sim a_4 \succ a_1 \sim a_5\}$, can be concisely represented as (234)(15).

The condition (4) is necessary as series of elements $\{a_i\}$ of the set A is *strongly monotonic*, i.e., $a_i < a_{i+1}$ for all $i \in \infty$. The class $A_k \subseteq A$ can include merely sequential subseries of elements of A without breaks, that is, indexes of these elements must constitute a *segment of the natural series*. It means that the difference of indexes for any two neighbor elements a_i and a_j in A_k cannot be larger than 1, i.e., $j \equiv i + 1$.

A concrete preference profile always consists of inrankings taken from the space of all possible ones. Hence, now we try to answer the question how many different inrankings exist for given n.

Let us denote the space of weak orders via Ω_0 and the space of inrankings via Ω_2. We need also to consider the space Ω_1 elements of which satisfy the conditions (1)–(3) but violate the condition (4). In fact, the members of Ω_1 are just those having a single symbol of strict order $\succ$. Clear that

$$\Omega_2 \subseteq \Omega_1 \subseteq \Omega_0, \tag{5}$$

for which the following expressions are valid:

$$\Omega_1 = \Omega_2 \cup (\bar{\Omega}_2 \cap \Omega_1),\ \Omega_1 \cap \bar{\Omega}_1 = \varnothing, \Omega_2 \cap \bar{\Omega}_2 = \varnothing,$$
$$|\Omega_1| = |\Omega_2| + |\bar{\Omega}_2 \cap \Omega_1|. \tag{6}$$

The space Ω_0 plays the part of universal set for the sets Ω_1 and Ω_2. The space Ω_1 consists of the space Ω_2 of inrankings and the space of forbidden rankings $\Omega_f = \bar{\Omega}_2 \cap \Omega_1$.

Cardinalities of the space $|\Omega_0|$ and its subspaces $|\Omega_1|$ and $|\Omega_2|$ are connected to Stirling numbers of the second kind $S_{n,q}$ that defines number of unordered partitions of n-element set into q non-empty subsets [11]:

$$|\Omega_0| = \sum_{q=0}^{n} q! S_{n,q}\ , \tag{7}$$

Journal of Physics: Conference Series **1379** (2019) 012052 doi:10.1088/1742-6596/1379/1/012052

Composition of the spaces Ω_1, Ω_2 and $\bar{\Omega}_2 \cap \Omega_1$ for $n = 1, \ldots, 5$, at corresponding cardinalities of Ω_0 and A_k, is shown in table 1. It is clear from table 1 that cardinality of the inranking space $|\Omega_2|$ in dependence on n follows the series 1, 3, 6, 10, 15, 21, 28, 36, 45, 55, …, elements of which are called *triangle numbers* T_n [12]. A triangle number and, hence, the cardinality $|\Omega_2|$ are defined by formula:

Table 1. Composition of spaces Ω_1, Ω_2 and Ω_f for $n = 1, \ldots, 5$.

$\|\Omega_0\|$	$\|A_k\|$	Ω_1	$\|\Omega_1\|$	Ω_2	$\|\Omega_2\|$	Ω_f	$\|\Omega_f\|$	$\|\Psi_k\|$	$n - \|A_k\| + 1$
				$n = 1$					
1	1	1	1	1	1	∅	0	1	1
				$n = 2$					
3	1	12, 21	3	12, 21	3	∅	0	1	2
	2	(12)		(12)		∅		2	1
				$n = 3$					
13	1	1(23), 2(13), 3(12)	7	1(23), 2(13), 3(12)	6	∅	1	2	3
	2	(12)3, (23)1, (13)2		(12)3, (23)1		(13)2		2	2
	3	(123)		(123)		∅		6	1
				$n = 4$					
75	1	1(234), 2(134), 3(124), 4(123)	15	1(234), 2(134), 3(124), 4(123)	10	∅	5	6	4
	2	(12)(34), (23)(14), (13)(24), (24)(13), (14)(23), (34)(12)		(12)(34), (23)(14), (34)(12)		(14)(23), (13)(24), (24)(13)		4	3
	3	(123)4, (134)2, (124)3, (234)1		(123)4, (234)1		(124)3, (134)2		6	2
	4	(1234)		(1234)		∅		24	1
				$n = 5$					
541	1	1(2345), 2(1345), 3(1245), 4(1235), 5(1234)	30	1(2345), 2(1345), 3(1245), 4(1235), 5(1234)	15	∅	16	24	5
	2	(12)(345), (13)(245), (14)(235), (15)(234), (23)(145), (24)(135), (25)(134), (34)(125), (35)(124), (45)(123)		(12)(345), (23)(145), (34)(125), (45)(123)		(13)(245), (14)(235), (15)(234), (24)(135), (25)(134), (35)(124)		12	4
	3	(123)(45), (124)(35), (125)(34), (134)(25), (135)(24), (145)(23), (234)(15), (235)(14), (245)(13), (345)(12)		(123)(45), (234)(15), (345)(12)		(124)(35), (125)(34), (134)(25), (135)(24), (145)(23), (235)(14), (245)(13)		12	3
	4	(1234)5, (1235)4, (1245)3, (1345)2, (2345)1		(1234)5, (2345)1		(1235)4, (1245)3, (1345)2		24	2
	5	(12345)		(12345)		∅		120	1

$$|\Omega_2| = T_n = \frac{n(n+1)}{2} = \binom{n+1}{2} = \binom{n+1}{n-1} = (S_{n+1,n}) = \sum_{i=1}^{n} |A_k|_i . \tag{8}$$

The cardinality $|\Omega_f|$ of the space of forbidden rankings is defined by equation:

$$|\Omega_f| = |\bar{\Omega}_2 \cap \Omega_1| = F_n = 2^n - 1 - n(n+1)/2 . \tag{9}$$

Evidently, the cardinality $|\Omega_1|$ is defined by

$$|\Omega_1| = T_n + F_n = 2S_{n,2} + 1 = 2^n - 1 . \tag{10}$$

It follows from equations (8)–(10) that numbers T_n can be found among Stirling numbers of the second kind and in Pascal triangle [11, 13].

4. Inranking Generating Sets

Inrankings, as well as any weak orders, are shaped of sets of strict orders. Let us demonstrate this by a simple example.

Example 3. Given two strict orders $\rho_1 = 213$ and $\rho_2 = 231$. Then lying between them (generated by them) weak order is $\lambda_k = 2(13)$ or in extended notation $\lambda_k = \{a_2 \succ a_1 \sim a_3\}$.

Designate the permutations set of elements of A_k through $\Pi = \{\pi_1, ..., \pi_{|A_k|!}\}$, and the permutations set of elements of $\bar{A}_k$ through $\Pi' = \{\pi'_1, ..., \pi'_{|\bar{A}_k|!}\}$.

The set of strict orders $\Psi_k = \{\rho^k_g\}$, $g = 1, ..., |A_k|! \cdot |\bar{A}_k|!$, will be called a *generating set* for an inranking $\lambda_k = A_k \succ \bar{A}_k$ if $\Psi_k = \Pi \times \Pi' = \{(\pi_u, \pi'_v) \mid \pi_u \in \Pi,\ \pi'_v \in \Pi'\}$, where the Cartesian product's element (π_u, π'_v) is deemed as a result of concatenation operation $\rho = \pi_u \div \pi'_v$ of two permutations π_u and π'_v.

Example 4. Let $\lambda_k = (234)(15)$. Then the generating set Ψ_k consists of $|\Psi_k| = 3!\cdot 2! = 12$ following strict orders ρ_g, $g = 1, \ldots, 12$: 23415, 23451, 24315, 24351, 32415, 32451, 34215, 34251, 42315, 42351, 43215, 43251.

Given cardinality $|A_k|$, at any n, the number of generating sets is defined by the expression $(n - |A_k| + 1)$. Hence, the total number of generating strict orders for all possible inrankings can be calculated as

$$N(A_k) = |A_k|! \cdot |\bar{A}_k|! \cdot (n - |A_k| + 1) . \tag{11}$$

One can claim that for given n a union of generating sets for $|A_k| = 1$ coincides with the single generating set for $|A_k| = n!$. All possible cardinalities $\Psi_k = |A_k|! \cdot |\bar{A}_k|!$ and $(n - |A_k| + 1)$ for different $|A_k|$ and $n = 1, \ldots, 5$ are shown in the ninth and tenth columns of table 1 correspondingly.

The introduced notations can be illustrated using permutohedra. Figure 3 shows the space of all strict orders for $n = 5$ and the space Ω_2 of all corresponding inrankings. For clarity, we use five permutohedra which are truncated octahedra with central nodes corresponding to inrankings 1(2345), 2(1345), 3(1245), 4(1235) and 5(1234). There are shown only links between inrankings and facet centers of permutohedra containing weak orders. Inranking designations are shown with bold font. Each of the truncated octahedra represents a generating set for inrankings with $|A_k| = 1$. In color there are shown subsets of the generating sets belonging to the permutohedron for inrankings at $|A_k| = 2, 3, 4$.

Figure 4 shows the generating set for the inranking $\lambda_k = (234)(15)$ discussed in examples 1–4. One can see that centers of edges and facets of permutohedra correspond to weak orders. As above, there are shown only links between facet centers of permutohedra and the inranking (234)(15). The number of permutohedra is defined by the cardinality $|A_k| = 3$.

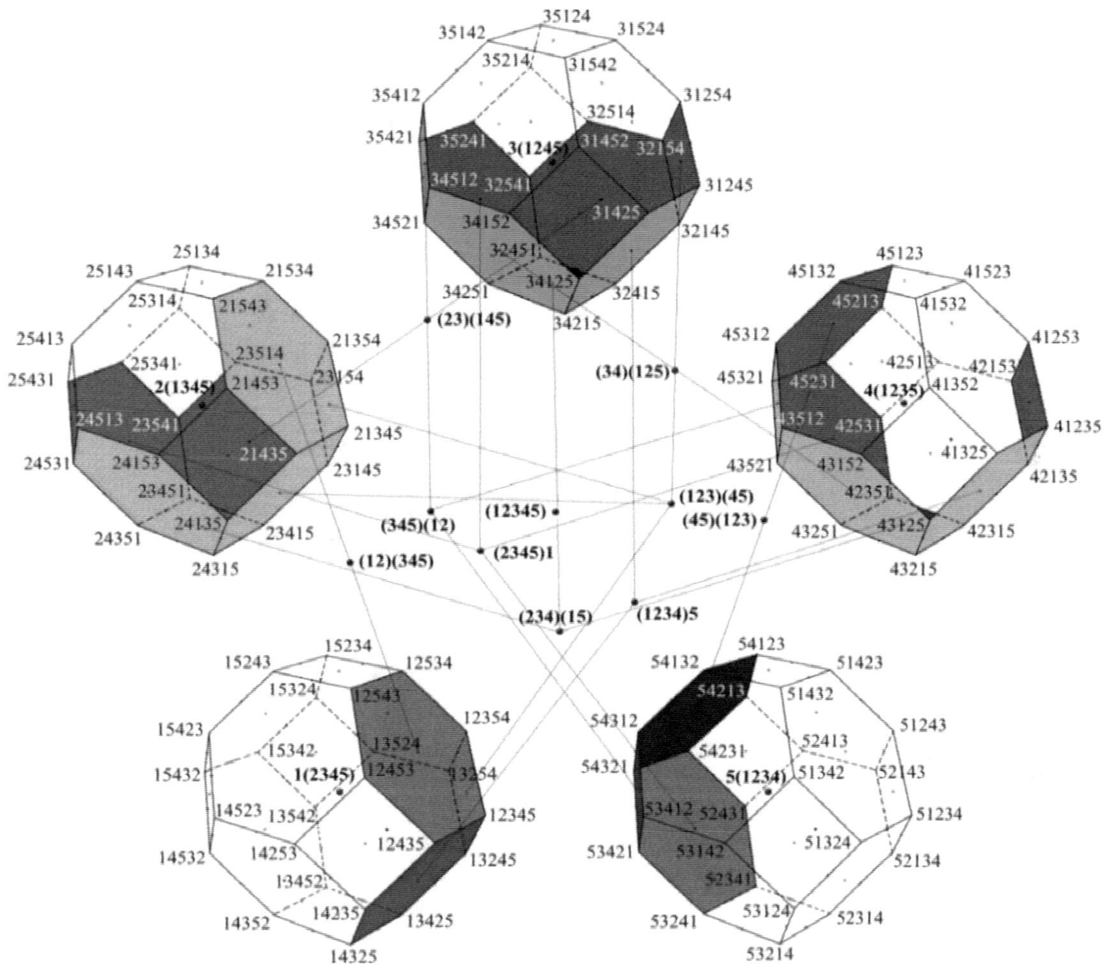

Figure 3. The space of all strict orders for $n = 5$ and the space Ω_2 of all corresponding inrankings.

5. Conclusion

It is shown that produced by the IF&PA procedure inranking space is a subset of the set of all weak orders with a single symbol of strict order. The cardinality of inranking space is defined by the triangle number for the given number of the discrete elements of the RAV. Knowing properties of the spaces can help to recommend an effective mode of the interval data fusion applications. The properties are particularly important for improvement of search techniques of the consensus ranking for a given preference profile consisting of the inrankings.

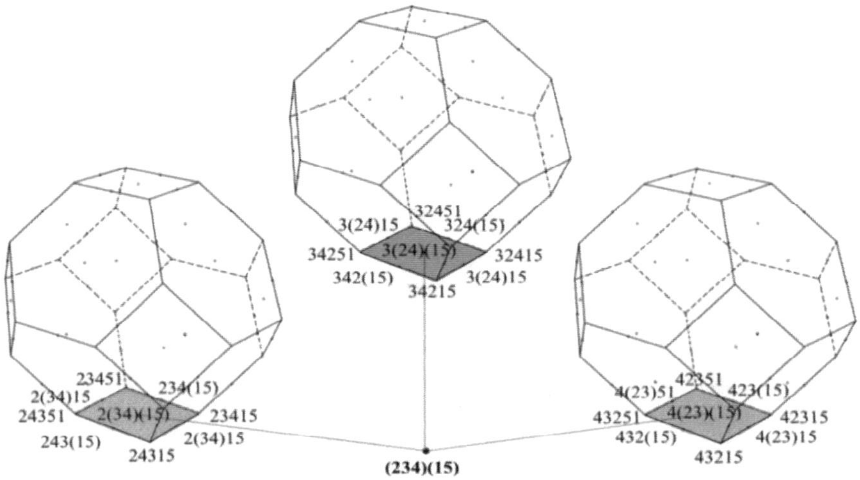

Figure 4. Generating set for the inranking $\lambda_k = (234)(15)$.

Acknowledgment
This work was supported by the Russian Science Foundation, project 18-19-00203, in part of the IF&PA method development, and by the Russian Ministry of Education and Science, basic part of the state task "Science", project 2.5760.2017/8.9, in part of the IF&PA method applications.

References
[1] Hall D L and McMullen S A H 2004 *Mathematical Techniques in Multisensor Data Fusion* (Norwood: Artech House Publishers)
[2] Meeker W Q, Hahn G J, and Escobar L A 2017 *Statistical Intervals: A Guide for Practitioners and Researchers* (New York: John Wiley & Sons)
[3] Tofallis C 2009 Least squares percentage regression *J. Mod. Appl. Stat. Methods* Available at SSRN: https://ssrn.com/abstract=1406472
[4] Brams S J and Fishburn P C 2007 *Approval Voting* (New York: Springer)
[5] Muravyov S V, Khudonogova L I and Emelyanova E Y 2018 Interval data fusion with preference aggregation *Measurement* **116** 621–630
[6] Kemeny J G 1959 Mathematics without numbers *Daedalus* **88** 571–591
[7] Muravyov S V 2013 Ordinal measurement, preference aggregation and interlaboratory comparisons *Measurement* **46**(8) 2927–35
[8] Muravyov S V and Marinushkina I A 2011 Largest consistent subsets in interlaboratory comparesons: preference aggregation approach *Proc. 14th Joint Int. IMEKO TC1+TC7+TC13 Symp. (August 31 – September 2, 2011, Jena, Germany)* pp 69–73
[9] Khudonogova L I and Muravyov S V 2018 Interval data fusion with preference aggregation in wireless sensor network: Energy-accuracy trade-off in presence of outliers *J. Phys. Conf. Series* **1065** 072016
[10] García-Lapresta J L and Pérez-Román D 2011 Measuring consensus in weak orders In: *Consensual Processes. Studies in Fuzziness and Soft Computing* **267** (Berlin Heidelberg: Springer)
[11] Aigner M 1997 *Combinatorial Theory* (Berlin Heidelberg: Springer-Verlag)
[12] Sloane N J A The On-Line Encyclopedia of Integer Sequences. The OEIS Foundation, https://oeis.org
[13] Anderson J A 2003 *Discrete Mathematics with Combinatorics* (Upper Saddle River: Prentice Hall)

Journal of Physics: Conference Series **1379** (2019) 012053 doi:10.1088/1742-6596/1379/1/012053

How to transform all multiple solutions of the Kemeny Ranking Problem into a single solution

S V Muravyov, P F Baranov and E Y Emelyanova

National Research Tomsk Polytechnic University
Pr. Lenina, 30, Tomsk, 634050, Russian Federation

E-mail: muravyov@tpu.ru

Abstract. Preference aggregation as a problem of a single consensus ranking determination, using Kemeny rule, for m rankings, including ties, of n alternatives is considered in the paper. The Kemeny Ranking Problem (KRP) may have considerably more than one optimal solutions (strict orders or permutations of the alternatives) and, hence, special efforts to deal with this phenomenon are needed. In the paper, there is proposed an efficient formal rule for convolution of the N multiple optimal permutations, the output profile B(N, n), into an exact single final consensus ranking, which can include ties. The convolution rule is as follows: in the final consensus ranking, alternatives are arranged in ascending order of their rank sums (total ranks) calculated for the output profile B; some two alternatives are considered to be tolerant if they have the same rank sums in B. The equivalent convolution rule can be also applied as follows: in the final consensus ranking, alternatives are arranged in descending order of row sums (total scores) calculated for a tournament table built for B; some two alternatives are deemed to be tolerant if they have the same row sums. It is shown that, for any alternative, its total rank and total score are equal in sum to the output profile dimension $N \times n$. The convolution rules are validated using Borda count.

1. Introduction

In this paper we consider a preference aggregation procedure in the form of Kemeny rule. Being essentially the multiple heterogeneous properties measurement in an ordinal scale, preference aggregation belongs to a circle of problems relevant to measurement science [1]. Additionally, in the recent few years, the preference aggregation and the processing of ranked data have attracted considerable interest within a much broader consideration covered by such a rapidly developing area of artificial intelligence as *machine learning* [2]. Appropriate methods are used to solve newly emerging problems, such as crowd sensing/labeling, sentiment analysis, meta-search engines, interval data fusion and others [2–7].

The preference aggregation is a single consensus ranking determination for m rankings (voters), possibly including ties, of n alternatives (candidates). This is a classical problem that has been intensively investigated firstly as a Voting Problem in the framework of Social Choice Theory [8].

Let we have a set $A = \{a_1, a_2, ..., a_n\}$ of n alternatives and a set $\Lambda = \{\lambda_1, \lambda_2, ..., \lambda_m\}$ of *rankings* defined over the set A, where each of m rankings (also called *preference relation* or *weak order*) has a form of chain of size n, e.g. $\lambda = a_2 \succ a_1 \succ ... \succ a_s \sim a_t \succ ... \succ a_p \sim a_q$, and may include $\succ$, a *strict preference* relation ρ, and $\sim$, a *tolerance* relation (or *tie*) τ, so that $\lambda = \rho \cup \tau$. The set $\Lambda(m, n)$ is called a *preference profile* for the given m rankings of n alternatives.

We will use the following obvious *short notation* for a ranking:

$$\lambda_k = a_2 \succ a_1 \succ ... \succ a_s \sim a_t \succ ... \succ a_p \sim a_q \Leftrightarrow 21...(st)...(pq). \quad (1)$$

For example, $a_2 \succ a_5 \sim a_1 \succ a_3 \sim a_4 \succ a_6 \Leftrightarrow$ 2(51)(34)6 or $a_3 \succ a_1 \succ a_5 \succ a_2 \succ a_6 \succ a_4 \Leftrightarrow$ 315264.

Let Π_n be a set of all $n!$ strict order relations $\succ$ on A; each strict order $\rho \in \Pi_n$ is in one-to-one correspondence with a *permutation* of first n natural numbers $\infty_n = \{1, 2, ..., n\}$; in this sense, we will treat notions of strict (or linear) order and permutation as synonyms.

Preference aggregation goal is to determine a *consensus ranking* β that is a single preference relation that would give an integrative characterization of the profile $\Lambda(m, n)$.

There are many rules to find the consensus ranking [9]. We use the *Kemeny rule* [10] consisting in determination of such linear order (*Kemeny ranking*) $\beta \in \Pi_n$ of the alternatives of A that the distance $D(\beta, \Lambda)$ (defined in terms of the *number of pairwise inconsistencies* between rankings) from β to the rankings of the initial profile $\Lambda(m, n)$ is minimal for all possible strict orders ρ (permutations), that is

$$\beta = \arg\min \sum\nolimits_{i<j} p_{ij}, \quad (2)$$

where $p_{ij} = \sum\nolimits_{k=1}^{m} [1 - \mathrm{sgn}(a_i^k, a_j^k)],\ i, j = 1,...,n,$ is an element of the $(n \times n)$ *profile matrix* $[p_{ij}] = P$, rows and columns of which are labeled by the alternatives' numbers (indexes); $\mathrm{sgn}(a_i, a_j) = \begin{cases} 1, & a_i \succ a_j \\ 0, & a_i \sim a_j \\ -1, & a_i \prec a_j \end{cases}$ is a function that reveals the sign (or direction) of the pair $(a_i, a_j) \in \lambda$.

The Kemeny rule (2) is an optimization problem, called the *Kemeny Ranking Problem (KRP)*, which means the determination of such a transposition of the profile matrix rows and columns that the sum of elements of its upper triangle submatrix, i.e. $D(\rho, \Lambda) = \sum\nolimits_{i<j} p_{ij}$, is minimal.

Clearly since the set Π_n is finite, the Kemeny ranking always exists, however it is *not always unique* as the optimal distance value may correspond to not the only transposition of the profile matrix rows and columns, that is $D(\beta_1, \Lambda) = \sum\nolimits_{i<j} p_{ij}^1 = D(\beta_2, \Lambda) = \sum\nolimits_{i<j} p_{ij}^2 = ... = D(\beta_N, \Lambda) = \sum\nolimits_{i<j} p_{ij}^N$, where N is a number of Kemeny rankings (optimal permutations) [2,3,11,12]. Furthermore, it was shown in [1] that in certain situations the number of multiple optimal permutations may exceed 10^7 even for small $m = 4$ and $n = 15$ and also the solutions may rank the alternatives in significantly different ways what produces much more ambiguity than that inherent in the initial profile Λ.

Example 1. Let the following profile be given: $\Lambda(3,3) = \{1{\sim}32;\ 31{\sim}2;\ 2{\sim}31\}$. Then the KRP has two solutions: $\beta_1 = 312$ and $\beta_2 = 321$, both having minimal distance to the profile $D(\beta_1, \Lambda) = D(\beta_2, \Lambda) = 5$. The considered profile $\Lambda(3,3)$ is shown in figure 1 as belonging to the weak order space for $n = 3$. □

In order to solve the KRP we use the recursive algorithm of our own design, RECURSALL, implementing the branch & bound technique, allowing to find *all possible* Kemeny rankings for a given preference profile [4]. It is convenient to term the initial preference profile $\Lambda(m, n)$ as *input profile* and the multiple optimal permutations as output profile $\mathrm{B}(N, n) = \{\beta_1, \beta_2, ..., \beta_N\}$, $\mathrm{B} \subset \Pi_n$ (see figure 2).

Journal of Physics: Conference Series **1379** (2019) 012053 doi:10.1088/1742-6596/1379/1/012053

Clear that special efforts to deal with the phenomenon of the KRP multiple optimal solutions, where $N_s >> m$, are needed. The rest of the paper is devoted to development of an efficient formal rule for convolution of the output profile into a single final consensus ranking β_{fin}.

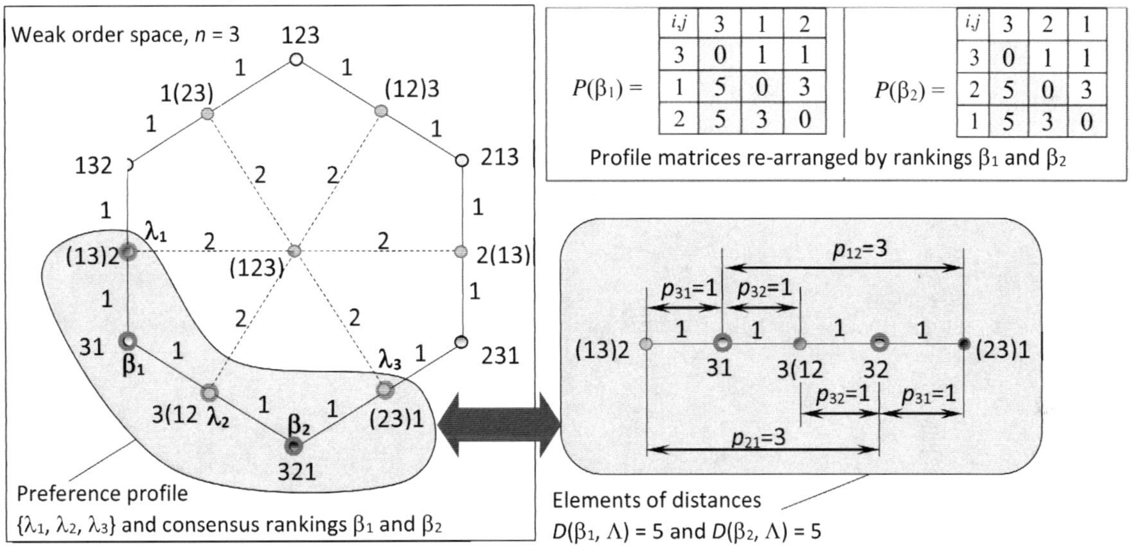

Figure 1. Toward s multiplicity of the KRP solutions: preference profile $\{\lambda_1, \lambda_2, \lambda_3\}$ and consensus rankings β_1 and β_2 in the weak order space for $n = 3$.

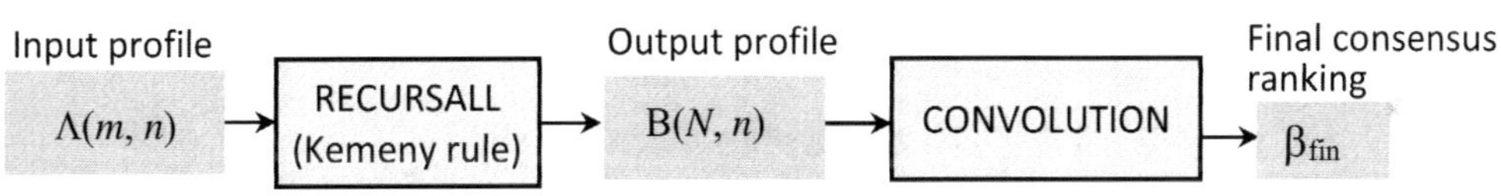

Figure 2. Stages of the final consensus ranking determination.

2. Convolution rule for the KRP multiple solutions

A rule for convolution of the set of optimal permutations can be formulated as follows. Let $B(N, n) = \{\beta_1, \beta_2, \ldots, \beta_N\}$, $B \subset \Pi_n$, be an output profile of the KRP applied to an input profile $\Lambda(m, n) = \{\lambda_1, \lambda_2, \ldots, \lambda_m\}$ given over some set of alternatives $A = \{a_1, a_2, \ldots, a_n\}$ and *rank* r_i^k be a position of an alternative a_i in the consensus ranking $\beta_k \in B$, $k = 1, \ldots, N$. Let a *total rank* r_i of the alternative a_i is defined as $r_i = \sum_{k=1}^{N} r_i^k$. Then, for all $i < j$, $i, j = 1, \ldots, n$,

$$r_i < r_j \Rightarrow a_i \succ a_j \text{ and } r_i = r_j \Rightarrow a_i \sim a_j, \qquad (3)$$

where both relations $\succ$ and $\sim$ are in the single final consensus ranking β_{fin}.

Thus, the final consensus ranking may include tolerances as well as an arbitrary $\lambda_k \in \Lambda$, i.e. generally speaking, $\beta_{fin} = \rho \cup \tau$ and $\beta_{fin} \notin \Pi_n$.

Example 2. Table 1 and figure 3 show an example of convoluting two optimal permutations into a single β_{fin}. One can see that alternatives 1, 3, 5, and 6 have different positions in two optimal solutions β_1 and β_2, however, their distances to the initial profile are $D(\beta_1, \Lambda) = D(\beta_2, \Lambda) = 58$. The convolution rule (3) allows to find their true positions in β_{fin}, namely as in the ranking 421(35)6. □

Table 1. Example of the convolution rule (3) application.

Input profile Λ	Output profile B	Ranks of alternatives in B							Final solution β_{fin}
λ_1: 1 2 6 4(3 5)			a_1	a_2	a_3	a_4	a_5	a_6	Short form:
λ_2: 4 5 1 2 3 6	β_1: 4 2 1 3 6 5	r_i^1	3	2	4	1	6	5	4 2 1 (35) 6
λ_3: 2(51)(34)6	β_2: 4 2 5 1 3 6	r_i^2	4	2	5	1	3	6	Extended form:
λ_4: (63)4 2(15)		$r_i^1 + r_i^2$	**7**	**4**	**9**	**2**	**9**	**11**	$a_4 \succ a_2 \succ a_1 \succ a_3 \sim a_5 \succ a_6$
λ_5: 3 4 (26) 51									

The convolution rule (3) guarantees determination of an exact single optimal permutation β_{fin} for any output profile B.

	1	2	3	4	5	6
1	0	6	4	6	6	4
2	4	0	4	6	2	3
3	6	6	0	5	5	3
4	4	4	5	0	2	4
5	4	8	5	8	0	6
6	6	7	7	6	4	0

Initial profile matrix *P*

	4	2	1	3	6	5
4	0	4	4	5	4	2
2	6	0	4	4	3	2
1	5	6	0	4	4	6
3	5	6	6	0	3	5
6	6	7	6	7	0	4
5	8	8	4	5	6	0

The profile matrix re-arranged by consensus ranking β_1

	4	2	5	1	3	6
4	0	4	2	4	5	4
2	6	0	2	4	4	3
5	8	8	0	4	5	6
1	6	6	6	0	4	4
3	5	6	5	6	0	3
6	6	7	4	6	7	0

The profile matrix re-arranged by consensus ranking β_2

Figure 3. Profile matrices from Example 2.

3. Convolution rule justification

Let the output profile B(N, n) be represented by the (n×n) *tournament table* $S = [s_{ij}]$, where $s_{ij} = \sum_{k=1}^{N} \mathbf{I}_\rho(a_i^k, a_j^k), i, j = 1,\ldots,n$, is the matrix element and $\mathbf{I}_\rho(a_i,a_j) = \begin{cases} 1 \text{ if } (a_i,a_j) \in \rho \\ 0 \text{ if } (a_i,a_j) \notin \rho \end{cases}$ is the indicator function[1].

A sum of elements of i-th row of the table S is termed as the *score* z_i of the alternative a_i, i.e. for $i = 1, \ldots, n$, $z_i = \sum_{j=1}^{n} s_{ij}$. Then multiple permutations can be also convoluted using the scores z_i as follows: for all $i < j$, $i, j = 1, \ldots, n$,

$$z_i > z_j \Rightarrow a_i \succ a_j \text{ and } z_i = z_j \Rightarrow a_i \sim a_j, \tag{4}$$

where both relations ≻ and ~ are in the single final consensus ranking β_{fin}.

The convolution rule (4) guarantees determination of an exact single optimal permutation β_{fin} for any output profile B, and the single optimal permutation is the same as that found using the rule (3). In fact, the rule (4) coincides with the preference aggregation rule well known as Borda count [13].

[1] For the indicator function, the following statements are hold: $(a_i, a_j) \in \rho \Leftrightarrow a_i \succ a_j$ and $(a_i, a_j) \notin \rho \Leftrightarrow a_i \prec a_j$.

For any output profile B(N, n) represented by a corresponding tournament table [s_{ij}], the following equation is valid:

$$r_i = N \cdot n - z_i, i = 1,...,n. \quad (5)$$

Example 3. Table 2 shows the values of scores and rank sums calculated for the profile alternatives from Example 2. The alternatives are sorted in accordance with rules (3) and (4). It can be seen the fairness of expression (5) for these data.

Table 2. Scores and rank sums of the output profile B from Example 2.

Alternatives	a_4	a_2	a_1	a_3	a_5	a_6
Total score z_i	10	8	5	3	3	1
Total rank r_i	2	4	7	9	9	11
$N \cdot n = z_i + r_i$	12	12	12	12	12	12

The expression (5) can serve as a basis for justification of the convolution rule (3) validity. Indeed, let a *social score function* f assigns a nonempty choice set $C \subset A$ to a profile B. It was proved by H.P. Young in [13] that "for any fixed number n of alternatives, there is one and only one social score function that is neutral, consistent, faithful, and has cancellation property – namely, Borda's rule". The properties of the function f listed in the Young's statement are deemed as the set of axioms necessary and sufficient in order to have a fully characterized preference aggregation rule [14]. Thus, the scores z_i, implementing Borda rule, guarantee that the rule (4) leads to convolution of the output profile B(N, n) into the exact and single consensus ranking β_{fin}. It follows from (5) that social score functions f coincide for the rules (3) and (4).

4. Conclusion

It was shown in the paper that the multiple optimal permutations, i.e. the output profile B(N, n) = {β_1, β_2, ..., β_N}, as solutions of the KRP for m rankings, including ties, of n alternatives can be efficiently transformed into an exact single final consensus ranking β_{fin}, which can also include ties. The proposed convolution rule is as follows: in the final consensus ranking β_{fin}, alternatives are arranged in ascending order of their rank sums calculated for the output profile B; in β_{fin} some two alternatives are tolerant to each other if they have the same rank sums in the output profile. The equivalent convolution rule can be also applied as follows: in the final consensus ranking β_{fin}, alternatives are arranged in descending order of row sums (total scores) z_i calculated for a tournament table S of the output profile B; in β_{fin} some two alternatives are tolerant to each other if they have the same row sums of the tournament table.

It was also shown that, for any i-th alternative, its total rank r_i and total score z_i are equal in sum to a constant value $N \cdot n$, that is the dimension of the output profile B. The convolution rules validity has been shown using such well known preference aggregation rule as Borda count.

Acknowledgment

This work was supported by the Russian Science Foundation, project 18-19-00203, in part of the preference aggregation algorithms development, and by the Russian Ministry of Education and Science, basic part of the state task "Science", project 2.5760.2017/8.9, in part of the preference aggregation method applications.

References

[1] Muravyov S V 2014 Dealing with chaotic results of Kemeny ranking determination *Measurement* **51** 328–334
[2] Korba A, Clemencon S and Sibony E 2017 A learning theory of ranking aggregation *Proc. 20th Int. Conf. on Artificial Intelligence and Statistics*, PMLR, 54, pp 1001–1010
[3] Ali A and Meilă M 2012 Experiments with Kemeny ranking: What works when? *Math. Soc. Sci.* **64**(1) 28–40
[4] Grandi U, Loreggia A, Rossi F and Saraswat V 2016 A Borda count for collective sentiment analysis *Ann. Math. Artif. Intell.* **77**(3–4) 281–302
[5] Chen X, Bennett P N, Collins-Thompson K and Horvitz E 2013 Pairwise ranking aggregation in a crowdsourced setting *Proc. 6th ACM Int. Conf. on Web Search and Data Mining* (February 04 – 08, 2013 Rome, Italy) pp 193–202
[6] Muravyov S V and Marinushkina I A 2011 Largest consistent subsets in interlaboratory comparisons: preference aggregation approach *Proc. 14th Joint Int. IMEKO TC1+TC7+TC13 Symp. (August 31 – September 2, 2011, Jena, Germany)* pp 69–73
[7] Muravyov S V 2013 Ordinal measurement, preference aggregation and interlaboratory comparisons *Measurement* **46**(8) 2927–35
[8] Young H P 1988 Condorcet's theory of voting *Am. Polit. Sci. Rev.* **82**(2) 1231–44
[9] Young H P 1995 Optimal voting rules *J. Econ. Perspect.* **9**(1) 51–64
[10] Kemeny J G 1959 Mathematics without numbers *Daedalus* **88** 571–591
[11] Davenport A and Lovell D 2005 *Ranking Pilots in Aerobatic Flight Competitions*, IBM Research Report RC23631 (W0506-079) (NY:T J Watson Research Center)
[12] Muravyov S V and Marinushkina I A 2013 Intransitivity in multiple solutions of Kemeny Ranking Problem *J. Phys. Conf. Series* **459** 012006
[13] Young H P 1974 An axiomatization of Borda's rule *J. Econ. Theory* **9** 43–52
[14] Arrow K J 1963 *Social Choice and Individual Values* 2nd ed (New York: Wiley)

The metrologically reasonable clustering of measurement results

K K Semenov [1], **V A Bolschikov** [1]

[1] Department of measuring informational technologies, Institute of computer science and technologies, Peter the Great Polytechnic University, 29, Polytechnicheskaya str., St. Petersburg 195251, Russia.

E-mail: semenov.k.k@iit.icc.spbstu.ru, vitaliy.bolschikov@gmail.com

Abstract. This paper is dedicated to the metrological aspects of clustering procedure. It is shown that taking into account the information on uncertainty of the data to be clustered allows easily making reasonable decisions that are metrologically supported. The paper presents that the problems that are usually difficult to solve in practice of the clustering become simpler: the proposed approach allows determining the number of maximum recognizable clusters and thus protects from the unreasonable conclusions and allows performing the preconditioning of data to be clustered simpler.

1. Introduction

The clustering of inaccurate data is a problem that usually occurs in practice. Indeed, when some complex object is studied, its different characteristics are measured. The obtained results are used to predict its future behaviour. Examples of such objects can be different: turbines, engines, power units and other complex technological installations, distributed systems with numerous components, human health, environmental processes etc. In many cases, there is no theory that can describe the laws, under which the object of interest acts, or it is no sufficient amount of information to build the mathematical model that exhaustively describes the object. So, it is treated as "black-box" and we only can cluster its different states that can be recognized from the obtained data. After that, for any new observation, we can classify the current object state by assigning to one of the clusters. Such multi-stage cognitive procedure is usually used when the object under study is extremely difficult to be described – like human health, medicine conditions and others.

The inaccuracy of used initial data imposes restrictions on the possible conclusions that can be made: the accurate measurements results of characteristics of object under study allow differing much more object states than if the uncertainty is large. This paper proposes a quantitative method for describing this circumstance and fills the gap in the literature dedicated to clustering: there are enough papers on clustering the inaccurate data – like interval data [1-4], fuzzy variables [5-8], Dempster-Shafer sets [9-11] etc – and there is almost no papers on the determination the number of clusters and the agreement of the final results of clustering with the inaccuracy degree of the initial data [12, 13].

2. The metrological aspects of clustering procedures

Studying different objects, processes, phenomena in practice, all the objective data are obtained from the measurements results. So, all decisions based on it are performed from the uncertain initial grounds and that's why they may be unreasonable. The clustering procedure isn't the exception – some of the discovered clusters can be parts of the one cluster in reality and look separate because of random errors in initial data; some of the identified clusters may consist several unrecognizable smaller subclusters we aren't able to see – they merged into one since their boundaries are blurred and overlap because of errors of the clustered results of measurement. To detect such situations is extremely difficult problem in practice and is supposed to be solved using the only data to be clustered, which usually doesn't contain any corresponding information in real-life situations. That's why to accompany any decision with a characteristic of its quality, we need to take into account the metrological characteristics of data we process. Accounting the information on the initial data uncertainty brings benefits in many cases: we can easily answer some questions that are difficult if we deal with data treated as absolutely accurate. Applying to clustering, the algorithm shown below allows estimating the maximum reasonable quantity of recognizable clusters. To get such number is a hard problem in practice that is often solved subjectively based on some a priori information, but if we have information on metrological characteristics of data to be clustered then this quantity can be easily obtained and used as restriction or reinforcement of conclusions and decisions.

Traditional approach to clustering consists of the following main stages.

1) Collecting the initial data for clustering.

The array $\mathbf{\Omega}=\{\omega_1, \omega_2, \ldots, \omega_n\}$ of states of studied object, process or phenomenon of our interest is composed at this stage. Each element ω_i is described by a vector of values $\boldsymbol{x}_i^{\mathrm{T}}= (x_{i1}, x_{i2}, \ldots, x_{im})$ having the one structure for all i. Its elements are a set of measured characteristics of studied object and are disturbed with measurement errors. All measured vectors $\boldsymbol{x}_i$, $i = 1, 2, \ldots n$ form a set $\mathbf{X}$ of observations on studied object, process or phenomenon.

2) Preparing the data before clustering.

The vectors $\boldsymbol{x}_i$ can be inconvenient to deal with during clustering. So, they can be transformed to vectors $\boldsymbol{y}_i = (y_{i1}, y_{i2}, \ldots, y_{ir})$ of may be the smaller dimension or that are more comfortable to deal with. The elements of $\boldsymbol{y}_i$ inherit the uncertainty from the measurements results x_{ij} and are inaccurate too. All vectors $\boldsymbol{y}_i$, form a set $\mathbf{Y}$ consisting of principal components of quantitative description of the studied object, process or phenomenon state.

The preconditioning of data can use different and extremely complex algorithms – including for example neural networks and self-organized maps [14].

3) Choosing measure of distance between elements of the set $\mathbf{Y}$.

The measure $\rho(\boldsymbol{y}_i, \boldsymbol{y}_j)$ that allows quantifying degree of proximity of vectors $\boldsymbol{y}_i$ and $\boldsymbol{y}_j$ towards each other is chosen at this phase. Since vectors $\boldsymbol{y}_i$ and $\boldsymbol{y}_j$ are associated with elements of the array $\mathbf{\Omega}$ by the chain of transformations $\mathbf{\Omega} \rightarrow \mathbf{X} \rightarrow \mathbf{Y}$, the value $\rho(\boldsymbol{y}_i, \boldsymbol{y}_j)$ allows making a conclusion on the proximity degree of the corresponding elements ω_i and ω_j.

4) Choosing clustering algorithm.

On this stage, a rule $f(\boldsymbol{y}_i)$ should be chosen that assigns to each vector $\boldsymbol{y}_i$ a number from array $\mathbf{C} = \{1, 2, \ldots \mathrm{K}\}$, which determines the cluster that contains $\boldsymbol{y}_i$ and, as a consequence, ω_i. Results of clustering by particular algorithm essentially depend on measure $\rho(\boldsymbol{y}_i, \boldsymbol{y}_j)$ that is used for calculations (one algorithm can produce different results for different measures ρ – just like different methods of clustering do that use the same measure ρ). The choice of a particular clustering algorithm may depend in practice on number of analyzed vectors n and their length m.

5) Clustering

The clustering procedure in introduced terms has the meaning of the mapping f: $\mathbf{Y}\rightarrow\mathbf{C}$. For each cluster its center's coordinates $\mathbf{c}_k$, $k = 1, 2, .., K$, should be determined. Number of clusters K is given by user and, as a rule, is based on available a priori information. The quantitative expression of quality of performed clustering is usually the sum of measures of proximity between vectors $\boldsymbol{y}_i$ and the closest cluster centers:

Journal of Physics: Conference Series **1379** (2019) 012054 doi:10.1088/1742-6596/1379/1/012054

$$R(K) = \sum_{i=1}^{n} \rho\left(\boldsymbol{y}_i, \mathbf{c}_{f(\boldsymbol{y}_i)}\right).$$

The listed stages are the main for clustering procedure; in concrete situations it may differ only in some particular details.

We can conclude that the value $R(K)$ computed from the measurements results x_{ij} inherits its uncertainty and that's why is uncertain too. This circumstance creates the possibility for determining the maximum possible number $K_{\max}$ of clusters that can be recognized from the initial data. To achieve this goal, it is necessary to metrologically accompany all preformed calculations from the list above: for any computation, we should obtain not only the calculation results but also the metrological characteristic of its accuracy. Then we will be able to connect the uncertainty of value $R(K)$ as the quantitative quality estimation with the accuracy of the initial data – the measurement results x_{ij}. For this, it is necessary at the first stage of clustering procedure to involve information on the possible absolute errors Δx_{ij} of values x_{ij}. In the most frequent practical case this will be the maximum possible error Δ_{ij}: $|\Delta x_{ij}| \leq \Delta_{ij}$, $i = 1, 2, \ldots n$ and $j = 1, 2, \ldots m$. If values x_{ij} are measured then the values of Δ_{ij} are known from technical documentation for used measuring instruments. If not only maximum possible error values are available from the manuals, then all other metrological characteristics should be taken into account.

To estimate the metrological characteristics of the results of the calculations with values x_{ij}, we can use one from many approaches and methods that are developed to the present moment to propagate uncertainty of measurement results through the computational procedure. These methods can be grouped in two big sets: methods that use the randomization (like Monte-Carlo approach) and methods that perform automatic analysis by the overloading operations performed during calculations (a linearization of calculated functions with automatic differentiation or interval arithmetic). All of these techniques can be applied to any computational problem; the clustering isn't the exception. We propose to use the approach of paper [15] for propagating the uncertainty of inaccurate initial data through all the computations because this approach has a metrological orientation, but any other suitable method can be also chosen.

If it is no need in initial data preliminary processing before clustering, then a computationally cheap simple procedure can be used in the case of small uncertainties of initial data. It can be described as follows. Let for each vector $\boldsymbol{x}_i^{\mathrm{T}} = (x_{i1}, x_{i2}, \ldots, x_{im})$ of initial data the vector $\boldsymbol{\Delta x}_i^{\mathrm{T}} = (\Delta_{i1}, \Delta_{i2}, \ldots, \Delta_{im})$ of maximum possible absolute errors of its components be known (from the technical documentation). Since no processing is provided, $\boldsymbol{y}_i = \boldsymbol{x}_i$ and $\boldsymbol{\Delta y}_i = \boldsymbol{\Delta x}_i$. Otherwise, we need to use one of the mentioned methods to propagate information on errors $\boldsymbol{\Delta x}_i$ during calculation the values $\boldsymbol{y}_i$ and, as a result, estimate their maximum possible absolute errors $\boldsymbol{\Delta y}_i$. Then the chosen clustering algorithm f should be applied to the set of vectors $\boldsymbol{y}_i$ and, as a result, the set of cluster centers coordinates $\mathbf{c}_k$, $k = 1, 2, 3, \ldots K$, should be obtained and the index $f(\boldsymbol{y}_i)$ of corresponded cluster should be determined for each vector $\boldsymbol{y}_i$. The area of possible actual values of vector $\boldsymbol{y}_i$ can be described as $\boldsymbol{y}_i \pm \boldsymbol{\Delta y}_i$ and represents the polygon $\mathbf{P}_i$ in space $\mathbf{R}^r$. To estimate the error bound $\Delta_{R(K)}$ for the value $R(K)$, we should compute the interval $[R_{\min}, R_{\max}]$ of possible values of $R(K)$ using the following algorithm:

1) Assign a zero value to $R_{\min}$ and $R_{\max}$.

2) For each i from the set 1, 2, 3, … n, check if $\mathbf{c}_{f(\boldsymbol{y}_i)} \in \mathbf{P}_i$. If it is so, assign $\rho\left(\boldsymbol{y}_i, \mathbf{c}_{f(\boldsymbol{y}_i)}\right) = 0$ and go to the next value of index i. Otherwise, add to $R_{\min}$ the minimum value of measure ρ between $\mathbf{c}_{f(\boldsymbol{y}_i)}$ and the vertices of $\mathbf{P}_i$.

3) For each i from the set 1, 2, 3, … n, add to $R_{\max}$ the maximum value of measure ρ between $\mathbf{c}_{f(\boldsymbol{y}_i)}$ and the vertices of $\mathbf{P}_i$.

The obtained bounds $[R_{\min}, R_{\max}]$ allow easily computing the value of $\Delta_{R(K)}$:

So, using the presented algorithm or using other suitable technique of propagating the uncertainty through the computational procedure, we will automatically obtain the value $R(K)$ together with the upper bound $\Delta_{R(K)}$ for absolute value of its error inherited from initial data on each step of the clustering.

To automatically determine the maximum possible number of clusters $K_{\max}$, the following iterative procedure should be applied.

1) Assign a zero value to the number of clusters K and assign $R(0)=+\infty$.
2) Assign K the value $(K+1)$, perform the clustering procedure by chosen method and get the value of $R(K)$ with the limits $\Delta_{R(K)}$ of its possible absolute error inherited from the initial data.
3) If $|R(K)–R(K–1)|>\Delta_{R(K)}$, then go to step 2. Otherwise, the procedure should be finalized and $K_{\max} = K$.

Thus, the idea of the proposed approach is as follows. If the clarification of clustering results caused by increase in number of clusters turns out to be more significant than the results' error, inherited from inaccuracy of initial data, then the number of clusters should be increased. Otherwise, the clustering procedure should be stopped because of reaching the maximum number of clusters that can be recognized in such an inaccurate data. This idea is a particular case of a more general idea: if iterative procedure deals with inaccurate initial data, then it should be stopped earlier in accordance with data accuracy [16].

The obtaining the value $K_{\max}$ protects us from unreasonable conclusions on clusters structure. The dealing at any step of clustering procedure with uncertainty characteristics creates the opportunity for better applying clustering results for classification of further observations of the state of the studied object, process or phenomenon: we will get the error characteristics $\Delta\mathbf{c}_k$ for clusters center's coordinates $\mathbf{c}_k$ and use it to find out if there are reasons to include the new observation $\boldsymbol{x}_{n+1}$ to several clusters at once. The usual practice is to attribute the vector to the closest cluster, but in reality the distances to clusters' centers are distorted by errors inherited from initial data used during clustering and this brings us in some cases to situation when we aren't find out what cluster is the closest one. Taking into account the uncertainty of measurement results allows us to detect this situation and draw right conclusion.

The other great possibility that is the consequence from the taking into account the uncertainty of initial data is that the proposed approach produces the reasonable and objective way to compare different measures ρ (or clustering algorithms at whole) and to choose the most suitable. Really, the best measure ρ from the set $\{\rho_1, \rho_2, \rho_3...\}$ is one that for the same data to be processed produces the largest value $K_{\max}$: this means that this measure allows recognizing the maximum number of clusters. At the moment, the choosing of measure ρ is poorly formalized procedure and usually is based on some subjective grounds or is ill-founded. The proposed approach produces the objective reasons for choice. The presented method can be applied for any dataset that should be clustered and allows individual determining the measure of proximity what is the new word in clustering.

3. Using example

Let us examine some metrological situation as an example of using the proposed approach. Let the physical process $\boldsymbol{\Omega}$ be studied that takes place during electrochemical reaction in substance. Let the flow of this process can be described only with two electrical potentials x_1 and x_2 that can be measured directly. Different modes of process $\boldsymbol{\Omega}$ flow can be recognized varying the external conditions, which x_1 and x_2 depend on. The measurements of x_1 and x_2 are performed independently but using the same voltmeter. All obtained values form the vectors $\boldsymbol{x}_i^{\mathrm{T}} = (x_{i1}, x_{i2})$ that are processed in the following manner: the measurements results are scaled to fit the diapason [–1; 1] V. So, using the notation defined in previous section of the paper, the vectors $\boldsymbol{y}_i^{\mathrm{T}} = (y_{i1}, y_{i2})$ are calculated from preliminary data processing performed before clustering procedure applying to determine different states of process $\boldsymbol{\Omega}$. Let the quantity of measurements results be equal to $n = 40$ and, as before, index $i = 1, 2, \dots n$.

Let the described physical process $\boldsymbol{\Omega}$ tend to have $K_{\mathrm{true}} = 4$ separate modes of its flow. In figure 1a, the all obtained vectors $\boldsymbol{y}_i$ are shown. Different colors are used to mark data that are corresponded to different states of the process $\boldsymbol{\Omega}$ and consequently to different clusters. In practice, we almost always

Journal of Physics: Conference Series **1379** (2019) 012054 doi:10.1088/1742-6596/1379/1/012054

don't know the actual value K_{true}, so we should assign the maximum quantity of clusters K_{max} and its value potentially can be equal to any number from 1 to n. The concrete value depends on additional information and on some subjective factors – intuition and personal experience of the researcher. If we can't take them into account, the only way to produce the reasonable results of clustering is to use information on the accuracy of the data to be processed. The figure 1b contains the graphical representation of data and its potential absolute errors' bounds inherited from used measurement instruments.

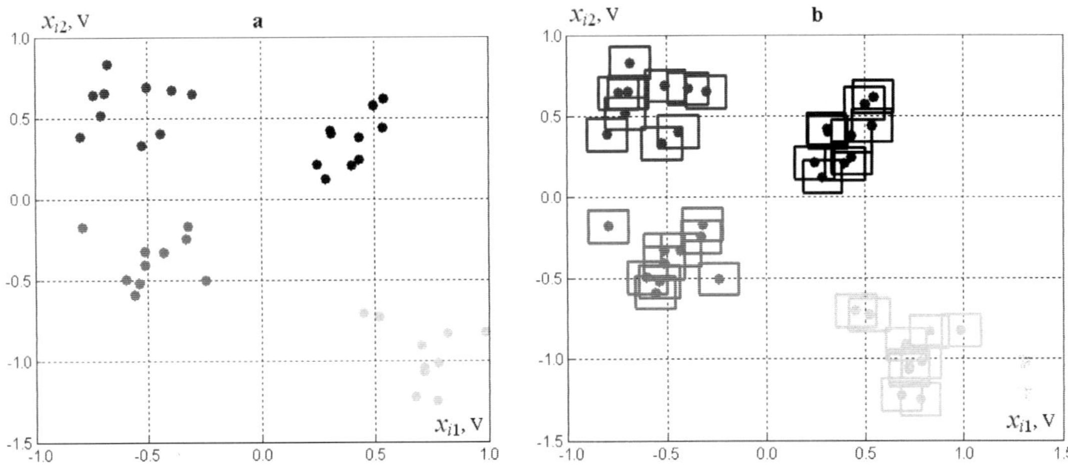

Figure 1. Measured initial data to be clustered:
a – without taking into account information on measurements errors
b – with bounds of possible values in accordance with data uncertainty characteristics.

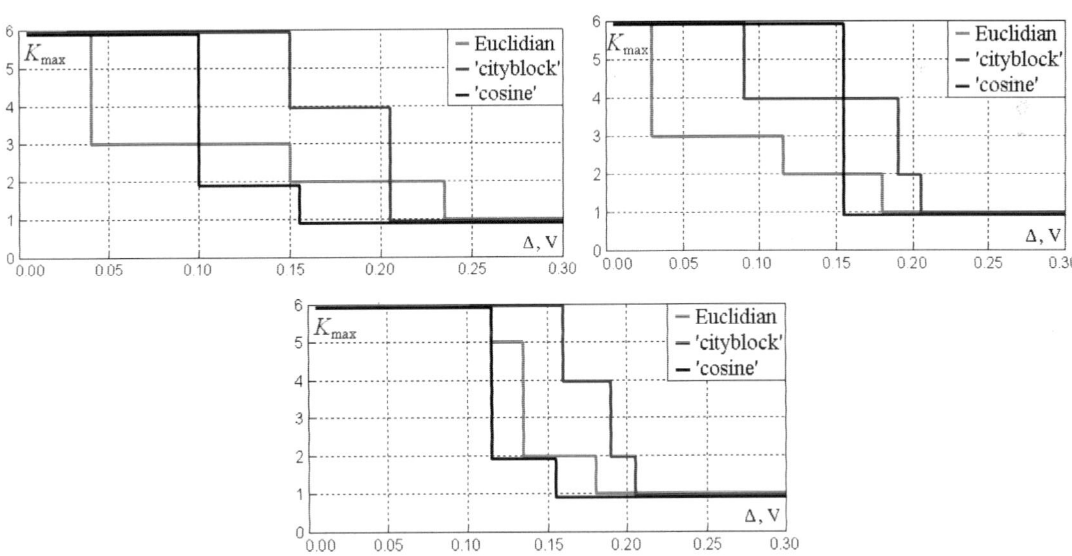

Figure 2. The dependence of maximum number of recognizable clusters K_{max} on used metric in clustering procedure obtained in three different launches of 'kmeans' method.

Let us apply the proposed approach to the data shown in Fig. 1 and take into account measurements uncertainties. As the clustering procedure, the 'kmeans' method was chosen because of its popularity in practice. Earlier we suppose that there is no additional information on the nature of the process $\mathbf{\Omega}$. So, we need to use the random guesses as the initial approximation of clusters centers' positions. In the performed modelling, we use the uniform distribution when covering the data to be clustered with such start points. In this situation, the results are to a certain extent random. For simplicity, the procedure was limited to the number of clusters – no more than 6 could be returned.

In the figure 2, three different results of applying proposed method are shown for different initial guesses for clusters centers depending on the maximum possible error of used voltmeter recalculated to the scale of y_{i1} and y_{i2} values (marked Δ in figure the 2). Three different metrics of proximity for clustered vectors were used: in the figure 2, the red line is corresponded to popular Euclidian metric (L_2), the blue line – to 'cityblock' metric (the sum of absolute values of components of the comparing vectors difference, L_1), the black line – to 'cosine' metric (the proximity degree is calculated as one minus the cosine of the included angle between comparing vectors).

We see that 'cityblock' metric (L_1) better corresponds to the data to be clustered: it allows recognizing larger quantity of clusters that can be potentially highlighted in the data. The metric 'cosine' produces the lowest sensitivity to find out clusters. The important feature of the proposed approach and this model application is that performed analysis is very simple, it doesn't require significant amount of computations and can produce results that are individual for each concrete practical situation.

4. Conclusions.

The presented paper is dedicated to the metrological aspects of clustering procedure and the consequences of outlined viewpoint. It is shown that taking into account the information on uncertainty of the data to be clustered allows easily making reasonable decisions that are metrologically supported: it becomes possible to determine the maximum number of recognizable clusters; the indicator arises that can be used for reasonable comparing the different metrics used in clustering methods or for comparing clustering algorithms at whole. The paper presents that the problems that are usually difficult to solve in clustering practice become simpler in this case.

References

[1] De Souza R M C R and De Carvalho F A T 2004 *Pattern Recogn. Lett.* **25** 353
[2] Asharaf S, Murty M N and Shevade S K 2006 *Pattern Recogn. Lett.* **27** 515
[3] Peng W, Li T 2006 *Proc. ICTAI'06 – 18th IEEE Int. Conf. on Tools with Artificial Intelligence* **11** 355
[4] Irpino A and Verde R 2008 *Pattern Recogn. Lett.* **29** 1648
[5] De Carvalho F A T and Tenório C P 2010 *Fuzzy Set. Syst.* **161** 2978
[6] D'Urso P and Leski J M 2016 *Pattern Recogn.* **58** 49
[7] Ji Z, Xia Y, Sun Q and Cao G 2014 *Fuzzy Set.Syst.* **253** 138
[8] Hwang C and Rhee F C H 2007 *IEEE Trans. on Fuzzy Systems* **15** 107
[9] Le Hegarat-Mascle S, Bloch I and Vidal-Madjar D 1997 *IEEE T. Geosci. Remote Sensing* **35** 1018
[10] Denoeux T A 1995 *IEEE T.Syst. Man Cyb.* **25** 804
[11] Schubert J 2000 *Information, uncertainty and fusion* 419
[12] Benderskaya E N 2017 *Proc. 2017 XX IEEE Int. Conf. on Soft Computing and Measurements (SCM)* 162
[13] Parschutin S V 2007 *Integrated models and soft computing in artificial intelligence: collection of scientific papers [Integrirovannye modeli i myagkiye vychisleniya v iskusstvennom intellekte: sbornik nauchnykh trudov]* 465 (in Russian).
[14] Yakovenko A A and Malykhina G F 2018 Procedia Comput. Sci. **123** 547
[15] Semenov K K and Solopchenko G N 2010 Meas. Tech. **53** 592
[16] Semenov K K 2015 *Advanced Mathematical and Computational Tools in Metrology and Testing X (AMCTM X)* (F Pavese, W Bremser, A G Chunovkina, N Fischer, A B Forbes eds) **10** *Series on Advances in Mathematics for Applied Sciences* **86** (Singapore: World Scientific) 320-9

Journal of Physics: Conference Series **1379** (2019) 012055 doi:10.1088/1742-6596/1379/1/012055

Music and growl of a lion: anything in common? Measurement model optimized with the help of artificial intelligence will answer

K Sapozhnikova[1], S Hussein[1,2], R Taymanov[1], Iu Baksheyeva[3]

[1] D.I. Mendeleyev Institute for Metrology, 19 Moskovsky pr., St.Petersburg, 190005, Russia
[2] Computer Engineering and Systems Department, Mansoura University, 60 Elgomhoria Street, City, 35516, Egypt
[3] St. Petersburg State University of Aerospace Instrumentation, 67 Bolshaya Morskaya str., St. Petersburg, 190000, Russia

E-mail: k.v.s@vniim.ru

Abstract. Adaptive possibilities of population, to a significant extent, are conditioned by the capability to transform and receive information. Music has overcome information borders of bioacoustic signals but it represents their organic development. At present, automatic interpretation of information hidden in emotionally coloured bioacoustic signals of non-human animals not to mention music, has not been realized. The paper deals with topical tasks, the solution of which relies upon the opportunity to decode the flows of nonverbal information of such a kind. The prospects of developments in this field based on new measurement methods using artificial intelligence, are shown.

1. Introduction

N. Wiener and S. Lem showed that the development of technical systems and the evolution of biological systems lead to the appearance of many similar structural and functional solutions. However, these two types of systems differ significantly in system complexity and durability of the constituent processes that take place. Therefore, the study of the emergence and improvement of the methods and means of obtaining, transmitting and processing information, which were used in the course of biological evolution, may be of interest to professionals who deal with measurement technology.

This approach to solve the problem of increasing the reliability of measurement results has proved its efficiency at VNIIM in developing methods for organizing metrological self-check of various measuring instruments and measurement systems [1].

Application of such an approach to emotions, in particular, transmitted by acoustic signals, opens up new perspectives in measuring technique.

2. The evolution of communication between living creatures

In the course of evolution, nonverbal acoustic signals have originated long before the appearance of humans. These signals accompanied the estimation reactions of living creatures to the situations, which were important for population vital activities, and influenced the behaviour of other members of the population. As evolution was progressing, the number of different signals increased.

Life has originated in the ocean. When marine animals needed to transmit information about the occurrence of danger, an unexpected situation with unknown consequences, or the necessity to be ready for active actions, infrasound signals proved to be efficient. The need for estimation reactions to changing situations and transmission information that is important for the population to its other members has been preserved at the next stage of the evolution when living creatures left water to occupy dry land.

Terrestrial animals preserved the main neurophysiological reactions that determine the emotions of marine animals. Later, these reactions characterized by the infrasound frequency range developed. If one considers mammals, this range corresponds to delta- and theta- brain biorhythm frequencies.

However, the density of air is many times lower than that of water. This feature has not allowed animals to apply infrasound frequencies for transmitting information by oscillations of the medium around animals on land.

To transmit signals that were vital for population development, in the course of evolution, to create a corresponding mechanism has become necessary. Amphibians, reptiles, and mammals (much later) gained a respiratory organ with the function of reproducing sound signals, modulation of these signals by neurophysiologic reactions within the infrasound range being realized.

To receive these acoustic signals (hereinafter, they are called emotiogenic ones), it was necessary to have a special organ of hearing and realize a structure that demodulates the received signals by the part of a nervous system connected with this organ. These new features enabled animals to select infrasound frequency oscillations within the biorhythm range and stimulate the neurophysiologic reactions that are adequate to them.

Hereinafter, these neuro-physiological reactions are considered as emotions, regardless of whether they are formed, e.g., in crabs (as proto-emotions) or humans.

The beginnings of social relations inside animals' population and the formation of their packs, caused the need to enrich transmitted information to coordinate their actions. This need resulted in the evolution of the sound apparatus aimed at improving the controllability of sound reproduction.

The maximum distance that enables an animal to hear the signal, depends mainly on its loudness. If an acoustic power is constant, the loudness grows with the increase of sound frequency. To transmit information to other animals being at different distances, the animals' ability to change the frequency and power of audio signals became necessary.

The use of sound oscillations of a relatively high frequency as a carrier frequency contributes to the growth of the sound loudness to the level close to its limit, which facilitates the transmission of a bioacoustic signal of danger to a greater number of the population members. It is typical that the high frequency sounds are typical for small animals.

If a predator transmits bioacoustic signals using a relatively low-frequency carrier but the carrier is modulated by the signal with the same biorhythm frequency, this can be perceived as a threat ("anger"). As the simultaneous appearance of other predators is undesirable, the loudness of such a signal is wittingly limited.

At the lowest level of controllability of the sound apparatus, there were poliharmonic bioacoustic signals reflecting only the simplest basic emotions. At the next level, there appeared a possibility of initiating or slowing down such signals as well as using their combinations (the simplest emotional images). The highest level is characterized by the ability to control the acoustic signal parameters, which opened the way to onomatopoeia (sound imitation), and later to intonated speech.

The appearance of a human ability to transmit and receive verbal information led to the increase in the meaningfulness of communication, including the use of logical reasoning, but did not suppress the ability to transmit and perceive emotional information. Considering the special value of imaginative perception and associative thinking for adaptation to changing conditions, the possibilities of receiving and transmitting emotional information have been enriched by elements of musical and visual arts.

Journal of Physics: Conference Series **1379** (2019) 012055 doi:10.1088/1742-6596/1379/1/012055

As for music coming to life, its development can be traced since the time of primitive man. The enrichment of verbal communication with variations of pitch and rhythm, using harmony and group singing, contributed to the success of human evolution [2].

Following a similar trend, at comparatively recent stages of human society development, a relative increase in the number of connections between members of certain groups of society and, accordingly, growth in the diversity of information that they exchange, also have resulted in the progress of these groups [3].

3. Opportunities for measuring emotions

The simplest emotions and simplest emotional images in total are encompassed in scientific literature by the term "basic emotions". Basic emotions in animals can be quite clearly identified by the situations that contribute to their occurrence. At present, in the literature, there is no generally accepted conception of basic emotions, their names as well as the number, some authors often including feelings (comparatively long-term reactions) in the list of emotions (short-term reactions).

To study the relationship of the spectrum of bioacoustic signals that transmit emotions, and the neuro-physiological reactions giving rise to emotions when these signals are perceived, a concept of an appropriate mechanism was developed. Then, based on it, a model was formed for measuring the expected emotions arising from certain acoustic impacts [4], which was supplemented and developed later [5, 6]. The first step of the measurement model consisting of 3 steps, is shown in figure 1.

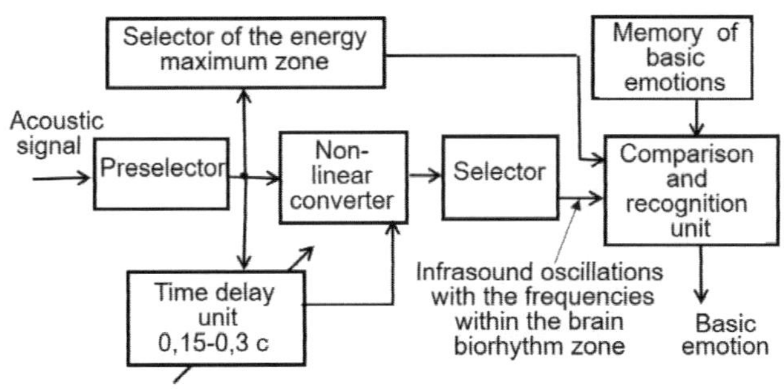

Figure 1. The first step of the measurement model [6].

To test the effectiveness of the measurement model, records of bioacoustic signals of animals are expedient to apply. The advantage of such a choice is that the emotions reflected in the signals of animals, in contrast to the emotions in the sound of a human's voice, are not under non-human animals' voluntary control, i.e., they are "honest" ones [7, 8]. In zoologists' investigations, records of emotional bioacoustic signals are often accompanied by characterization of the situation that caused them.

In [9], there are given the results of the first experiments with recognizing emotions in the records of bioacoustic signals of 14 animals, which we based on the measurement model [6]. Artificial intelligence was used as a recognition means. Such emotions as "fear" and "happiness" were identified with 100 % confidence, while for other emotions, the level of confidence lied within (67-87) %.

These data have confirmed the link of certain modulating infrasound oscillations with emotions being transmitted by bioacoustic signals, but raised the issues of both a basic emotion definition and reason for a noticeable share of unspecified emotions.

The analysis enabled the assumption that mainly, the revealed difficulties were due to variations of basic emotions during measurement time.

To test this assumption, bioacoustic signals of mammals (growl, hiss, yelp, purr, etc.) used in [9] after nonlinear conversion and selection of infrasound oscillations were processed with the help of continuous wavelet transform (CWT). An example of visualization of the processing results is given in figure 2. The analysis has shown that for the considered cases, the first revealed components with the frequencies in the zones of delta-, theta- and alpha- brain biorhythms, distinguished by the maximum values, as a rule, have the duration up to 0.3 s. Sometimes, it is only one maximum, and sometimes, in some cases, are two or three (in different zones).

When CWT was applied to relatively long records of acoustic signals (longer than 1.5 c) after the first maximum, previous maximums could disappear, and new ones could arise. It shows that emotional colour of mammals' bioacoustic signals can change noticeably even during a few seconds. It seems that this fact can explain the main reason of vagueness in terms of emotion recognition according to the procedure applied previously, which relied upon obtaining integral evaluations for a few second intervals [9]. Additional CWT calculations of signals after nonlinear conversion, using time delays 0.1 s, 0.2 s, and 0.3 s have confirmed that in all the cases, the basic emotion is identified most clearly for the time delay 0.2 s with the uncertainty 0.02 s.

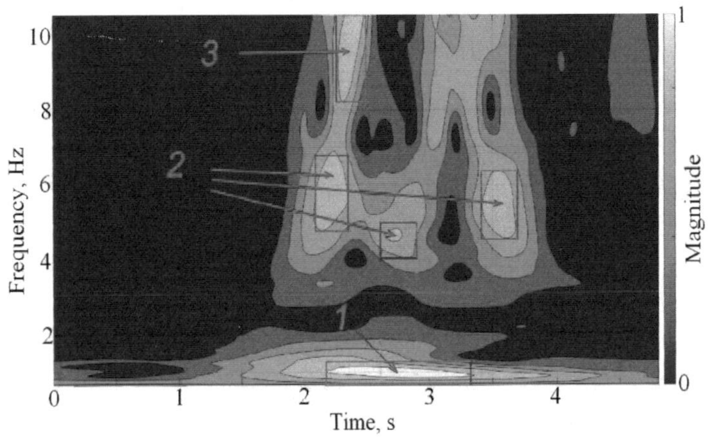

Figure. 2. Lion growl sounds. The result of wavelet transform after nonlinear conversion (2.8 s duration; 0.2 s time delay). The colour coding corresponds to relative units.
1: maximum is in the delta-biorhythm zone;
2: maximum is in the theta-biorhythm zone;
3: maximum is in the alfa-biorhythm zone.

The concept of emotion origination mechanism enabled considering the task of identification of basic and more complicated emotions using quantitative estimates. Let us suppose that for living creatures that breathe air, the signals corresponding to basic emotions include only signals modulated by oscillations with frequencies within delta- and theta- rhythm zones and their combinations. Let us also assume that if the level of modulating oscillations in the zone of any of two biorhythm types is significantly greater than both the level of noise and level of the signal in the zone of the second biorhythm type, the value of the first level is assigned "1" and that of the second one is assigned "0". If these levels are close, depending on the signal-to-noise ratio, it is possible to attribute them "11" or "00". Accordingly, provided the carrier frequency of bioacoustic signal $f > 500$ Hz, "1" can be assigned to it, while provided $f \leq 500$ Hz, "0" can be assigned. Then, for each emotiogenic bioacoustic signal, decoding of its meaning can be proposed (see table 1). The total number of basic emotions in table 1 is limited to 7. (It is natural that marine animals breathing by means of gills are characterized by no more than 3 emotions). The table has been formed taking into consideration the correlation between activated biorhythms and the animal's state.

Basic emotions causing the signals included in table 1 do not have generally accepted names. For higher non-human animals, including marine ones, and especially for humans, emotions corresponding to the signals from the list given in table 1, can have various nuances taking into consideration a greater variety of levels of signals with the frequencies of delta- and theta- rhythms. The nuances also depend on the levels of oscillations with the alfa- and beta-rhythm frequencies. The former characterizes visual sensibility and muscular activity, while the latter reflects brain stress related to a task being solved.
The meaning of signals based on the above-mentioned factors can correspond to different emotions. For example, signal "100" characterizes "anger", "threat", or " disgust", while signal "010" reflects emotion "interest" or "surprise", signal "110" concerns with various manifestations of tenderness while signal "111" testifies sadness or happiness.

Subsequent analysis of bioacoustic signals has shown that in records, to detect structured fragments with the durability less than 1 s is possible. In many cases, they are repeatable or have minimal variations. The sequence of such fragments seems to cause a complicated listener's emotion. Taking into consideration the previous experiments [9], 81 typical fragments were chosen from the records of animals' voices. Further, they were processed according to the measurement model [6].

Binary codes of signals (ratios) were experimentally checked with the help of these records. Not all 7 types of signals (see table 1) were found in them. Nonetheless, the concept of basic emotions taking into consideration the decoding given becomes more certain being characterized by numbers. Later it can be refined.

Table 1. Decoding of the meaning of emotiogenic bioacoustic signals.

Delta	Theta	Frequency	Signal meaning
1	0	0	Discontent, e.g., at meal or meeting with a specific animal
0	1	0	Warning! Here we have something important (to a partner that is nearby)
1	1	0	Pleasure of comfort (to a partner that is nearby)
0	0	1	No grounds for anxiety (to a community).
1	0		Threat for a community has originated. Call for help.
			Warning! Something important takes place (to a community
			Call for interaction (to a distant partner)

4. Prospects for the development of instruments aimed at measuring (recognizing) emotions

Not only does cognitive interest cause the studies of the evolution of emotions and methods providing their transmission. If we limit ourselves to acoustic signal features, the following fields of research and development can be mentioned:

a) investigation of animals' language (In particular, it is necessary to decode their communication, which would enable monitoring of ecological situation in certain places);

b) automatic interpretation of speech from any language to another one considering emotional meaning of intonation used while pronouncing words and sentences (Such a demand was repeatedly mentioned by philologists. One of the purposes is to facilitate studies of foreign languages);

c) development of the scientific foundations of the theory of music as emotion language based on numerical assessments (The prospects of this work are associated with the development of methods for enhancing the emotional impact of various musical forms typical for various national and social cultures);

d) development of the theory, methods, and means focused on the designing an optimal sound scale (Basing on this work it is planned to improve musical instruments, primarily, synthesizers and other keyboard instruments;

e) development of methods for designing technical instruments that will enable helping a composer and performer in advance to take into consideration the influence of sociocultural features of supposed listeners on their probable reaction (This field was formulated by professional musicians as one of the actual);

f) diagnostics of characteristics of the brain and psyche based on voice sounding features [10] (Diagnostics of infants' brain development in the first months of life is a subject of particular interest to the authors of the present paper).

In the listed fields, the authors of the present paper are carrying out certain investigations together with specialists from other organizations. To provide progress in most of these fields, automatic recognition of emotion information contained in emotionally coloured soundings, is necessary.

The aforementioned model used for measuring expected emotions caused by acoustic signals [6], can be the basis for recognition of such information.

At the first stage of the development of such an automatic device, recognizing basic emotions with a rather high degree of confidence should be provided. To achieve this result, it will be necessary to define the limits of assumed "1" and "0" in each bit, receive more information on the influence of spectrum components within alfa- and beta- biorhythm frequencies as well as refine the border being used for dividing carrier frequency into two ranges.

Success at the first stage will provide a possibility to proceed to the study of human complex emotions and feelings as well as to that of the connection of emotional perception with national culture and culture of certain social strata of modern society.

For such studies, the sociological surveys will be required being carried out within the framework of restrictions [11, 12]. However, the research should be supplemented by an analysis of the correlation of survey results with records of sequences of expected basic emotions, obtained based on the aforementioned measurement model. This is a qualitatively new approach.

The solution to the applied tasks that concern measuring emotions, e.g., speech interpretation, usually requires automatic recognizing emotiogenic signals within tenths of a second. In some cases, simultaneous automatic analysis of the sequence of these signals, is necessary to allow recognizing a feeling being formed. It should be added that for a number of applications, while identifying emotions, emotional nuances should be considered, but they are determined by ratios that at present, are insufficiently known. Therefore, one of the most important tasks is recognizing emotions with the help of artificial intelligence [6, 9].

However, a continuation of studies of artificial intelligence possibilities [9] carried out in this field, have shown that it is critical to supplement emotion definitions. Using Radial Basis Function Neural Networks (RBFNNs), accompanied by k-means clustering method, proved insufficiently effective. From 35 fragments of bioacoustic signals chosen by an expert (each of the fragments represented one of 7 basic emotions: anger, disgust, fear, happiness, interest, sadness, and tenderness) only 29 were identified by RBFNN. After excluding wrong signals using k-means clustering, 28 fragments were identified but with 100 % confidence. Thus, for original bioacoustic signals, improving the method has resulted in the increase in the identification confidence to 80 % only, i.e., 20 % of signals were taken away. It is specific that only "fear" was identified in all the cases, while recognition of "interest" was the worst. To investigate complicated emotions and processes of feeling formation, to take away a part of signals is undesirable.

In our view, a future prospect could be related to clustering associated with determination of codes for any cluster and evaluation of the limits of assumed "1" and "0" in each bit for signal levels. Corresponding clustering seems to be a multistep process taking into consideration the necessity to preserve the binary codes and duration of bioacoustic signal fragments no longer that 1 c.

Certainly, for some clusters there will be no analogy in the available set of bioacoustic signals. However, if the number of binary bits influencing the emotional colour is known and permissible limits of variation of infrasound oscillation levels are established, an opportunity will open to form a full set of emotions based on binary codes.

5. Conclusions

The level reached by population in the course of evolution can be characterized by the level of methods and means for transmitting and receiving information inside the population, which enable it to adapt to the changing environment. Successful spreading of mammals on the Earth, to a significant extent, is due to the development of their vocal organs, qualitative leap in the development of primitive human being linked with origination of singing and musical instruments. During many centuries, the process of human society civilization was accompanied by the improvement of music theory and musical instruments, development of methods for verbal communication (speech and written language) as well as information transmission instruments.

At present, the devices that transmit and receive information are practically available to anyone. Automatic speech interpretation and written text translation from any language and to any other language is also possible.

However, nowadays, the requirement has originated for automatic interpretation of the emotional content of non-verbal information hidden in acoustic signals including music. Opportunities for such interpretations are provided by the modern level of measuring instruments and computer technologies.

References

[1] Taymanov R and Sapozhnikova K 2018 *Smart Sensors and MEMS for Industrial Applications* (2ed) ed S Nihtianov and A L Estepa (Woodhead Publishing, Elsevier Limited) chapter 1 pp 1-22
[2] Mithen S 2009 *The Neurosciences and Music III — Disorders and Plasticity: Ann. N.Y. Acad. Sci.* **1169** pp 3–12
[3] Taymanov R and Sapozhnikova K 2016 *J. Phys.: Conf. Ser.* **772** 012040 6 p
[4] Taymanov R and Sapozhnikova K 2010 *Meas. Sc. Rev.* **10** 3 pp 78-88
[5] Sapozhnikova K, Taymanov R, Baksheeva Iu, Kostromina S, Gnedykh D and Danilova I 2017 *18th Int. Congr. of Metrology, EDP sciences, Web of Conf.* (Paris, France, September 19º- 21) ed C Corleto 10006
[6] Sapozhnikova K, Hussein S, Baksheeva Iu and Taymanov R 2018 *J. Phys.: Conf. Ser.* **1065** 072041
[7] Juslin P N and Laukka P 2003 *Psychol. Bull.* **129** 5 pp 770–814
[8] Briefer E F 2012 *J. Zool.* pp 1-20
[9] Sapozhnikova K, Hussein S, Taymanov R and Baksheeva Iu 2018 *Proc. 28th Int. Sc. Symp. Metrology and Metrology Assurance 2018 (September 10-14 2018, Sozopol, Bulgaria)* pp 297-302
[10] Baksheeva Iu, Sapozhnikova K, Taymanov R and Danilova I Proc. *27th Int. Sc. Symp. Metrology and Metrology Assurance 2017 (Sept. 8-12, 2017, Sozopol, Bulgaria)* pp 292-6
[11] Mari L and Wilson M 2014 *Measurement* **51** pp 315-27
[12] Pendrill L and Fisher W P Jr 2015 *Measurement* **71** pp 46-55

Journal of Physics: Conference Series **1379** (2019) 012056 doi:10.1088/1742-6596/1379/1/012056

Radiometric extension of a measurement arrangement in accordance with the EMVA 1288 standard for camera characterization in UV to NIR wavelength range

David Rebhan[1], Paul-Gerald Dittrich [1,2*], Maik Rosenberger[1], Gunther Notni[1]

[1] Technische Universität Ilmenau,
Faculty of Mechanical Engineering,
Group for Quality Assurance and Industrial Image Processing,
Gustav-Kirchhoff-Platz 2, 98693 Ilmenau, Germany
[2] SpectroNet c/o Technologie- und Innovationspark Jena GmbH,
Hans-Knöll-Straße 6, 07745 Jena, Germany

E-mail: paul-gerald.dittrich@tu-ilmenau.de

Abstract. For the objective comparison of different cameras or image sensors, it is necessary to describe their image acquisition channel by suitable parameters. The EMVA 1288 standard, developed by camera manufacturers and research institutes, distinguishes itself from other standards by considering the camera as a linear model. The camera is treated as a black box of which only pixel size and exposure time must be known. The recording of standardized test images is also omitted, allowing the camera to be described without optics. The only input variable of the linear camera model of the EMVA 1288 standard is the number of photons that hit a pixel of the image sensor during the exposure time. Therefore, the correct determination of the photon count is essential to calculate important camera parameters from the linear camera model, such as quantum efficiency or signal-to-noise ratio. To determine the number of photons, the irradiance of the radiation incident on the image sensor must be measured. This is usually done using a radiometer instead of the camera or image sensor. The number of photons per pixel during the exposure time can then be calculated from the irradiance, considering constants like the wavelength of the incident radiation, the area of the pixel and the exposure time of the camera. Calibrated radiometers were procured which are sensitive in the UV-A, VISNIR and NIR wavelength ranges in order to be able to perform camera or image sensor characterizations according to the EMVA 1288 standard in these wavelength ranges in the future.

1. Introduction

The EMVA 1288 standard was published in 2005. The reason for the publication of this standard by the European Machine Vision Association (EMVA) was, among others, the difficult comparability of the data sheets of the camera manufacturers. Often different camera parameters or different units were used. This made it difficult to select the right camera for a particular application. The EMVA 1288 standard defines an accurate and reliable measurement procedure and guidelines for the presentation of the measurement results. This standard makes it much easier to compare cameras and image sensors from

[*] To whom any correspondence should be addressed.

different manufacturers. The EMVA 1288 standard was developed by a consortium of companies and research institutes in Europe [1]. The current version of the standard is 3.1. The calculations of the EMVA 1288 standard are based on a linear camera model (figure 1) [2].

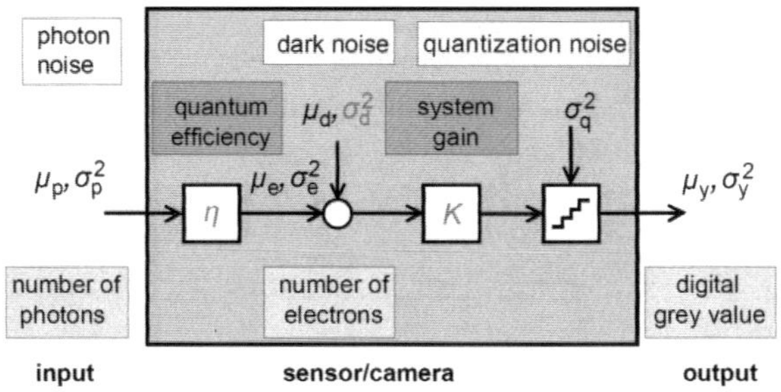

Figure 1. Schematic representation of the linear camera model of the EMVA 1288 standard [2]. During the exposure time an average number of photons μ_p hits on the surface of a pixel. The input value μ_p is converted into the output value of the digital grey value μ_y in the camera model.

The linear camera model contains only three parameters, the quantum efficiency η, the dark noise σ_d and the camera gain K. When photons hit the pixel, only a certain part of them is absorbed and generates charge carriers μ_e. The proportion of generated charge carriers μ_e to incident photons μ_p is referred to as quantum efficiency η (1).

$$\eta(\lambda) = \frac{\mu_e}{\mu_p} \tag{1}$$

The quantum efficiency η refers to the entire area of a pixel, not only to the photosensitive part. The quantum efficiency is dependent of the wavelength λ of the incident light [2]. The average number of photons μ_p that hit one pixel of the surface A during the exposure time t_{exp} can be calculated using the irradiance E on the sensor surface (2).

$$\mu_p = \frac{AE_{t_{exp}}}{hc_0/\lambda} \tag{2}$$

Using Planck's constant $h = 6{,}6260755 \cdot 10^{-34}[Js]$ and the speed of light $c_0 = 2{,}99792458 \cdot 10^8[m/s]$, the average number of photons μ_p can be simplified calculated (3).

$$\mu_p[photons] = 50{,}34 \cdot A[\mu m^2] \cdot t_{exp}[ms] \cdot \lambda[\mu m] \cdot E\left[\frac{\mu W}{cm^2}\right] \tag{3}$$

The digital gray value μ_y results from the sum of the photon-induced charge carriers μ_e and the dark signal μ_d integrated over the exposure time t_{exp}, multiplied by the absolute gain factor K (4).

$$\mu_y = K(\mu_d + \mu_e) = \mu_{y.dark} + K\mu_e \tag{4}$$

The absolute gain factor K has the dimensionless unit number of electrons per digital value $[e^-/DN]$. However, the relationship between equation (1) and (4) cannot be determined directly, since the number of charge carriers μ_e is unknown. Only the relationship between the photon number per pixel μ_p and the image signal μ_y can be determined by inserting equation (1) into equation (4) [2] (5).

$$\mu_y = \mu_{y.dark} + K\eta(\lambda)\mu_p \tag{5}$$

This is the linear camera characteristic whose axis section forms the dark signal $\mu_{y.dark}$ and whose gradient is the product of the absolute amplification factor K and the quantum efficiency η. The gradient of the camera curve is also referred to as sensitivity R. However, the unknown camera parameters K, η and σ_d, marked red in figure 1, can only be determined unambiguously from the camera characteristic after an analysis of the noise. The accumulated electrons are assumed to be Poisson distributed thus their variance is equal to the mean value (6).

$$\sigma_e^2 = \mu_e \tag{6}$$

The same applies to the variance of the photon noise σ_p^2. Due to the linear camera characteristic, all noise sources resulting from the transport or amplification of the charge carriers can be combined into a single noise source, the dark noise σ_d^2 [2]. Another noise source that must be included in the consideration is the so-called quantization noise, due to the final analog-digital conversion. Their variance can be described as follows: $\sigma_q^2 = 1/12\ DN^2$ [2]. Thus, including equations (4) and (5), total noise σ_y^2 is obtained (7).

$$\sigma_y^2 = K^2\sigma_d^2 + \sigma_q^2 + K(\mu_y - \mu_{y.dark}) \tag{7}$$

With known K, the quantum efficiency η can be calculated from the gradient of the camera curve R from equation (8). This method is called photon transfer method (figure 2) [2].

$$\eta = \frac{R}{K} \tag{8}$$

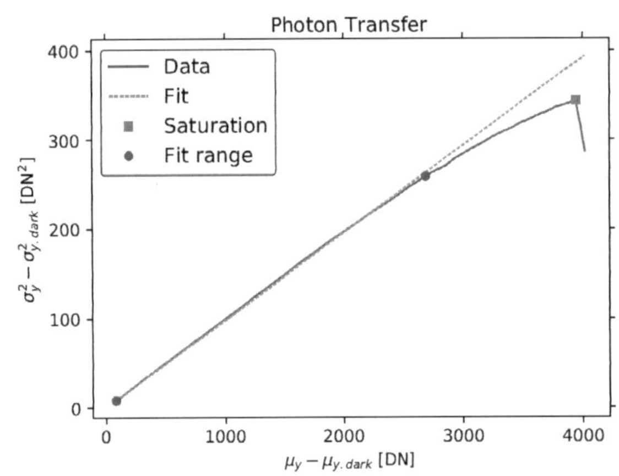

Figure 2. Exemplary photon transfer curve [2]. The absolute gain factor K can be read directly from the slope of the photon transfer curve. The variance of the dark noise σ_d^2 results from the axial section of the photon transfer curve after deduction of the quantization noise.

2. Adding radiometer to the EMVA 1288 test system

In accordance to the manufacturer Gigahertz Optik, the radiometers to be connected are pn photodiodes. However, the radiometers differ in the semiconductor materials of their photodiodes. For example, GaP (Gallium Phosphide) is used as the semiconductor material for the UV radiometer. In the NIR radiometer the photodiode is based on the semiconductor material InGaAs (Indium Gallium Arsenide), the VISNIR radiometer is based on conventional Si (Silicon) (figure 3).

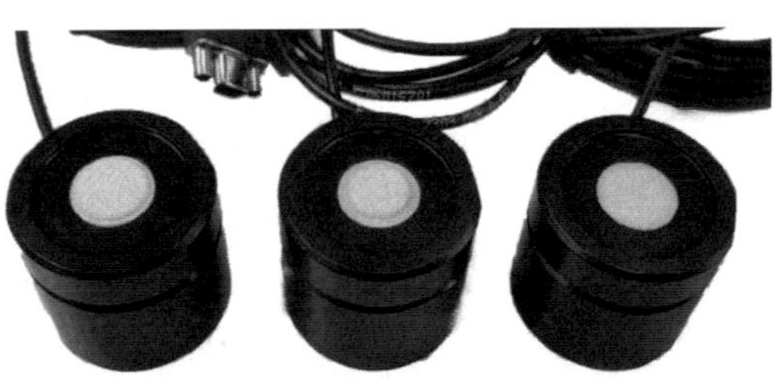

Figure 3. UV, VISNIR and NIR radiometers by Gigahertz-Optik. Each radiometer is equipped with a diffuser in order to minimize the directional dependence of the irradiance by generating a diffuse transmission [3].

Journal of Physics: Conference Series **1379** (2019) 012056 doi:10.1088/1742-6596/1379/1/012056

All radiometers have the same geometric proportions, which simplifies mechanical coupling to the EMVA 1288 test system. By combining color glass filters with the optimized diffuser, the sensitivity curve of each sensor is approximated to a rectangle function of the respective spectral range (figure 4) [4].

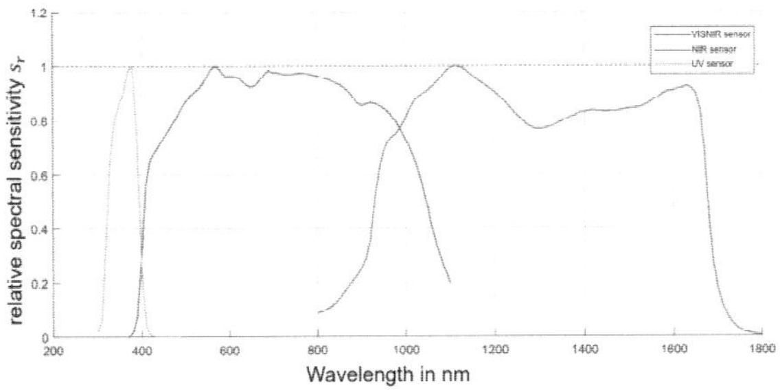

Figure 4. Relative spectral sensitivity of the three radiometers. The adaptation of the relative spectral sensitivity to a rectangular function of the VISNIR and NIR measuring heads is clearly visible.

The measuring heads were set by the manufacturer to the integral irradiance of the respective rectangle function calibrated. This calibration factor is already set in the so called optometer module for each radiometer. However, this integral calibration factor cannot be used for (nearly) monochromatic measurements, because for a monochromatic measurement at least one irradiance value or calibration factor on the relative spectral sensitivity function must be available. In the following, this irradiance value is also referred to as the grid point. Therefore, the correct calibration factors for monochromatic measurements were determined on the basis of further calibration data from the manufacturer.

3. Example procedure for absolute calibration of radiometer for VISNIR

The relative spectral sensitivity s_r of the VISNIR radiometer was determined by the manufacturer by comparison with a reference standard BN-9102-118 IET 1008, whose spectral sensitivity is attributed to the national standards of the Physikalisch-Technische Bundesanstalt (PTB).

Irradiation was performed with the radiation of a halogen incandescent lamp [4] diffracted by a grating double monochromator. The relative spectral response can be shown (figure 5).

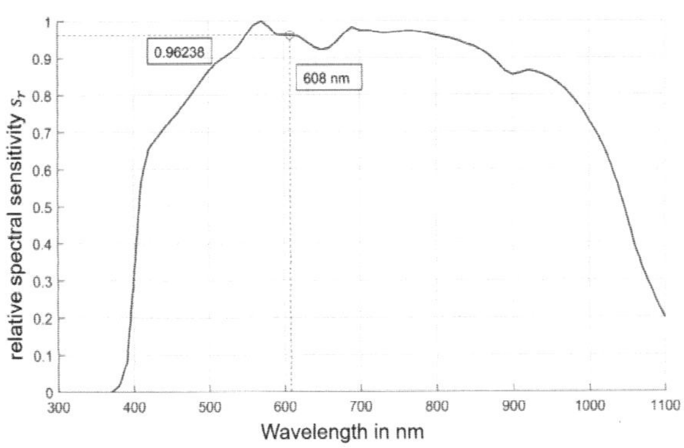

Figure 5. Grid point and relative spectral sensitivity of VISNIR radiometer. In addition to the relative spectral sensitivity function, the manufacturer measured a grid point. The irradiance was determined by a 1000 W lamp with a 608 nm filter IET 1899, FWHM 11 nm and a reference receiver whose spectral irradiance is attributed to the national standards of the PTB [5].

The manufacturer has determined relative measurement uncertainty of the relative spectral sensitivity measurement (table 1).

Table 1. Relative uncertainty of measurement of the relative spectral sensitivity calibration of the VISNIR radiometer [4].

Wavelength range in nm	Uncertainty of measurement in %
380 – 900	± 4
910 – 1070	± 4,5
1080 – 1100	± 5

In order to determine the correct calibration factors k_λ, the relative spectral sensitivities specified by the manufacturer in 10 nm steps were first interpolated linearly in 1 nm steps. This results in a relative spectral sensitivity s_r 0.96238 at the 608 nm grid point (see figure 5). Since the relative spectral sensitivity values are normalized to 1, the sensitivity function is first adapted so that it assumes the value 1 at the grid point (9).

$$s_k(\lambda) = \frac{1}{s_r(608)} \cdot s_r(\lambda) = \frac{1}{0{,}96238} \cdot s_r(\lambda) \tag{9}$$

The calibration factor k at the grid point is known and results from the photocurrent I of the radiometer and the irradiance determined by the reference receiver E (10).

$$k_{608}\left[\frac{A}{W/m^2}\right] = \frac{I[A]}{E[W/m^2]} \tag{10}$$

Thus the irradiance E can be determined depending on the sensitivity $s_k(\lambda)$ (11).

$$E = \frac{I}{k_{608}} \cdot \frac{1}{s_k(\lambda)} \tag{11}$$

The constant k_{608} can be combined with the sensitivity $s_k(\lambda)$ to a calibration factor, here referred to as spectral irradiance sensitivity k_λ (12).

$$k_\lambda = \frac{k_{608}}{\frac{1}{s_k(\lambda)}} = \frac{k_{608}}{\frac{0{,}96238}{s_r(\lambda)}} \tag{12}$$

The spectral irradiance sensitivity k_λ was calculated over the entire sensitivity range of the VISNIR radiometer in steps of 1 nm (figure 6).

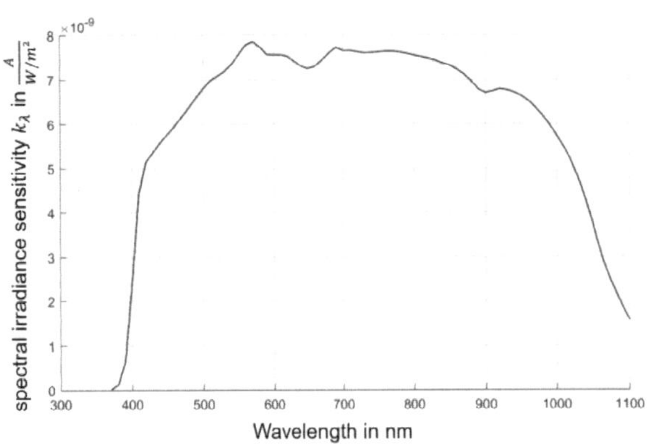

Figure 6. Determined spectral irradiance sensitivity of the VISNIR radiometer. The measurement program determines the spectral irradiance sensitivity k_λ from the given wavelength and divides the selected photocurrent I by k_λ. Thus the correct irradiance E is output for monochromatic measurements.

The course of k_λ reflects the course of the sensitivity function (see figure 6). The irradiance E can simplified be calculated (13).

$$E\left[\frac{W}{m^2}\right] = \frac{I[A]}{k_\lambda\left[\frac{A}{W/m^2}\right]} \tag{13}$$

By determining the absolute values of the spectral irradiance sensitivity over a measured sampling point, the relative measurement uncertainties of both measurements performed by the manufacturer add up. According to the manufacturer, the relative measurement uncertainty of the measurement of the sampling point is ± 5 % [5]. The relative measurement uncertainties of the relative spectral sensitivity calibration depend on the observed wavelength and ranges between ± 4 % and ± 5 % (see table 1). The calibration of the UV and NIR radiometers was carried out analogously to the calibration of the VISNIR radiometer.

4. Results and summary

This paper deals with the radiometric extension of a measurement arrangement in accordance with the EMVA 1288 standard for camera and or image sensor characterization in UV, VISNIR and NIR wavelength ranges. The measurement of the irradiance is of great importance here, in order to calculate the number of photons that hit the image sensor during the exposure time. The irradiance was previously measured using a non-calibrated photodiode from Texas Instruments whose spectral sensitivity is limited to the VISNIR range. In order to extend the measurable radiation range, three radiometers of the company Gigahertz-Optik where added to the measurement arrangement.

In this paper, a method was developed with which the spectral irradiance sensitivity of each radiometer can be determined. The absolute sensitivity values were calculated based on the relative spectral calibration and a grid point.

In summary, it can be said that the new radiometric measuring system is now usable in our EMVA 1288 measurement arrangement for extended wavelength ranges. On the one hand, the correct mechanical coupling was verified. Accordingly, the irradiance at the image sensor level differs by only 0.2 % from that at the radiometers. The determined inhomogeneity ΔE of the radiation source of approx. 2 % is within the maximum permissible inhomogeneity of the EMVA 1288 standard of 3 %. The correct calculation of the spectral irradiation sensitivity of the radiometers could be verified quantitatively by the measurement with a monochromator and the comparison with the spectrum of its Xe arc lamp. The correctness of the absolute measured values of the new radiometric measuring system was qualitatively confirmed by a comparison measurement of the VISNIR radiometer with a calibrated luxmeter. The irradiance of the VISNIR radiometer deviated from that of the luxmeter only by a maximum of 1 %, which is within the relative measurement uncertainty of the radiometer.

References

[1] Erz M and Jähne B 2010 *Forum Bildverarbeitung* ed Puente León F and Heinzmann M (Regensburg: KIT Scientific Publishing) pp 155–166
[2] European Machine Vision Association 2016 *EMVA Standard 1288 – Standard for Characterization of Image Sensors and Cameras (release 3.1)* www.emva.org
[3] Gigahertz-Optik GmbH, https://www.gigahertz-optik.de Access on 13.11.2018
[4] Gigahertz-Optik GmbH 2017 calibration certificate 1642353-05 (Türkenfeld)
[5] Gigahertz-Optik GmbH 2017 calibration certificate 1740347-02 (Türkenfeld)

Modelling and optimisation of design of non-conventional instrument transformers

Bojan Nikolić and Sanowar Khan

City, University of London, Northampton Square, London EC1V 0HB
United Kingdom

E-mail: bnikolic@ieee.org

Abstract. In this paper, we have proposed, modelled and optimised several designs of non-conventional instrument transformer (NCIT) for high voltage overhead transmission lines (400kV). We have discussed several parameters and investigated how they influence the sensitivity of our NCIT, consisting of magnetic shape memory (MSM) element, magnetic circuit and an LVDT (linear variable differential transformer). One of the most used conductors in these lines, 528-Al1/69-ST1A ACSR conductor (old code MOOSE), was modelled together with the MSM element and the magnetic circuit in ANSYS APDL. Based on the obtained results we have given suggestions on how NCIT could be designed taking into account a choice of the most appropriate material for this application. The way how the model was developed was presented as well as calculations of errors in the model in ANSYS APDL for electromagnetic problems.

1. Introduction

One of the main reasons that a need for so-called non-conventional instrument transformers (NCIT) has arisen recently is incompatibility of conventional instrument transformers with modern measurement equipment. Current and voltage transformers with an iron core are usually considered as conventional instrument transformers and their design has not changed significantly over the last few decades.

In our previous papers, we proposed a novel non-conventional instrument transformer based on magnetic shape memory (MSM) alloys for measuring electrical current in HV transmission lines and in this paper, that idea has been developed further [1, 2]. MSM alloys are relatively new materials which change their shape when subjected to external magnetic fields. The design of an NCIT, consisting of an MSM alloy element, a magnetic circuit and an LVDT, was optimised for 400kV high voltage overhead transmission lines. One of the most frequently used conductors in overhead transmission lines at high voltage level and above, 528-Al1/69-ST1A ACSR conductor (old code MOOSE), was modelled together with the MSM element and the magnetic circuit in ANSYS APDL [3, 4]. The model development was presented as well as calculations of discretisation error in ANSYS APDL for electromagnetic problems.

Our initial research showed that these sensors were not sensitive enough to be used in overhead transmission lines. By adding a magnetic circuit, in order to concentrate as much as possible magnetic field around the current-carrying conductor and focusing it towards the MSM element, this problem can be solved.

2. Designs of NCIT

A particular magnetic material called Hiperco 50 was proposed as a material for the magnetic circuit around the MSM element in our previous paper, taking into account several aspects, such as saturation flux density, resistivity, skin depth, availability in the market and others [2]. Since the skin depth of magnetic circuit is very small, a laminated magnetic circuit should be used.

The NCIT was modelled around the ACSR conductor 528-Al1/69-ST1A which consists of 54 aluminium strands and 7 steel strands in its core. Both, aluminium and steel strands, have equal diameters of d=3.53 mm. Similar models can be easily obtained for other types of conductors by slightly changing the code that we developed.

A 2D model of three different designs of the proposed NCIT can be seen in figure 1. The MSM element is placed between the poles. The dimensions of the MSM element that can be found in the market and are suitable for the proposed NCIT are 10 mm x 3 mm x 1 mm.

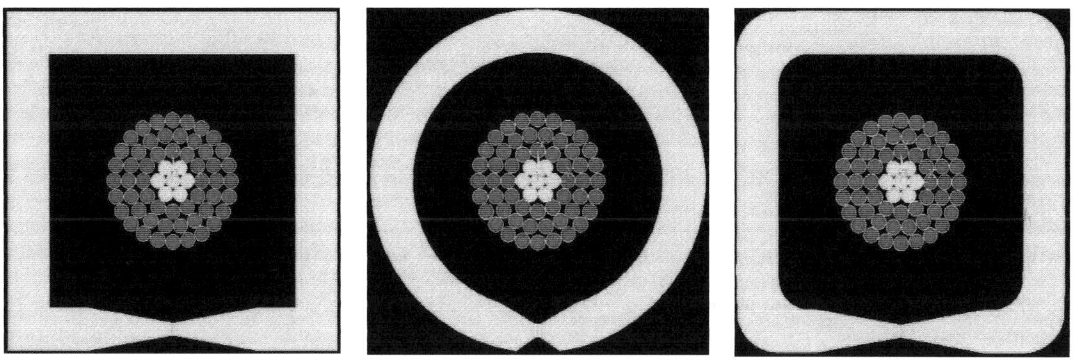

Figure 1. 2D model of the three different designs of the proposed NCIT around a 528-Al1/69-ST1A conductor – from left to right, rectangular, circular and rectangular magnetic circuit with rounded corners.

By analysing the strain-magnetic field relationship of MSM crystals (ETO Magnetic) at different pre-stress levels (loads) [5], it can be seen that the best option for our NCIT would be the MSM element which follows the 'ETO curve' at a load of 0.5 N/mm^2. This curve has the lowest threshold for triggering the MSM element (reversibly) and it has the largest strain. The MSM element would be triggered in this case by an approximately 0.085 T magnetic field, but higher values of the field are needed in order that the MSM element starts elongating linearly. MSM element consists of small parts called twin variants which start to reorient when magnetic field exceeds a certain threshold. This threshold is characterised by the minimum value of the external stress or magnetic field required to overcome the crystal's twinning stress and initiate reorientation. When an MSM element is not elongated it consists only of so-called hard variants. The relative magnetic permeability of MSM element in this situation is μ_r=2 what was used in this model. When the MSM element is fully elongated it consists of only so-called easy variants and its relative magnetic permeability is μ_r=50 in that case.

3. Optimisation of the magnetic circuit geometry

In order to optimise the magnetic circuit geometry for the application of NCIT for high voltage overhead transmission lines, we have analysed several parameters: size of the airgap (the airgap between the MSM element and the poles), position of the magnetic circuit relative to the conductor (distance between the centre of the conductor and the inner side of the magnetic circuit, r), magnetic circuit's width w and the distance from the airgap to the point where the poles start to taper, t. Comparisons have been done for the rectangular Hiperco 50 magnetic circuit.

Table 1 shows the variation of magnetic field amplitude on the surface of the MSM element, $B_{surface}$, and the maximum value of magnetic field inside the magnetic circuit, B_{max}, when the size of the airgap has been changed while keeping the other parameters constant.

Table 1. Variation of magnetic flux density on the surface of the MSM element and in the magnetic circuit with the airgap (Current I=500 A, f=50 Hz).

w [mm]	t [mm]	r [mm]	w [mm]	$B_{surface}$ [T]	$B_{surface}$ [%]	B_{max} in mag. circuit [T]	B_{max} in mag. circuit [%]
10	20	30	10	0.33	0	2.213	0
10	20	30	10	0.306	-7.27	2.195	-0.81
10	20	30	10	0.285	-13.64	2.179	-1.54

It can be seen that even a small increase in the airgap significantly decreases the magnetic flux density on the surface of the MSM element. In order to increase the sensitivity of the NCIT, it is better that its size is as small as possible. As the further decrease in the airgap is limited by technological factors, the best choice for the airgap was found to be 0.1 mm.

As discussed in [2], we modelled the 528-Al1/69-ST1A ACSR conductor (old code MOOSE). The overall diameter of this conductor is 31.8 mm which means that the magnetic circuit around the conductor can be placed only in a way that r>15.9 mm. The magnetic circuit geometry should be carefully chosen as it affects its saturation, NCIT's sensitivity to high temperatures of the conductor and, finally, possibility to trigger the MSM element.

Table 2 shows the variation of magnetic flux densities $B_{surface}$ and B_{max}, with thc position of the magnetic circuit relative to the conductor.

Table 2. Variation of magnetic flux density on the surface of the MSM element and in the magnetic circuit with distance, r between the conductor and the magnetic circuit (I=500 A, f=50 Hz).

w [mm]	t [mm]	r [mm]	$B_{surface}$[T]	$B_{surface}$[%]	B_{max} in mag. circuit [T]	B_{max} in mag. circuit [%]
10	20	25	0.354	100.00	2.357	100.00
10	20	30	0.33	-6.78	2.213	-6.11
10	20	40	0.293	-17.23	2.091	-11.29
10	20	50	0.264	-25.42	1.987	-15.70
10	20	60	0.241	-31.92	1.849	-21.55

Since the flux density on the surface of the MSM element decreases significantly as the magnetic circuit is placed away from conductor, it should be placed as close as possible to the conductor, but yet far enough so that the heat around conductor does not affect the magnetic circuit and the temperature-sensitive MSM element. A distance of r=25-30 mm could be a solution.

Tables 3 and 4 show the change of $B_{surface}$ and B_{max}, when the width of the magnetic circuit, w and the above geometric parameter of the poles, t have been changed.

There needs to be trade-off with respect of the width, w. A narrow magnetic circuit, although more sensitive, saturates more easily. On the other hand, a wide magnetic circuit is not very practical as it will increase the size of the NCIT as well as its cost. We have found that a w=10 mm magnetic circuit is a reasonable trade-off.

From table 4, it can be seen that tapering of poles does not affect significantly the magnetic flux density on the MSM surface, but still it increases. Thus, it is recommended to taper pole ends as much as its geometry allows. For example, for r=30 mm, this value could be chosen around t=20 mm.

Finally, we have compared sensitivity of all three NCIT's designs in figure 1 for similar geometric parameters. Results are shown in table 5.

As expected, the rounding of corners of rectangular magnetic circuit has improved the sensitivity of the NCIT ($B_{surface}$ is comparable to that of the circular design). Although any of the three proposed designs can be used, due to practical reasons of handling and mounting of the NCIT around the conductor, the preferred design would be the rectangular one with rounded corners.

Table 3. Variation of magnetic flux density on the surface of the MSM element and in the magnetic circuit with the width, *w* of the magnetic circuit (*I*=500 A, *f*=50 Hz).

w [mm]	*t* [mm]	*r* [mm]	*B*surface [T]	*B*surface [%]	*B*max in mag. circuit [T]	*B*max in mag. circuit [%]
5	20	30	0.338	0	2.31	0
8	20	30	0.333	-1.48	2.213	-4.20
10	20	30	0.33	-2.37	2.213	-4.20
15	20	30	0.321	-5.03	2.231	-3.42
20	20	30	0.312	-7.69	2.217	-4.03
25	20	30	0.302	-10.65	2.227	-3.59

Table 4. Variation of magnetic flux density on the surface of the MSM element and in the magnetic circuit with the distance, *t*, (*I*=500 A, *f*=50 Hz).

w [mm]	*t* [mm]	*r* [mm]	*B*surface[T]	*B*surface [%]	*B*max in mag. circuit [T]	*B*max in mag. circuit [%]
10	10	30	0.322	0	2.231	0
10	15	30	0.327	1.55	2.245	0.63
10	20	30	0.33	2.48	2.213	-0.81

Table 5. Magnetic flux density on the surface of the MSM element and in the magnetic circuit for three different magnetic circuit designs of NCIT (I=400 A, f=50 Hz).

Shape of magnetic circuit	*w* [mm]	*t* [mm]	*r* [mm]	*B*surface [T]	*B*surface [%]	*B*max in mag. circuit [T]	*B*max in mag. circuit [%]
Circular	10	20	40	0.237	0	1.9981	0
Rectangular	10	20	40	0.224	-5.49	1.8926	-5.28
Rectangular with rounded corners	10	20	40	0.23	-2.95	2.047	2.45

4. Model development and errors in the model

The model of the conductor, MSM element and magnetic circuit was developed in ANSYS APDL. It consists of 5 different materials (aluminium, steel, air, Hiperco50 and MSM alloy). The element Plane53 was used for modelling. This is an eight nodes element which is based on the magnetic vector potential formulation and has a nonlinear magnetic capability for modelling B-H curves.

The mesh was obtained using ANSYS's Mesh Tool where the level of fineness can be set between 1 (the finest mesh) and 10 (the roughest mesh). Additional refinements were required, especially in the most sensitive areas such as in magnetic circuit and in the air gap between the MSM element and the magnetic circuit's poles. The level of refinement can be chosen in the range from 1 to 5. The choice of level 1 means that the selected part of mesh will be the least refined while by choosing the level 5 the most new elements will be added.

In an electromagnetic analysis, ANSYS APDL allows calculating the mesh discretisation error for the parts of the model using macro EMAGERR. Two parameters are used to describe its value: B_{ei} - relative error for the magnetic flux density (magnitude) for element *i* and its normalised value, B_{nei}. They can be calculated using the equations (4.1) and (4.2), respectively.

$$B_{ei} = \frac{1}{n}\sum_{j=1}^{n}\left|B_j - B_{ij}\right| \tag{4.1}$$

where:

B_j - nodal averaged magnetic flux density (magnitude),
B_{ij} - magnetic flux density (magnitude) of element i at node j

$$B_{nei} = \frac{B_{ei}}{B_{max}} \tag{4.2}$$

where:
B_{max} - maximum nodal averaged magnetic flux density (magnitude)

The airgap between the MSM element and the poles of the magnetic circuit is the most delicate part of the model due to its geometry and presence of several different materials, and at the same time, the one which is of the particular interest for obtaining the simulation results. Therefore, B_{ei} and B_{nei} errors were calculated specifically for that part of the model.

The calculation is valid within a material boundary and does not consider the error in continuity of fields across dissimilar materials. Therefore, in addition to the values of B_{ei} and B_{nei}, in order to control the error in the model, it is required to check the values of the normal component of B on the border between two materials (air in the airgap and Hiperco 50 in the magnetic circuit). Ideally, there should be no difference in their values, but in FEM simulations that never happens. It is convenient to observe the relative error due to this discontinuity which is given by equation (4.3):

$$B_{nerr}\ [\%] = \frac{B_{nair} - B_{nmag}}{B_{nair}} \cdot 100 \tag{4.3}$$

where:
B_{nair} – normal component of B in the airgap on the border with the magnetic circuit,
B_{nmag} – normal component of B in the magnetic circuit on the border with the airgap

The values of the above-mentioned parameters will obviously depend on the chosen geometry, mesh fineness, loads, and so on. Several models with different levels of mesh fineness were developed and tested.

Figure 2 shows a mesh of one of the suggested designs – a rectangular magnetic circuit with rounded corners (w=10 mm, t= 20 mm, r=30 mm, I=500 A) as well as the mesh of the part including the MSM element, the airgaps around it and the poles of the magnetic circuit.

The maximum value of B_{nei} for the observed part was $B_{nei\ (max)}$=0.06832 and discontinuity on the border was B_{nerr}=0.63 %. The total number of elements in the model was 329,794 out of which 121,299 were in the airgap between the MSM element and the pole of the magnetic circuit. This mesh was chosen due to its precision of obtained results and yet still not a long time to obtain results on the computer that was used for simulations. Increasing the number of elements in the model particularly affects the value of B_{nerr}. As an example, it can be mentioned that this parameter took values from 17.15 % for the mesh consisting of 13,486 elements (136 elements in the airgap) to 0.63 % for the presented case. At the same time, the range of B_{nei} was 0.033-0.093. The number of elements in the presented model is relatively high, but it is worth it to mention that, for example, B_{nerr}=1.96 % was achieved with 47,424 elements out of which 14,561 were in the airgap. Furthermore, it should be said that a convergence tolerance in the model was 0.001.

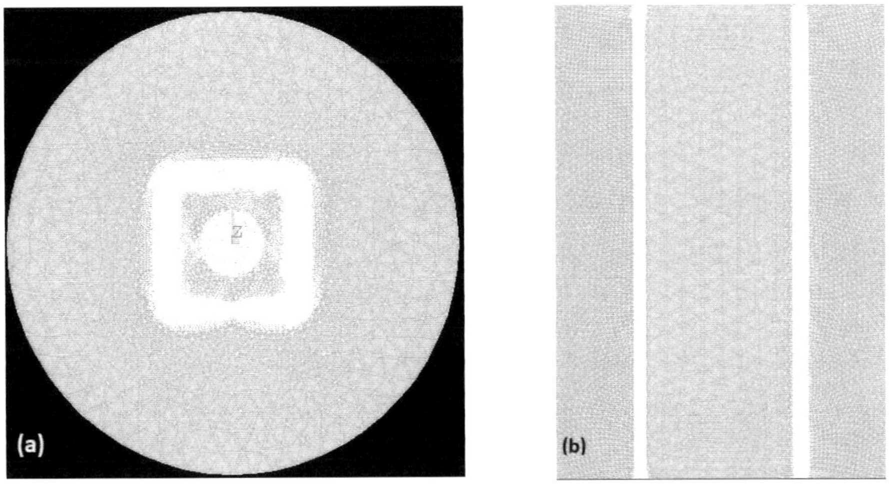

Figure 2. Mesh of one of the suggested designs – a rectangular magnetic circuit with rounded corners (w=10mm, t=20mm, r=30mm) a) full model b) MSM element, airgaps and poles of magnetic circuits.

5. Conclusions

In this paper, we have analysed parameters that affect the sensitivity of the proposed NCIT designs in order to optimise its magnetic circuit geometry for high voltage overhead transmission lines (400kV). One of the most used conductors in these lines, the 528-Al1/69-ST1A ACSR conductor, was modelled together with the MSM element and the magnetic circuit in ANSYS APDL.

We have proposed and modelled three designs of magnetic circuit and presented how variation of different parameters (e.g. airgap size, placement of magnetic circuit in relation to the conductor, and its various dimensions) affect, for example magnetic flux density on the surface of the MSM element and in the magnetic circuit.

The way how the model was developed was shown as well as calculations of errors in the model in ANSYS APDL for electromagnetic problems. The obtained results for one of the suggested designs were presented.

Based on the results obtained so far, we have recommendations on NCIT design taking into account a choice of the most appropriate material for this application. Before a viable prototype can be designed, more NCIT modelling and analyses of other parameters will be done in our future work.

6. References

[1] Nikolić B, Khan S and Gabdullin N 2016 Development of non-conventional instrument transformers (NCIT) using smart materials *J. Phys.: Conf. Ser.* **772** 012065

[2] Nikolić B and Khan S 2018 Modelling of non-conventional instrument transformers (NCIT) by FEM *J. Phys. Conf.: Ser.* **1065** 072046

[3] British Standards Institution 2013 BS EN 50182:2001 Conductors for Overhead Lines - Round Wire Concentric Lay Stranded Conductors (London, England: BSI) pp 27-74

[4] Lings R 2005 Overview of transmission lines above 700 kV *IEEE Power Engineering Society Inaugural Conference and Exposition in Africa*, 2005 pp 11–15.

[5] EtoMagnetic: Magnetic Shape Memory Technology (MAGNETOSHAPE) http://etogroup.com/MAGNETOSHAPE_EN.html

Group representative brain connectivity model of episodic encoding using large fMRI dataset

Supat Saetia, Natsue Yoshimura, Yasuharu Koike

Department of Information Processing, Interdisciplinary Graduate School of Science and Engineering, Tokyo Institute of Technology, 4259 Nagatsuta-cho, Midori-ku, Yokohama, Kanagawa 226-8503, JAPAN

E-mail: saetia.s.aa@m.titech.ac.jp

Abstract. Episodic memory is autobiographical memory crucial to human cognitive function. Brain connectivity modeling is an essential tool for describing brain function of interest. While a large body of research has focused on analyzing functional brain connectivity, most studies has focused on a connectivity of an individual. Brain connectivity studies are usually based on functional Magnetic Resonance Imaging (fMRI) which often involved many subjects. Only a few researches have focused on group level connectivity. We employ a framework called Tigramite to estimate causal interaction and model episodic memory connectivity on Human Connectome Project (HCP) BOLD data. Regions of interest are identified by performing fMRI univariate group subtraction analysis on working memory task-fMRI data using subsequent memory test paradigm. From each resulting individual connectivity graphs, we construct a group representative graph by performing median aggregation on each corresponding edges and nodes of each individual graphs. However, there is no established method to measure a quality of the group representative graph. So, we propose the use of Levenshtein distance as a measurement to validate the group representative graph if it represents general characteristics of each individual graphs in the population.

1. Introduction

Writing is one of the greatest invention of humanity, it enables us to physically record our ideas or experiences we have encountered, then to pass it on to others or to the next generation. Modern day human has several means of storing information either physically, or even digitally. However, for most of humanity's existence, it has been only a short period of time since the invention of writing. Before our means of recording information was invented, all human had was their brain, a biological storage which stores information called memory. The brain is our most primitive means of information storing, yet slightly we know of how information, our memories, are organized inside our brains.

Episodic memory is a memory of personal events times, places, associated with emotion and other contextual knowledge (autobiographical events). With episodic memory, a person can remember personal experienced events and consciously aware of the certain situation at a certain time of that events [1]. It is a part of a bigger category of long-term memory called declarative memory [2]. The episodic memory is different from other types of memory in that it often represented in form of visual information in order of event occurrence on personal timeline, and it needs to have perspective observer of the memory owner. The importance of episodic memory is that it enables human to project oneself backward in time and recall many aspects of one previous experience. This ability becomes the source of self-awareness and induces intelligence which set human apart from other animals.

Observing how information are organized in human brain is a challenging task. However, in recent years, the progress in functional Magnetic Resonance Imaging (fMRI) technology has allowed the study of brain connectivity, the statistical dependencies between localized regional brain activity. In this context, we can model how information flows in the brain by analyzing brain connectivity during the memory-related cognitive task. Brain connectivity can be visualized in a graphical model [3] or time series graph [4].

Granger causality [5] is a mathematical framework commonly employed to model causality of the neuronal activity from fMRI BOLD data. The underlying assumption of this framework is that if $X \rightarrow Y$ if and only if a change in X has an effect on Y [6]. However, all we can imply from observational data are statistical dependencies. It is controversial to infer effective connectivity (directed connectivity) due to the low temporal resolution nature of the BOLD signal as Granger causality is prone to under-sampling signals. To avoid that issue in this study, we utilize time-lagged causal discovery framework called Tigramite (time-series graph-based measures of information transfer) [7]. This framework relies on different set of assumptions to identify a causal graph. It performs conditional independence testing using the assumptions of time-order, Causal Sufficiency, the Causal Markov Condition, and Faithfulness, among others [8].

While most conventional fMRI studies rely on pool of data collected across individuals, modeling of the connectivity from data across individuals is controversial [9]. Nonetheless, a method to establish a general model that represents brain connectivity across individuals is essential. We propose a method of constructing a representative graph based on graph structure by using median aggregation of the weight of corresponding edges and nodes of each individual graphs.

2. Method

2.1. Human Connectome Project

Human Connectome Project is a project conducted by the Washington University-University of Minnesota Human Connectome Project Consortium (WU-Minn HCP) [10]. We utilize this dataset because it provides access to large exceptional spatiotemporal resolution of well- characterized samples of healthy individuals.

This study made use of 376 subjects available in HCP data. Of all available 511 subjects, 52 subjects are excluded due to data incompletion, 83 subjects are excluded from the study due to biased responses in picture sequence memory test (subjects always respond to the stimuli in the test only either remember or know). Of the 376 subjects included in this study, 54 are between the ages of 22-25, 179 are between the ages of 26-30, 141 are between the ages of 31-35, and 2 are above 35 year-old.

2.2. Subsequent memory paradigm

The HCP dataset does not include a specific task related to episodic memory, hence, we employ a subsequent memory paradigm by combining HCP working memory task-fMRI and corresponding picture sequence memory test. Subsequent memory paradigm is a study of neural activity during an encoding phase of the stimuli that are later subsequently remembered in contrast to the stimuli that are forgotten [11].

In the encoding phase, a subject is given a chance to study a series of stimulus while their brain activity data are being collected. Later after the encoding phase is completed, the subject then performs a recollection test. In this test, the subject is shown a series of stimulus consists of both old stimuli presented during encoding phase and completely new unseen stimuli. During this phase, the subject is allowed to respond to the stimulus by indicating that they either remember, know, or seeing new stimulus.

The response is collected and evaluated for memory recollection quality. The correction and the confidence of the response is evaluated. If remember or know responses are given to the old stimuli, it indicates the subject's correct recollection. The remember response indicates subject's confidence in their conscious recollection, and the know response indicates subjects doubt in their recollection, in which they may feel familiarity with the stimulus but fails to consciously recall it precisely. Then if the subject gives the new response to the old stimuli, it is considered that the stimuli have subsequently been forgotten by the subject, and if remember or know response is given to the new stimuli, that response is considered as a false recollection. If the new response is given to the new stimulus, it indicates that the subject remembers that the presented stimulus is not included in the set of stimuli previously shown during encoding phase.

2.3. Univariate group subtraction

The scanned images are processed and analyzed using FSL 5.0.7 (FMRIBs Software Library, www.fmrib.ox.ac.uk/fsl) software suite. The HCPs structural MRI and fMRI were preprocessed using FSL 5.0.6 software suite. The preprocessing pipeline is done by HCP using the pipeline derived from [12]. FEAT (FMRI Expert Analysis Tool, v6.00) was used to analyze first- and second-level analysis. The General Linear Model was used to model the event duration. The event-timing were specified in FSLs 3-column Explanatory Value file format. The explanatory file describes time when the interested events occurred, the duration of the events, and the value of the input during that time. The trial on which the stimuli are identified by the subjects as remember and know are selected for the analysis because this study focuses on comparing the condition where the subject encodes the memory of the stimuli that later be successfully recalled with one that failed to be recalled.

Additional Psychophysiological Interaction (PPI) Analysis is done using parahippocampal gyrus as a seed to identify additional memory encoding related region.

2.4. Group causal modeling

We model causal relation between 4 regions of interest (table 1) for each individual subjects using Tigramite causal time series analysis software package. Tigramite is a time-lagged causal discovery frameworks [8]. There are 2 free parameters for the algorithm, the maximum time lag τ_{max}, and the significance threshold α in the condition- selection step. To determine maximum time lag τ_{max}, we estimate lagged unconditional dependencies of the BOLD time-series and found the dependencies diminish beyond a lag of 8. The significance threshold α is set to 0.1. The α is a regularization parameter in model-selection techniques, and should not be seen as significance test level in the condition-selection step [13].

Then we construct a group representative causal model by median aggregation of connection weight of corresponding link of each individual connectivity graph. We use median to reduce the effect of outliers. However, interpreting a generalized representative connectivity model from data across individuals is controversial. So, we propose and perform a validation process to establish that the representative connectivity reflects common characteristics of connectivity in each individual subjects.

We use Levenshtein distance [14], a classic graph edit distance (GED), to measure similarity between connectivity graphs. The edit distance compares only structural aspect of the graphs and completely omits link weight. We validate the representative graph by comparing it with a representative graph of a sub-group of randomly selected 100 individual for 1,000 iteration. Then, we average the resulting distance from all iteration.

3. Results

3.1. Univariate group subtraction and regions of interest

The peak activated region yielded by fMRI group subtraction analysis is large (17,529 voxels) and cover wide area of the brain (figure 1). Harvard-Oxford cortical and subcortical structural probability atlases are used to identify the activated region. The major activated regions are selected as regions of interest. Additionally, the PPI shows that lingual gyrus has significant correlation with parahippocampal gyrus. Therefore, 4 ROIs shown in table 1 are selected for connectivity modeling.

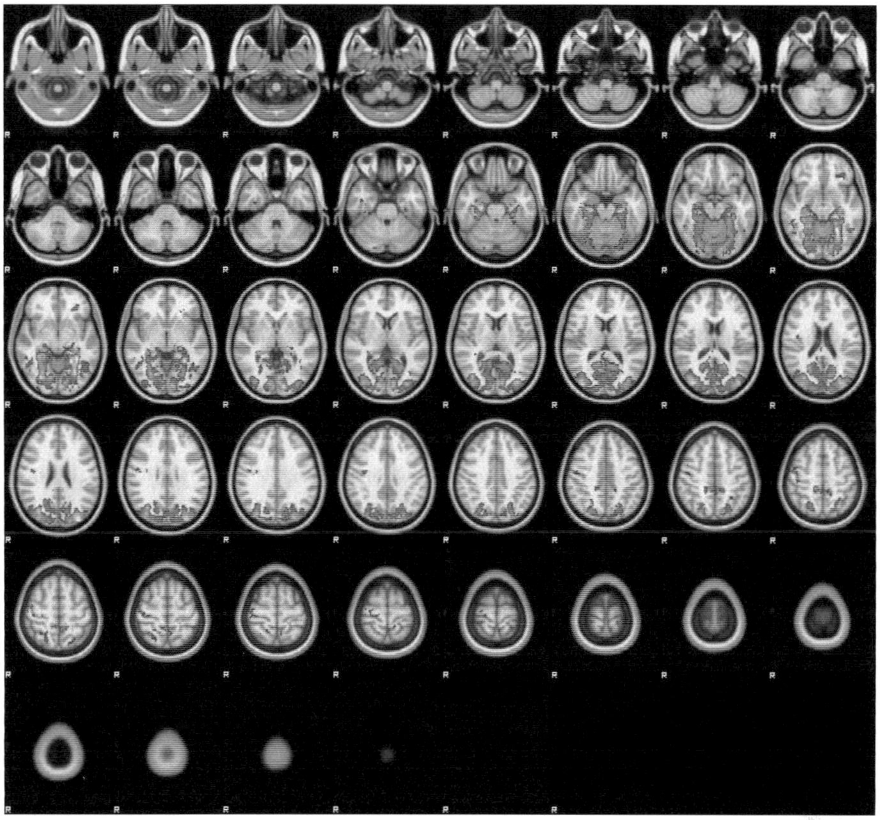

Figure 1. Activated cluster from univariate group subtraction analysis of all 376 subjects on Successful recollection > Failed recollection (threshold at $Z > 2.3$, $P < 0.05$).

Table 1. Selected regions of interest (Harvard-Oxford cortical structural probability atlases).

Regions	Hemispheres
parahippocampal gyrus	Left and right
temporal occipital fusiform cortex	Left and right
occipital gyrus	Left and right
lingual gyrus	Left and right

3.2. Causal modeling and group representative model

The resulting group representative graphical model and time series graph are presented in figure 2 and figure 3, respectively. In the representative model, only 2 significant connections, a connection from occipital gyrus to temporal occipital fusiform cortex, and from lingual gyrus to parahippocampal gyrus, are presented. Its average Levenshtein distance from our test is 6.35 (SD: 0.28).

The main concern of connectivity measure derived from fMRI data is that movement and physiological noise sources, which vary from subject to subject and session to session, can potentially induce spurious correlations between ROIs, which could confound the interpretation of the results by increasing chance of false positive [15]. The effect of aforementioned spurious correlations may contribute to the variation among individual graphs. The number of connections in the individual graph varies from only 1 to as many as 12 connections. Group model is calculated using median aggregation to reduce the effect of outlier. The Levenshtein distance is introduced to measure the reliability of the resulting group model by showing that the model is in the average distance among the population.

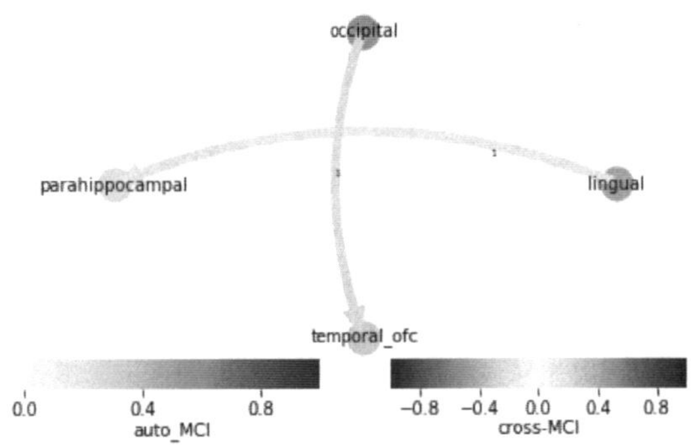

Figure 2. Representative connectivity graphical model of subsequent memory task.

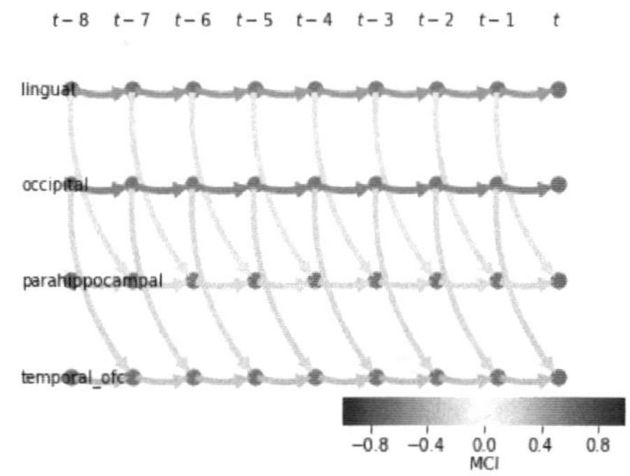

Figure 3. Representative connectivity resulting time series graph of subsequent memory task.

4. Discussions and conclusions

4.1. Episodic memory encoding

In this study's subsequent memory paradigm, brains activity is observed only during the encoding phase. Cognitive functions related to memory beyond memory encoding, such as, memory organization or memory recollection are not addressed.

According to previous studies, the resulting peak activated regions are known to have roles in memory formation. A study by Ofen et al. [16] found that middle temporal gyrus, fusiform gyrus, and parahippocampal gyrus play significant role in development of declarative memory formation development from child to adult. Occipital area and fusiform gyrus are known to play important roles in face recognition [17] and declarative memory forming [16]. The protocol used by HCP may contribute to the activation of this region since images of human face are used as stimuli for working memory task fMRI [18]. Lingual gyrus is a part of declarative memory system [19] and also contributes to visual memory forming [20].

The results of this study agree with previous studies and affirm that the prior knowledges hold in large population since most of the previous fMRI studies included only small number of subject due to technical difficulty of fMRI experiment protocol.

4.2. Causal modeling and group representative model

The goal of brain research is to describe biological mechanic of the brain that responsible for any particular cognitive function. Most researches identify brain function in forms of active region and pathway. Sensory pathways is one of the well-known example. However, the knowledge of higher cognitive function's pathway, such as, consciousness or memory organization is limited since these functions lack physical interaction that can be observed. Causal interaction analysis on fMRI BOLD time-series is a common way to study said brain functions.

Granger causality [5] is a popular approach among numerous techniques developed to analyze causal interactions in fMRI data [21]. However, inferring causal relationship (effective connectivity or directed connection) using this framework on fMRI BOLD data is controversial because low temporal resolution nature of the BOLD signal may confound its statistical dependencies implication. Without the temporal order integrity, the framework shows only correlation (functional connectivity or undirected connection). Consequently, we use Tigramite framework [8] for our study to avoid temporal sensitivity issue of Granger causality framework.

Journal of Physics: Conference Series **1379** (2019) 012058 doi:10.1088/1742-6596/1379/1/012058

In fMRI study, the localization of specific brain function is observed across individuals using statistical inference, and therefore the location can be generalized across population. However, fMRI statistical framework does not account for temporal order of activation. We have no statistical basis that guarantee temporal order of activation or activation time lag across individual to be able to infer general causal model of a group of individuals. Nonetheless, the general representative model is needed for drawing a general conclusion.

To address this issue, we compare difference between the resulting representative model with model of random sub-groups using Levenshtein distance. The average distant is 6.35 (SD: 0.28), with the maximum distance at 11, and the minimum at 1. While structure of individual subject's model varies, ranged from a graph with only one connection to a complete graph, our test shows that the representative graph is balanced.

Therefore, we conclude that the graph is a general representative model of this sample population. However, it is important to emphasize that this method only concerns structural similarity, the connection weights are ignored.

References

[1] Tulving E 1993 *Curr. Dir.Psychol. Sci.* **2** 67–70
[2] Ullman M T 2004 *Cognition* **92** 231–270
[3] Lauritzen S L 1996 *Graphical Models* (Oxford: Oxford University Press)
[4] Eichler M 2012 *Probab. Theory Rel.* **153** 233–268
[5] Granger C W J 1969 *Econometrica* **37** 424
[6] Pearl J 2009 *Causality: Models, Reasoning and Inference* 2nd ed (New York, NY, USA: Cambridge University Press) ISBN 052189560X, 9780521895606
[7] Runge J 2015 *Phys. Rev. E* **92** 062829–18
[8] Runge J 2018 *Chaos: An Interdisciplinary Journal of Nonlinear Science* **28** 075310–21
[9] Duggento A, Passamonti L, Valenza G, Barbieri R, Guerrisi M and Toschi N 2018 *Sci. Rep.* **8**(1) 5571
[10] Van Essen D C *et al.* 2012 *NeuroImage* **62** 2222–2231
[11] Dolcos F, LaBar K S and Cabeza R 2004 *NeuroImage* **23** 64–74
[12] Glasser M F *et al.* 2013 *NeuroImage* **80** 105–124
[13] Runge J, Nowack P, Kretschmer M, Flaxman S and Sejdinovic D 2017 *arXiv.org* (Preprint 1702.07007v2)
[14] Levenshtein V I 1965 *Soviet Physics. Doklady* **10** 707–710
[15] Whitfield-Gabrieli S and Nieto-Castanon A 2012 *Brain connectivity* **2** 125–141
[16] Ofen N, Kao Y C, Sokol-Hessner P, Kim H, Whitfield-Gabrieli S and Gabrieli J D E 2007 *Nat. Neurosci.* **10** 1198–1205
[17] Fusar-Poli P *et al.* 2009 *J. Psychiatr. Neurosci.* **34** 418–432
[18] Marcus D S *et al.* 2013 *NeuroImage* **80** 202–219
[19] Menon V, Boyett-Anderson J M, Schatzberg A F and Reiss A L 2002 *Cognitive brain res.* **13** 261–265
[20] Schankin C J, Maniyar F H, Sprenger T, Chou D E, Eller M and Goadsby P J 2014 *Headache* **54** 957–966
[21] Margulies D S *et al.* 2010 *Magma (New York, N.Y.)* **23** 289–307

Assessment of the generic problem-solving construct across different contexts

A K L Nguyen[1], C T K Nguyen[1], and R J Adams[1,2]

[1]Assessment Research Centre, The University of Melbourne, Vic 3010, Australia
[2]Australian Council for Educational Research, VIC 3124, Australia

E-mail: lananh289@gmail.com

Abstract: Problem-solving, one of the generic skills in the so-called '21stcentury' skillset, has been found to be critical to the personal and professional success of learners. This study evaluates the existence of conceptual meaning of students' generic problem-solving (GPS) competency in three distinct contexts, namely mathematics, reading, and creative problem-solving using the PISA Computer-Based Assessments (CBA) 2012 data. The analyses were framed in the context of a content validity exploration from Messick's validation framework to reveal the GPS conceptual meaning underlying PISA CBA 2012 measurements. Specific tasks carried out include: (a) conducting skill auditing of PISA CBA 2012 items to understand their measured skill content and relevant to and representative of the generic problem-solving construct's level; (b) using the outcome of the skill auditing stage to interpret the GPS competency levels measured by the domain-specific subject tests before comparing them with corresponding levels together.

1. Background and objective of the study

Over the last 40 years, economic and technological changes have drastically reshaped education systems throughout the world. Problem-solving (PS) is arguably one of the most pertinent due to its inherent ability to deal with complexity and its mutual relationship with other skills. Numerous studies have been conducted worldwide to assess students' PS in a specific range of contexts. However, none of these studies has specifically identified PS competencies that are likely to transfer across more than three different PS contexts. This leads to the question of whether there is a generic problem-solving (GPS) construct that can apply in a wide range of fields. The answer to this question has not been adequately justified. This study investigated the extent to which a single GPS construct can be invariant across mathematics, reading and creative PS. The objectives of the research were (a) to identify the measured skills in PISA 2012 computer-based assessments (CBA) relevant to and representative of the GPS construct's level and (b) to compare the interpretations of the measured GPS competency levels in the domain-specific areas together.

2. Theoretical framework

In educational assessment, the assessment for a GPS construct's existence across fields has received a lot of interest due to an increased emphasis on preparing students for 21st century skills. A shared interest for researchers, policymakers, educators and the public is whether a GPS construct exists across different domain-specific areas and can thus be taught in the classroom [1]. Instances of students performing similarly in solving problems in different contexts, for example, in analytic PS and interactive PS [2], and the existence of GPS skills which are independent of any specific context, make imperative the need for assessing the existence of a GPS construct. To explore the existence of a GPS construct, most studies have evolved by comparing different PS constructs to reveal the similarities and differences between those constructs [1, 2]. However, it seems impossible to find any valid research in the literature focusing only on a comparison of three or more measured constructs in PS tests.

This paper intends to provide a validity judgement approach by adapting the content validity process from Messick [3]'s validation framework to reveal the GPS conceptual meaning measured by PISA CBA 2012 instruments. Messick [3] acknowledged the importance of expert judgement to evaluate content relevance, by which researchers look for evidence of content relevance and representativeness of the tests. Specifically, to justify the claim that GPS skills underlie three domain-specific PS constructs, the evaluation of this claim would begin with assessing the evidence that the PISA 2012 item test content is relevant to description of GPS construct by using skill auditing. This stage leans on the *skill-auditing process,* which uses expert review to support judgements of the relevance and representativeness of the content upon which items are based. A skill-auditing process has been applied to the content validity in different studies. For example, Tran, Griffin and Nguyen [4] undertook a skill-auditing process as the evidence for content validity on the interpretation and use of the 2008 University Entrance Examination English test scores.

The research questions to be addressed in this paper are as follows:

- To what extent are the skills that underlie the PISA CBA 2012 released items relevant to and representative of the levels in the GPS construct?
- To what extent do the interpretations of item ordering in the construct of GPS competence measured by the PISA CBA 2012 released items match with each other?

3. The secondary PISA CBA 2012 data and the participants

This study used both previously collected data and direct correspondence from human participants. The secondary data, including released items and analysis reports from PISA CBA 2012, were downloaded from the OECD website (www.oecd.org/pisa/pisaproducts/database-cbapisa2012.htm). Data from the participants of 32 countries/economies, with 117,933 students, were used in this study. These released computer-based items can be seen at www.oecd.org/pisa or http://erasq.acer.edu.au (username: public; password: access). No new reading material was released after the 2012 survey administration. There were 30 released items including both multiple-choice and partial-credit formats. There are two reasons why this study focused only on computer-based PISA 2012 data. First, for an equal comparison between constructs, the method of delivering the tests needs to be the same in three assessments. The interactive PS assessment was administered in a computer environment. As a result, the computer-based mathematics test and the digital reading test in PISA 2012 were chosen for a similar administered approach. There were also a science test and a financial test in the PISA 2012 cycles; however, they were only administered in paper-and-pencil form. Thus, neither was included in this research. Second, a requirement of this study was that one student had to answer three assessments at a similar time. Thus, only PISA 2012 CBA data could satisfy this condition.

In collecting subjects' experts at the skill-auditing stage, the targeted participant group comprised six adults 22 years or older who were competent to give consent. They were senior testing developers and specialists in the field of testing, assessment and evaluation in Australia. They were also selected based on their familiarity with the PISA tests, mathematics assessment, reading assessment and PS assessment. Individual opinions from the experts regarding the skills needed to answer the questions in the PISA CBA 2012 tests were collected via an item analysis form. To record the discussion, as well as using participants' completed forms, the researcher employed both audio recording and note-taking. These data and materials were collected at the University of Melbourne site.

4. Research design

4.1. Using individual analysis and group discussion to explore the skill measured by the PISA 2012 computer-based released items relevant to and representative of the levels in the GPS competence

This analysis stage adopted the skill-auditing procedure suggested by Tran et al. [4] as this would provide high methodological clarity and the fundamental developmental PS (DPS) construct [5-7] as the reference framework. The description of the six levels of DPS is given in [5]. Like many taxonomies, the levels in this framework display a hierarchy in which learners have more probability of displaying the performance at a higher level as their PS competence increases. The purpose of this analysis was to describe the measured continuums in term of GPS skills and levels based on the item's intended measured skills so as to compare the derived continuums' descriptions in the next analysis. Skill auditing was evaluated in three different content areas, mathematics, reading, creative PS, varying in knowledge content. Thereby, the skill auditing of each item and level description of the GPS in each domain were analysed in depth.

4.2. Comparing the interpretations of item ordering in terms of GPS skills measured by the PISA 2012 computer-based tests

The second research question aimed to compare the interpretations of the GPS constructs measured by the PISA CBA 2012 released items to evaluate the extent to which the three derived constructs were similar. In this study, the similarities represented the existence of a GPS construct underlying three domain-specific PS constructs. To arrive at such an interpretation for the measured GPS constructs in three PISA CBA 2012 tests, the expert judgement-based approach was used. As Mach, Mastrandrea, Freeman and Field [8] pointed out, applying rigorous judgement is an important dimension of assessment. The results of the previous skill auditing helped to inform whether there was a common substantive interpretation of the underpinning construct. The released PISA CBA 2012 items were placed in descending order of levels, according to the experts' judgement. The set of items in each level was then examined to interpret different measured GPS levels using the fundamental DPS proficiency and referring to the European Qualifications framework for generic competencies. Then, to validate these substantive interpretations, the researcher emailed these level descriptions to the expert participants and asked them to judge if these levels were meaningful, informative and useful. Finally, the interpretations of the GPS levels measured in PISA CBA 2012 released item tests were compared with the descriptions of corresponding levels. The purpose was to inform the similarities and differences between the three measured GPS levels in three domain-specific areas.

5. Results and discussion

5.1. The skill measured by PISA CBA 2012 released items relevant to and representative of the levels in the GPS construct

At the skill-auditing stage, expert participants identified the GPS skills measured by each PISA 2012 CBA test item (domain-specific areas) and located the item in the GPS levels. Differences could be seen in their individual analyses, but these were few and were mainly due to the varying extent of detail in their descriptions rather than a fundamental dissimilarity in their perceptions of what GPS skills were needed to succeed. With many years of testing experience, all the participants responded elaborately to the researcher's questions of clarification and came to relatively uniform conclusions. Based on their individual analyses, the expert participants joined the group discussion to express their views. The judgement they provided at this stage was generally consistent with the points they made in the completed individual forms. Figure 1 presents the relations between PISA CPS 2012 levels (dash line) versus the skill-auditing GPS level (solid line) of released items across creative PS, computer-based mathematics and digital reading domains. Two main findings can be drawn from the skill-auditing results.

Journal of Physics: Conference Series **1379** (2019) 012059 doi:10.1088/1742-6596/1379/1/012059

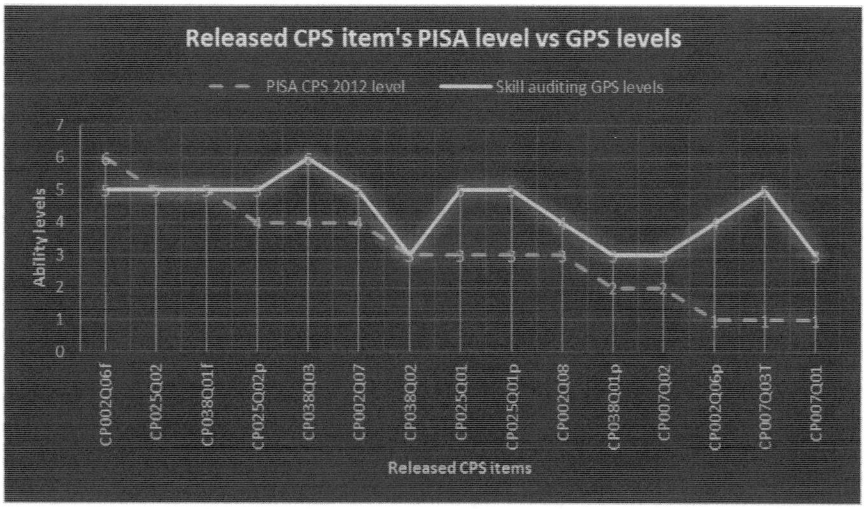

Figure 1. The relative between PISA CPS 2012 level versus skill auditing GPS level of released items across creative problem-solving area.

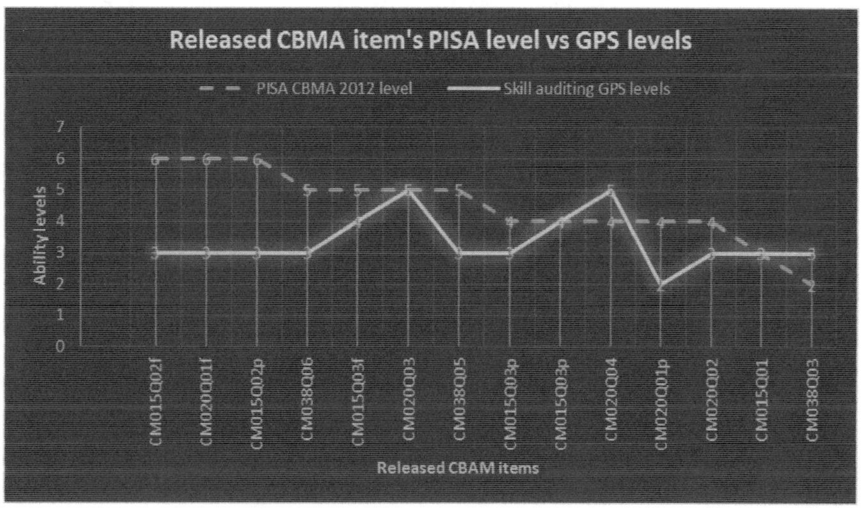

Figure 2. The relative between PISA CPS 2012 level versus skill auditing GPS level of released items across computer-based mathematic area.

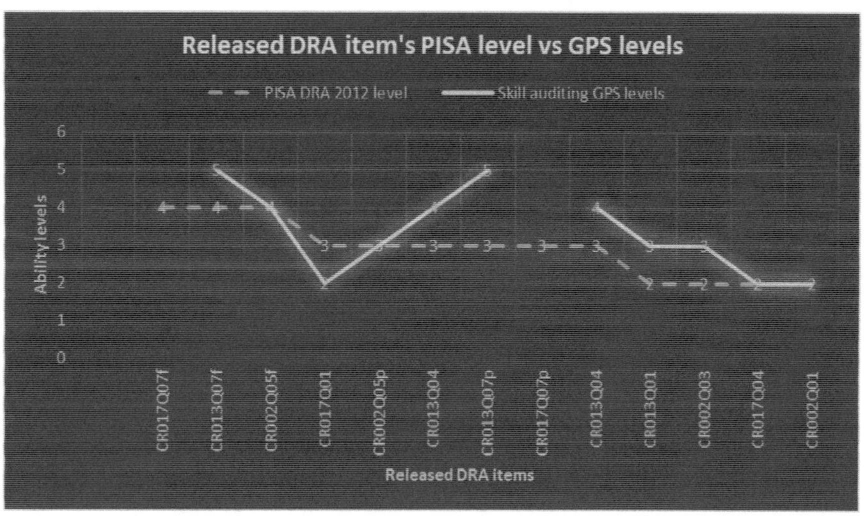

Figure 3. The relative between PISA CPS 2012 level versus skill auditing GPS level of released items across digital reading area.

In the first finding, all the experts asserted that while it was difficult to classify the PISA CBA 2012 released items into GPS levels, the test content was highly relevant to the GPS skills descriptions in the six levels of the fundamental GPS construct. These comments were reasonable, and it was fair to say that the PISA CBA 2019 items were developed to measure literacy competencies such as mathematics, reading, or science constructs rather than the GPS construct. This can be observed in figure 1, figure 2 and figure 3, in which most of the released items have two different levels according to the PISA CBA 2012 scale and skill-auditing results. An example of such items is as follows. For item CR017Q01, the participants found that students would need to locate explicitly provided information using a conventional symbol system and a text label. Thus, strictly speaking, the item should be classified under both reading competence and GPS competence. From their experience, though, the participants judged that candidates would be more challenged in reading than in the GPS. Hence, it was agreed that item CR017Q01, which was classified at Level 3 according to the PISA CBA 2012 scale, should be classified at Level 2 of GPS.

For items CP007Q02, CP002Q08, CP002Q06, CP025Q01, CP025Q02, CP038Q02 and CP038Q01 in released PISA CPS 2012 tasks, CR013Q04 and CR013Q07 in released PISA DRA 2012 tasks and CM015Q03, CM020Q02, CM020Q03, CM020Q04 and CM038Q06 in released PISA CBMA 2012 tasks, though expert participants' detailed descriptions of the underlying knowledge and skills were similar, their individual level classifications differed significantly. Each item was classified under at least three or four levels. During the group meeting, they agreed that the classified level was the highest one, and success in these items required students to use most of the GPS skills at that level. Thus, these items were finally categorised under only one level (see figure 1, figure 2, and figure 3).

For items CR017Q07 full credit and CR017Q07 partial credit, especially, the expert participants explained that these items first tested students' ability to explore the individual elements within the task in a random manner. Because these skills were classified under the GPS competence, two experts classified them under Level 2, two classified them under Level 4, one classified them under Level 3 and one classified them under Level 5. However, when they came to analyse item CR017Q07, they found that these items strongly tested students' ability to recognise the message as potentially misleading or risky, and to evaluate the content of a text, which was classified under reading ability. After a long discussion, the experts agreed that items tested purely reading ability rather than GPS competence. Thus, the items CR017Q07 full credit and CR017Q07 partial credit were omitted from this study. This is why, in table 1, there is no solid line for these items when they belong to Level 4 and Level 3, according to the PISA CBA 2012 scale.

In the second finding, while the views on GPS skills relevance were highly complementary, those on level representativeness were not as positive. The results of skill auditing showed that the released PISA CBA 2012 items measured students in only 4 levels of GPS competence. On the one hand, the released items covered moderate GPS levels such as Level 3. It tested this level extensively, given that, of 30 items, 16 required students to explore and then recognise patterns. On the other hand, all the expert participants found that the measures for Level 6 and Level 1 were not represented by the released PISA CBA 2012 items. The measures of remainder of the levels (5, 4, and 2) were only limitedly in all of three domain-specific areas by several released items. Thus, the PISA CBA 2012 released test items were deemed able to measure a limited range of GPS competencies.

On a more positive note, the results of the inquiry showed that the released test items and the marking key were appropriate for rendering evidence of students' GPS ability. All the experts acknowledged that the PISA CBA 2012 released items were in close accordance with the GPS skills. They asserted that, though these PISA CBA 2012 tests more generally measured the literacy abilities, students needed to have skills and abilities related to GPS competence to answer the items correctly. During the discussion, all the experts agreed with the skills needed to answer items correctly. The expert participants further shared that they and other colleagues had developed these PS tasks in several areas such as mathematics, reading or social science. Therefore, according to the experts, the test tasks were relevant to the fundamental GPS construct.

5.2. *Comparing the interpretations of item ordering in terms of GPS skills measured by the PISA 2012 computer-based tests*

To interpret the measured GPS levels in the PISA CBA 2012 tests and compare these interpretations with the descriptions of corresponding levels, the fundamental GPS construct was used as the reference framework. This study also referred to the illustrative descriptors from eight levels of the European Qualifications framework for generic competencies, elaborating descriptive progressions from UNESCO [9] to synthesise elaborated descriptors for the dimensions of the GPS skills to increase the amount of information for interpretive purposes.

Table 1 presents the interpretation of GPS levels in three domain-specific areas. As can be seen from table 1, the GPS interpretations in three domain-specific areas were compared with corresponding levels. The GPS Level 1 had no data in all three domains, and Level 6 was not present in mathematics and reading because there were no data in these domains. Beyond those mismatches, there were many similarities across the rest of the corresponding GPS levels in the three domains.

For example, the descriptions of GPS Level 2 (explore patterns) in the mathematics domain were equal to the descriptions of this level in the reading domain. Students in this low level in both scales had mastered the basic knowledge of domains and were starting to explore, search, locate, retrieve and interpret information. There were no data at this level in the CPS domain.

The similarities between GPS levels across three domains were also revealed in the comparison between Level 3, Level 4 and the highest Level 5 (testing rules and forming hypotheses). The highest level in the three areas involved formulating rules by identifying the relevant factors in the problem and their interrelationships, formulating rules to solve the problem according to the given constraints, and synthesising information and forming an opinion with evidence to support reasons. In other words, those students who could represent the rules were considered to have the highest GPS competency in all specific domains. Due to these commonalities, it can be concluded that the hierarchies of released items in the three domain-specific areas greatly resemble each other, which is another indicator that GPS competence exists across three domain-specific areas.

6. Conclusion and recommendations

Given the results presented and discussed so far, the findings for the study are clear. The content of the released PISA CBA 2012 items was relevant to the GPS construct's levels. Nevertheless, since the released PISA CBA 2012 items targeted only a limited range of GPS skills, the released PISA CBA 2012 items had moderate representativeness of the measured GPS construct. On a brighter note, the results of the inquiry show that the released PISA CBA 2012 test items and the marking key are appropriate for rendering evidence of students' GPS ability.

Given the experts' rich experience of working with testing and assessment, their professional judgement presented so far has been an illuminating source of evidence. Because the four-level scales obtained were interpretable and meaningful, it is reasonable to conclude that there are some similarities in the score interpretations of the PISA CBA 2012 in the light of the GPS competency across three domain-specific areas. These were important pieces of evidence for the content aspects of validity. However, to achieve higher validity standards, more secured items should be included in future analyses as replacements for items within higher or lower GPS levels.

Special attention should be given to devising items that discriminate students in the ability range such as Level 6, Level 5, and Level 1 so that the competency of students in the GPS ability levels can be interpreted with greater precision. Izard [10] pointed out that items should also be designed in a way that they can be spread over different difficulty ranges so that evidence can be obtained for students of all ability levels.

Table 1. The GPS levels' interpretations in three domain-specific areas measured by PISA CBA 2012 released items.

GPS's level based on experts' judgments	Derived GPS levels measured by PISA CPS 2012 released items	Derived GPS levels measured by PISA CBMA 2012 released items	Derived GPS levels measured by PISA DRA 2012 released items
Level 5: Testing rules and forming hypotheses	Students at this level can organize and critically evaluate information, to find the (relevant) relationships between problem elements. This enables them to formulate complex hypotheses to predict the behaviour of the system. In controlling the execution of a solution strategy, they can plan ahead and monitor progress systematically to avoid errors and ensure actions contribute to the solution.	Students at this level can understand a mathematical concept and apply it in a new situation. They employ previous knowledge to form rules to solve the problem according to the given constraints.	Students at this level can access and retrieve information, combine, interpret, compare and evaluate a series of descriptions with a set of requirements. They synthesize information from several sources and form a conclusion supported by evidence and reasons.
Level 4: Forming rules	Students at this level can explore an unfamiliar system and identify patterns in different aspects of its behaviour. They can form and confirm a simple hypothesis as to the rules governing a system.	Students at this level can recognize and understand the relationship between sets of variables, to identify and reflect on a proposed incorrect solution. They execute a strategy to solve the task and give the correct solution.	Students at this level can use multiple steps to locate, search and retrieve required information. They evaluate, compare, integrate and interpret information and use the information to make a decision. They form rules, reflect on them and justify the suitability for a given situation with evidence.
Level 3: Recognizin g patterns	Students at this level can explore a problem situation or system, recognise patterns in its components and infer simple relationships between these components. They can compare different plans or strategies and execute the best one.	Students at this level can recognize and understand relationships/connections/pat terns in data and calculate correctly within them. They can manipulate a representation and apply it successfully. They can critically evaluate a statement about a representation, identifying flaws in it.	Students at this level can explore, extract, and retrieve information to identify a common criterion or find some connection between elements in the task.
Level 2: Explore patterns	N/A	Students at this level can understand a new, simple mathematical concept and apply it, but only partially successfully.	Students at this level can use background knowledge to explore elements of the task to find information that is explicitly stated. They search for patterns and connections between elements and may find them by trial and error.

Journal of Physics: Conference Series **1379** (2019) 012059 doi:10.1088/1742-6596/1379/1/012059

The results of this research have also raised important points regarding the test design of these measured GPS tests. The PISA CBA 2012 tests seem to have the features of a proficiency test. While the content of the PISA CBA 2012 released tests covered areas of domain-specific literacy, they also tested some areas of GPS skills such as Level 3 to Level 5. Though the GPS skills tested were relevant to the fundamental GPS construct, the reality is that not all the levels had been tested by the released items. On the other hand, it must be acknowledged that the designers of PISA CBA 2012 tests had done a good job in assembling items to measure PS in different subject areas. Although it seems no explicit assumption of a latent trait with specific levels of GPS ability had been nominated by OECD, most of the literacy domains tend to measure this GPS ability.

References

[1] Baird L L 1983 Review of problem solving skills. *ETS Research Report Series* **1**

[2] Fischer A, Greiff S, Wüstenberg S, Fleischer J, Buchwald F and Funke J 2015 Assessing analyticand interactive aspects of problem solving competency *Learn. Individ. Differ.* **9** pp 172–9

[3] Messick S 1989 Validity, ed R L Linn *Educational Measurement* (New York: American Council on Education) pp 13–103

[4] Tran H P, Griffin P and Nguyen C 2010 Validating the university entrance English test to theVietnam National University: A conceptual framework and methodology *Proced. Soc. Behv.* **2** pp 1295–304

[5] Rahman S, Griffin P and Zang Z 2016 Development and validation of the construct of problem solving *Prof. Australian Association for Research in Education Conf. 2016* (Melbourne, Australia: AARE)

[6] Callingham R and Griffin P 2000 Towards a framework for numeracy assessment. *23rd Annual Conf. of the Mathematics Education Research Group of Australasia Incorporated* pp 134-141

[7] Callingham R and Griffin P 2001 Beyond the basics: Improving indigenous students' numeracy. *24th Annual Conf. of the Mathematics Education Research Group of Australasia Incorporated* pp 128-136

[8] Mach K J, Mastrandrea M D, Freeman P T and Field C B 2017 Unleashing expert judgment in assessment *Sci. Tech. (Global Environmental Change)* **44** pp 1–14

[9] UNESCO 2015 *Level-setting and Recognition of Learning Outcomes*, ed J Keevy and B Chakroun (Paris, France: The United Nations Educational, Scientific and Cultural Organization)

[10] Izard J 2005 Overview of test construction *Quantitative Research Methods in Educational Planning* ed K N Ross (Paris: UNESCO International Institute for Educational Planning) pp 1–51

Legal metrological verification in health area in Brazil

A N Soratto, L L Nunes and G Cassol

National Institute of Metrology, Quality and Technology (INMETRO), Rio de Janeiro, Brazil.

E-mail: ansoratto@inmetro.gov.br

Abstract. Presents the method used by Inmetro to measure the coverage of periodic metrological verification of sphygmomanometry and clinical scales in Brazil in 2018. It was used the Metrological Verification Coverage Ratio, that represents the percentage of healthcare facilities that received the periodic metrological verification service on the total number of healthcare facilities in the country. The results reveal great opportunity for growth of the instrument verification in several sectors, helps to size of the workforce and provides strategic information for the planning and execution of mandatory metrological verification in Brazil.

1. Introduction

According to data from the Brazilian Ministry of Health, there are more than 300,000 health facilities in the country, most of which make use of various measuring instruments in their day-to-day activities.

Clinical scales and sphygmomanometers (blood pressure meters) are examples of important measuring instruments regulated by Inmetro - National Institute of Metrology, Quality and Technology, because blood pressure and body mass are fundamental information for the good medical diagnosis of a patient.

Therefore, sphygmomanometers and clinical scales are subject to legal metrological control, so that they present an adequate level of confidence in their measurements [1]. When in use, these instruments must receive a legal and annual metrological verification, performed by the 26 Public Bodies that integrate the Brazilian Network of Legal Metrology and Quality (RBMLQ). It is estimated that there are over 200,000 clinical scales and more than 500,000 sphygmomanometers in use in Brazil, in a various kind of health facilities [2, 3].

A major problem for Inmetro in the field of legal metrology was to find out how much of this great universe of measuring instruments was receiving the annual metrological verification of the RBMLQ.

In this sense, Inmetro has carried out studies in order to expand the periodic metrological verification in clinical scales and sphygmomanometers [4, 5], and to improve the quality of measurement of these instruments [6].

In one of these works, Inmetro and RBMLQ developed the MVCR - Metrological Verification Coverage Ratio, an indicator that shows the capacity of Inmetro and each RBMLQ Public Body to meet the demand for mandatory metrological verification in measuring instruments [6].

The MVCR is the ratio of the number of facilities that received RBMLQ services, like metrological verification in clinical scales, and the number of facilities subject to RBMLQ services.

Journal of Physics: Conference Series **1379** (2019) 012060 doi:10.1088/1742-6596/1379/1/012060

The objective is presents the MVCR calculation report and its results in some sectors in health area in Brazil, like hospitals, health clinics and doctors' office, among others. The paper will be presents the MVCR ranking of all 26 RBMLQ Public Bodies, in a composition of sectors in health area.

2. Legal metrological control in clinical scales and sphygmomanometers in Brazil

The consolidation of the legal framework of metrological control in health in Brazil began in the 90's with the establishment of Technical Metrology Regulations for non-automatic weighing instruments - weighing scales for adults and children, thermometers and sphygmomanometers.

Currently, health scales and sphygmomanometers are subject to the following controls specified in regulations:

a) Technical model assessment: this is the approval procedure for the prototype of the instrument, comprising systematic tests of the performance of the measurements.

b) Initial verification: this is the first verification of the instrument after it has been manufactured to confirm that it retains the metrological characteristics of the approved model in the technical assessment.

c) Periodic verification: it is the subsequent verification, carried out annually on the instruments in use to contact if it maintains the reliability of the measurements.

d) Verification after repair: is the verification performed after the instrument has been repaired.

e) Inspection: is the verification conducted on a special inspection operations or for the care of complaints or legal expertise.

Figures 1 to 6 present traditional examples of mechanical and digital scales for weighing adults and children, and mechanical and digital sphygmomanometers.

Figure 1. Mechanical scale – adult.

Figure 2. Digital scale – adult.

Figure 3. Pediatric digital scale.

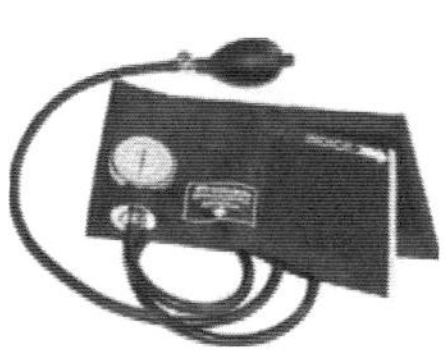

Figure 4. Mechanical aneroid sphygmomanometer.

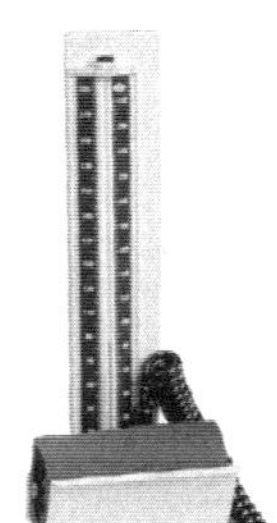

Figure 5. Mechanical Sphygmomanometer mercury column.

Figure 6. Digital Sphygmomanometer.

This study focuses on the scope of Periodic Verification in the universe of these instruments in use in Brazil, by calculating the index of coverage of this service.

3. Metrological Verification Coverage Ratio (MVCR)

One of the most important tools used by Inmetro to follow the metrological verification services in Brazil is the SIMCS - System for Monitoring the Coverage of RBMLQ Services, whose main result is the MVCR - Metrological Verification Coverage Ratio [6].

The MVCR shows the percentage (%) of facilitie that the RBMLQ Public Bodies has the capacity to visit, within the universe of existing facilities that make use of measuring instruments regulated by Inmetro [6].

In this way, the MVCR is a percentage calculated by the ratio between the visited facilities and the total of facilities subject to visit:

$$MVCR_{i;j} = (VF/UF)_{i;j},$$

where:

MVCR is metrological verification coverage ratio;

VE is visited facilities by RBMLQ public bodies;

UE is universe of facilities engaged in economic activities using regulated measuring instruments;

i is index that represents any sector on commerce, industry or service providers;

j is index that represents any geographical areas (city, state, region or country).

The data sources of the VE and AE are, respectively, the Inmetro's Integrated Management System (SGI) and the Annual Social Information Report (RAIS) of the Brasilian Ministry of Labor and Employment (MTE).

One important SIMCS filter is the code and the description of the economic activities, that are selected by association with the use of measuring instruments regulated by INMETRO.

SIMCS is a systematized database that monitors the coverage of periodical metrological verification services in Brazil, in several geografic areas and several economic activities. It can calculate MVCR in any sector on commerce, industry and service providers, in geographical areas that can range from a municipality or state, to even a big region or the entire country.

For this paper, it was extracted information related to some kind of health facilities that make use of clinical scales and sphygmomanometers in a hole country and in each of the 26 states of Brazil.

4. Results and discussions

In this section, the Metrological Verification Coverage Ratio (MVCR) reveals the percentage of establishments visited by RBMLQ, to perform periodic metrological verification services on clinical scales and sphygmomanometers.

Table 1 presents the Coverage Ratio of periodic verification in some kind of health facilities in Brazil in 2018: hospitals, health clinics, doctors'office, pharmacies and fitness center. It is important to note that this study does not contemplate the Municipal Health Posts, because these establishments were not included in the database used.

The results show a great potential for growth in the number of facilities visited and the consequent number of new periodic checks on clinical scales and sphygmomanometers.

Although Brazil has a robust framework of regulations for legal metrological control of clinical scales and sphygmomanometers, there is a long way to go to establish this control periodically of the instruments that are in use.

Table 1. Metrological Verification Coverage Ratio in health area - Brazil/2018.

Economic Activity	Nº Establishments	MVCR 2018
Hospitals	8,868	7.3 %
Health clinics	16,612	1.3 %
Doctors'office	40,823	0.7 %
Pharmacies	60,092	2.7 %
Fitness center	15.731	0.7 %

The MVCR reveal that there is a large gap in the periodic metrological control in Brazil in all health sectors studied. It helps in directing investments to better meet such demand.

Table 2 presents the MVCR ranking of all 26 RBMLQ Bodies (plus DF – Capital of Brazil) in 2018 in a group formed by the just following medical facilities: hospitals, health clinics and doctors'office.

Table 2. MVCR Ranking – medical scales and sphygmomanometros – Brazil/2018.

Position	RBMLQ Body	Number of Establishments	MVCR	Position	RBMLQ Body	Number of Establishments	MVCR
1	SE	518	15.6 %	15	PB	788	3.1 %
2	AL	488	11.3 %	16	SC	2,888	1.9 %
3	AM	482	7.9 %	17	RS	5,264	1.8 %
4	CE	1,496	6.9 %	18	MG	7,765	1.7 %
5	PE	1,686	6.1 %	19	SP	18,397	1.6 %
6	AP	90	5.6 %	20	RO	390	1.5 %
7	MS	956	4.5 %	21	MT	1,056	1.5 %
8	PI	563	4.4 %	22	DF	1,434	0.9 %
9	RN	740	4.3 %	23	RJ	7,240	0.7 %
10	AC	149	4.0 %	24	GO	1,986	0.5 %
11	ES	1,668	3.8 %	25	PA	1,097	0.4 %
12	MA	703	3.7 %	26	TO	411	0.0 %
13	BA	3,236	3.6 %	27	RR	84	0.0 %
14	PR	4,728	3.3 %	-	**BRAZIL**	**66,303**	**2.3 %**

The average MVCR of RBMLQ in hospitals, clinics and doctor's offices in 2018 was only 2.3 %. Even in the State of Sergipe (SE) that had the highest MVCR, with 15.6 %, there is a great possibility of expanding coverage of periodic verification in health facilities. This results denotes the need for structural and management actions throughout Brazil, such as:

a) Expand and train the number of metrological verification teams in all RBMLQ Bodies to work in the health area.

b) Establish goals for this service in the Annual Work Plan that the RBMLQ Bodies agrees and sign with Inmetro.

c) To make the owners of health establishments aware of the importance of periodic metrological verification of scales and sphygmomanometers for the correct diagnosis and medication of patients.

Knowing the MVCR is fundamental to dimension the quantitative of the labor force and of the vehicles and equipment needed to meet the real demand for periodic verification. It allows the comparison of performance between the RBMLQ Bodies, and is also important to define strategic planning scenarios and goals.

5. Conclusion

This survey represents a very close picture of the periodic metrological verification coverage of clinical scales and sphygmomanometrs use in the health facilities in Brazil.

The Metrological Verification Coverage Ratio (MVCR) is an importante indicator to monitoring the legal metrology control. In 2018, the MVCR average of clinical scales and sphygmomanometers in hospitals, health clinics and doctors'office was only 2.3 %, revealing great opportunity for growth of these services.

By showing the gap between the visited facilties and the total volume of facilities subject to the visit, the MVCR becomes valuable information in the adjustment of the current capacity of the RBMLQ, so that it can meet the demand for its periodic verification services.

Some potential impacts of MVCR use in Brazil are:

a) reveals an estimate of demand for periodic verification services at each location in the country;
b) provides information for dimensioning the number of verification teams, vehicles and equipment;
c) allows the comparison of performance between the RBMLQ Bodies;
d) supports strategic planning and goals.

With a simple calculation memory, the MVCR can be replicated in any country that has basic information about the healthcare facilities and metrology services performed.

Acknowledements

The authors thank Inmetro, Imetro/SC and Mr. Ricardo Basso Giordani, Specialist in Database Programming.

References

[1] Rodrigues Filho 2018 *Clinical methods and pathophysiology* **23** n.3 p 141
[2] Paiva M R, Pohlmann Filho O and Soratto A 2015 *J. Phys.: Conf. Ser.* **575** 012047
[3] Rodrigues Filho BA, Soratto A N and Gonçalves R *Advances in Production Management Systems* **459** p 121
[4] Soratto A N, Rodrigues Filho B A and Nunes L L 2018 *J. Phys.: Conf. Ser.* **1044** 012061
[5] Soratto A N, Pohlmann Filho O, Paiva M R and Nunes L L 2015 *IV Congr. Brasileiro de Metrologia* (Bento Gonçalves/RS)
[6] Soratto A N, Pohlmann Filho O, Paiva M R, Giordani R B and Bringhenti C 2014 *International J. Metrol. Qual. Eng.* **5**(3) p 5

Journal of Physics: Conference Series **1379** (2019) 012061 doi:10.1088/1742-6596/1379/1/012061

Measuring the oscillometric method contribution to the measurement uncertainty of automated sphygmomanometers using a patient simulator

B A Rodrigues Filho[1], **R F Farias**[1] **and W E dos Anjos**[2]

[1]National Institute of Metrology, Quality and Technology – Inmetro, 50 Nossa Senhora das Graças Av., Xerém, Duque de Caxias, RJ, Brazil
[2]Weights and Measures Institute of São Paulo – IPEM-SP, 154 Muriaé St., Ipiranga, São Paulo, SP, Brazil

E-mail: bafilho@inmetro.gov.br

Abstract. The sphygmomanometer is a widely used device for non-invasive blood pressure measurement. Systematic measurement errors and uncertainties can affect measurements, consequently, impacting diagnosis or medication. In order to maintain the conformity of the instruments, sphygmomanometers are usually under control of legal metrology demanding periodical verifications. Automated sphygmomanometers are based on the oscillometric method, measuring small changes of arterial blood pressure occurring in the cuff. Once pressure working standards are unable to generate the oscillometric pulses, verifications are performed on the sphygmomanometer on its manometer mode, disregarding the contribution of the oscillometric mode to the uncertainties. The present study investigates the contribution of the oscillometric method to the expanded measurement uncertainty. We used a patient simulator in four sphygmomanometers available in the market for consumers. The results show that the contribution of the oscillometric mode to the expanded uncertainty can be significant, varying from 0.16 to 1.56 mmHg and from 0.12 to 1.55 mmHg, for 100 and 150 mmHg, respectively. We also observed different contributions to the uncertainty for systolic and diastolic blood pressure.

1. Introduction

The sphygmomanometer is the device used for non-invasive blood pressure (BP) monitoring, widely used in primary care and health care, by practitioners, physicians and doctors in order to diagnose hypertension or monitor patients for medication. It is a simple procedure conducted also by individuals at home to monitor their condition. However, systematic deviations due to measurement errors and uncertainties can affect the results, consequently, impacting the diagnosis. For example, a healthy patient may be treated as a hypertensive one or a hypertensive one may not receive the proper treatment. Literature shows that 32.95 % of healthy patients would be diagnosed as hypertensive for a + 5 mmHg measurement error associate to ± 5 mmHg uncertainty, considering the diastolic pressure [1].

Sphygmomanometers are usually under the legal metrology regulation due to their impact on health, demanding verifications according to national requirements. The International Organization of

Legal Metrology (OIML) issues international standards, used as a recommendation applied by members and associate countries. Consequently, the instrument is subjected to type approval, verification and surveillance, i.e. the legal metrology control activities [2].

Automated sphygmomanometers have an increasing percentage in the market when compared to aneroid devices and mercury-based ones, which are being discontinued due to the poisonous aspect of mercury.

The measurement of automated sphygmomanometers is based on the oscillometric method, where an algorithm developed by each manufacturer, using empirical data, is responsible to recognize the small changes of arterial blood pressure occurring in the cuff. Those oscillations are used to identify both systolic and diastolic pressure [3]. Consequently, the oscillometric-based devices rely on previous clinical investigations in order to validate the manufacturer methodology [4].

Although the sphygmomanometers are submitted to verification after being placed on the market, the periodic verification tests do not represent the real use of the instrument. Since a working pressure standard is unable to generate small oscillations in pressure, simulating the blood pressure measurement, automated sphygmomanometers have a "testing mode", where the oscillometric algorithm is disabled, testing only the manometer, and not all the aspects of the instrument. Consequently, an import metrological aspect regarding the oscillometric algorithm is not being tested during metrological verifications.

Moreover, for diagnostic purposes, since BP is affected by external factors, such as anxiety, several measurements are usually carried out in order to mitigate these factors [5]. So, the number of measurements taken increases the correctness in categorization the patients as healthy or hypertensive according to the treatment threshold [6]. Consequently, the sphygmomanometer repeatability plays an important aspect in this BP measurement methodology.

In order to verify the contribution of the oscillometric method to the sphygmomanometer performance, the present study carries out a test using a patient simulator in order to measure the contribution of the oscillometric component to the expanded measurement uncertainty.

2. Methodology

In order to check the metrological accordance of sphygmomanometers in-use conditions, we randomly purchased four different brands easily found on the Brazilian market. Despite the wide selection of brands available, we chose only those manufacturers whose metrological assurance is under the metrological supervision process. All tested instruments have a type approval certificate issued by Inmetro, the National Metrology Institute in Brazil. Two instruments take measures from the wrist and two from the upper arm, representing the most common sphygmomanometers found on the market for consumers. Sample characteristics are shown in table 1. For impartibility and confidentiality, the manufacturers are not disclosed.

Table 1. Sample of tested sphygmomanometers.

Sphygmo.	Measurement Taken	Measuring Interval (mmHg)
1	Upper arm	30 – 280
2	Upper arm	0 – 299
3	Wrist	40 – 250
4	Wrist	30 – 280

Both accuracy of the electromechanical transducer and repeatability were tested in order to check the behavior of the sample. The applied measurement procedure for the accuracy test is in accordance with the OIML Recommendation [7], for automated sphygmomanometers verification. The contribution of the oscillometric method is evaluated by the repeatability of sphygmomanometers using a Dynatech Nevada Inc., model Non-Invasive Blood Pressure Analyser Cuff Link, whose repeatability = 0.6 mmHg is appropriate to test automated sphygmomanometers [8-9]. The patient simulator is a device that simulates the oscillation of blood pressure enabling metrological testing of

Journal of Physics: Conference Series **1379** (2019) 012061 doi:10.1088/1742-6596/1379/1/012061

oscillometric automated sphygmomanometers in conditions representing a real individual. Figure 1 shows the experimental set-up.

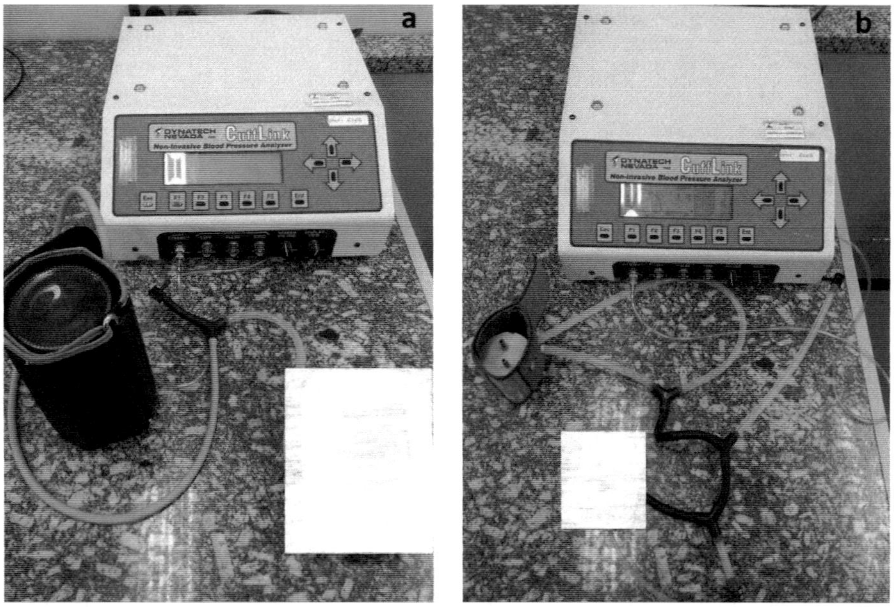

Figure 1. Experimental set-up for upper arm (a) and wrist (b) sphygmomanometers. Devices are blurred for confidentiality.

The measurements were performed within the recommended environmental conditions which range from 15° to 25° C for temperature, and 20 % to 85 % for relative humidity [7]. The measurement uncertainty is computed for both manometer and oscillometric mode according to the procedure described by the ISO GUM [10].

3. Results

Table 2 shows the comparison of the expanded measurement uncertainty regarding manometer ($U_{mano.}$) and the oscillometric contribution for systolic ($US_{Oscill.}$) and diastolic ($UD_{Oscill.}$) BP. As each sphygmomanometer (sphygmo.) presents different repeatability for systolic and diastolic blood pressure, using the patient simulator, we observed distinct expanded uncertainties for either systolic or diastolic BP. The individual contribution to the combined uncertainty, due to the repeatability of the oscillometric mode, is seen in figure 2.

Table 2. Expanded measurement uncertainty for manometer-only and the manometer plus oscillometric mode.

Standard Value	**Sphygmo. 1 (upper arm)**			**Sphygmo. 2 (upper arm)**			**Sphygmo. 3 (wrist)**			**Sphygmo. 4 (wrist)**		
(mmHg)	$U_{mano.}$	$US_{Oscill.}$	$UD_{Oscill.}$	$U_{mano.}$	$US_{Oscill.}$	$UD_{Oscill.}$	$U_{mano.}$	$US_{Oscill.}$	$UD_{Oscill.}$	$U_{mano.}$	$US_{Oscill.}$	$UD_{Oscill.}$
50.00	0.59	-	0.67	0.84	-	0.91	0.59	-	0.66	0.16	-	0.33
80.00	0.92	0.96	0.97	0.84	0.89	0.88	0.59	0.67	0.62	0.16	0.16	0.16
100.00	0.84	-	0.90	0.84	-	0.84	0.59	-	0.59	0.16	-	1.56
120.00	0.59	0.69	-	0.84	0.89	-	0.59	0.73	-	0.16	0.16	-
150.00	0.59	1.16	1.15	0.59	0.69	0.65	0.59	0.65	0.74	0.12	0.12	1.55
200.00	0.59	0.88	-	0.84	0.95	-	0.59	0.87	-	0.11	0.11	-

Journal of Physics: Conference Series **1379** (2019) 012061 doi:10.1088/1742-6596/1379/1/012061

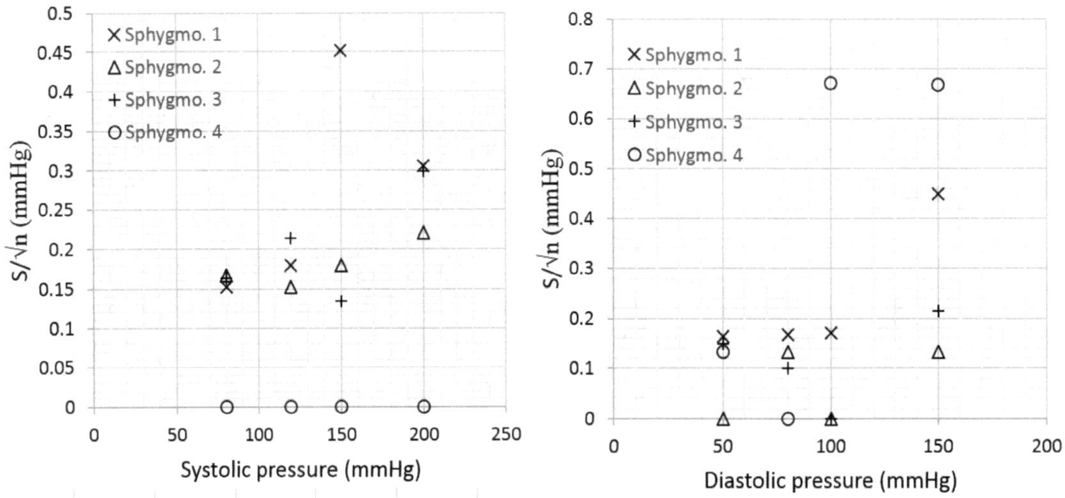

Figure 2. Contribution from oscillometric repeatability to the expanded uncertainty, where n=10.

The contribution due to the oscillometric method is important, especially for the sphygmomanometer 4, where the expanded uncertainty for diastolic BP = 100 and 150 mmHg varies from 0.16 to 1.56 mmHg and from 0.12 to 1.55 mmHg, respectively. Sphygmomanometer 1 presents a similar behavior for BP=150 mmHg, where the expanded uncertainty varies from 0.59 to 1.16 mmHg for systolic BP and from 0.59 to 1.15 mmHg for diastolic BP. For the sphygmomanometer 1 to 4, we also observed an increasing contribution of the oscillometric component in the higher BP points, for both systolic and diastolic BP, as observed in figure 2.

Based on the contribution of the uncertainties due to the repeatability of the oscillometric mode and expanded uncertainty we conclude that expanded uncertainty is independent of the measurement taken location (wrist or upper arm), for the tested sphygmomanometers. For example, the sphygmomanometer 4 presented the smallest contributions to the uncertainty for systolic and the greatest ones for diastolic BP, according to figure 2.

Although the observed discrepancies, it is not possible to conclude which specific component of the programmed oscillometric mode contributes the most for the divergences, such as frequency of data acquisition, air inflate and release rate. A further investigation is necessary in order to rank the contribution of each of those components individually, however, the oscillometric program is kept confidential by manufacturers.

4. Conclusion

The sphygmomanometer is the device used for non-invasive blood pressure measurement, and more specifically, to control hypertension. A proper measurement is a vital aspect for proper diagnosis and use of medication. Consequently, the present study showed the contribution of the oscillometric method to the expanded measurement uncertainty.

This study revealed that all expanded uncertainties of tested devices, considering systolic and diastolic pressure, were influenced by the oscillometric contribution, especially for sphygmomanometers 1 and 4. We also concluded that the contribution of the oscillometric method is not impacted by the measurement location (wrist or upper arm).

Acknowledgments

The authors acknowledge the support of FAPESP (grant 2017/50173-0), which enabled us to carry out this research.

Journal of Physics: Conference Series **1379** (2019) 012061 doi:10.1088/1742-6596/1379/1/012061

References

[1] Rodrigues Filho B A, Farias R F and dos Anjos W 2018 Evaluating the impact of measurement uncertainty in blood pressure measurement on hypertension diagnosis *Blood Press. Monit.* **23** 141-7

[2] Klenovsky P 2006 Legal metrological control over measuring instruments in use: current situation *OIML Bulletin* **XLVII** 22–33

[3] Amoore J N 2012 Oscillometric sphygmomanometers: a critical appraisal of current technology. [Review] *Blood Press. Monit.* **17** 80–8

[4] ISO 2013 *ISO 81060-2:2013(en) Non-invasive Sphygmomanometers — Part 2: Clinical Investigation of Automated Measurement Type*

[5] Padial L R, Fragoso A S, Moreno F J A, Arias M A, Castro A V and Roca G C R 2017 Blood pressure differences between one automatic measurement and the mean value of 3 automatic measurements. SPRINT trial *Med. Clin.-Barcelona* **149** 72–4

[6] Powers B J, Olsen M K, Smith V A, Woolson R F, Bosworth H B, Oddone E Z 2011 Measuring blood pressure for decision making and quality reporting: where and how many measures? *Ann. Intern. Med.* **154**(12) 781-8

[7] OIML 2002 *OIML 2002 R 16 - 2 Non-invasive Automated Sphygmomanometers* (International Organization of Legal Metrology)

[8] Farias R F and Neto J A G 2015 Evaluation of the repeatability of non-invasive blood pressure simulators *Proc. Metrologia 2015*

[9] Farias R F 2015 *Evaluation of the Effectiveness of Legal Metrological Control of Automated Sphygmomanometers* (Duque de Caxias, RJ: National Institute of Metrology, Quality and Technology - Inmetro)

[10] Joint Committee for Guides in Metrology 2008 *Guide to the Expression of Uncertainty in Measurement (GUM)* (Brookhaven National Laboratory)

Journal of Physics: Conference Series **1379** (2019) 012062 doi:10.1088/1742-6596/1379/1/012062

The feasibility of using an international common reading progression to measure reading across languages: a case study of the Vietnamese language

C Nguyen

Melbourne Graduate School of Education University of Melbourne, Vic 3010 Australia

E-mail: cuc@unimelb.edu.au

Abstract. In the last few decades, student achievement in many countries has been assessed and reported against a developmental learning continuum which underpins the knowledge and skills being measured. There have been efforts to place students from different countries on the same scale. However, those efforts have often been limited to individual projects and countries. There has been no common framework and no common metric for assessments to be compared across countries and levels. Recently, the UNESCO Institute for Statistics (UIS) has worked with the Australian Council for Educational Research (ACER) to develop UIS reporting scales for reading and mathematics with the aim of enabling countries to examine and report using a common framework and common quality control. This paper presents the results of an evaluation of the feasibility of adapting and applying the UIS reporting framework to the reading of Vietnamese language.

1. General description of research questions, objectives and theoretical framework

1.1. Introduction

In the past few decades, there has been a tendency for countries to assess and report student achievement at a national level against developmental continua. There have also been international and regional efforts in assessing and reporting student achievement among many countries using a common framework, such as the Trends in International Mathematics and Science Study (TIMSS), the Progress in International Reading Literacy Study (PIRLS) and the Programme for International Student Assessment (PISA). However, there is a lack of a unified international framework across projects enabling countries to examine and report the outcomes of these assessment activities [1].

Without a common framework, it is difficult to compare and monitor student learning outcomes at international levels, and especially difficult for developing countries to compare their student learning outcomes with those of other countries, as learning progressions may not be used in those countries or, if they are used, the validity and reliability of the scales in use may be questionable.

Yet, the United Nations Education 2030 agenda reinforces the need to monitor students' achievement in terms of proficiency levels at the end of educational cycles such as primary and lower secondary school [2]. With the aim of enabling countries to examine and report student outcomes using a common framework, the Australian Council for Educational Research (ACER) has been working with the UNESCO Institute for Statistics (UIS) to create common reporting scales for reading which can be applied in many languages.

1.2. The reading framework

In the framework of the UIS reading scales [1], reading concepts and knowledge are grouped into two broad domains: comprehension skills and constrained skills. Comprehension skills are considered unconstrained in their development. This means that comprehension skills include a set of skills that continue to develop before, during, and after formal schooling. Comprehension has been conceptualised to includes three strands: retrieving, interpreting and reflecting. [1]

Constrained skills are the decoding skills required to understand the symbolic representation of a language as written text. As such, the decoding skills are required to convert written symbols into words in a particular language. These skills involve smaller numbers of discrete things to be learned (e.g. the letters of the alphabet) and they are usually mastered completely in a few years [3]. They are referred to as 'constrained' because they are constrained in scope and time of acquisition. Constrained skills are fundamental and need to be fully mastered in order to become a proficient reader [1]. Constrained skills can also be referred to as 'pre-reading' skills.

For each of strand of the UIS reading framework, ACER experts worked together to map out important milestones of the "underlying developmental progression from simple to complex skills applied to familiar then progressively more complex texts" [1].

1.3. Research questions

As the objective of the UIS reading scales is to develop empirically derived common reporting scales in mathematics and reading that will support governments to effectively measure and monitor learning outcomes, there is a need to evaluate the extent to which the scales are compatible with teaching curriculum and assessments from a range of countries. This paper presents the evaluation of the extent to which it is feasible to translate or adapt the UIS learning and reporting framework for use with the Vietnamese language. The questions addressed by this paper are:

- To what extent does the UIS reading scale align with the Vietnamese curriculum?
- To what extent can the UIS reading scale be used in the Vietnamese context?
- If necessary, how can the UIS Reading scale be adapted to the Vietnamese context?

2. Methodology

The method used to conduct the evaluation of the extent to which the UIS reading scale aligns with the Vietnamese curriculum was document analysis. Vietnamese language and literature curriculum from early childhood to year 12 [4] were analysed to establish the extent to which the UIS reading scale aligns with the Vietnamese curriculum and whether the milestones along the UIS reading scale are comparable with the milestones in the developmental stages of reading in Vietnamese.

3. Results

3.1. Alignment of the UIS reading scale with the Vietnamese curriculum

In Vietnam, the two subjects that are related to the UIS reading scale are 'Vietnamese' for Primary Education Grades 1 to 5, and 'Phonetics and Literature' for secondary Education Grades 6 to 12. Reading is one of the four language skills taught in these two subjects: reading, speaking, writing and listening. The extent to which reading is covered at each of the grade levels in the 2006 Vietnamese curriculum is presented in table 1.

Comparing the Vietnamese curriculum [4].and the UIS reading scale, the following are noted:

- The strands of the UIS reading scale are not clearly stated in the Vietnamese curriculum. The structure of the reading section of the curriculum of the two subjects 'Vietnamese' and 'Phonetics and Literature' does not completely align with the structure of the UIS reporting scale in reading: 'Retrieving,' 'Interpreting' and 'Reflecting' are placed in the section 'Reading Skills'; Constrained skills such as 'Decoding and concept of print' and 'Vocabulary' are partly

Table 1. Vietnamese curriculum in reading [4].

Skills			Grade 1	2	3	4	5	6	7	8	9	10	11	12
Reading fluency	Read aloud	Accurate reading	*	*	*	*	*							
		Emotionally expressive reading				*	*	*	*	*	*	*	*	*
		Scanning and salience reading		*	*	*	*	*	*	*	*	*	*	*
Reading Understanding		Word and sentence meaning	*	*	*	*	*	*	*	*	*	*	*	*
		Content of paragraphs and documents		*	*	*	*	*	*	*	*	*	*	*
		Indirect meanings of words and implications of sentences				*	*	*	*	*	*	*	*	*
		Authors' ideas and emotions					*	*	*	*	*	*	*	*
		Artistic values and ethical meanings of a story/novel/book				*	*	*	*	*	*	*	*	*
		Meanings of documentation				*	*	*	*	*	*	*	*	*
		Links between reading documentations and individual life and practice			*	*	*	*	*	*	*	*	*	*
Application of reading skills		Rote Learning	*	*	*	*	*	*	*	*	*	*	*	*
		Research of documentation			*	*	*	*	*	*	*	*	*	*

placed in the section 'Knowledge of Vietnamese language' and partly placed in the section 'Reading Skills.' However, in general, the UIS reporting scale in reading is still applicable in the Vietnamese context.

- While there is not a clear progression continuum across the standards of Grades 1 to 12, the 'Decoding and concept of print' in the Vietnamese curriculum appears to have higher levels than level 7 of the UIS reading scale 'Decoding and concept of print' strand. It can be noted that, from Grade 4 to Grade 12, Vietnamese students continuously work to improve emotionally expressive reading skills.
- The speed of reading is one of the elements of the Vietnamese reading progression.
- There is a slight mismatch between the UIS reading scale and the Vietnamese curriculum in the 'Reflecting' and 'Interpreting' strands: the Vietnamese 'Reflecting' and 'Interpreting' strands focus heavily on moral and ethical implications;
- Reasoning levels are not a focus of Vietnamese reading;
- The scoring rules of the sample items drafted for 'Reflecting' and 'Interpreting' strands may

be a surprise to teachers and education officers in Vietnam as there are multiple correct options allowed. Students in Vietnam are likely to have been taught to think 'the same way,' and Vietnamese test developers and teachers would normally provide only one option as a correct answer. However, allowing multiple options can facilitate critical thinking and this approach is progressively being adopted in Vietnam as the country works with educators from other countries and encourages Vietnamese students to be critical and independent thinkers.

3.2. The extent to which the UIS reading scale can be used in the Vietnamese context

The Retrieving, Interpreting and Reflecting strands do not need adaptation. The level descriptions for these strands are appropriate in the Vietnamese context and can be directly translated into Vietnamese. The reading tasks can also be translated into Vietnamese. There is a minor point for revision in the Reflecting strand. The constrained skills, however, will need to be adapted due to the different features of Vietnamese in comparison to English.

3.3. Recommendations for further work for adaptation of the UIS reading scale to the Vietnamese context

The steps needed for adapting the UIS constrained skills into the Vietnamese context are:

- Step 1 Translation from English to Vietnamese
- Step 2 Workshop/focus group interviews with Vietnamese subject matter experts on selection of words for each level
- Step 3 Revision
- Step 4 Back-translation into English
- Step 5 Comparison of translation from step 4 with original English version and final revision if necessary.

The purpose of the workshops in Step 2 is to elicit the following:

- Are the descriptions of the levels of each of the elements of the constrained strand relevant for the Vietnamese context? Is there a need for modification? If so, what modifications are needed? (This exercise to be repeated with each of the elements to describe the continuum for all elements of the Constrained strand.)
- Are the elements in each level of the Constrained strand appropriate for that level?
- For each of the Constrained strand levels, what is the corresponding level in the Vietnamese standards?

4. Discussion

The study demonstrates, first, that it is feasible to apply the UIS reading scale in the Vietnamese language context, and, second, that the process of evaluating the feasibility of applying the UIS reporting scales in general can be applied to other languages and other disciplines such as mathematics and science. While rigorous document analysis and expert consultation are important for ensuring the validity and reliability of the conclusions drawn from the feasibility study, there is a need to develop international common items and Vietnam-specific items along the reading scale, and empirical data should be collected and analysed for this purpose. Re-interpretation of the reading scale, based on the location of items along the continuum, should be conducted [5,15]. The same process and data collection should be collected for all other languages if the countries where they are spoken want to adapt and use the UIS reporting scale. Comparison between the empirically derived scales from country data and the UIS reading scale would lead to a solidly validated UIS reporting scale in reading. There may be variation in constrained skills due to the uniqueness of languages. The extent to which constrained skills in other languages differ from those of English can be established. The Rasch model will play an important role in this process of validation [6] to compare item location on a scale and compare interpretations of an underlying continuum derived from data gathered from different languages.

The validated UIS reporting scales can facilitate international collaborations in teaching and assessment of reading skills. They potentially can also enhance the intercultural capability of students through exchange of reading materials and test items. Education in foreign languages will also benefit from the initiatives in the development of such a common framework in reading. For countries where competency-based education is being promoted and adopted, the common UIS framework in reading is a good starting point to inspire teachers to work collaboratively to elaborate progressions in the context of the grades that they teach. In this way, the UIS framework can enhance classroom assessment and not be limited to use in large scale surveys of student achievement and cross-country comparison.

References

[1] Turner R *et al.* 2018 *Development of Reporting Scales for Reading and Mathematics: a Report Describing the Process for Building the UIS Reporting Scales* (The Australian Council for Educational Research)
[2] United Nations 2015 *Transforming Our World: the 2030 Agenda for Sustainable Development* https://sustainabledevelopment.un.org/post2015/transformingourworld
[3] Paris S G 2005 Reinterpreting the development of reading skills *Read. Res. Quart.* **40** (2) pp 184–202
[4] Vietnamese Ministry of Education and Training 2006 *Curriculum Standards for Vietnamese Language Grade 1 to 5 Vietnam* (Hanoi: Vietnamese Ministry of Education and Training) p 4
[5] Vietnamese Ministry of Education and Training 2006 *Details of Teaching Curriculum for Vietnamese Language Grade 1 to 5*
[6] Vietnamese National Institute of Educational Sciences 2016 *Grand Curriculum for Vietnamese General Education* (Hanoi: Vietnamese Ministry of Education and Training)
[7] Vietnamese Ministry of Education and Training 2013 *Text Book for Teaching and Learning Spelling, Grade 1* (Hanoi: Vietnamese Ministry of Education and Training)
[8] Vietnamese Ministry of Education and Training 2015 *Standards for Early Childhood and Preschool Education* (Hanoi: Vietnamese Ministry of Education and Training)
[9] Nguyen T K C and Griffin P 2013 Developing of two instruments to measure attitudes of Vietnamese parents and students toward schooling. *Journal of Applied Measurement*, **14**(3) 282-298
[10] Nguyen T K C and Griffin P 2007 Development of a scale to measure economic status of students in rural Vietnam *Asia Pac. Educ. Rev.* **8**(2) 205-15
[11] Rasch G 1960 *Probabilistic Models for Some Intelligence and Attainment Tests* (Copenhagen: Danmarks Paedagogiske Institut)

Journal of Physics: Conference Series **1379** (2019) 012063 doi:10.1088/1742-6596/1379/1/012063

Methods to ensure the reliability of measurements in the age of Industry 4.0

E A Budylina[1] and A A Danilov[2]

[1] Moscow Polytechnic University, Moscow, Russia
[2] The State Regional Center of Standardization, Metrology and Tests in the Penza region, Penza, Russia

E-mail: aa-dan@mail.ru

Abstract. The methods of increasing the reliability of measurement results based on the use of structural redundancy in the implementation of measuring instruments and multidimensional models, calibration of measuring instruments (including remote) and methods of processing large arrays of measurement results are considered. It is noted that these methods will be demanded by mass production in the age of Industry 4.0.

1. Introduction

In an era of Industry 4.0 a lot of attention should be given to the reliability of primary measurements. To ensure the accuracy the measurement results can be used in various ways, for example:

– the use of structural redundancy in the implementation of measuring instruments and multidimensional models;

– the use of calibration of measuring instruments, in the perspective of remote calibration of measuring instruments using the Internet of Things technology;

– the use of techniques for processing a large array of measurement results (big data), etc.

2. State in the redundancy application field

The addition of redundancy in the measuring instruments structure gives an opportunity:

- to increase reliability due to the addition of redundancy in expanse by using several measuring channels of the same type functioning in parallel [1];

–to carry out self-calibration in order to reduce the systematic components of error (additive, multiplicative, from nonlinearity) by adding amendments to the measurement results due to the addition redundancy in expanse and time by using built-in calibrators (more accurate than the calibrated measuring instrument) and additional measurement cycles [2, 3];

– to reduce the systematic components of the error from external influencing quantities (temperature, humidity, etc.) by adding redundancy in expanse and time – through the use of additional measurement channels of external influencing quantities and the addition of appropriate amendments to the measurement results, etc. [3, 4];

– to carry out self-check and self-diagnostics of measuring instruments, including the evaluation of the critical component of the error [5] and evaluation of residual metrological resource [4, 5] due to the addition of redundancy in expanse and time.

3. The use of structural redundancy in the implementation of measuring instruments and multidimensional models

One of the ways to ensure the reliability of measurement results is the introduction of additional devices into the structure of measuring instruments (that is, the introduction of some redundancy). Additional devices may include tools for measuring operating conditions, such as ambient temperature, supply voltage, etc. This will allow for the time point t_i to get not only the value x_i of the measurement result of the measured quantity X, but also the value of $\xi_{1i};\ldots;\xi_{mi}$ measurement results of influencing quantities (operating conditions) $\Xi_1;\ldots;\Xi_m$. The measurement results of measurable and influencing quantities obtained at different points in time can be the basis for constructing multidimensional mathematical models. These mathematical models will introduce amendments in the measurements due to the change of impact values (operating conditions) of measuring instruments. For example, with minor deviations of the measurement results of the influencing quantities $\xi_{1i};\ldots;\xi_{mi}$ according to the values in which the measuring instrument was calibrated, i.e. $\xi_{10};\ldots;\xi_{m0}$, it is advisable to use a multidimensional model type

$$y_i = x_i + \alpha_1(\xi_{1i} - \xi_{10}) + \ldots + \alpha_m(\xi_{mi} - \xi_{m0}) \tag{1}$$

y_i – the value of the measured quantity X obtained in accordance with the model; x_i – the value of the measurement result of the measured value (reading of the measuring instrument) X; $\alpha_1;\ldots;\alpha_{mi}$ – coefficients of the influence of the influencing quantities on the reading of the measuring instrument of the measured quantity X.

In the case of a substantial systematic time drift of the characteristics of the measuring instrument, model (1) should be supplemented with one more component that takes into account the instability correction [4]:

$$y_i = x_i + \alpha_1(\xi_{1i} - \xi_{10}) + \ldots + \alpha_m(\xi_{mi} - \xi_{m0}) + \beta(t_i - t_c) \tag{2}$$

β – the drift rate; $(t_i - t_c)$ – time interval since the last calibration of measuring tools t_c.

As additional devices, built-in calibrators can also be introduced into the structure of measuring instruments [3]. As part of a measuring instrument, they can be used as multi-valued measures in the implementation of additional measurement cycles. The introduction of the results of additional measurement cycles into multidimensional models will make it possible to introduce corrections in measurement results due to instrumental uncertainty of measuring instruments.

For example, in the case of a linear transformation function of the measuring instrument X a multidimensional model can be constructed using two additional measurement cycles:

$$\begin{cases} y_{i1} = \varepsilon + (1+\eta)x_i \\ y_{i2} = \varepsilon + (1+\eta)x_{s1} \\ y_{i3} = \varepsilon + (1+\eta)x_{s2} \end{cases} \tag{3}$$

ε and η – the values of the additive and multiplicative components of the error of the measuring instrument; x_{s1}, x_{s2} – the values of X, reproduced by the built-in calibrator.

The solution of the system (3) relatively x_i gives:

$$x_i = \frac{y_{i3} - y_{i2}}{y_{i1} - y_{i3}} x_{s1} + \frac{y_{i1} - y_{i2}}{y_{i1} - y_{i3}} x_{s2}$$

that allows to make changes amendments to the measurement results due to the instrumental uncertainty of the measuring instruments, i.e. take into account the values of the additive and multiplicative components of the error of the measuring instrument.

Thus, the introduction of structural redundancy in the implementation of measuring instruments together with the use of multidimensional models will improve the reliability of measurement results.

4. Using calibration of measuring instruments

Calibration of measuring instruments is a traditional way to increase the reliability of measurement results obtained with its help. However, in an era of Industry it 4.0 should be further developed in the calibration of measuring instruments remotely. Preferably, when implementing the remote calibration of measuring instruments, the procedures namely without human intervention, provided for by the Internet of Things should be used:

– the analysis and identification of the need for calibration of measuring instruments (through the establishment of intervals between calibrations);

– to identify the moments in those the measuring instrument is not used for the purpose;

- work was carried out to search the network for necessary standards that meet the characteristics that would be available at the required time points (not occupied by other consumers) for calibrating the measuring instrument;

– work is carried out for the calibration of measuring tools, etc.

In contrast to the recommendations of ILAC-G24 / OIML D 10: 2007 [6] a new approach is offered below for determining the intervals between calibrations of measuring instruments, based on taking into account the component of measurement uncertainty caused by instability of the measuring instrument in the expanded measurement uncertainty, and comparing the later with some previously accepted maximum permissible measurement uncertainty is offered.

When calibrating the measuring instrument, the calibration certificate, together with the ratio between the values provided by the standards and the corresponding readings of the measuring instruments, also indicates the expanded uncertainty U, which is usually obtained by multiplying the summary standard measurement uncertainty u by the coverage factor k (corresponding to the accepted probability P):

$$U=ku \tag{4}$$

It should be noted that the value of the summary standard measurement uncertainty u is obtained in the calibration laboratory and belongs to the time of the calibration. At the same time, during the operation of the measuring instrument (after a certain time interval t from the moment of calibration), in addition to this, it is necessary to take into account the component of measurement uncertainty $u_{dritf}(t)$ due to the instability of the measuring instrument.

Then the summary standard measurement uncertainty u^*, taking into account, in addition to a summary of the standard measurement uncertainty u, obtained in the calibration laboratory, also the component of the uncertainty $u_{dritf}(t)$ from the instability of the measurement, will be equal to:

$$u^* = \left[u^2 + u^2_{drift}(t)\right]^{1/2} \tag{5}$$

Supposedly that after a certain time interval t from the moment of calibration, the component of measurement uncertainty due to the instability of the measuring instrument $u_{dritf}(t)$ will be $b(t)$ times greater than the summary standard measurement uncertainty u, i.e:

$$u_{dritf}(t) = ub(t) \tag{6}$$

where a certain function of time is $b(t)$ (in the simplest case, linear).

After the substitution (6) into (5) we get:

$$u^* = u\left[1 + b^2(t)\right]^{1/2} \tag{7}$$

In this case, the expanded uncertainty of measurements U^*, taking into account the component of uncertainty from the instability of the measuring instrument, will be equal to:

$$U^* = k^* u^* \tag{8}$$

where the coefficient of coverage is k^* (the value of which may be different from k).

After the substitution (7) into (8) we obtain:

$$U^* = k^* u\left[1 + b^2(t)\right]^{1/2} \tag{9}$$

or subject to (4):

$$U^* = U\left(k^*/k\right)\left[1 + b^2(t)\right]^{1/2} \tag{10}$$

Certainly, that the calibrated measuring instrument can continue to be used until, over time, the extended measurement uncertainty does not exceed some pre-accepted maximum allowable uncertainty *MPU*.

In particular, for determining the *MPU* may be used recommendations OIML G19 [7]:

$$MPU = f\,MPE \tag{11}$$

where f is the coefficient equals to 0.2 or 0.33; *MPE* is the maximum admissible error.

Then the right side of (10) should not exceed *MPU*, i.e.

$$U\left(k^*/k\right)\left[1 + b^2(t)\right]^{1/2} \le MPU \tag{12}$$

Using a new approach will allow us to objectively determine the need for remote calibration of measuring instruments. Expressive examples of measuring instruments in respect of which there is a huge demand for remote calibration of measuring instruments are:

– most of the measuring instruments for medical purposes, used by patients of medical institutions at home (primarily patients with disabilities);

– multichannel geographically distributed measuring systems, etc.

To perform remote calibration of measuring instruments, many issues will have to be resolved, such as:

– financially justified complication of the structure of measuring instruments in terms of the implementation of the possibility of calibrating remotely (for example, the introduction of a switch for connecting a measured quantity or a value realized using a standard, as well as the introduction of elements implementing information exchange protocols between devices complying with the Internet of Things standard);

– implement for the use as standards only of those standards, with the help of which calibration can be carried out remotely, for example, frequency (time) standards, etc.

The later will lead to the need for a significant change in the structure of the currently used built-in calibrators (referred to in paragraph 3 of this article), which should be based on frequency converters (time) to other quantities necessary for the built-in metrological control of measuring instruments.

5. Using methods for processing a large array of measurement results

The presence of a large array of measurement results allows us to solve many problems of increasing the reliability of measurement results, mainly when using multi-channel measurement systems:

– the use of a multidimensional model of the object of measurements together with an array of measurement results obtained using a multi-channel measurement system will not only clarify the type and parameters of the multidimensional model of the object, but also can serve as the basis for the construction of various criteria for checking the reliability of measurement results [8];

– the results of calibration of a set of similar measuring channels of a measuring system can be the basis for developing a methodology for assessing the reliability of values assigned to a standard during its calibration, etc.

When constructing these techniques, it should be remembered that the measurement result is not just a number, but a number with the corresponding measurement uncertainty.

6. Conclusions

The age of Industry 4.0 opens up new opportunities for improving the reliability of measurement results, which can be solved by using the following promising methods:

– use of structural redundancy in the implementation of measuring instruments and multidimensional models;

– the use of calibration of measuring instruments, in the long-term perspective, using the Internet of Things technology;

– the use of techniques for processing a large array of measurement results (big data), etc.

References

[1] Czichos H, Saito T and Smith L 2011 *Handbook of metrology and testing* (Heidelberg Dordrecht London New York: Springer) p 963
[2] Danilov A A and Kucherenko Yu V 2013 *Meas. Tech.* **55** 1447-50
[3] Danilov A A, Kucherenko Yu V, Berzhinskaya M V and Ordinartseva N P 2014 *Meas. Tech.* **57** 228-30
[4] Berzhinskaya M V and Danilov A A 2009 *Meas. Tech.* **52** 220-22
[5] Pronin A N, Saposhnikova K V and Taymanov R E 2015 *T-Comm* **9**(3) 32-37 (In Russian)
[6] ILAC-G24/OIML D 10: 2007 *Guidelines for the Determination of Recalibration Intervals of Measuring Equipment Used in Testing Laboratories*
[7] OIML 2017 *OIML G19:2017 The Role of Measurement Uncertainty in Conformity Assessment Decisions in Legal Metrology*
[8] Sheinin E M 2013 *Meas. Tech.* **55** 1442-46

Journal of Physics: Conference Series **1379** (2019) 012064 doi:10.1088/1742-6596/1379/1/012064

Synergy and completeness of simple A/D conversion and simple signal processing

P Sovilj[1], L Zupunski[1], B Vujicic[1], A Radonjic[2], V Vujicic[3] and D Kovacevic[4]

[1]Faculty of Technical Sciences, Trg Dositeja Obradovica 6, Novi Sad, Serbia,
[2]Institute of Technical Sciences of the Serbian Academy of Sciences and Arts, Knez-Mihailova 35/IV, Belgrade, Serbia,
[3]Vladimir Vujicic - Entrepreneur Consultant in Electrical Engineering and Energetics, Mokranjceva 16, Novi Sad, Serbia
[4]Electrical Engineering Institute "Nikola Tesla", Koste Glavinica 8, Belgrade, Serbia.

E-mail: vujicicv@uns.ac.rs

Abstract. This paper strictly defines a simple A/D conversion and simple signal processing within the stochastic digital measurement method (SDMM). It is shown that their synergy and completeness optimizes the measurement instrument or a measurement system as a whole. As a result, the instrument or system then has: 1) extremely simple, robust and reliable hardware, 2) extreme accuracy, 3) extreme wide frequency range, 4) the ability to perform extremely simple parallel measurements and processing.

1. Introduction

In [1, 2] a substantially different paradigm of digital measurements from the standard sampling method (SSM) was proposed. It is generally described in [3], and a few years later it is called the stochastic digital measurement method (SDMM). The essential difference between SSM and SDMM is that the latter does not presuppose a negligible quantization error. In the SDMM, a quantization error may be even 100 % of full scale [1], but the final result is still precise and accurate. This is impossible to achieve by using SSM. The key consequence of this fact is that the SDMM uses an extremely low resolution A/D converter - one-bit and two-bit resolution. Such A/D converters (flash A/D converters) have extremely simple hardware and, consequently, a small number of systemic error sources (which can be easily identified and eliminated [4]). As a result, instruments using one-bit or two-bit SDMM are extremely accurate. This is the most important advantage of SDMM over SSM.

The processing of one-bit or two-bit data is performed extremely fast and accurate, because it uses integer arithmetic of short word lengths. This has been demonstrated in the first paper in which SDMM was applied [1]. Extremely fast and accurate stochastic A/D conversion allows very accurate measurements of the mean values of the signal and their parameters, primarily signal power and signal harmonics or signal spectrum. A similar approach, in terms of A/D conversion, can be seen in sigma-delta A/D converters. In their case, however, the stochastic occurs in analogue modulator that operates at multi-GHz frequencies. Besides this, measurement samples are taken at lower rates (in the order of MHz), have high resolution (16 bits or more) and moderate accuracy. For this reason, sigma-delta A/D converters use floating-point arithmetic or are based on oversampling sigma-delta modulators and digital decimation filters [5]. In the case of SDMM, the processing speed is identical to the speed of A/D conversion, i.e. it is two orders of magnitude greater than that of the sigma-delta A/D converters.

The only limiting factors are correlational characteristics of the random number generator, as well as resolution and accuracy of the applied D/A convertor.

On the other hand, there is the issue of reconstructing a signal which is digitized with a low-resolution A/D converter. Recently, this issue became important in measuring of the probing voltage of high-voltage isolation materials. In [6] it is shown how a signal that is previously measured with a two-bit SDMM can be reconstructed. In this way, the topic related to the measurement and reconstruction of a signal using a low-resolution SDMM is completely rounded off.

All the above mentioned indicates that there is a synergy between simple A/D conversion and simple processing, and their completeness in SDMM. The final results are extremely simple, accurate, fast, robust and reliable instruments that are used in laboratories and industry. In this paper they are strictly defined and elaborated.

2. Basic Definitions and Schemes

2.1. Simple A/D conversion

This section provides strict definitions of simple A/D conversion and simple processing within a simple SDMM, as well as appropriate hardware schemes.

One-bit A/D conversion. The basic condition is that the measured signal is non-negative. The input signal is an analog sum of the measured analog signal (AI) and an analog uniform random noise (*h*). The average value of *h* is zero, whereas its absolute value is less than or equal to half of the quantum 2g. The quantum, on the other hand, is equal to the range of one-bit A/D converter. As a result, the circuit generates two signals (Figure 1a): digital output signal (DO1) and overflow bit signal (OF).

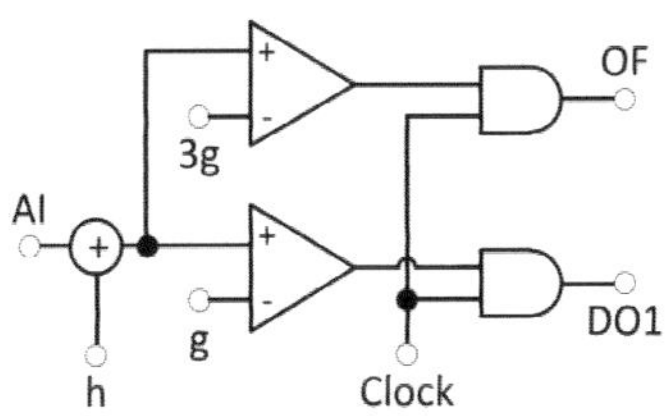

(a)

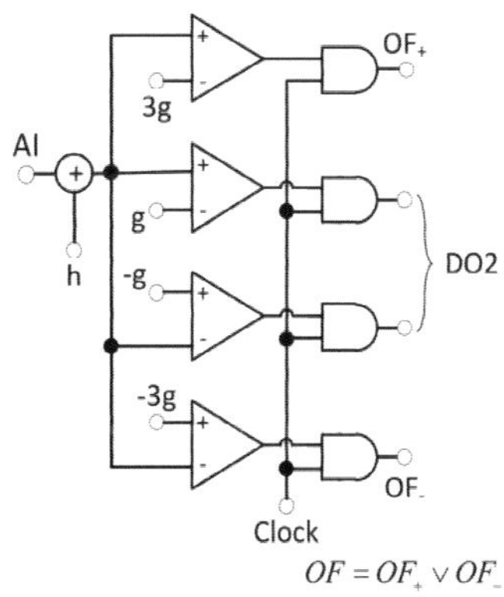

(b)

Figure 1. Simple A/D conversion: (a) one-bit and (b) two-bit.

Two-bit A/D conversion. In the case of a two-bit A/D conversion, the input signal is again an analog sum of the measured signal AI and an analog uniform random noise *h*. Both these signals must satisfy the same conditions as in the previous case (Figure 1b). The only difference is that, in this case, the quantum $2g$ is equal to the half of the range of a two-bit A/D converter, i.e. $\pm 2g$.

2.2. Simple signal processing

Simple signal processing is one-bit or two-bit processing. The basic processing operations are multiplications and accumulations. They are performed by the multiply and accumulate (MAC) block.

Time domain. In the time domain, both MAC operands are the samples of the input signals.

One-bit MAC scheme: As the digital outputs of an A/D converter are 0 or 1, their product is also 0 or 1. Hence, in this case, the multiplier is an ordinary AND-gate, while the accumulator is an ordinary binary counter. The basic scheme of one-bit time domain MAC block is given in figure 2a.

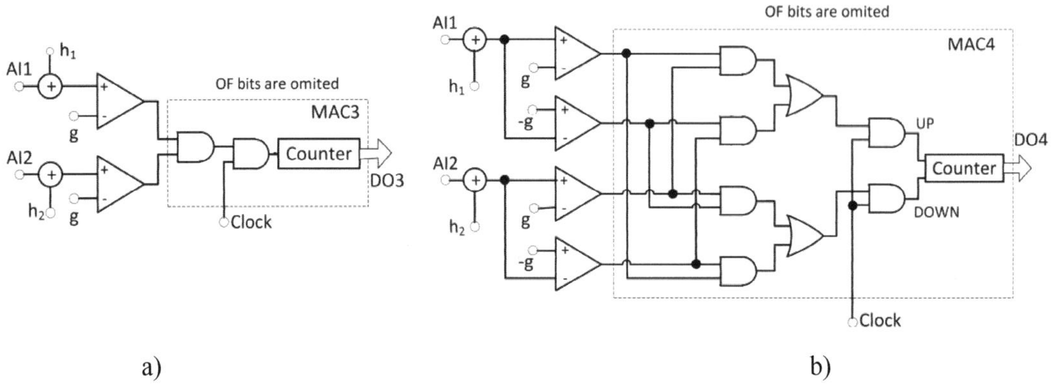

Figure 2. MAC scheme in time domain: (a) one-bit and (b) two-bit.

Two-bit MAC scheme: In this case, the digital outputs of an A/D converter are - 1, 0 or 1. As a result, their product is also - 1, 0 or 1. So, in this case, the multiplier consists of four AND-gates and two OR-gates, whereas the accumulator is a binary up/down counter (Figure 2b).

Transformation domain. In the transformation domain, one MAC operand is the sample of the input signal, while the other is the sample of a dithered base function (DBF) from an orthogonal normed function set (the most often from the Fourier set).

One-bit MAC scheme: The digital output of an A/D converter is 0 or 1, while the sample of a DBF may have values - 1, 0 or 1. Hence, in this case, the multiplier consists of one AND-gate, while the accumulator is a binary up/down counter. The basic scheme of one-bit transformation domain MAC block is given in figure 3a, [2], [7].

Two-bit MAC scheme: In this case, both the digital output of an A/D converter and the sample of a DBF may have values - 1, 0 or 1. Hence, their product is also - 1, 0 or 1. So, in this case, the multiplier consists of four AND-gates and two OR-gates, while the accumulator is a binary up/down counter as it is shown in figure 3b and [7].

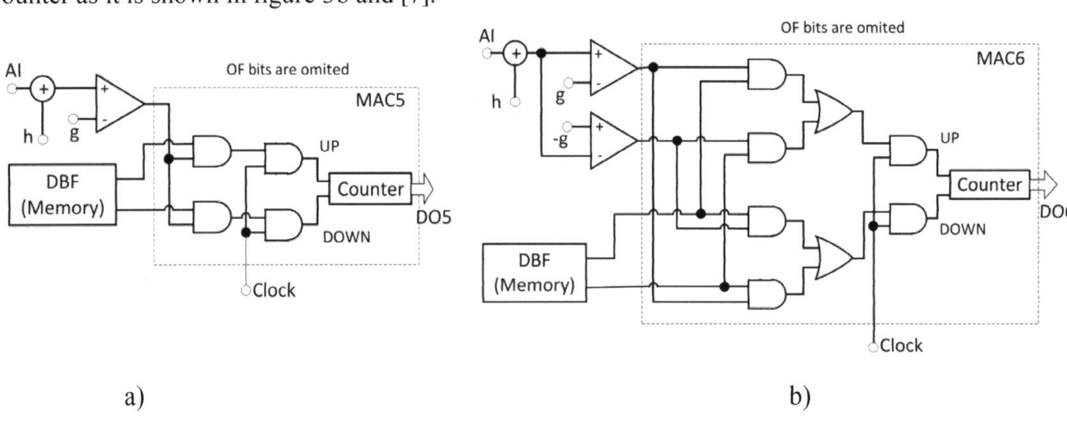

Figure 3. MAC scheme in transformation domain: (a) one-bit and (b) two-bit.

3. Current Application of Simple SDMM

In the last 20 years, simple SDMM is intensively applied in the area of measurements in a power grid. The following quantities are measured by simple SDMM-based instruments: voltage RMS, current RMS, power grid frequency, active power, and reactive power (Budeanu's, Fryze's and that recommended by IEEE Std. 1459-2010). Very difficult conditions in the substations were the best test for the reliability and robustness of all SDMM-based instruments. One of them is the quadruple three-phase power analyzer called MM4 (Figure 4). It can measure electric power and energy according to IEEE Std. 1459-2010. The heart of this instrument is two-bit MAC scheme in time domain (Figure 2b) and two-bit MAC scheme in transformation domain (Figure. 3b). The main processing block is a two-bit stochastic digital DFT processor [7], which is realized by employing several two-bit MAC schemes in parallel. Despite being extremely simple, this instrument has an accuracy of 0.2 % of full scale. In the following areas, the authors and associates have also applied a simple SDMM: measurement of biomedical signals, temperature measurement, dose measurement, wind energy measurement, and measurement on the graphene.

Figure 4. Two-bit quadruple three-phase power analyzer (MM4).

4. Discussion - Synergy and Completeness

The usual way of considering A/D conversion and signal processing is their complete mutual independence. This approach leads to the optimal design of each of them. However, is the instrument or measuring system, based on these, also optimal? The answer is no! This is indicated by the statement of the famous Pontryagin theorem: "The optimum of the system never corresponds to the optimum of any of its subsystems." [8]. Practical generalization of this theorem is as follows: "Optimizing the outcome of a subsystem will in general not optimize the outcome of the system as a whole" [9]. This means that the instrument or measuring system optimization has no alternative, and it is a central problem in the science of measurements. Because of this, a simple SDMM treats at the same time both A/D conversion and signal processing. The outcome is impressive: 1) extremely simple, robust and reliable hardware, 2) extreme accuracy, 3) extremely wide frequency range, 4) extremely simple parallel measurements, 5) extremely simple parallel processing and 6) extremely simple ASIC design verification. In other words, the application of a simple SDMM leads to optimal performance, i.e. to the optimization of the instrument as a whole. This has been confirmed in the previous research and implementation of a simple SDMM [1-4, 7]. This completes the implementation of a simple SDMM in the instrumentation. In addition, it confirms the synergy and completeness between simple A/D conversion and simple signal processing. The most recent and most complete evidence of all these claims is given in detail in recently published paper [7].

5. Conclusion

This short essay presents the synergy and completeness of simple A/D conversion and simple signal processing within the simple stochastic digital measurement method. It was shown that the application of this method allows multiple optimization of the measuring instrument or measuring system as a whole. Specifically, the instrument or measuring system then has: 1) extremely simple, robust and reliable hardware, 2) extreme accuracy, 3) extreme wide frequency range, 4) the ability to perform extremely simple parallel measurements and processing. These features have been confirmed in all our studies, especially in the most recent ones.

Acknowledgment

This research is a part of the project III 43011 (Joint research of measurements and effects of ionizing and UV radiation in medicine and environmental protection). The authors are sincerely grateful to Ministry of Education and Science, Republic of Serbia, for supporting this study.

References

[1] Vujicic V, Milovancev S, Pesaljevic M, Pejic D and Zupunski I 1999 *IEEE Trans. Instrum. Meas.* **48** 467

[2] Santrac B, Sokola M, Mitrovic Z, Zupunski I and Vujicic V 2009 *IEEE Trans. Instrum. Meas.* **58** 3434

[3] Vujicic V, Zupunski I, Mitrovic Z and Sokola M 2009 *Proc. XIX IMEKO World Congress Fundamental and Applied Metrology (Lisbon)* pp 1128-32

[4] Urekar M, Pejic D, Vujicic V and Avramov-Zamurovic S 2017 *Measurement* **98** 139

[5] Janssen E and van Roermund A 2011 *Look-ahead based Sigma-Delta Modulation* (New York: Springer Science + Business Media)

[6] Vujicic V, Zupunski L, Sovilj P and Radonjic A 2017 *Proc. 17th IEEE Int. Conf. Smart Technologies-EUROCON 2017 (Ohrid)* pp 829-31

[7] Pejic D, Naumovic-Vukovic D, Vujicic B, Radonjic A, Sovilj P and Vujicic V 2018 *Measurement* **124** 494

[8] Pontryagin L, Boltyanskii V, Gamkrelidze R and Mishchenko E 1962 *The Mathematical Theory of Optimal Processes* (New York: John Wiley)

[9] Heylighen F 1999 *Principia Cybernetica Web* (Principia Cybernetica, Brussels) URL: http://pespmc1.vub.ac.be/SUBOPTIM.html

Direct measurement of stationary objects' dimensions with Michelson type incremental laser interferometer

T Stejskal[1], M Dovica[1], S Ďuriš[2], Z Ďurišová[3], J Palenčár[2]

[1] Technical University of Košice, Faculty of Mechanical Engineering, Slovakia
[2] Faculty of Mechanical Engineering, Slovak University of Technology, Bratislava, Slovakia
[3] Slovak Institute of Metrology, Bratislava, Slovakia

E-mail: miroslav.dovica@tuke.sk

Abstract. The paper deals with a design of an original length measuring equipment for a measurement by Michelson-type laser interferometer. Laser measuring operate on the principle of relative movement of two components, where one component is fitted with an interference unit and the other one with a reflector. Such system is capable of measuring the length of displacement post motion, however, it is unable to measure the length of a stationary object. Functionality of the measuring equipment is ensured by a measuring board and a ruler. At a relatively low additional cost, the measuring configuration is capable of measuring the lengths of motionless objects achieving the precision of laser. The paper describes the measurement methodology verified in an experiment. Expanding verification tests of the device may facilitate possible commercial production of such auxiliary device for the Michelson-type laser interferometer, which would make the symposium motto "The future glimmers long before it comes to be" sound especially true.

1. Introduction

The laser interferometer is structurally adapted to measure the positioning accuracy of coordinate systems. The coordinate system comprises a moving part in the form of a guideway in either of the axes. The positions are programmable and the laser interferometer can verify the degree to which the programmed value has been reached [1]. Should the coordinate system lack its own measuring ruler, even then the movement of the moving part can be measured accurately.

The solution is based on a two-carrier measuring equipment for interferometer components. The carriers have a fixed arm with a three-point contact for the rulers, a spot for fitting the interferometer and the reflector, respectively. The carriers can move freely along the ruler and their function is to limit the contact of the arm with the object measured. At the end of both arms, co-linear contact sensors are placed, facing each other. All other movements related to handling the object in preparation for the measurement are performed by the object. This ensures high accuracy and reliability of measurement.

To verify the measurement methodology, a granite surface plate and a precise knife straight edge were used. The carriers and the arms were made of balsa wood. A high accuracy digital sensor was subject to verification as it served as a contact sensor.

Length measurement is principally done in many ways, ranging from utilizing mechanical systems to laser metering. The authors are not aware of any commercial application using the proposed principle of measurement linked to Michelson-type laser interferometer. The experiments below confirm that this system could expand the portfolio of measuring products produced and sold by laser interferometer manufacturers and sellers, respectively.

2. Theory

It is commonly known that there are 21 sources of kinematic inaccuracy accompanying spatial movement of a body in three axes [2-4]. In linear motion, this is reduced to 9 sources of kinematic errors. When measurement is done with the arms, the primary source of errors, in accordance with the Abbe principle, is the measuring arms' length. Therefore, the most accurate part must be the guide ruler and the flatness of the base surface. The most beneficial arrangement based on this principle is when the arms are non-reconfigurable. All other moves required to measure the dimension must be done by the object. These moves are another source of errors, but they are second-order errors. These are mostly cosine errors. The nature of the measuring contact is a force which acts not only on the object but also on the end of the measuring arm. This force must be minimized to avoid arm deformation [5]. Deformation propensity increases with the increase of the arm length. On the other hand, the arm length determines the range in which the object can be measured. Ideally, a proximity sensor should be used. However, there is a risk that such sensors will not determine a clear distance from the surface due to roughness and type of the material. A compromise solution is to use a special contact sensor with a small and precisely defined contact force. For the methodology proposed, a high sensitivity sensor should be used. Linearity requirements are smaller, because repeated measurements must always be made to the same load of the sensor. However, this last requirement gives rise to special demands on positioning accuracy of the interferometer component carrier [6].

Naturally, thermomechanical sources of errors may also occur. In this respect, the proposed measurement methodology poses an advantage, because the measurement starts with the system reset. If the whole system is thermally stabilized, this error source is also minimized by resetting the system before the measurement, and the error increases only with the increased time of measurement duration. This even has an advantage over conventional CMM systems, where the temperature compensation needs to be ensured by means of a complex system design.

3. Measurement procedure

The first step in the measurement is resetting the measuring system. Measuring contact sensor heads at the end of the measuring arms are aligned co-linearly (Figure 1). Resetting is done when both sensors touch by gently bringing the interferometer to the reflector. The laser system is reset and the values of the contact sensors, the accuracy of which corresponds to the accuracy required, are recorded.

The second step is the separation move of the arms, which creates space for the object to be measured in. The object is inserted into this space so that the desired dimension is measured exactly in the axis of the contact sensor.

The third step is moving the reflector arm to touch the object. At this step, the value of the contact sensor recorded at reset has to be obtained.

The fourth step is the same as the third step but the contact is achieved with the interferometer arm. When the correct contact is achieved, the laser display shows the dimension measured.

When determining a different size, the system does not need to be reset. All that needs to be done is moving the arms away from the object, and after setting a new position or inserting a new object, the procedure can be repeated from step three. If there is considerable thermal expansion between the measurements, it is advisable the system be reset before starting a new measurement.

The accuracy and reliability of the measurement can be increased by eliminating random error resulting from dynamic motion of the arms. Position verification in this case can be performed by a set of repeated micro-arm movements in the alternating direction around the contact point. The most frequent value obtained by the measurements is considered to be the correct value.

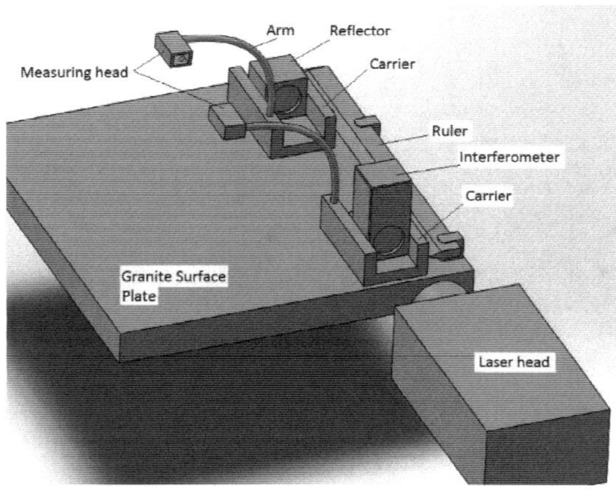

Figure 1. Measurement arrangement.

4. Experiments to verify contact impact during measuring

The most important part of drafting the measurement methodology is to verify the contact of the measuring arm with the object measured. This task gives rise to a lot of possible influences, which will definitely affect the design of the measuring equipment's correct rendering. Problems arise in precision positioning, in setting the arm stiffness and the in selecting the correct contact sensor.

4.1. Arrangement of the contact-measuring experiment

The highly accurate digital GT2-P12KL sensor from Keyence was used to verify the contact force. The sensor accuracy is 1 µm and its resolution is 0.1 µm. The XL80 laser from Renishaw was used as a laser system. The laser system precision is ± 0.5 µm/m. A 500 mm, DIN 874/00 knife straight edge was used to guide the carriers. The system is set on a granite measuring slab 5000x400x100 mm. Figure 2 describes the experiment arrangement. The principal arrangement is shown in figure 3.

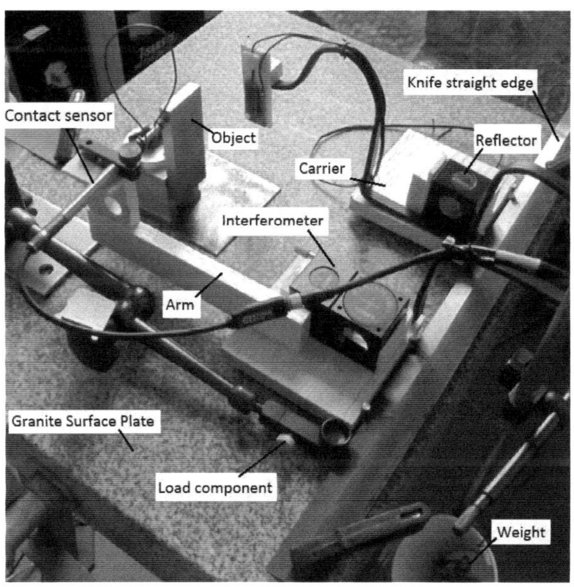

Figure 2. Measuring of contact accuracy.

Journal of Physics: Conference Series **1379** (2019) 012065 doi:10.1088/1742-6596/1379/1/012065

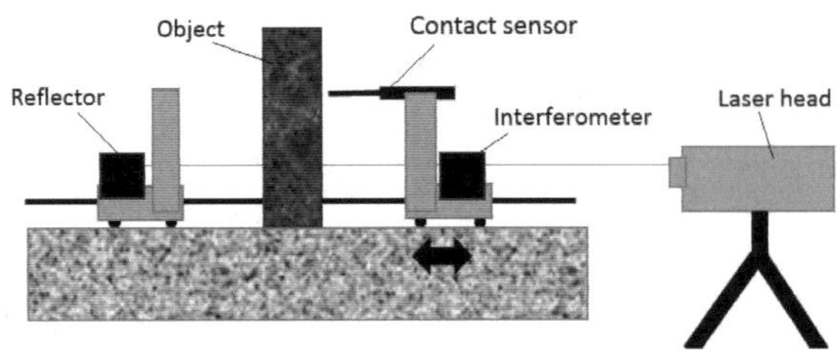

Figure 3. Experiment layout.

The length of the arm from the laser beam axis is 180 mm. The arm carriers have a three-point contact with the measuring board and a two-point contact with the knife ruler. Two-sided arm was used to balance the carrier stability. To eliminate the backlash in the motion along the ruler, weight on a rope transmission was applied.

4.2. Contact counterbalance spring and measuring the contact stiffness

To minimize the contact force, the sensor was counterbalanced by a preloaded spring. The magnitude of contact force dropped from about 0.14 N to 0.007 N. The principle of negative spring preload allows for a significant reduction in the contact force (Figure 4).

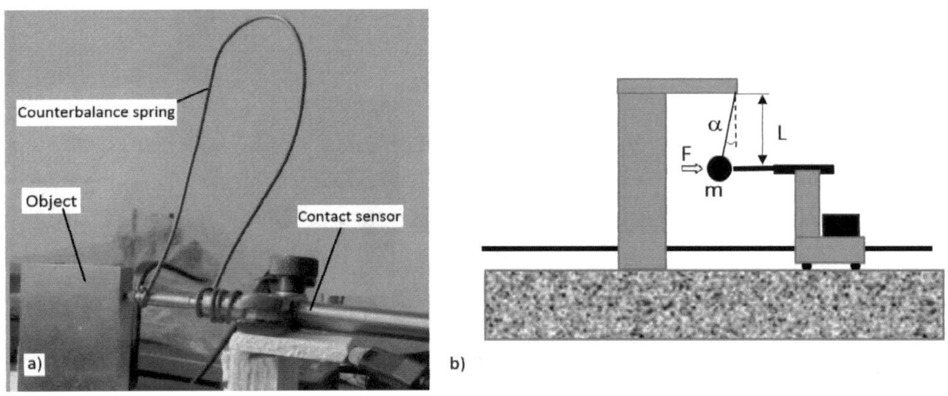

Figure 4. Measuring contact.
a) counterbalance spring b) principle of measuring contact stiffness.

4.3. Repeated contact measurements

Repeated contact measurements enable to determine the accuracy of the measurement under the given conditions. Figure 5 shows 32 repeated measurements. The biggest problem was encountered in positioning the carrier arm. Manual position tuning brings considerable scatter of measured positions to the system. For practical purposes, it is worthwhile to use a motor drive in this respect.

Based on the evaluation of deviations in the logs from two positions, the scattering of the sensor contact is about ± 9.5 μm (the confidence interval is in the 4 sigma range). This is insufficient for precise measurement. The accuracy of the contact sensor is considerably higher. The measurement error is affected mostly by dynamic phenomena occurring during changes in position. The magnitude of the contact force acting on the carrier arm is assumed to have less influence. Motorically driven micro-motions can greatly improve accuracy. This has been proven by preliminary experiments measuring the contact of the carrier with the object of measurement. The contact was repeated by the contact force relief by means of motoric drive. The contact force was generated by the weight through rope transmission. The results were significantly better than those achieved by manual positioning. Figure 6 shows the results of repeated measurement of the contact of the carrier with the measured object. Only

Journal of Physics: Conference Series **1379** (2019) 012065 doi:10.1088/1742-6596/1379/1/012065

values that were repeated several times were selected to determine the actual dimension. In this case, precision is determined in the order of a tenth of a micrometer.

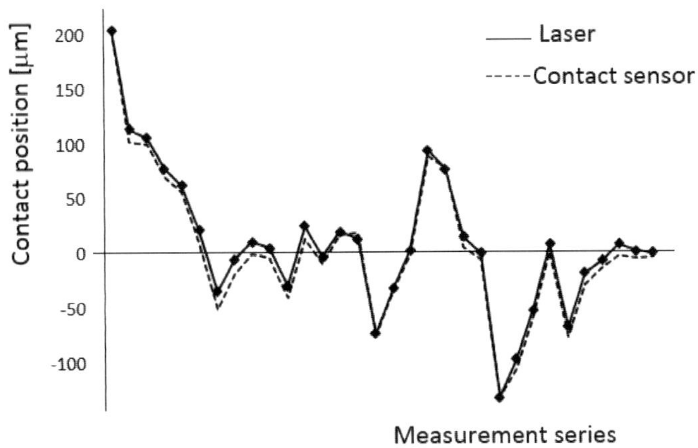

Figure 5. Logs of repeated measurements by the sensor contact.

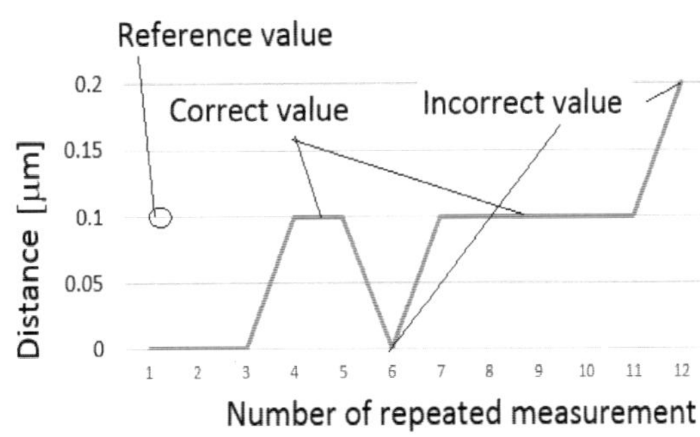

Figure 6. Log of repeated measurements of the carrier contact by means of motoric drive.

5. Design of the future device

Device design is based on experimental experience. The first proposal's idea was that the measuring arms would be adjusted during measurement. Such intervention, however, introduces a first-order error into the measurement. Therefore, the best length measurement philosophy is that the arms are not reconfigured and that all the adjustments needed to measure the size are carried out by the measured object. In this case, moving the object causes a second order error. In this respect, the measuring configuration consists of two separate parts. The first part consists of a measuring ruler with the laser interferometer carrier and the arms fitted with the contact sensors. The second part is a measuring jig handling the measured object. The measuring jig ensures that the correct position is achieved relative to the measuring arms, that the object is removed from the measuring zone after a change in the measured dimension or before the system is reset. If a rotary table is located on the measuring jig, this is an advantage.

Measuring surface plate is not required for measurement purposes. Precise guideway of the carriers can be ensured by two parallel knife edge. This part requires the most accurate execution. Any deviation from linearity causes angular rotation of the measuring arm, which results in the greatest contribution to the measurement error. For this reason, it is suitable for the arms to be as short as possible. On the other hand, by shortening the arms, the measure range is limited. The compromise solution is to use exchange arms of different lengths. The guideways should be in horizontal position, which is properly achieved by a three-point arrangement.

To ensure accuracy, it is also necessary to deal with the axial alignment of contact sensors. The sensors should terminate in a sphere shape. Co-ordination could be set up with a measuring instrument. A better solution is to use the laser interferometer itself.

The contact sensors usually have a high contact force. Proximity sensors are highly sensitive to the object changing the environment and the position. Therefore, it is suitable to produce a combined sensor. The sensor measuring head operates on the proximity principle. For example, the capacitive sensor achieves very high accuracy [7]. The end of the head is the contact part, which has a small stroke, very low stiffness to the axis of measurement.

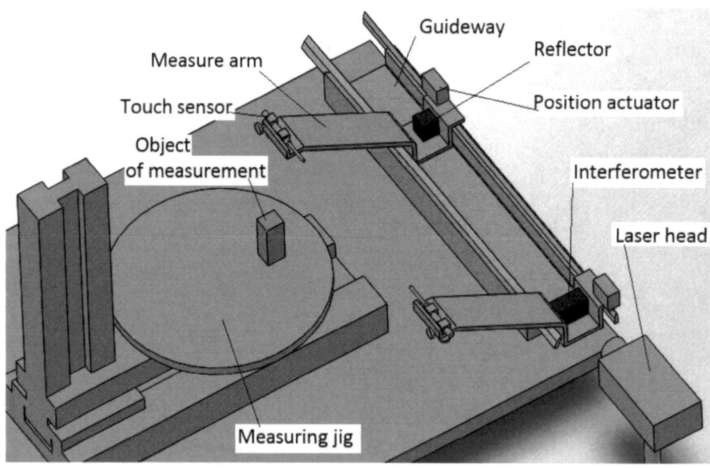

Figure 7. Design of the future measurement configuration.

6. Conclusion

The laser interferometer measuring equipment does not replace conventional measuring systems. The system can only be used where a Michelson laser interferometer is already available. On the other hand, the measuring equipment expands the possibilities of its use. The measurement methodology itself is progressively designed to eliminate the effect of the thermal expansion of the measuring system. This achieves a very high accuracy of measurement. The method provides the ability to measure the length dimensions of static objects at a given level of accuracy. But it is a less accurate method than the direct measurement of moving objects by laser interferometer.

Acknowledgments

This contribution has been supported by research grants: APVV-15-0149, APVV 15-0164, APVV-18-0413 and APVV-15-0295, VEGA 1/0224/18, VEGA 1/0098/18, VEGA 1/0556/18 and VEGA 1/0437/17, KEGA 006STU-4/18, KEGA 025TUKE-4/2019 and KEGA039STU-4/2017.

References

[1] Hillard R L and Shepherd G G 1966 Wide-angle Michelson interferometer for measuring Doppler line widths *JOSA* **56**(3) pp 362-69

[2] Schwenke H *et al.* 2008 Geometric error measurement and compensation of machines—an update *CIRP Annals* **57**(2) 660-675

[3] Abbe E 1890 Messapparate fur Physiker *Zeitschrift für Instrumentenkunde* **10** pp 446-447

[4] Dassanayake K M; Tsutsumi M and Saito A 2006 A strategy for identifying static deviations in universal spindle head type multi-axis machining centers *Int. J. Mach. Tool. Manu.* **46**(10) pp 1097-106

[5] Cauchick-Miguel P A and King T G 1998 Factors which influence CMM touch trigger probe performance *Int. J. Mach. Tool. Manu.* **38**(4) pp 363-374.
[6] Stamenkovic M 2007 Method and device for the three-dimensional measurement of objects *U.S. Patent No 7 310 889*
[7] Dean T J, Bell J P and Baty A J B 1987 Soil moisture measurement by an improved capacitance technique, Part I. Sensor design and performance *J. Hydrol.* **93**(1-2) pp 67-78

Journal of Physics: Conference Series **1379** (2019) 012066 doi:10.1088/1742-6596/1379/1/012066

A seismocardiography system and a possibility of its use for diagnosis of internal organs diseases using seismocardiogram information analysis

V M Uspensky[1], V M Achildiev[2,4], N A Bedro[2], V A Soldatenkov[2], Y K Gruzevich[2,3], Y N Evseeva[2], M N Komarova[2], A D Levkovich[2]

[1] Branch of Kirov Military Medical Academy, Moscow, Russia

[2] Scientific Production Unity B«GEOPHIZIKA-NVB», Moscow, Russia

[3] Bauman Moscow State Technical University, Moscow, Russia

[4] Mytischi branch of Bauman Moscow State Technical University, Moscow, Russia

E-mail: medddik@mm.st, yukg@mail.ru

Abstract. The work is devoted to the development of a cardioseismometer system for obtaining a seismocardiogram for informational analysis of seismocardiographic signals and diagnosis of internal organs diseases on their basis. The cardioseismometer system allows measuring angular velocity projections and the vector of apparent acceleration recorded on the chest due to the mechanical movement of the heart, calculation of heartbeat acceleration vector - magnitude and direction relative to the reference coordinate system. The most informative and available for high-precision measurement parameters of seismocardiographic signals are selected, which are compared to the parameters of the electrocardiograms measuring model used to diagnose diseases. The conclusion was made about the possibility of using the dynamics of the seismocardiographic signal main parameters for informational analysis to diagnose diseases of internal organs. Experimental research results are given.

1. Introduction

The accumulated experience of studying the variability in the main parameters of serially registered electrocardiosignals indicates that electrocardiosignals have all the properties of the signals, and the heart has all the signs of information organ. In this regard, V. M. Uspensky first substantiated and proposed the theory of information function of the heart [1, 2], and then on its basis, a technology of electrocardiosignals information analysis that has no analogues in the world practice was developed [2, 3]. Through experience in the use of technology, it was established that at the time of electrocardiosignals generation by the sinus node the programs of norms and various diseases are laid in them by amplitude, frequency and phase modulation of the main ECG parameters. Specific code templates for more than 30 of the most common and life-threatening diseases of the internal organs were obtained. On their basis, a diagnostic system "Screenfax" was created and tested, that has a long experience of more than 15 years, the use of which is detailed in the monograph [4].

The most important of this experience is the fact that the information analysis of electrocardiosignals allows for the first time to carry out with high reliability the diagnosis of internal organs diseases at any stage of development, including the initial formation stage. Diagnosis of diseases at the initial stage of formation opens the possibility of organizing and conducting primary personalized prevention and preventive treatment to prevent the final stage fraught with life-threatening complications development. The absence of primary targeted specific prevention of the internal organs diseases is current importance problem of modern medicine. In particular, that fact explains the high reserving frequency of these diseases among the population despite the evident successes of diagnosis, but only at the final stage of their development, when therapeutic measures are not so effective, and complications are life-threatening.

An essential element of "Screenfax" diagnostic system is electrocardiograph providing electrocardiogram (ECG) capturing acceptable for the information analysis. Electrocardiographs that are widely used in medical practice are unsuitable for this purpose. Biophysical explanation of the electrocardiograph requirements allows formulating of necessary technical electrocardiograph parameters to implement technologies for electrocardiosignals information analysis [4, 5]. Further improvement of the electrocardiograph should be associated with the improvement of the element base.

Taking into account that the heart generates cardiosignals of different physical nature (electrical, magnetic, hydrodynamic and mechanical) which carry the same information improvements in their registration devices are of interest. In particular, seismocardiography system that is capable of registering seismocardiogram (SCG) reflecting the mechanical effect of the heart on the chest during its contraction in the form of seismic cardiopulses. We have assumed that the mechanical heart action with pulse nature may have the properties of signals (seismocardiosignals) with the possible use of them for information analysis.

The purpose of the study: the development of a seismocardiography system with state-of-the-art components and software for its capturing and processing, methods of main seismocardiogram parameters measuring, comparison with the dynamics of the same ECG signal parameters and reveal the possibility of their use for the diagnosis of internal organs diseases.

While a person moves and breathes, the orientation of body changes relative to the plane of the local horizon and accordingly the orientation of seismocardiography sensor measuring axes changes, that leads to distortion of the accelerometer signals due to the influence of free fall acceleration on the accelerometer data. So there is a task of measuring the orientation angles of the accelerometers sensitivity axes relative to the plane of local horizon and carrying out algorithmic compensation of the body movement.

The coordinate system associated with the seismocardiography system is shown in the figure 1, it is formed by the following axes: X-axis is directed from right shoulder to left, Y-axis is directed to the head, Z-axis is perpendicular to the XY plane and complements the coordinate system to the right.

2. Seismocardiogram signal processing algorithm

A seismocardiogram signal processing algorithm is based on a measurement model for processing an electrocardiogram that was proposed by Uspensky [6, 7].

In the process of research, a rational scheme for constructing a seismocardiogram capture was chosen and experimental samples were made [8, 9]. Figure 2 shows a picture of a seismocardiogram capture device.

The method for determining the single heart beat cycle in the seismocardiogram consists of the following sequence of accelerometers and gyroscopes signals processing. The measurements of apparent acceleration projections on the axis of the instrumental coordinate system, analog-to-digital conversion of signals and primary filtering, compensations of zero shifts and accelerometers sensitivity errors are performed. In addition, projections of the angular velocities on the axis of the instrumental coordinate system are calculated and zeros offset and gyroscopes sensitivity errors are compensated. The rotation matrix of the instrumental coordinate system relative to the reference one is calculated and then an estimation of orientation angles (roll, pitch, yaw) is performed.

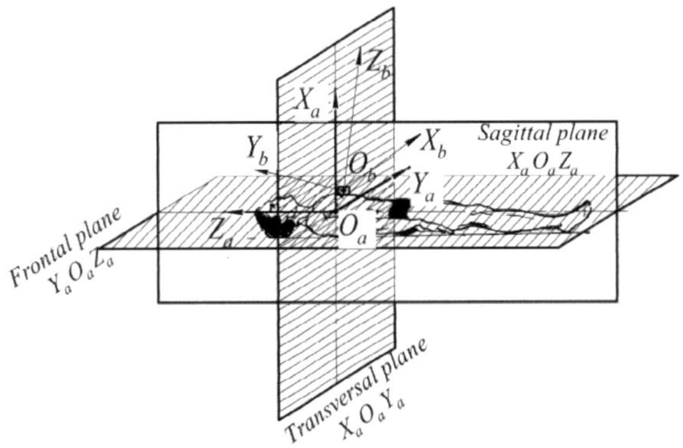

Figure 1. Seismocardiography coordinate systems

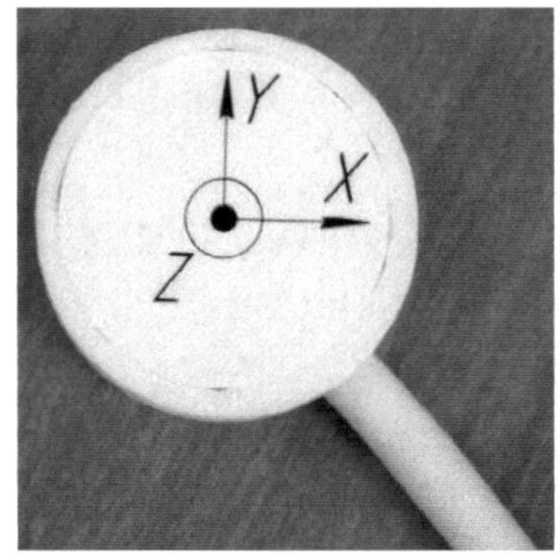

Figure 2. MEMS-based cardioseismometer

This calculation is based on the angular velocities projections and apparent accelerations. After that, the value of heartbeat self-acceleration vector, its modulus and angles α (the angle of heartbeat acceleration vector deviation fromthe horizontal plane) and β (angular orientation of the heartbeat acceleration vector relative to the Y_b axis, that is directed to the head) are calculated.

The algorithm of seismocardiogram signal processing takes into account the interrelation of coordinate systems: instrumental and reference.

The result of processing of the apparent accelerations measured projections and angular velocities in the instrumental coordinate system are the modulus of the self-acceleration vector of the heartbeat, that is registered on the chest, and the angles of its orientation relative to the orthogonal coordinate system associated with the human body.

3. Experimental research

An experimental simultaneous ECG and SCG capturing was performed. Time synchronisation between signal capturing was not less than 25 microseconds.

Accelerations spectrum measured on the chest is shown on the figure 3. The results have shown that micromechanical accelerometer, in addition to low-frequency microvibrations, measures the entire spectrum of acoustic, infra-sound and infra-low frequencies. The primary sources of noise and microvibrations are compressors, ventilation, vehicles, pipelines, working machines, equipment. Production and transport noises contain infrasonic components that are not recorded by conventional measuring instruments, are not audible and have high levels of sound pressure. At the same time, the frequency spectrum of the seismocardiogram is very close to the spectrum of low-frequency sound and low-frequency microvibrations. This phenomenon complicates the task of precise seismic cardiac cycle measuring and its selection against the background of external disturbances. The appearance of a seismocardiogram does not depend strongly on the presence of additional noise components, however, as our research have shown, the amplitude variability depends on the presence of infrasonic noise during the measurement.

Figure 5 shows an electrocardiogram and a seismocardiogram (SCG), recorded with special software for the diagnosis according to the combined method of electro-seismocardiography [10]. As can be seen from the figure, a regularity is observed between ECG and SCG signals amplitude, so a seismocardiogram can act as an additional source of cardiac signal parameters for information analysis.

Figure 6 shows a segment of seismic cardiac cycle acceleration module that corresponds to the time when the mitral valve is closed with the points: MC – mitral valve closure, IM – isovolumic movement, AO – aortic valve opening, IC – isotonic contraction, RE – peak of rapid ejection.

An estimation of ECG and SCG parameters variability was performed both for amplitude R_n and time intervals T_n between peaks corresponding to the heart contraction as shown on the figure 4. The upper plot on the figure 7 shows a comparison of time intervals T_n variability calculated as $T_{n-1} - T_n$, solid plot represents $T_n^{SCG} - T_{n-1}^{SCG}$ for SCG along the Z-axis, while dashed line represents

$T_n^{ECG} - T_{n-1}^{ECG}$ for ECG. As can be seen from the graph, there is a coincidence between ECG and SCG time intervals increments, that confirms the assumption made in the introduction about the same nature of mechanical and electrical heart activity.

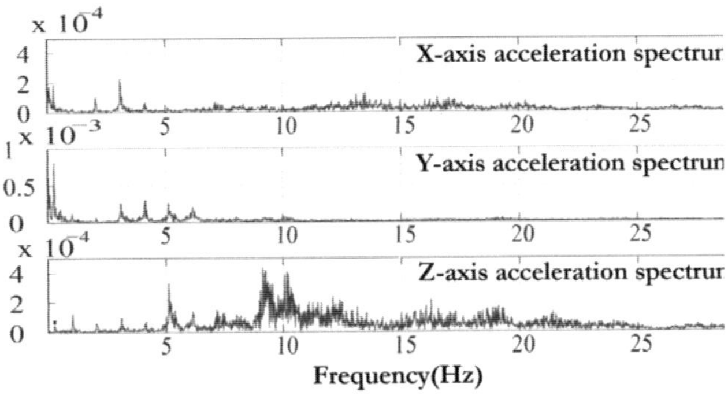

Figure 3. Seismocardiogram spectrum

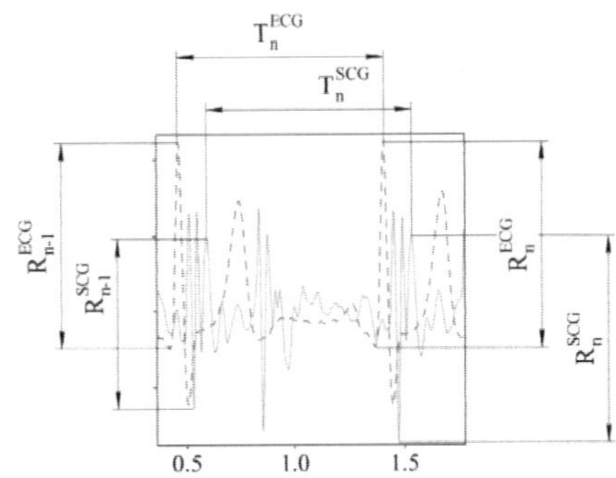

Figure 4. Intervals T_n, R_n

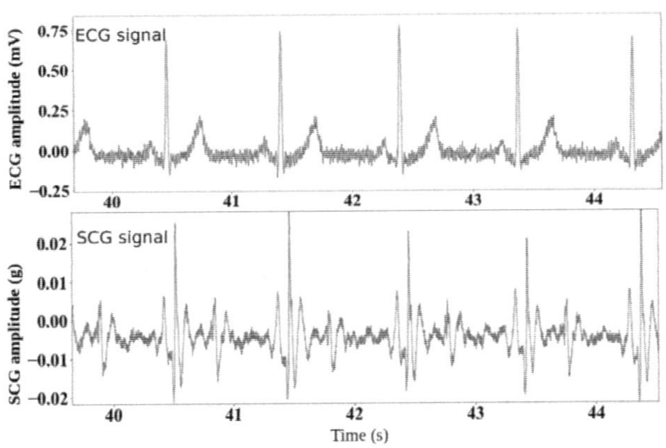

Figure 5. An example of simultaneously recorded ECG and SCG

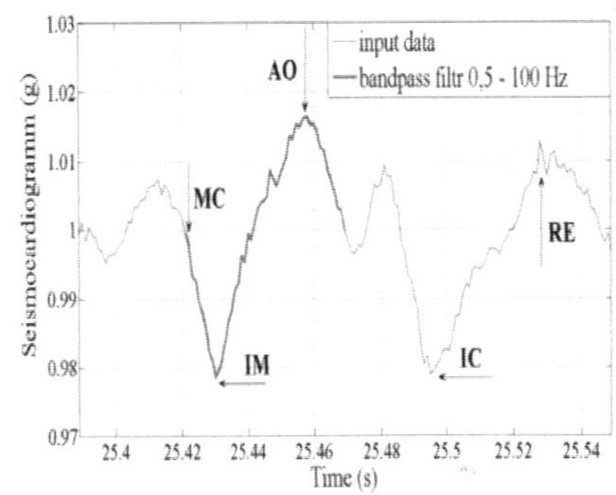

Figure 6. Acceleration module corresponding to a single cardiac cycle

The lower graph on the figure 4 shows an example of ECG and SCG amplitude R_n variability calculated as $R_{n-1} - R_n$. As can be seen from this graph, the coincidence of these variabilities is not so evident as for $T_n - T_{n-1}$, this problem is under our investigation now. One of the reasons of that may be errors in automatic SCG peaks detection (SCG signal is different from ECG: it has many extremums, the algorithm for their detection is not so simple as for ECG), but for a large number of intervals, the dynamics of ECG and SCG variability in R_n increments coincides.

Diagrams of T_n and R_n increments were obtained by the representation of their values in the form of a flat curve connecting the points with coordinates ($T_n - T_{n-1}$, $R_n - R_{n-1}$). Figures 8 and 9 show examples of such diagrams. They are different for different people, the authors suggest that their appearance also carries some information about the health state.

4. Conclusion

The application of electrocardiogram information analysis technology in medical practice can be the technical basis for diagnosing diseases of internal organs at the stage of their initiation and the initial stage of development. Prevention and preventive treatment at this stage will prevent the development of final diseases stage that is dangerous for human life with its complications and help to improve the health of the population. The results of comparative analysis of main electrocardiograms and

Journal of Physics: Conference Series **1379** (2019) 012066 doi:10.1088/1742-6596/1379/1/012066

seismocardiograms variability parameters indicate that they have the same character. Moreover, the coincidence is not only in the values of parameters: amplitude, repetition rate and phase dynamics but also in their dynamics in sequential mode.

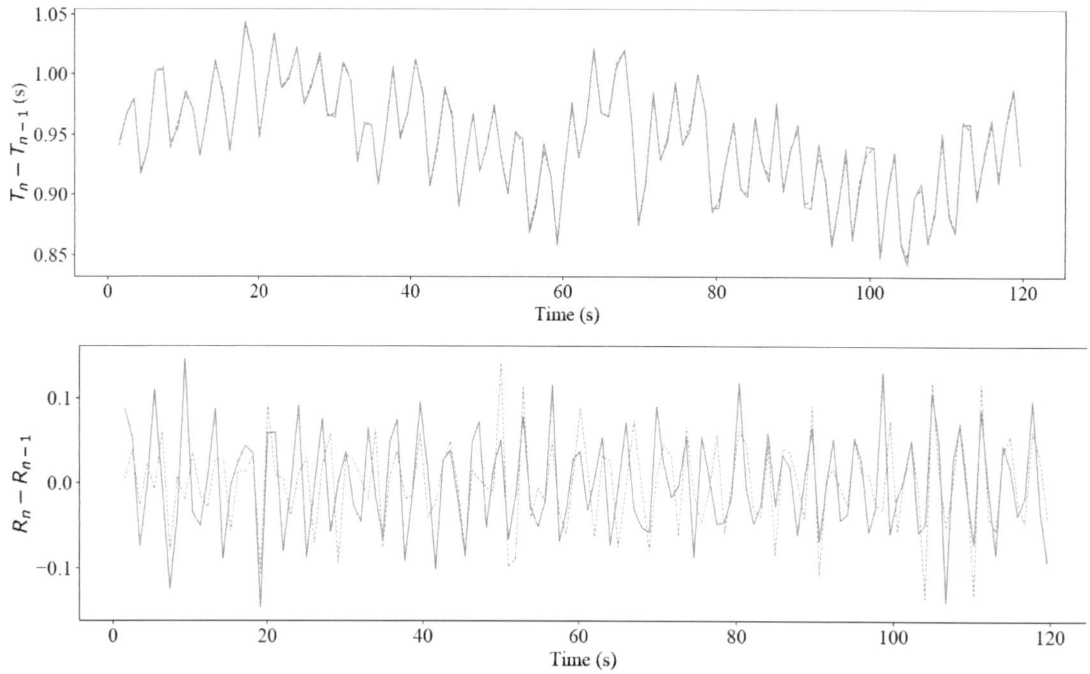

Figure 7. An example of T_n and R_n increments variability for ECG (- - - -) and SCG (—)

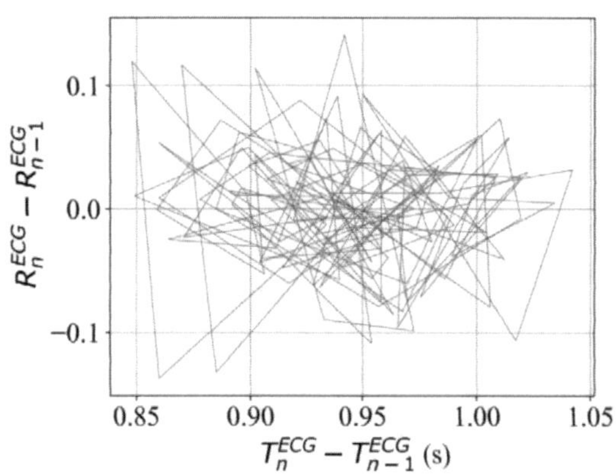

Figure 8. An example of T_n; R_n increments diagram for ECG

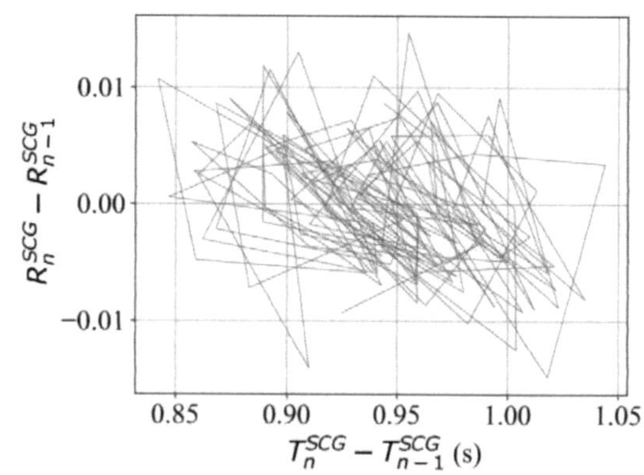

Figure 9. An example of T_n; R_n increments diagram for SCG

The conducted research has shown that:

- It is possible to register the variability of the main seismocardiographic pulse, corresponding to the contraction of the heart ventricles, time intervals between these pulses and the dynamics of acceleration vector due to cardiac contraction.
- There is a parallelism in the dynamics of similar parameters for electrocardiogram and seismocardiogram signals, that indicates the prospect of seismocardiogram usage in the diagnostic systems based on information analysis of cardiovascular signals of different physical nature.

- The absence of electrodes and electrode cables for seismocardiogram registration gives it a certain advantage over the electrocardiograph if it is necessary to register the cardiogram in non-standard conditions.

Experimental studies have confirmed the correctness of chosen cardioseismometer schematic design solutions and algorithms of its work.

Acknowledgments
This work was supported by the Russian Foundation for Basic Research, project no. 18-29-02019.

References

[1] Uspensky V M 2008 *Information Function of the Heart. Theory and Practice of the Internal Organs Diseases Diagnosis Using the Electrocardiogram Information Analysis* (Moscow: Economics and information) (In Russian)
[2] Uspensky V M 2008 Information function of the heart *Clin. Med.* **86**(5) pp 4-13
[3] Uspenskiy V M Information function of the heart. A measurement model 2011 *Proc. of the 8-th Int. Conf. Measurement* (Slovakia) pp 383-386
[4] Uspensky V M 2016 *Information Function of the Hheart. Theory and Practice of the Internal Organs Diseases Diagnosis Using the Electrocardiogram Information Analysis* (Moscow: Planeta) (In Russian)
[5] Uspensky V M 2011 Information function of the heart. Biophysical substantiation of technical requirements for electrocardioblock registration and measurement of electrocardiosignals parameters acceptable for information analysis to diagnose internal diseases *Proc. Joint Int. IMEKO TC1+ TC7+ TC13 Symp.* (August 31–September 2, 2011, Jena, Germany) urn:nbn:de:gbv:ilm1-2011imeko-024:6
[6] Patent RU2407431 (C1) — 2010-12-27 Uspenskij V M *A Method for Internal Organs Diseases Diagnosis* https://worldwide.espacenet.com/publicationDetails/biblio?II=0&ND=3&adjacent=true&locale=en_EP&FT=D&date=20101227&CC=RU&NR=2407431C1&KC=C1#
[7] Patent RU2159574 (C1) — 2000-11-27 Uspenskij V M, Kravchenko Ju G, Pavlovskij K P and Averbakh Ju I *Device for Performing Express Diagnostic of Internal Organ Diseases and Oncopathological Disorders* https://worldwide.espacenet.com/publicationDetails/biblio?II=0&ND=3&adjacent=true&locale=en_EP&FT=D&date=20001127&CC=RU&NR=2159574C1&KC=C1#
[8] Achildiev V M, Basarab M A, Bedro N A, Soldatenkov V A, Evseeva Y N, Gruzevich Y K, Konnova N S and Levkovich A D 2018 Cardioseismometer unit based on micromechanical sensors *Proc. of the 25-th Saint Petersburg Int. Conf. on Integrated Navigation Systems* (ICINS) pp 1-10
[9] Levkovich A, Achildiev V, Soldatenkov V, Basarab V, Bedro N, Gruzevich Y, Evseeva Y, Konnova N and Komarova M 2018 Seismocardiography system based on micromechanical sensors *Proc. Int. Conf. IERI on Medical Physics, Medical Engineering and Informatics* (Basic Clin. Pharmacol. Toxicol.) 123 (Denmark:Wiley) pp 9-10
[10] Achildiev V M, Basarab M A, Bedro N A, Soldatenkov V A, Gruzevich Y K, Evseeva Y N, Konnonova N S, Levkovich A D and Uspensky V M 2018 Electroseismocardiography system consisting of a high resolution electrocardiograph and a seismocardiography system based on micromechanical sensors *Proc. 4th Int. Conf. "Electronic component base and microelectronic modules" of the Int. Forum "Microelectronics-2018"* (Moscow: Technosphera) pp 514-6 (In Russian)

Design and implementation of a digital mixer with digital logic

Leonidas Niyonkuru, Gerd Vandersteen and Leo Van Biesen

ELEC Department, Vrije Universiteit Brussel, Pleinlaan 2, 1050 Brussels, Belgium

E-mail: niyonkuru.leonidas@vub.be, gerd.vandersteen@vub.be, lvbiesen@vub.be

Abstract. A mixer is a device whose output frequency is a sum or a difference of input frequencies. With f1 and f2 as input frequencies, the output frequency is f1+f2 or |f1-f2|. The mixer is generally an analogue device. In this work, we suggest a digital mixer, designed and implanted using digital logic. A phase locked loop is added at the output of the mixer acting as a pass band filter. The input and output digital nature implies that there is no conversion loss and the isolation between output port and input ports was no less than 45 dB. The mixer can be used for downconversion of the signal to be measured in spectrum analyzer.

1. Introduction

Frequency mixers are widely used in communications systems for signals up conversion or down conversion. Usually they are analogue devices. Digital mixers have been also proposed in the literature. In [1] EXOR gates were used for multiplication of the input signals and some of the intermodulation spurs were cancelled through multi-phasing of input signals [2]. However multiple phases (up to 7-phase in the example given in the paper) were necessary to obtain satisfactory results. In [3] a nonlinear Digital-to-analog conversion was used in addition to multi-phasing. This nonlinear DAC add more complexity to the method. An FPGA implementation of a frequency mixer was also proposed in [4], in which VLSI (Very large Scale Integration) is used for frequency conversion. The last method appears to be too complex, in view of to the hardware involved. We suggest a simple digital mixer whose operation theory is based on trigonometric identity (1). The novelties of the method are:

- The use of the model of trigonometric identity for intermodulation cancelation of half of intermodulations spurs,
- The use digital logic for the operation of "ADDITION" of two square wave signals as shown in paragraph 2.2.
- Obtaining square wave at the output of the frequency mixer. The method used in [2] and [3] lead to obtaining sinusoidal signal.
- The suggested mixer does not need any calculation as in [2] and [3] and therefore can be used for high frequencies.

The limitation of proposed method is the presence of spurs which could not be removed by the filter as they are near the fundamentals and the harmonics of the square wave. The proposed mixer is digital in the sense that digital logic gates and digital signals are used instead of analogue components and analogue signals. We begin with design methodology which is based on mathematical trigonometric identity. The simulation and hardware implementation will follow. From the simulation, the hardware implementation and the measurements that have been made, it will be shown, that a pass band filter at the output is necessary for the attenuation of spurious components.

2. Design methodology

The functional diagram of the suggested digital mixer is shown on figure 1.

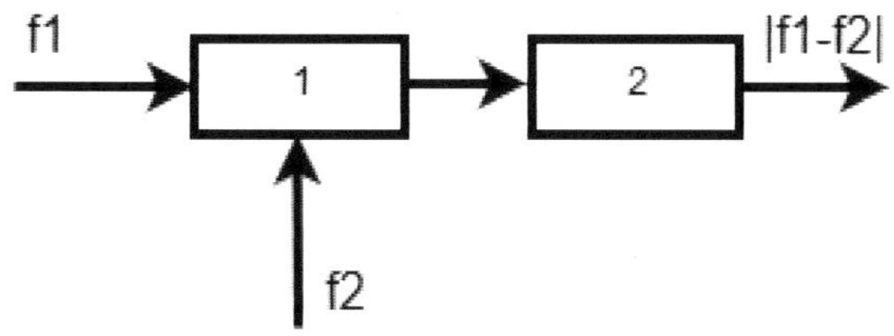

Figure 1. Functional diagram of the suggested mixer
1: digital multiplier
2:PLL based pass band filter
f1, f2: input frequencies
|f1-f2|: output frequency

The block 1 represents a digital multiplier whose role is to generate the difference of the input frequencies and the block 2 is a pass band filter added at the output of the digital mixer. The digital mixer operation theory is elaborated based on trigonometric identity (1). Square wave signals are used as digital inputs and ADDITION and MULTIPLICATION are replaced by digital logic. Simulation is done using Simulink and hardware implementation is achieved using High speed CMOS logic gates.

2.1. The theory of the suggested mixer

We will focus on digital mixer with the difference frequency at the output. However, the proposed methodology can be equally applied to get the sum frequency at the output. An ideal analogue mixer with the difference frequency at the output can be expressed by the following trigonometric identity:

$$\cos(\omega_1 t) * \cos(\omega_2 t) + \sin(\omega_1 t) * \sin(\omega_2 t) = \cos[(\omega_1 - \omega_2)\, t],$$

where ω_1 and ω_2 are the input signals pulsations.

$$\cos(\omega_1 t) * \cos(\omega_2 t) + \cos\left(\omega_1 t - \frac{\pi}{2}\right) * \cos\left(\omega_2 t - \frac{\pi}{2}\right) = \cos[(\omega_1 - \omega_2)\, t] \qquad (1)$$

A simple multiplication of two cos would have given:

$$\cos(\omega_1 t) * \cos(\omega_2 t) = \frac{1}{2}[\cos[(\omega_1 - \omega_2)\, t] + \cos[(\omega_1 + \omega_2)\, t]] \qquad (2)$$

Equation (1) contains only difference of frequencies while equation (2) contains both the sum and the difference of input signals. In case of square wave signals, this method allow cancellation of the intermodulation of the frequencies $(n\omega_1 + m\omega_2)$ with n and m integers. We will use the model (1) for the digital mixer, with the difference that the signals at the input are digital square waves now and that the ADDITION and MULTIPLICATION in the analogue domain are replaced by their equivalent in the digital domain. It is well known that MULTIPLICATION in that case is equivalent to XOR [1].
In this paragraph, we explain how the ADDITION is implemented. Let A and B be the inputs to the digital mixer and let 1 and -1 be the low and high signal level respectively. The “ADDITION A+B” is shown on figure 2.

A and B are 2-level square wave signals while A+B is a 3-level signal. Based on figure 2, A+B can be implemented using digital logic. The first row in the figure 2 will be transformed in the low level, while the last row will be transformed into the high level. The first row is “NEITHER A nor B (logic NOR)” and it will reset the RS trigger while the last row “A and B (AND)” will set the RS trigger. The figure 3 shows the digital mixer based on the above mentioned principle. Therefore, equation (1) can be implemented by the digital circuit represented on figure 3 provided the shift $-\frac{\pi}{2}$ is done before the signal is fed to XOR gate.

Journal of Physics: Conference Series **1379** (2019) 012067 doi:10.1088/1742-6596/1379/1/012067

A	B	A+B
-1	-1	-2
-1	1	0
1	-1	0
1	1	+2

Figure 2. The "ADDITION A+B".

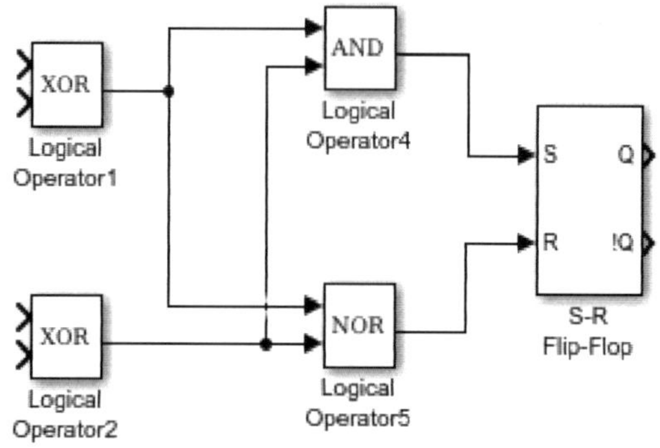

Figure 3. Digital mixer with digital logic.

2.2. Simulation

The figure 4 shows the simulation model using Simulink. The digital mixer is fed with two clock signals with frequencies f1 and f2. The gate NOT shifts the input signal by 180° and each D flip flop divides by 2 the frequency of the signal at his input. This results in a 90° shift between the signals at the output of "D flip flop" and "D flip flop1". The same 90° shift exists between the "D flip flop2" and "D flip flop3" outputs. Due to the division by 2 when using D flip flops, the output of mixer is $|f_1-f_2|/2$. Simulation showed that whenever one of the frequencies is a multiple of the other, the mixer works perfectly while spurious components appear whenever this is not the case (figure 5 and figure 6). An output filter is therefore necessary as early mentioned.

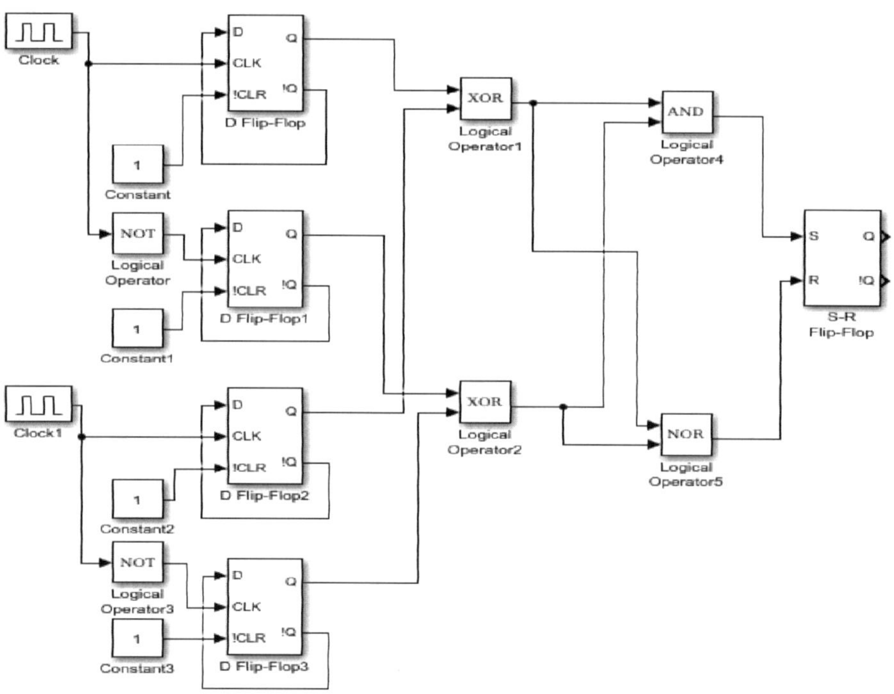

Figure 4. Simulation model using Simulink.

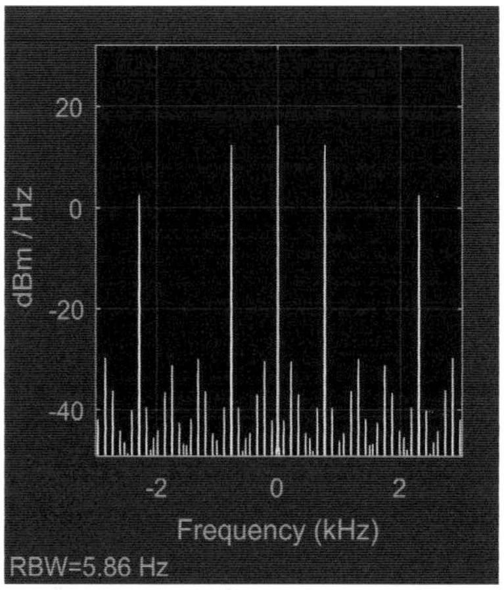

Figure 5. Output power spectrum f1=1530 Hz, f2=3060Hz, fout=765 Hz.

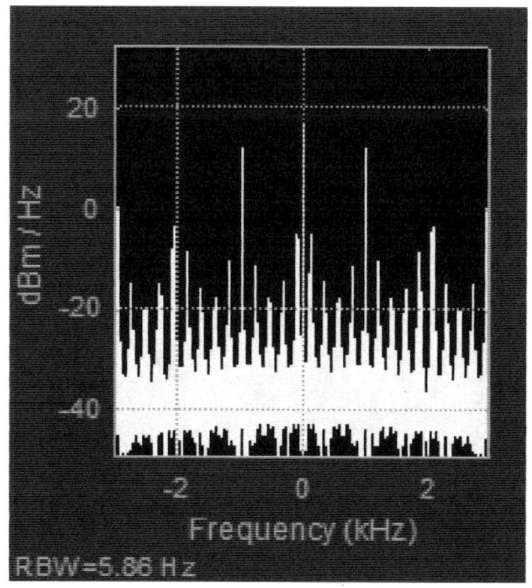

Figure 6. Output power spectrum f1=1530 Hz, f2=3500Hz, fout=985 Hz.

2.3. Hardware implementation and measurements

For the bloc 1 on figure 1, the hardware implementation is straightforward from simulation functional diagram by replacing different blocs with corresponding logic gates. The following integrated circuits were used: D flip flop CD74HC74E, NOT gate CD74HC04E, XOR gate CD74HC86E, NOR gate CD74HC02E and AND gate CD74HC08EE4. The RS flip flop was implemented using two NOR gates s shown on the figure 7.

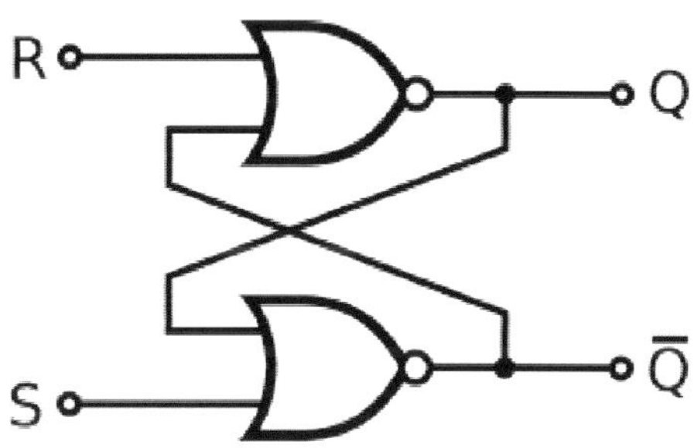

Figure 7. RS flip flop implementation from NOR gates.

For the bloc 2 on figure 1, a passband frequency filter was implanted using PLL (phase locked loop). The chip CD74HC4046AE was used for this purpose. For the illustration of the mixer operation 2 para of frequencies where used. The figure 8 shows the output power spectral density with and without the output filter when f_1=56464 Hz, f_2=37918 Hz and f_{out}=9273 Hz. The effect of the PLL based pass band filter is also shown on figure 8. Figure 9 shows the output power spectral density when f_1=9536 Hz, f_2=25893 Hz and f_{out}=8178.5 Hz. Isolation of output from input is shown on figure 10 and figure 11, where the output power spectral density is measured at input frequencies.

Journal of Physics: Conference Series **1379** (2019) 012067 doi:10.1088/1742-6596/1379/1/012067

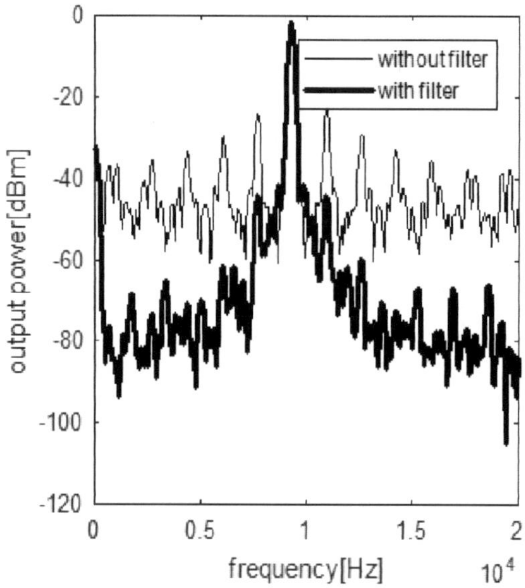

Figure 8. Output power spectrum with and without filter f_1=56464 Hz, f_2=37918 Hz, f_{out}=9273 Hz.

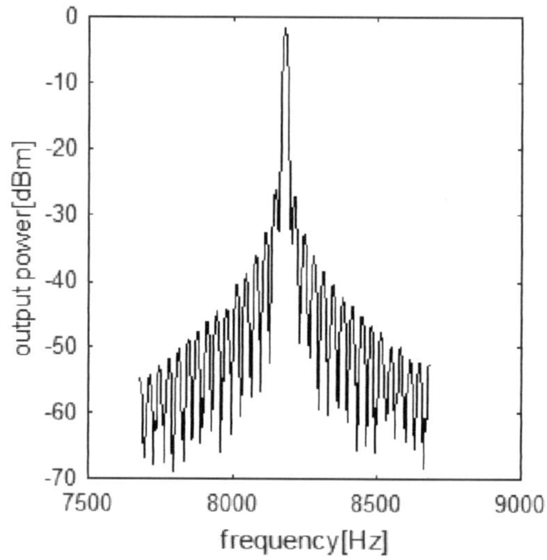

Figure 9. Output power spectrum at $|f_2f_1|/2f_1$=9536 Hz, f_2=25893 Hz, f_{out}= 8178.5 Hz.

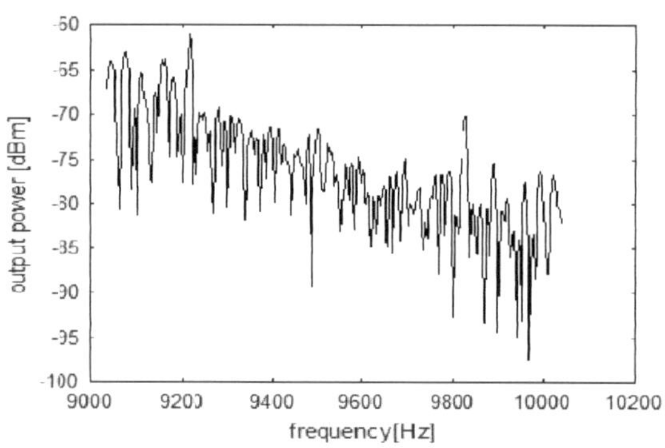

Figure 10. Output power spectrum measured around input frequency f_1=9536 Hz.

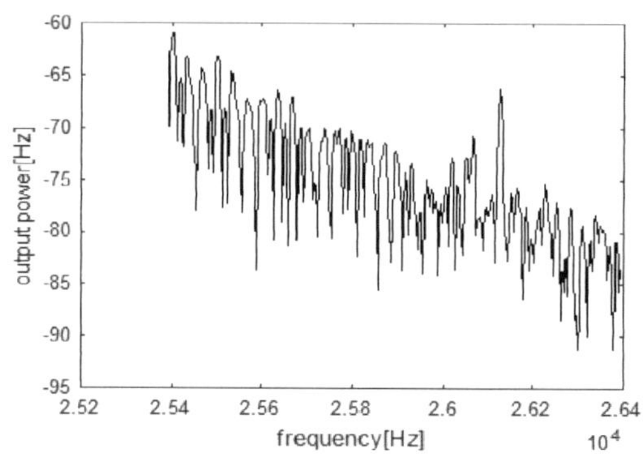

Figure 11. Output power spectrum measured around input frequency f_2=25893°Hz.

3. Results and discussion

Both the simulation and hardware implementation show that the suggested method for digital frequency mixer can be implemented. It is simplicity is attractive, if we compare it to the methods described section 1. However, the measurement show that more effort is needed for spurs attenuation. It's main advantage, compared to its analogue equivalent, is that there is no conversion loss and that good isolation of the output from inputs is assured.

Acknowledgment

We thank the Vlaamse Interuniversitaire Raad (VLIR) for providing a PhD scholarship in the framework of which this research is conducted.

Journal of Physics: Conference Series **1379** (2019) 012067 doi:10.1088/1742-6596/1379/1/012067

References

[1] Sotiriadis P P and Ling W A 2009 Multiphase EXOR frequency mixers in *Proc. IEEE Radio Wirel. Symp.* pp 260–3

[2] Ling W A and Sotiriadis P P 2009 Almost-digital multiphase frequency mixing in *Proc. IEEE Sarnoff Symp.* pp 1–4

[3] Ling W A and Sotiriadis P P A 2011 Nearly All-Digital Frequency Mixer Based on Nonlinear Digital-to-Analog Conversion and Intermodulation Cancellation *IEEE Transanctions on Circuits and Systems* **58** (8) pp 1695–704

[4] Cardells F and Valls J 2002 High Performance Quadrature Digital Mixers for FPGA *12th Int. Conf. FPL (2-4 September, 2002, Montpellier, France)*

Impact of enzymatic methods for glucose measurement on the performance of clinical laboratories in proficiency testing

Diogo José da Silva Jerônimo[1,2] **and Elisabeth Costa Monteiro**[1]

[1]Postgraduate Programme in Metrology, Pontifical Catholic University of Rio de Janeiro (PUC-Rio), Rio de Janeiro, 22451-900, Brazil
[2]Control Lab - Controle de Qualidade para Laboratórios, Rio de Janeiro, Brazil

E-mail: beth@puc-rio.br

Abstract. The present study compares performance scores obtained by clinical laboratories participating in Proficiency Testing schemes for measuring three concentration items of glucose in lyophilized serum, according to the analytical method employed (Oxidase-Peroxidase or Hexokinase-Dehydrogenase). Two distinct scoring criteria guided the evaluation: one based on the consensus value from the participants' results; and another one that uses parameters exclusively based on the assigned value defined by the reference material. The results indicate that the measuring method employed in laboratory practice significantly affects the uniformity of PT participants' outcomes. The larger dispersion of data provided by participants using the Oxidase-Peroxidase enzymatic method caused a higher index of evaluation's inadequacies for this group when the laboratory performance score applies the consensus approach.

1. Introduction

Efforts toward reliable clinical laboratory results, aiming at guaranteeing the appropriate diagnoses and treatments, are continuously increasing [1-4]. An essential tool for quality assessment of measurement capability has been widely employed utilizing participation in proper proficiency testing (PT) schemes [5,6]. Medical laboratories apply a wide variety of techniques to quantify a large number of parameters, so the results may not be comparable across labs [7].

The same analyte can be measured by different analytical methods, according to the routine employed by each medical laboratory. The non-uniformity of the analytical procedure used by participants may affect their performance results from PT interlaboratory comparisons. If the performance of the different methods is not uniform, it would probably impact on the proficiency test evaluation, mainly when the consensus value derived from participants' results is the basis for the assigned value, which is the most commonly employed procedure.

PT performance of clinical laboratories has been demonstrated to be affected by the scoring approach employed for their measurement capability evaluation [1]. Discrepancies in the PT participant score resulted from bias introduced by the use of statistics incorporating elements associated with the consensus value obtained from the outcome of participants when compared to strategies entirely based on the reference material value for determining assign value and acceptable ranges [1]. Characteristics inherent to the different analytical methods employed for the estimative of the measurand may play a role in the discrepancies produced by participants' consensus in the target value and the standard deviation for the proficiency assessment.

Glucose measurement is among the most customary analyzes of the clinical laboratory's routine. Measurement methods employed are based on enzymatic reactions, especially Hexokinase-Dehydrogenase (H-D) and Oxidase-Peroxidase (O-P). Both the approaches encompass spectrophotometric assays. In the H-D approach, the hexokinase facilitates the phosphorylation of glucose, being followed by an oxidation reaction by glucose-6-phosphate dehydrogenase, which produces reduced nicotinamide adenine dinucleotide (NADPH) proportionally to the amount of glucose in the sample. NADPH is, then, measured spectrophotometrically. Regarding the O-P method, glucose is oxidized to gluconic acid and hydrogen peroxide, followed by a reaction catalyzed by peroxidase that results in a product quantifiable by spectrophotometric detection [8].

The present study compares the proficiency testing performance of biomedical laboratories for detecting and quantifying glucose in human serum, which used the two different enzymatic methods: Oxidase-Peroxidase and Hexokinase-Dehydrogenase. The analysis compares all PT participants' results according to the ***Z score***, based on the consensus value obtained from the outcome of participants; and to an approach named ***Z'_{FL} score$_{(ref)}$*** in [1], which comprises parameters exclusively based on the assigned value defined by the reference material.

2. Materials and Methods

The study evaluates the results reported by 1737 clinical laboratories taking part on a proficiency testing round for glucose determination in samples of lyophilized human serum with three different glucose concentrations (91.08 mg/dL, 301.96 mg/dL, 143.33 mg/dL) [1]. The data were provided by an accredited PT provider. All PT participants reported the employment, in their routine glucose measurements, combinations of analytical systems based on two possibilities of enzymatic methods: Oxidase, with Peroxidase reaction (O-P); or Hexokinase, with Dehydrogenase reaction (H-D).

The comparative analysis of performance evaluation was based on two statistical designs. One of them, the ***Z score*** described in [9], is frequently employed by proficiency testing procedures (equation 1).

$$z\text{-}score = \frac{x - X_{cons}}{\sigma_{cons}} \tag{1}$$

where x is the result reported by a participant; X_{cons}, the assigned value obtained from the interlaboratory consensus amongst participants; and σ_{cons}, the standard deviation for proficiency assessment derived from the results reported by the participants in the same round.

The second approach for evaluating the clinical laboratories performance, proposed in [1], is based on parameters entirely independent of the participants' measurement capability (equation 2).

$$z'_{FL} - score_{(Ref)} = \frac{x - X_{ref}}{(\sigma^2_{refFL} + u^2_{X_{ref}})^{1/2}} \tag{2}$$

where X_{ref} is the assigned value derived from the reference value of the provided PT item; σ_{refFL}, the standard deviation for proficiency, determined using a fixed limit (FL) of 13 % of the assigned value defined by the reference value of the PT item ($\sigma_{refFL}=X_{ref}*(FL/2)$) [10]; and u_{XRef} is the standard uncertainty for reference values (X_{ref}) associated with each round item [9, 11], determined by:

$$u_{X_{ref}} = \frac{\sqrt{\frac{1}{n_h - 1}\left(\sum_{i=1}^{n_h} x_{ih}^2 - n_h X_{ref}^2\right)}}{n_h} \tag{3}$$

where x_{ih} is the i_{th} value obtained from the sample being analysed for the demonstration of sufficient homogeneity; and n_h is the sample size employed for the homogeneity assessment.

The standard deviation for proficiency assessment using the $\boldsymbol{Z'_{FL}}$ ***score***$_{\boldsymbol{(Ref)}}$ approach was established on a fit for purpose value, which is recommended by the International Union of Pure and Applied Chemistry (IUPAC) and described in the Harmonized Protocol for the PT in Analytical Chemistry [1, 9-11].

Outliers may be responsible for differences between means and variances of results provided by the two groups of participants [8], subdivided according to the analytical methods they used for measuring glucose in human serum (O-P or H-D). The Grubbs' test evaluated the statistical outlier of data reported for each item of the PT round by the two groups of participating laboratories. After eliminating the outliers, the Kruskal-Wallis' test (K-W) analysis investigated the difference between means of the two groups of laboratories. Under the same conditions, using the Levene's test, the variances of results for each group of analytical method were checked for equality. Differences between assigned values designated by consensus or by the reference, associated with each PT item, were investigated by applying the one-sample *t*-test.

3. Results and Discussion

The study evaluates PT performance of clinical laboratories for quantifying glucose in lyophilized serum using two frequently employed enzymatic methods, Oxidase-Peroxidase (O-P) and Hexokinase-Dehydrogenase (H-D). The criteria for PT assessment derived from the participants' consensus (***Z score***), or were entirely independent ($\boldsymbol{Z'_{FL}}$ ***score***$_{\boldsymbol{(Ref)}}$) and based on the assigned value defined by the reference material. The reference values (X_{ref}) for each round item, and their standard uncertainties (u_{Xref}), were obtained from the data provided by the assessment of homogeneity of the reference material, corresponding to items of round PT.

The *t*-test indicated differences between X_{cons} and X_{ref} for PT items CHM001 and CHM003, at a significance level of 5%. Table 1 presents the results of tests to investigate the differences between outcomes from the two distinct analytical methods (O-P and H-D) used by participants for quantification of glucose.

Table 1. Statistics and corresponding p-values obtained by the K-W and Levene's tests for comparison between the results of the analytical methods O-P and H-D, applied by PT participants for glucose measurement.

Item	test	Statistics	*p*-value
CHM001	**K-W**	19.07900	0.00001
	Levene	116.86000	0.00000
CHM002	**K-W**	48.76600	0.00000
	Levene	101.89000	0.00000
CHM003	**K-W**	0.10081	0.75090
	Levene	97.08300	0.00000

The Levene's test was significant at the 5% level for all three items of the round (table 1), indicating that there was a difference between the variances of the two sets of analytical methods employed by the participants. These differences, observed for the three PT items, impact on the standard deviation for proficiency when it derives from the participants' consensus.

In turn, the K-W test indicated that there was a significant difference between the means of the analytical methods for items CHM001 and CHM002 (table 1). For these two items, one expects that the differences between the means and variances of the analytical systems O-P and H-D will result in inconsistencies in the performance evaluation of the laboratories when using consensus-based approaches. For item CHM003 (table 1), discrepancies are expected to arise only from the differences between the variances of criteria.

Figure 1 presents the laboratory outcomes for CHM001 item, according to the analytical method used by participants of the proficiency test round of glucose dosing. From the total number of 1737 results reported, after applying the Grubbs' test for identifying and removing outliers, this number was reduced to 1392 results (19.86% sample reduction). The great majority of data originated (1038 laboratories) employing Oxidase-Peroxidase (O-P), which consists of the most commonly used method. Participants employing Hexokinase-Dehydrogenase (H-D) method provided 354 outcomes. In figure 1, the dashed blue line and the gray region represent the acceptable ranges for outcomes of glucose concentration according to ***Z score*** and $\boldsymbol{Z'_{LF}}$***-score*** $_{(Ref)}$, respectively. The solid red and blue lines indicate, respectively, the $\boldsymbol{X_{ref}}$ and $\boldsymbol{X_{cons}}$ assigned values (figure 1).

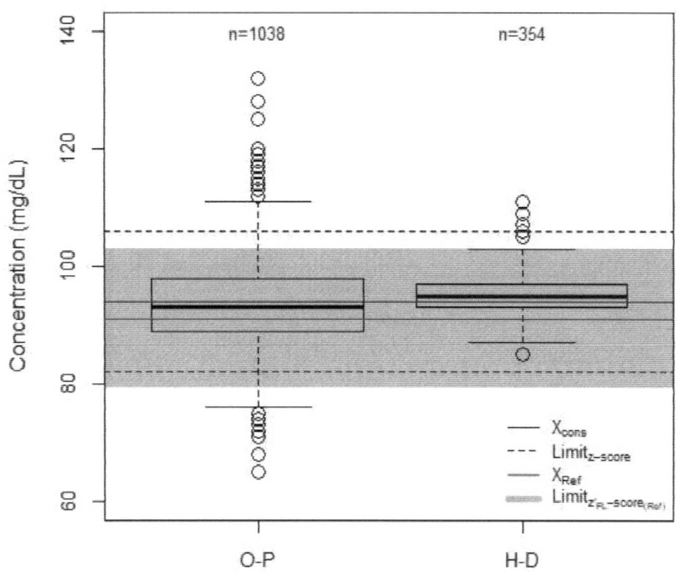

Figure 1. Distribution of reported results for glucose concentration of item CHM001, according to the analytical method employed by laboratories (O-P and H-D). The region in gray shows the acceptable range for $\boldsymbol{Z'_{LF}}$***-score*** $_{(Ref)}$, and dashed blue lines indicate the acceptance limits for ***Z score***.

The Oxidase-Peroxidase method shows a larger dispersion of the participants' results when compared to the outcomes reported by laboratories using Hexokinase-Dehydrogenase method. As a result, the great majority of participants using H-D enzymatic method presented a satisfactory classification when laboratory performance considered the independent criterion $\boldsymbol{Z'_{LF\text{-}Score(Ref)}}$, proposed in [1]. In turn, a significant number of results derived from O-P measurements were outside this acceptable range, which is exclusively based on the reference value ($\boldsymbol{X_{ref}}$), limited by the gray region. Even though there are differences between the O-P and H-D measurements' means (table 1 and figure 1), both shift away from the $\boldsymbol{X_{ref}}$. This difference is responsible for the discrepancy between $\boldsymbol{X_{cons}}$ and $\boldsymbol{X_{ref}}$ assigned values. Despite the higher adequacy of the results' set obtained using H-D, with reduced dispersion of outcomes, their average value is even more distant from $\boldsymbol{X_{ref}}$ than the bias associated with the O-P results (figure 1).

The discrepancy between the acceptable range for the ***Z score*** (dashed blue lines of figure 1), and for the $\boldsymbol{Z'_{LF\text{-}Score(Ref)}}$ reference approach (gray region) indicates a potential risk of an incorrect decision of clinical laboratories' classification when based on ***Z score***. For the proficiency testing with the CHM001 item, 6 % of participants received improper classification by ***Z score***, being 5.7 % from the laboratories employing the O-P method, and 0.3 % from the participants using the H-D approach. These results demonstrate the significant impact of the wider dispersion of O-P results, risking their users of an inappropriate classification when laboratory performance evaluation employs scoring approaches strongly dependent on the participants' consensus.

The *t*-test of both the PT items CHM001 (table 1 and figure 1) and CHM003 indicated differences between X_{cons} and X_{ref}, at a significance level of 5%. However, the impact on the metrological compatibility of PT results was more evident for the item with the lowest concentration (CHM001), for which both K-W and Levene tests showed significant differences between analytical systems. In turn, item CHM003, whose K-W indicates a good correlation between the enzymatic methods, the PT compatibility remained in the edge of an acceptable limit.

For the item with the highest glucose concentration (CHM002), although the *t*-test indicated an agreement between X_{cons} and X_{ref}, both the K-W and Levene tests pointed toward significant differences between the means and variances of the analytical systems O-P and H-D. Considering the fewer performance discrepancies and the satisfactory PT compatibility observed for the CHM002 item, differences between methods appear to be less relevant then agreement between consensus and reference assigned values.

According to literature, O-P and H-D approaches present good correlation [12], as also observed in this study for the CHM003 item measurements, but differences between their consensus and the reference values generate the need of applying independent criteria for performance evaluation.

The significant difference observed between the results obtained from these two distinct analytical methods, and their impact on the PT performance evaluation, particularly for the CHM001 item (table 1 and figure 1), corroborates with [7], which shows that laboratory results are strongly dependent on the selected method to perform measurements, with risks of making their results incomparable.

4. Conclusion

The present study characterizes the differences in performance scores obtained by clinical laboratories participating in Proficiency Testing schemes for measuring three concentration items of glucose in lyophilized serum, according to the analytical method employed (Oxidase-Peroxidase or Hexokinase-Dehydrogenase). The evaluation was guided by two distinct scoring criteria, the ***Z score***, based on the consensus value from the participants' results; and the Z'_{FL} ***score***$_{(ref)}$ [1], which considers parameters exclusively based on the assigned value defined by the reference material.

The results indicate that the measuring method employed in the laboratory practice significantly affects the agreement between outcomes provided by PT participants. The laboratories using the Oxidase-Peroxidase enzymatic method presented a much higher dispersion of reported data than Hexokinase-Dehydrogenase outcomes. This scattering of provided O-P results induced their high index of incorrect classification of performance (as satisfactory or unsatisfactory) when using the consensus approach for evaluation. On the other hand, due to the lower dispersion of the results provided by laboratories using the H-D method, the opposite was observed for this group of participants.

The widespread use of analytical methods with considerable variability of their results, coupled with a significant number of measurement systems with effectively unsatisfactory performance, generates an expanded distribution of the results reported in a proficiency test. This wide dispersion of results induces non-homogeneities of the performance of PT participants using these technologies, which increases their risk of unacceptable performance classification. Besides, the larger scattering of results directly affects the values of the parameters associated with the ***Z score*** evaluation. As a result, a bias in the acceptance range can occur when a proficiency test analysis employs parameters based on the consensus of the participants (***Z score***), generating a higher risk of incorrect decision regarding the performance of the laboratory.

This non-uniformity of the quality of laboratory results according to measurement methods employed emphasizes the relevance of using reliable targets to set the PT assigned value; ideally, traceable to SI [1, 3, 4, 13]. The use of proficiency testing criteria thoroughly independent of the results reported by the participants contributes to the appropriate decision of clinical laboratories' performance [1, 13].

Given the different combinations of instrumentation and reagents, which the two enzymatic approaches may employ, future work consists of extending the present investigation to evaluate the impact of the nuances of the complete set of analytical system's elements in the PT performance evaluation.

Journal of Physics: Conference Series **1379** (2019) 012068 doi:10.1088/1742-6596/1379/1/012068

Considering the possibility of a biased consensus due to the use of faulty methodology by participants, the score inadequacy can be prevented by applying reliable references to set the assigned value, ideally traceable to SI, utterly independent of the results reported by the PT participating laboratories.

Acknowledgment
The authors acknowledge the Control Lab management for providing the data analyzed in this study.

References
[1] Jerônimo D J S and Costa Monteiro E 2018 *J. Phys.: Conf. Ser.* **1065** 132003.
[2] Plebani M, Sciacovelli L and Aita A 2017 *Clin. Lab. Med.* **37** 187–205
[3] Costa Monteiro E 2007 *Metrologia e Instrumentação* **6** 6-12
[4] Costa Monteiro E and Leon L F 2015 *J. Phys.: Conf. Ser.* **588** 012032
[5] ISO/IEC 2010 *ISO/IEC 17043:2010 Conformity Assessment — General Requirements for Proficiency Testing* (International Organisation for Standardization)
[6] ISO 2012 *ISO 15189:2012 Medical Laboratories – Particular Requirements for Quality and Competence* (International Organisation for Standardization)
[7] Delatour V, Lalere B, Saint-Albin K, Peignaux M, Hattchouel J M, Dumont G, De Graeve J, Vaslin-Reimann S and Gillery P 2012 *Clin. Chim. Acta* **414** 182–184
[8] Moodley N, Ngxamngxa U, Turzyniecka M J and Pillay T S 2015 *J. Clin. Pathol.* **0** 1–7
[9] ISO 2015 *ISO 13528:2015 Statistical Methods for Use in Proficiency Testing by Interlaboratory Comparisons* (International Organisation for Standardization)
[10] Ehrmeyer S S and Laessig R H 1987 *Clin. Chem.* **33(10)** 1901–1902
[11] Thompson M, Ellison S L R and Wood R 2006 *Pure Appl. Chem.* **78(1)** 145–196
[12] Ayyanar K, Pichandi S, Janakiraman P 2018 *Int. J. Biotechnology and Biochemistry* **14** 51–58
[13] De Biévre P 2016 *Accredit. Qual. Assur.* **21** 167–169

Journal of Physics: Conference Series **1379** (2019) 012069 doi:10.1088/1742-6596/1379/1/012069

Readings correction and on-line monitoring of fluid level measuring channels at NPP

A A Kalashnikov

JSC «Rusatom automated control systems», Moscow, Russia

E-mail: AlAleKalashnikov@rasu.ru

Abstract. Presently the international program has gained momentum aimed at increasing the power of nuclear power plants (NPP) with WWER- and PWR-types reactors up to 107% and also at extending the automated process control system (APCS) repair intervals up to 18 months, which places higher requirements to the accuracy of measurements of NPP technological parameters. To enable the implementation of this program, an algorithm for processing the measurement signals has been designed that ensures high precision in readings correction of hydrostatic level sensors for all operating modes of technological equipment. Methods for correction of readings based on the developed algorithm have been implemented in Russian nuclear power industry. In addition to the correction of readings the algorithm also provides on-line monitoring to identify all possible defects of fluid level measuring channels including hidden defects that are not detected by existing facilities of NPP control system. The paper describes the algorithm and its application it to the tasks of readings correction and on-line monitoring of fluid level measuring channels at NPP. As result the article represents one of the first steps to methodological development of the correction reading and on-line monitoring to detection and minimization of hidden defect on NPP measuring channels of all thermotechnical parameters.

1. Introduction

The article summarizes the many years of experience in commissioning nuclear power plants (NPP). It includes research methodology from readings correction to on-line monitoring of fluid level measuring channels. Achieved research results allowed to revise methods of fluid level measuring in part of readings correction and after that they were introduced in nuclear industry. This paper includes these methods and opens up perspective to development of on-line monitoring systems for other industries.

Development of these methods in nuclear industry caused by international program to NPP power increasing. Presently the international program has gained momentum aimed at increasing the power of nuclear power plants with WWER- and PWR-types reactors up to 107 % and also at extending the APCS repair intervals up to 18 months [1]. The program places higher requirements in regard of safe operation, accuracy of thermotechnical parameters measuring, accuracy of automatic control and calculation of technical and economic indicators. To enable the implementation of the program we had to develop principles for processing measuring signals of hydrostatic level sensors that permit improved metrological reliability of their indications under different operating modes of technological equipment.

The paper presents main provisions on the development of a universal algorithm for processing the measuring signal of hydrostatic level sensors, which can be an instrument not only for high-precision correction of readings but also for on-line monitoring of measuring channels (MC) of fluid level at NPP.

Mathematical construction that form the basis of the algorithm are given for the level measurement scheme using a single-chamber leveling vessel [2] which is most common in nuclear industry and thermal power industry.

2. Level measurement using a single-chamber leveling vessel

A typical scheme for measuring the fluid level by hydrostatic method (by differential pressure sensors) using a single-chamber leveling vessel is shown in figure 1.

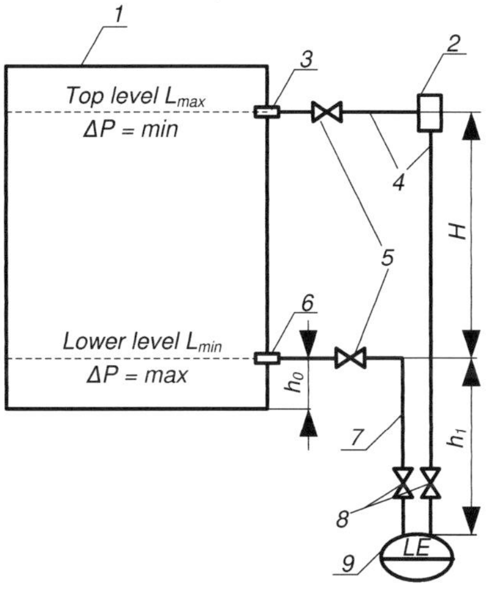

Figure 1. The level measurement in the storage vessel under pressure with a single-chamber leveling vessel

1 – storage vessel under pressure; 2 – single-chamber leveling vessel; 3 – tap (joint of plus sensing line); 4 – plus sensing line (instrument piping); 5 – main valves; 6 – tap (joint of minus sensing line); 7 – minus sensing line; 8 – shut-off valves; 9 – differential pressure sensor; ΔP – measured pressure difference; h_l – height of sensing lines until the lower tap; H – level sensor base; h_0 – distance from the storage vessel bottom to the lower tap; L_{max} – maximum measured level value; L_{min} – level value corresponding to the height of the lower tap.

In this measuring scheme the leveling vessel has the same pressure/vacuum as in storage vessel; the pressure differential on the sensor shows as:

$$\Delta P = P_{plus} - P_{minus} = P_{in} + \rho_1 \cdot g \cdot (h_1 + H) - P_{in} - (\rho_2 \cdot g \cdot h_1 + \rho \cdot g \cdot L) =$$
$$= \rho_1 \cdot g \cdot H + (\rho_1 - \rho_2)\, g\, h_1 - \rho \cdot g \cdot L,$$

where ΔP is the measured pressure differential;
P_{plus} is the overall pressure exerted on the side of the plus sensing line;
P_{minus} is the overall pressure exerted on the side of the minus sensing line;
P_{in} is the pressure/vacuum inside the storage vessel;
ρ_1 is the fluid density in the plus sensing line and the leveling vessel;
ρ_2 is the fluid density in the minus sensing line and the leveling vessel;
g is the local gravity acceleration;
L is the fluid level in the storage vessel;
ρ is the fluid density in the storage vessel.

When the level is measured using this method, the main causes of error are as follows: increase of temperature and pressure of the process fluid in the storage vessel, different temperature conditions in the plus sensing line (SL) and in the storage vessel; different phases of the fluid (water and steam) in the storage vessel; mismatch of the construction characteristics and MC settings. These causes result in the change of MC readings in general. Hence, to ensure the reliability of MC readings, the «hot» scale correction with proven the value sensor base is required.

When the process fluid temperature in the storage vessel equals to ambient temperature, the corresponding scale of measurement which provides reliable readings, is conventionally referred to as «cold» scale. When the process fluid temperature is higher than ambient temperature, the corresponding scale of measurement is conventionally referred to as «hot» scale.

In order to perform the correction of the MC readings in real time, which implies recalculation of their values from «cold» to «hot» scale, we shall derive formulae for the «hot» scale correction coefficients.

3. MC readings correction by the introduction of correction factors to the «hot» scale

For readings correction of fluid level measurement with hydrostatic sensors we derived the special formulae [3]:

$$L = \frac{L_{cs}\cdot(\rho_{cs}' - \rho_{cs}'') - H\cdot(\rho_{cs}' - \rho_{cs}'' + \rho'') + \left(\sum_{i=1}^{n} \rho_{1i}\cdot H_i + \sum_{j=1}^{m} \rho_{1j}\cdot h_j - \sum_{\gamma=1}^{k} \rho_{2\gamma}\cdot h_\gamma\right)}{\rho' - \rho''} \quad (1)$$

where L_{cs} is the water level in the storage vessel for the case of «cold» scale;
ρ_{cs}' is the density of water in the storage vessel corresponding to the case of «cold» scale;
ρ_{cs}'' is the density of water steam in the storage vessel corresponding to the case of «cold» scale;
H_i is the length of the i section of the plus SL within the actual base of the MC ($\sum_{i=1}^{n} H_i = H$);
h_j is the length of the j section of the plus SL within its height above the minus SL ($\sum_{j=1}^{m} h_j = h_1$);
h_γ is the length of the γ section of the minus SL within its height above the minus SL ($\sum_{\gamma=1}^{k} h_\gamma = h_1$);
ρ' is the density of water inside the storage vessel;
ρ'' is the density of steam inside the storage vessel.

To simplify the industrial calculation procedure it is convenient to use the resulting function transformed to the following form (2):

$$L = k \cdot L_{cs} + b, \quad (2)$$

where the correction factors k and b are calculated as follows:

$$k = \frac{(\rho_{cs}' - \rho_{cs}'')}{\rho' - \rho''},$$

$$b = \frac{\sum_{i=1}^{n} \rho_{1i}\cdot H_i + \sum_{j=1}^{m} \rho_{1j}\cdot h_j - \sum_{\gamma=1}^{k} \rho_{2\gamma}\cdot h_\gamma - H\cdot(\rho_{cs}' - \rho_{cs}'' + \rho'')}{\rho' - \rho''}.$$

Formulas (1) and (2) take into account a lot of fluid parameters and MC characteristic including uneven temperature distribution into the storage and sensing lines. It allow to realize the high-precision correction of MC readings in dynamic.

In order to implement the correction of the level readings to the nominal values of the process fluid, initially the correction factors K, B are calculated using the previously derived formulae. The density of water and steam required for the calculation are determined according to the reference book [4] for the nominal values of pressure and temperature of the process fluid. The calculated correction factors are entered in the software and hardware complex that is part of MC structure. This ensures a continuous recalculation of the level readings for the «hot» scale (Figure 2).

Figure 2 shows that this method of readings correction allows the control of level both at the «cold» and «hot» scales (at the nominal characteristics of the fluid) without additional resetting of the sensor. If necessary, it is possible to provide automatic bumpless switching of the level readings from the «cold» to «hot» scale in the software-hardware complex.

The single drawback to this method of correction is that the reliable range of level readings is limited for only two operational modes of technological equipment. During transient processes, when thermal and chemical characteristics of the fluid do not reach the nominal values, the reliability of the readings is not provided.

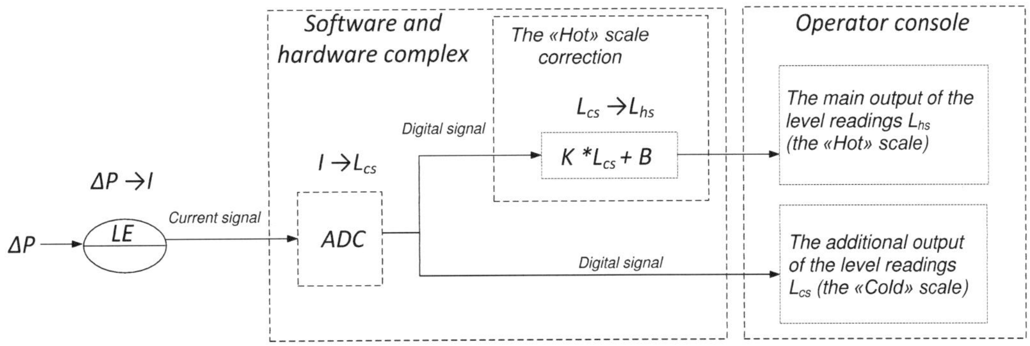

Figure 2. An explanatory diagram for the measurement signal processing with the correction at the nominal characteristics of the fluid
ΔP is the measured pressure differential; LE is a differential pressure sensor adjusted to the «cold» scale (according to GOST 22520-85); I is the output current sensor signal; ADC is analog-to-digital converter; L_{cs} is the «cold» scale level readings; L_{hs} is the «hot» scale level readings

In order to perform correction of the level readings in real time (including transient processes), it is necessary to arrange in the software-hardware complex the calculation of correction factors K, B taking into account the actual values of the density of water and steam.

In turn, the density of water and steam in real time may be determined on the basis of «International Formulation Equations IAPWS» [5]. In this case the initial data for the calculation of density are the related actual readings of the pressure and temperature sensors in the storage vessel (Figure 3).

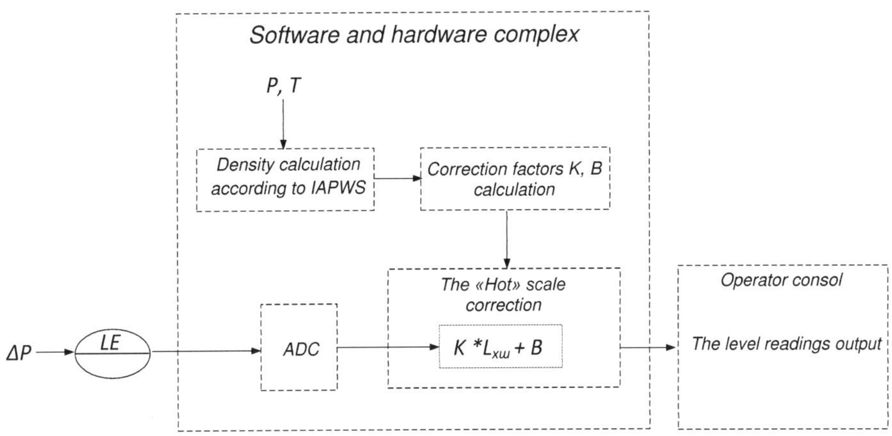

Figure 3. The explanatory diagram for dynamic correction
P is the actual pressure value in the storage vessel;
T is the actual temperature value in the storage vessel

This method of dynamic correction, as opposed to the previous one, ensures the reliability of readings in all possible operational modes of the equipment. "The other side of the coin" here concerns the higher requirements to the computing capacity of the software-hardware complex and the use of additional pressure and temperature measurement channels.

Depending on specific conditions, requirements to the accuracy of measurements/automatic control, and depending on the capabilities of APCS, the most appropriate of the suggested methods of correction can be chosen at NPP.

As a result we lift the readings correction procedure of hydrostatic level sensors from static to dynamic mode by algorithm with simplest realization. Dynamic error of MC readings was minimized until to 3 %. This 3 % error is minimum caused by pressure and temperature measurements accuracy [3].

Moreover it allow to realize on-line monitoring based on constructing of virtual sensor readings that will be present in the paper. The main purpose of this on-line monitoring is to detect the hidden defects of measuring channels without stopping the process. This is real step to increase the NPP safety.

4. Application of the developed algorithm for processing of measurement signals in the tasks of MC on-line monitoring

This algorithm for processing measurement signals of hydrostatic level sensors allows modeling of a virtual level sensor (mathematical model which uses the real MC readings and generates the level value) to perform on-line monitoring of MC metrological operability. Such a method of on-line monitoring is of particular relevance for existing NPP, where the implementation of the developed dynamic correction of readings is not possible due to absence of digital software-hardware complex in APCS structure. In such cases, modeling of the readings of virtual sensor provides opportunities for an independent high-precision control of level, including transient modes of technological equipment.

Mathematical modeling of a virtual sensor (Figure 4), similar to the developed correction methods, is based on calculating the level values taking into account the correction factors K and B, the difference being that the initial data are the measured values of differential pressure on the sensors rather than the readings of L_{cs} channel:

$$L_M = K \cdot \overline{\Delta P} + B \tag{3}$$

where:

$$K = (\rho'' - \rho')^{-1} \cdot g^{-1};$$

$$B = \frac{\sum_{i=1}^{n} \rho_{1i} \cdot H_i + \sum_{j=1}^{m} \rho_{1j} \cdot h_j - \sum_{\gamma=1}^{k} \rho_{2\gamma} \cdot h_\gamma - \rho'' \cdot H}{\rho' - \rho''} + h_0.$$

where L_M is simulated readings of the level sensor;
$\overline{\Delta P}$ is the average value of the measured pressure differential on the redundant level sensors;
H is a value of the MC level base found using results of specialized geodetic measurements.

The value of $\overline{\Delta P}$ included in the formula (3), is determined as follows:

$$\overline{\Delta P} = \rho_{1\mathrm{mc}}\, g\, H_{\mathrm{mc}} - \left(\rho'_{\mathrm{mc}}\, g \cdot \left(k^{-1} \cdot \sum_{i=1}^{k} L_i - h_0\right) + \rho''_{\mathrm{mc}} \cdot g \cdot \left(H_{\mathrm{mc}} - k^{-1} \cdot \sum_{i=1}^{k} L_i - h_0\right)\right) + \Delta P_0.$$

where $\rho_{1\mathrm{mc}}$ is the density of water in the plus sensing line, which is included in the settings of MC;
H_{mc} is the base value in the settings of MC;
ρ'_{mc} is density of water inside the storage vessel;
ρ'' is value of steam density inside the storage vessel;
L_i is the reading of the i measurement channel;
k is the total number of duplicating MC.

The factors K and B for the respective temperature and pressure values are calculated either with using the data from the reference book [4] or with using of IAPWS International Formulation Equations.

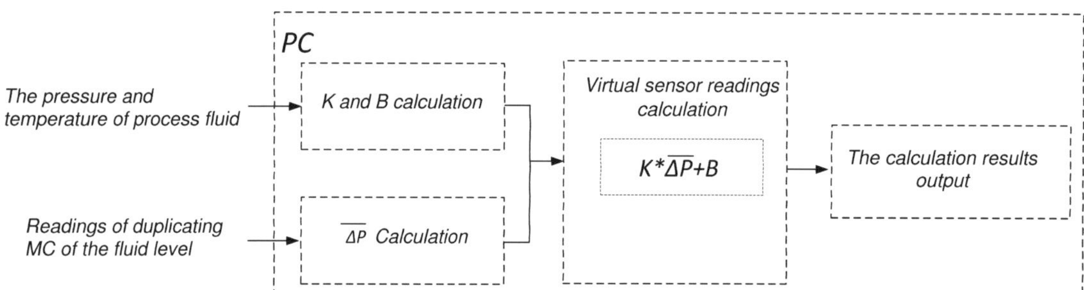

Figure 4. Mathematical modeling of virtual level sensor

The readings of each of the redundant MC of level are verified against modeled readings of the virtual sensor to realization the metrological diagnostics in real time. As a result we calculate the standard deviation:

$$\sigma = \sqrt{\frac{1}{N-1}\sum_{n=1}^{N}(L_n - M[L_M])^2} = \sqrt{\frac{1}{N-1}\sum_{i=1}^{N}\left(L_n - \frac{1}{N}\sum_{n=1}^{N} L_{M_n}\right)^2}$$

where σ is the standard deviation over time period N;

L_n is the current value of MC readings at the n time moment;

$M\ [L_M]$ is the expectation value of the modeled readings L_M for the entire time period;

$L_{M\,n}$ is the current readings of the modeled sensor at the n time moment.

The fact of MC fault is identified when the standard deviation exceeds the set limit of admissible deviation (Figure 5).

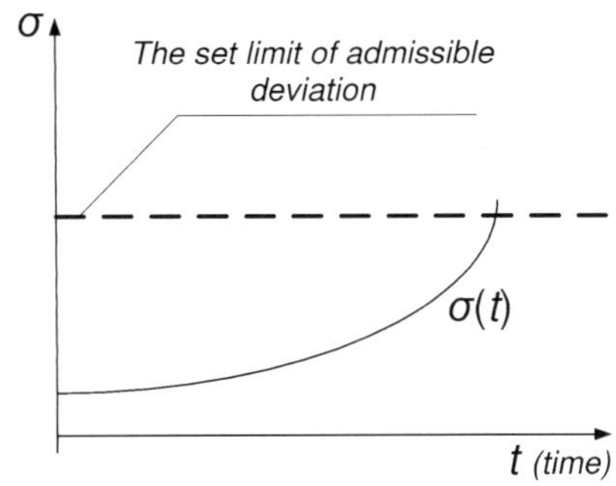

Figure 5. Standard deviation exceeding the preset limit

Analysis of the graph of the dependence of the standard deviation on time enables the identification not only of visible defects but also of invisible defects of MC. For example:

1) The constant standard deviation indicates a systematic measurement error, which most likely cause are concealed faults:

- The MC settings not matching the actual value of level sensors base;
- Correction of MC readings does not correspond to nominal thermal and physical characteristics of the process fluid.

2) The main causes for gradual and/or short-term growth of the standard deviation are:

- Partial emptiness of the sensing lines and the leveling vessel;
- "Zero drift" of the sensor;
- The deviation of sensor from the required metrological characteristics.

3) Instant significant increase of the standard deviation usually may be caused by the following main reasons: breakage of cable lines, deterioration of electrical contact, malfunction of any technical device in the measuring channel. These defects are visible, and most likely they would be identified with in-service standard self-diagnosis system of the APCS. Such defects are fixed by operational personnel immediately after they are detected.

4) When tracing the long history of standard deviation over many fuel loadings of the reactor plant, it is possible to identify the degradation of cable circuit and other elements that are part of MC structure.

The examples given above show that the proposed method of on-line monitoring solves the main task of maintenance, that is to timely diagnose a particular defect before it results in a critical failure/error of the process equipment. Application of this method opens up the possibility to monitor and ensure the metrological reliability of MC readings in the context of extended interrepair intervals of NPP control and measuring instruments. Known methods such as balance and neural networks, regression methods [6 – 9] did not reveal a lot of above reasons.

As a result, the introduction of represented readings correction and on-line monitoring methods allowed to revise all previously used methods at Russian NPP and allowed to reduce MC dynamic error to 3 %.

5. Conclusion

The described algorithm for processing of measuring signals for correction of MC level readings allowed to revise the approaches applied and to provide high accuracy of measurement results for different operational modes of technological equipment and for any technical characteristics of MC. The developed methods of readings correction were implemented during putting into operation of new power units of Russian NPP.

The developed mathematical model of the virtual level sensor for on-line monitoring enables diagnostics of visible faults as well as invisible faults of MC. The practical significance of this on-line monitoring method is the possibility of remote metrological diagnosis of MC during continuous technologic process. It is also possible to provide timely identification of the probable cause of the fault and, therefore, to inform the staff about the scope and volume of repair work, thus reducing the troubleshooting time (which is of particular relevance in the restricted area). Besides, the use of this method does not require any intervention into the functioning of in-service APCS equipment at NPP.

Developed methods improve NPP operational safety and provide opportunities for implementation of the program aimed at increasing the NPP power and extending interrepair intervals of APCS equipment

The positive results from implementation developed methods at NPP opens up perspective to development of on-line monitoring system for other thermotechnical parameters at the energetic industry.

References

[1] IAEA 2011 *NP-T-3.9 Power Uprate in Nuclear Power Plants: Guidelines and Experience* (Vienna: IAEA) p 106

[2] Kalashnikov A A 2017 *Guide to Setting up Industrial Hydroastatic Level Sensors* (Moscow: Infra-Engineria) p 194 (in Russian)

[3] Kalashnikov A A 2015 Correction of level measuring channels with differential pressure sensors at NPP *Kontrol'. Diagnostika 12* pp 69 – 75 (in Russian)

[4] Aleksandrov A A and Grigoriev B A 2006 *Tables of Thermal and Physical Properties of Water and Steam: Reference Book* (Moscow: MEI) p 164 (in Russian)

[5] 2007 *Revised Release on the IAPWS Industrial Formulation 1997 for the Thermodynamic Properties of Water and Steam* (Lucerne: IAPWS) p 49

[6] Subekti M 2006 Development of on-line monitoring system for nuclear power plant (NPP) using neuro-expert, noise analysis and modified neural networks. *Proc. 5th Int. Topical Meeting on Nuclear Plant Instrumentation Control and Human Machine Interface Technology (NPIC&HMIT) (Charlotte, NC)* pp 112 – 134

[7] IAEA 2008 *NP-T-1.1 On-line Monitoring for Improving Performance of Nuclear Power Plants/ Part 1: Instrument Channel Monitoring* (Vienna: IAEA) p 109

[8] O'Hagan R D and Hashemian H M 2015 Implementation of On-Line Monitoring to Optimize I&C Maintenance: A Case Study. *Proc. 9th Int. Topical Meeting on Nuclear Plant Instrumentation, Control and Human Machine Interface Technologies (NPIC & HMIT): (Charlotte, NC)* pp 65 – 74

[9] Hashemian H M 2006 *Maintenance of Process Instrumentetion in Nuclear Power Plants* (Springler-Verlag) p 304

[10] Trunov N B, Logvinov S A and Dragunov Ye G 2001 *Hydrodynamic and Thermochemical Processes in Steam Generators at NPP with WWER Reactors* (Moskow: Energoatomizdat) p 316 (in Russian)

The light and the dark sides of the moon: possibilities of Industry 4.0 for early non-invasive diagnosis of pathology and health-risks

M V Vasilieva[1], V A Glukhova[1], E V Grednovskaya[1], A A Dydrov[1], I I Klebanov[1], I A Nikolaeva[2], K Yu Nikolskaya[1], R V Penner[1], V E Tseilikman[1]

[1] South Ural State University (National Research University), Chelyabinsk, Russia, 76, Lenin prospekt, Chelyabinsk, Russia, 454080
[2] Kurgan State University, Kurgan, Russia, 63, Sovetskaya St., Kurgan, Russia, 640020

E-mail: tceilikmanve@susu.ru

Abstract. Industry 4.0 is an innovation infrastructure based on global digital connectivity to create smart plants and provide global automation in science and society. In Industry 4.0, the Internet of Things is used to collect big data from embedded sensors into live and inanimate objects. " The health calculator is positioned as a universal objective measure of human health, since it can be used not only in any disease, but also in preventive medicine, in particular, in examining astronauts and athletes and in life-risk occupations. However, even such a deeply thought-out software as health calculator requires its further development, that is, the transition to a second-generation device in which health is assessed from other methodological points of view. We have tried to summarize Industry 4.0 into medicine. We began to construct "Digital Twins" with the psychological "twins organisms". In the future, this structure will serve as a bridge from psychology to psychopathology. To conclude, Industry 4.0 can bring to humanity unique opportunities in maintaining health and preventing diseases and at the same time engaging it in qualitatively new stressful situations. Therefore, now it is worthwhile to look for ways to increase human resistance to new versions of informational stress.

Introduction

Industry 4.0 is a global innovation phenomenon based on the ability to digitize large amounts of all sorts of data. Such global digitization makes measurable the most unexpected things in the most diverse areas of our life. Global digitization allows real-time measurements as well as past experience measurements. When we add real-time data analysis to the Internet of Things, a state of global communication or “Quantitative Planet” is created [1]. Medicine is based on the results of past experience more than any other field of activity and the ability to measure the huge accumulated knowledge of medicine opens up a completely new era for medicine. Here we discuss some of the positive and negative effects of Industry 4.0 with respect to medicine. Currently, there are great expectations for revolutionary changes in personalized medicine due to Digital Twins.

Personalized medicine uses ultra slim information in the deviations from the normal on individual persons [2]. Digital Twins in engineering provide a conceptual base for measurements in therapy and preventative care. Digital Twins stand for a specific engineering paradigm, where individual physical artifacts are paired with digital models that dynamically reflects the status of those artefacts. When applied to humans, digital twins can be viewed as a fundamentally new technology based on human perceptions that dynamically reflect measurements of molecular status, physiological status, and lifestyle over time. Their use will allow an individual assessment of the long-term effects of experienced immediate events. In addition, the "twin organisms" will help doctors to develop personalized therapy for each specific case. It also seems possible to expand the software product by incorporating psychometric data, which, despite being "non-physical quantities", closely correlates with changes in the level of neurotransmitters and hormones. If it is possible to reproduce these behavioural disorders in Digital Twins, then they can be used in experimental psychotherapy. In general, the use of twin organisms for this purpose in the future will dramatically reduce the use of laboratory animals and cell cultures and thereby reduce the time of preclinical and clinical trials of pharmacological preparations.

The main indicator of physiological health is the maintenance of homeostasis or constancy of the internal environment of the body. However, under the shell of visible health pathological processes may form that give rise to the development of the disease. They can flow for a long time in a latent form and make themselves known when it is too late to take any measures. There is no doubt that sooner these adverse factors will be identified, the more chances there are to stop the development of fatal diseases in the bud. In our opinion, the solution of this important biomedical problem requires a large-scale use of the capabilities of Digital Twins. The mechanistic use of global digitalization can give rise to "A Demon of the Second Kind" from the short story by Stanisław Lem, to something that was able to work at a speed of "three hundred million information per second" and have an efficiency factor of zero. Therefore, it is crucially important at the start to determine the methodological approaches. There are two types of models for Digital Twin. The first model is creating through analytical techniques. The second model is being creating through the machine learning and artificial intelligence techniques. Neuronal Networks, which are, represents the mathematical models, as well as their software or hardware implementations, built on the principle of organization and functioning of biological neural networks – networks of nerve cells of a living organism successfully used for created Digital Twins in the both types of models.

The Digital Twins in psychology

Currently, there have been a few attempts to use Digital Twins in cardiology and in oncology [3, 4]. However, there are no data on the use of this approach for psychiatry, although the psychosomatic diseases are widely presents in the world. Therefore, the psychological fragments are quite necessary in the Digital Twin of those diseases. Here we have tried to summarize the further prospects of translating Industry 4.0 into medicine. We have taken the first steps. We began to construct the Digital Twins of the human psychological profile. In future, this structure will serve as a bridge from psychology to psychopathology. A further perspective for the development of this approach is the use of artificial intelligence for psych diagnostics. The next step to translate art intelligence will be building a bridge from psychology to recognizing psychosomatic pathology. The axial structure for this approach may be Jung's theory of psycho-types.

According to the theory of C G Jung [5], all people can be divided into eight groups, depending on their psychological attitude and basic psychological function, which together determine the psychological type of the individual. Thus, this typology is a multiparameter value and can serve both as a method for arranging the diverse psychological experience, and as a psychologist-practice tool for selecting a specific approach to each patient. Also, each psycho type has its own markers of stress. The selection of such markers on the video will allow you to make a more detailed picture of the patient's disease. The task of recognizing a psych type is the classification problem, which may be solved with a large amount of machine learning tools, e.g. neural networks. Neural network are able to extract high-level properties of abstraction from input data, therefore the range of problems solved by different typed of neural networks covers tasks from the definition of handwritten symbols [6] to human gait recognition [7], where theses how significant results.

The neural network, trained on bright, indisputable representatives of every type, will allow a combination of experience of many psychologists that leads to exact and impartial typing. At the moment, a data preprocessing model has been developed. Here the module preprocessing the source data. A video length of up to 10 seconds is used as the system input data. На вход подаются исходное видео и параметры, позволяющие регулировать размер видеофрагмента, порог бинаризации, а также размеры эрозии, наращивания и масштабирования. The video features interviews with people of various psycho-types. These videos are created by professional psychologists, so the accuracy of the types presented in the video is an absolute variable. The pre-processing of this video consists of several stages. Below is a description of each stage, as well as algorithms for their implementation.

1. Fragment extraction. If you need to use a video clip shorter than the original, you need to cut it to a specified duration. Using shorter fragments will speed up the work of the program, and longer ones will increase the recognition accuracy.

2. Binarization. To determine the moving areas in the video, background subtraction algorithms are used, which consist in the threshold binarization of the image sequence. In this case, the first frame of the sequence is memorized, which is then subtracted from the following, thus coloring all the pixels in one of two colors, signaling its movement or static state.

In this work we use the ViBe background subtraction algorithm, which is as follows. Denote by v (p) the value of the attribute, for example, the RGB component of a pixel located in on the image, and vi is the background sign with the subscript i. Then each pixel of the background in p is modeled by a set of N background segments:

$$\mu(\rho) = \{v_1, v_2, \ldots, N\} \quad (1)$$

taken from previous frames. To classify the pixel characteristic v (p), it is compared with all values contained in the μ (p) model relative to the distance function (a metric that is usually specified as Euclidean) and the distance threshold.

3. Framing (dividing into frames). Since the neural network processes the image sequence, the binarized video fragment needs to be divided into frames.

4. Erosion (cleaning noise). After binarization, noises in the form of single pixels may remain on the frames. Simple cleaning can be performed using morphological transformation – erosion. The erosion algorithm consists in applying a certain structural element to each pixel. If at any position every single pixel of a structural element coincides with a single pixel of a binary image, then a logical addition of the central pixel of the structural element with the corresponding pixel of the output image is performed. As a result of applying the erosion operation, all objects smaller than the structural element are erased, and the objects connected by thin lines are separated, and the sizes of all objects are reduced.

Erosion of the binary image A by the structure element B is denoted by $A \ominus B$ and is given by the expression:

$$A \ominus \mathrm{B} = \{z \in A | Bz \subseteq \mathrm{A}\} \quad (2)$$

5. Dilation (capacity). Since erosion can separate thinly connected parts of a single object, after erosion it is necessary to conduct a build-up operation that will increase the size of the moving areas and connect these parts. The transformation algorithm also consists in applying to each pixel an image of a certain structural element, but logical addition is performed over the entire structural element and provided that any single pixel coincides with the origin of its coordinates.

Erosion of the binary image A by the structural element B is denoted by $A \oplus B$ and is given by the expression:

$$\mathrm{A} \oplus \mathrm{B} = \mathrm{U\ b} \in \mathrm{B\ A} \quad (3)$$

6. Skeleton building. The skeleton building method is an erosion of the image until its thickness is equal to one pixel. In the Zong-Sun algorithm used, a 3x3 matrix is introduced. This algorithm summarized at the figure 1.

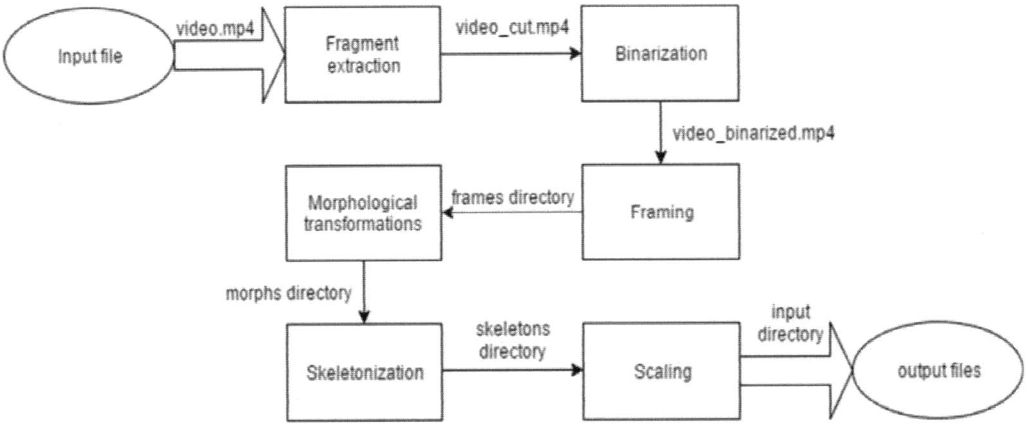

Figure 1. Transition from input measurement of video recording data to output Jung's Psychological Types Scale via binarization, framing and skeletonization.

This algorithm allows to obtain a set of specific data an image with a set of specific points and with skeletonization of the human body, for example. These data will be transmitted to the input of another neural network, which will allow to create the measurement model of the psycho-type of person, according to Jung.

Next part of the work is neural network topology adjustment a recognition task solution. Recurrent LSTM, convolutional networks will be applied to preprocessed video frames. Training set granted by experts and preprocessed with implemented utility will be used to adjust weights of network. Our next steps will be related to the formation of a biological passport for each psychological type which will include their physiological, metabolic and genetic features, as well as their susceptibility to stress what will allow in the future to prevent the stress-related diseases. In this study, we created a "Digital Twin" of in healthy people. Based on the obtained results conventional reference scale of Jung's Psychological Types (JPTS) was to construct. Exploratory factor analysis found three significant measurand pairs: extraversion-introversion (E-I), thinking-feeling (T-F), and sensation-intuition (S-N). The result showed that the three factors of E-I, T-F, and S-N corresponded to Extraversion, Agreeableness, and Openness. In the healthy people the references value for each measurand pairs were obtained. In the future, we plan to study the characteristics of the manifestation of psychosomatic diseases in individuals with different psychological types based on the reference values of this scale. In particular, we began to model the "Digital Twin" of an autoimmune psychosomatic psoriasis disease. We trace the features of the course of psoriasis in individuals with different psychotic types according to Jung.

Hypothetical risks of Industry 4.0 for human health

Despite the obvious benefits of Industry 4.0, it carries some dangers to human health. First of all, this is due to the fact that the activity of an employee at the enterprises of Industry 4.0 requires unlimited plasticity and readiness for change. Meanwhile, with age the majority of people become more and more conservative and any changes lead to violations in the psychological zone of comfort, which is fraught with the development of severe psycho-emotional stress. In addition, Industry 4.0 implies an increase in the importance of artificial intelligence, which leads to the development of information stress associated with the replacement of workers by automated control systems [8]. The indicated problems require complex measures of solution at the physiological levels. Countering information stress and disorientation in the context of high-speed technological progress is one of the priority transdisciplinary tasks. Obviously, it must be solved both with the help of medicine and with the help of other humanitarian technologies. These technologies fit organically into the strategy of futurization (including futurization of continuing education), which implies purposeful preparation of a person for the adoption and introduction of technical-technological and other innovations [9]. Futurization is associated with the expansion of contextual knowledge of employees, informing about new and, including, hypothetical technologies, as well as systematic participation in forecasting seminars and trainings. With proper implementation of the strategy, the employee will not be distracted from production and will adapt more effectively to change.

Accordingly, employees will be engaged in tasks that require responsible decisions in the interaction between human and machine and which may expose them to health risks as well as to the dangers associated with automated tools and greater psychosocial stress. Psychosocial stress can trigger the development of comorbidity, the nature of which remains a mystery.

Conclusion

Industry 4.0 can bring to humanity unique opportunities in maintaining health and preventing diseases and at the same time engaging it in qualitatively new stressful situations for which there are no recipes for adaptation. Therefore, now it is worthwhile to look for ways to increase human resistance to new versions of informational stress. With a successful solution to this problem, Industry 4.0 will undoubtedly revolutionize medicine.

Acknowledgments

The study was supported by the Russian Science Foundation (grant#17-15-013418) and by the Government of the Russian Federation according to Act 211 (contract 02.A03.21.0011).

References

[1] Özdemir V 2018 The Dark Side of the Moon: The Internet of Things, Industry 4.0, and the quantified planet *Omics* **22**(10) pp 637-641
[2] Bruynseels K, Santoni de Sio F, van den Hoven J 2018 Digital twins in health care: ethical implications of an emerging engineering paradigm *Front. Genet.* Feb 13 doi: 10.3389/fgene.2018.00031. eCollection 2018
[3] Miller W and Jaffe A 2016 Biomarkers in heart failure: the importance of inconvenient details *ESC Heart Fail.* **3**(1) pp 3–10 doi: 10.1002/ehf2.12071
[4] Kummar S, Williams P, Lih C-J, Polley E, Chen A P, Rubinstein L V, Zhao Y, Simon R M, Conley B A, Doroshow J H 2015 Application of molecular profiling in clinical trials for advanced metastatic cancers. *J. Natl. Cancer Inst.* **107**(4) doi: 10.1093/jnci/djv003
[5] Jung 1921 Psychological Types [*Psychologische Typen*] (Zurich: Rascher Verlag) (in German)
[6] Le Cun Y, Bottou L, Bengio Y and Haffner P 1998 Gradient-based learning applied to document recognition *Proc. IEEE* **86**(11) pp 2278-324
[7] Phung S L and Bouzerdoum A 2007 A Pyramidal Neural Network for Visual Pattern Recognition *IEEE Transactions on Neural Networks* **18**(2) pp 329-343
[8] Leso V, Fontana L and Iavicoli I 2018 The occupational health and safety dimension of Industry 4.0 *Med. Lav.* **110**(5) pp 327-338
[9] Ursul A D and Ursul T A 2017 Vectors of achieving sustainable future *Philosophical Sciences* 7 pp 139-149 (In Russian)

Gait rehabilitation monitor

P Leite[1,2], O Postolache[1,2], J M Dias Pereira[1,3] and G Postolache[4]

[1] Instituto de Telecomunicações, Lisboa, Portugal
[2] ISCTE – Instituto Universitário de Lisboa, Lisboa, Portugal
[3] Escola Superior de Tecnologia de Setúbal, Instituto Politécnico de Setúbal, Setúbal, Portugal
[4] CBIOS Universidade Lusofona, Lisboa, Portugal

Email: opostolache@lx.it.pt

Abstract. This paper presents a simple wearable, non-intrusive affordable mobile framework that allows remote patient monitoring during gait rehabilitation by doctors and physiotherapists. The system includes a set of 2 Shimmer3 9DoF Inertial Measurement Units (IMUs), an Android smartphone and a developed app for collecting, primary processing of data and for persistence of data in a remote PostgreSQL database, which is available in a remote server and where further data processing is performed. This framework provides gait features classifier by invoking an implemented REST API available in the remote server. Low computational load algorithms based on Euler angles and filtered signals were developed and used for the classification and identification of several gait disturbances. These algorithms include the alignment of IMUs sensors data by means of a common temporal reference as well as heel strike and stride detection algorithms. After segmentation of the remotely collected signals for gait strides identification relevant features were extracted to feed, train and test a classifier for prediction of gait abnormalities using supervised machine learning type and Extremely Randomized Trees method.

1. Introduction

Gait analysis support recognition of normal or pathological patterns of walking, for medical diagnosis and treatment [1], physical therapy [2] and sports training [3]. The traditional scales used to analyze gait parameters in clinical conditions include variable degree of subjectivism during assessment. Progress in new technologies has given rise to devices and techniques which allow objective evaluation of different gait parameters, resulting in more efficient measurement and providing specialists with a large amount of reliable information on patients' gaits. A simple and accurate technological system would be helpful to increase efficiency and affordability of motor rehabilitation. Gait analysis using wearable wireless sensors can be an economical, convenient and effective way to provide diagnostic and clinical information for various health-related issues [4]. In section II of the paper several related works are presented. Section III summarizes the proposed method and system. Section IV describes the results and discussion and Section V the conclusions.

2. Related Works

Several gait analysis systems have been proposed over the years. Alcaraz et. al [5] proposed a system using an IMU Shimmer2R and an Android application for mobile gait performance evaluation and feedback. In this system the extraction of important features from accelerometer raw data in both time and frequency domain was performed for gait classification, and each peak within a data window was processed to distinguish between normal and abnormal gait. A method for measuring human gait posture using wearable sensors was proposed by Takeda et al. [6]. The acquisition of the signals was realized with a tri-axial acceleration sensor and three gyro sensors aligned on three axes. These were placed at abdomen level and at lower limbs (both thighs, both shanks and both feet) to measure acceleration and angular velocity during walking [6]. Bonnet and Jallon presented a technique for recognition of different gait conditions from a body-worn sensor data. A sensor located at subject's shank was a combination of a 3-D accelerometer and a 3-D magnetometer [6]. Charry et al. [7] work have shown that inertial sensors have good performance comparable to video-based systems. In Nukala, et al. [4] work a custom designed low-cost wireless gait analysis based on inertial measurement unit (IMU) was used to collect the gait data for four patients diagnosed with balance disorders and from three normal subjects. Each subject has executed Dynamic Gait Index (DGI) tests while wearing the custom wireless gait analysis sensor (WGAS) [4]. Previous researches have shown that machine learning methodologies are effective for classification of different activities from sensor data [8].

3. System Description

The implemented measurement system for gait classification is a distributed computation system, which includes embedded sensing device that are Bluetooth compatible, edge computation platform and user GUI expressed by an Android OS device and by advanced processing and classification in a remote server. IMU as part of sensing device allow acquisition of the signals for gait monitoring. The signals from measurement channels are primary processed and Wi-Fi transmitted by the Android OS Platform that also assures the Internet connectivity. An Android app was developed for smartphone that invokes a REST API in a remote server. Java and Python programming languages in a server support data processing and classification using machine-learning. The developed system includes two Shimmer3 devices and a database. In figure 1 the system architecture including two IMU units, an Android OS computation platform (smartphone) and a remote server is presented.

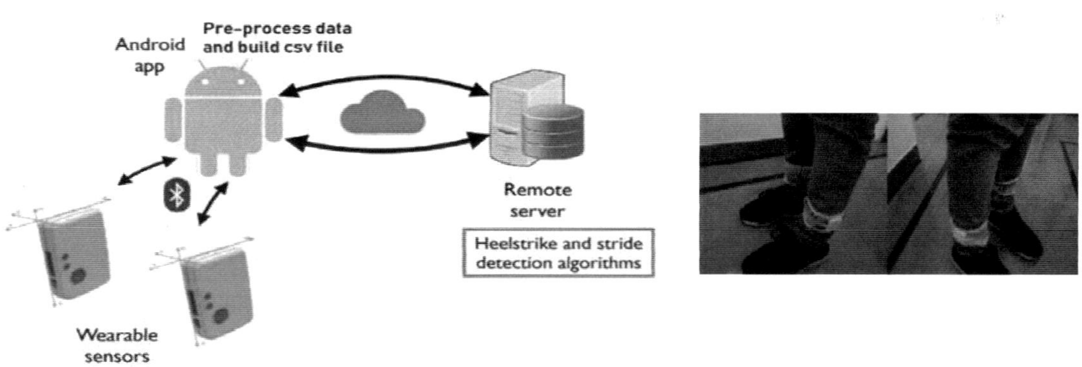

Figure 1. Gait Rehabilitation Monitoring System Block Diagram.

The software that runs in the Android OS device performs several tasks related to setup, data processing and packing, data transmission and some self-test functions that monitor the proper working of the system. Complex functions that require higher data processing capabilities are performed remotely by computation units as Laptop or remote server.

3.1. System Description - Hardware – IMUs

The hardware of the system includes 2 Shimmer3 IMUs. Shimmer3 is a small wireless sensing device, well suited for wearable applications. The integrated kinematic sensors, the storage and low-power standards-based communication capabilities of Shimmer3 IMUs are proper for development of motion capture applications, long-term data acquisition and real-time monitoring. 9DoF Calibration Application from Shimmer [9] was used to calibrate Shimmer3 devices.

3.2. System Description - Software

The APP for Android OS was developed using Multi Shimmer Template and the Shimmer drivers that provide Bluetooth connectivity to Shimmer3 devices. The *ShimmerDriver* and *ShimmerAndroidInstrumentDriver* that hold *ShimmerObject* (threads running connections between app and devices), *ShimmerBluetooth* (streaming over Bluetooth), *ObjectCluster* (that represents data from devices) and Shimmer (that represents a device) Java classes were used. Multiple Shimmer devices can be connected simultaneously to a device with Android OS, using multiple instances of the Shimmer Class object [10]. Shimmer drivers manage the connection with the app and data streaming over Bluetooth. Interactions with the remote server are executed using SSH2 implemented via JSch, Java Secure Channel and port forwarding and SFTP. Android *Fragments* were developed to implement the APP functionalities: *LoginFragment* and *RegisterFragment* for login and registration over SSH2. *WalkFragment* and *TrainFragment* have handlers to manage threads running the multiple Bluetooth connections with Shimmer3 devices, and to receive messages sent from the devices carrying the gait information. The received messages are synchronized using a common System timestamp reference, pre-processed and sent in a .csv to the remote server over SFTP. These two fragments layouts have the option to plot signals and YY Euler angles. *TrainFragment* has an *AlertDialog* to choose the classification label associated with each gait type and *WalkFragment* has an *AlertDialog* to show the test results. Figure 2 depicts the APP functionalities: Plot of the signals during the session, upload the data file on the server, and the display of classification results. Train classifier feature enables the APP to collect data and train a classifier model. The algorithm for gait classification is described in figure 3 and figure 4. Smartphone connects to the system devices using Shimmer drivers and Bluetooth, collecting tri-axial data from both devices that integrate accelerometers, gyroscopes and magnetometers. Data from the two Shimmer3 devices are pre-processed by Android device and transformed by the APP into a .csv file containing IMU Bluetooth address, device sampling-rate, gait classification labels, collected raw data, aligned timestamps and Euler angles. IMU signals alignment algorithm was developed and implemented. A common referential time was considered for IMUs and the obtained data from both devices was aligned to this common reference. A common temporal reference was considered as the instant at which the samples are received by the Android Smartphone [11] and the timestamp of arrival of each sample in the smartphone is collected.

Each IMU sample is associated with a time stamp according to the following relation:

$$ct = ws - sat \tag{1}$$

where *ct* is current timestamp; *ws* is start walking timestamp; *sat* is sample timestamp arrival in the smartphone.

Samples with timestamp prior to the moment gait starts are ignored. Euler angles are a convenient set of generalized coordinates for describing the orientation of a body-fixed coordinate frame [12]. In this application the Euler angles (yaw, pitch, and roll) were used to describe the legs movement based on the data in quaternion [13] format delivered by IMU Shimmer3 devices attached to the legs. On the APPs level a conversion is made from quaternion to Euler angles that are stored in .csv file that is sent to the system remote server by SFTP for off-line processing. Additionally, digital filtering has been considered. Thus a 4th order Butterworth bandpass digital filter characterized by 0.7 Hz low-cut off and 35 Hz high-cut off frequencies has been applied to the accelerometer's signals [7] to increase the signal-to-noise ratio which will help gait classification.

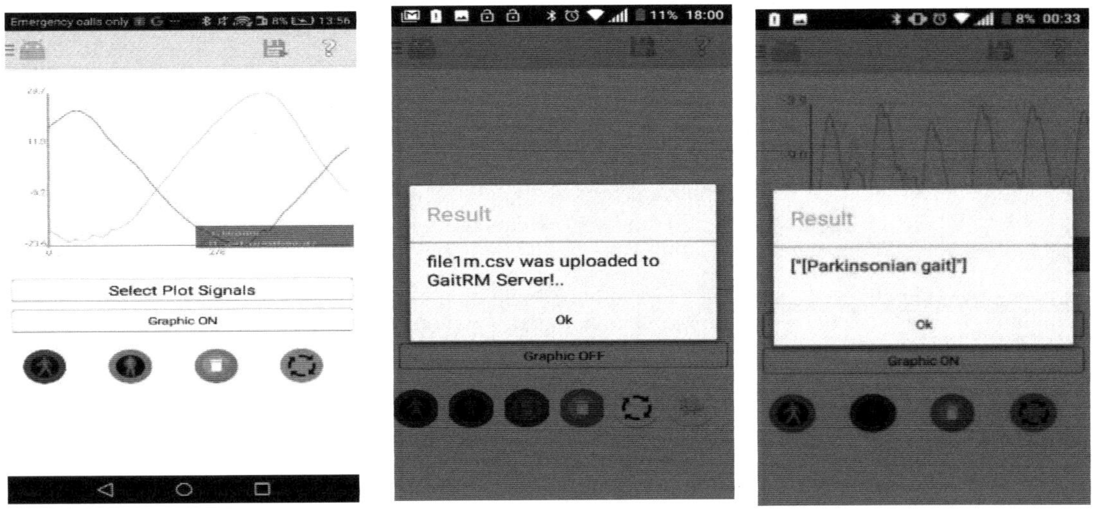

(a) (b) (c)

Figure 2. APP functionalities: (a) plot of the IMU signals; (b) upload the data as .csv file; (c) display classification results.

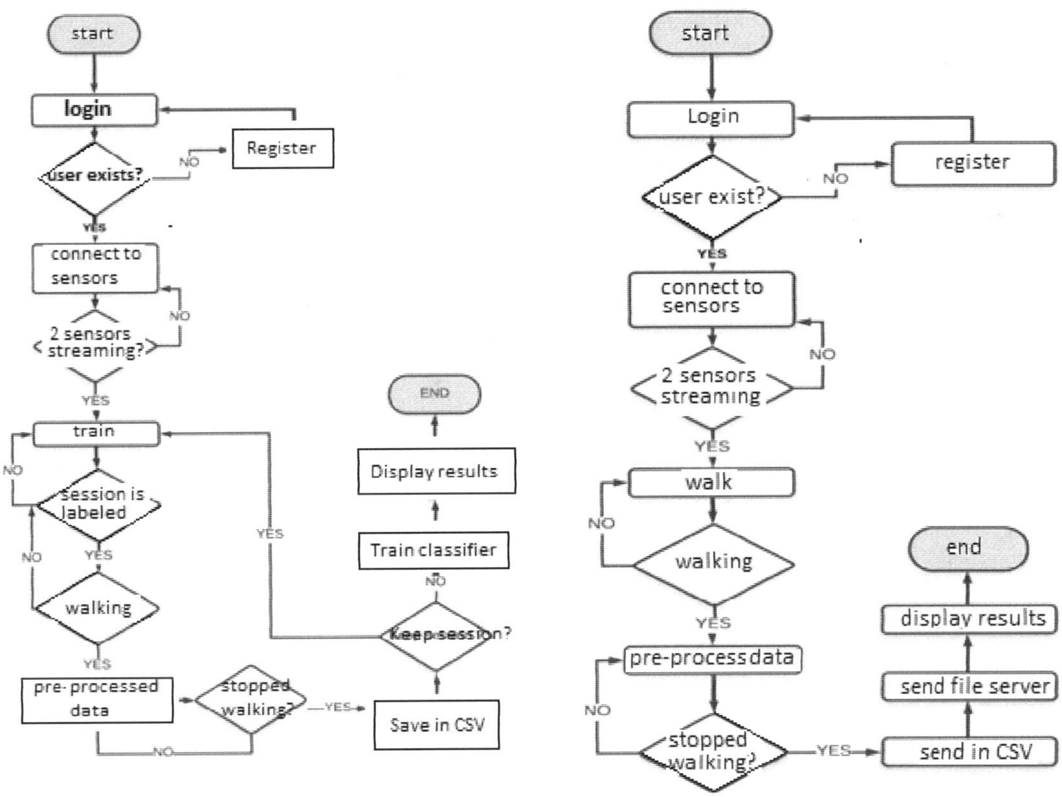

(a) (b)

Figure 3. Algorithms flowchart: (a) gait classifier training flow; (b) gait session flow.

Journal of Physics: Conference Series **1379** (2019) 012071 doi:10.1088/1742-6596/1379/1/012071

Additionally, 1st order Butterworth low-pass digital filter characterized by 10 Hz cut-off frequency has been applied to magnetometer signals and obtained Euler angles. These play a very important role in the detection of the heel strikes that occur during the gait cycles, which are useful to help segmenting the signal for stride identification and features extraction for training the classifier model. Euler angle from YY axis was used in this algorithm. The YY axis was chosen in this calculation considering the calibration procedure. A heel strike detection algorithm includes the following steps: i) calculation of the mean value of the discrete YY axis Euler signal; ii) determination of the signal indexes where values are below the mean value of the signals; iii) heel strikes detection, by calculating the maximum value between the lower and upper limits of the signal samples in noisy signals or in signals with many peaks, and elimination of peaks that may be misidentified as heel strikes (if two identified samples below the mean have more than two samples above the mean - initial adaptive distance = two samples, then there might exist a heel strike between them; at each iteration through samples determined in the second step, is performed the distance adjustment if the number of samples is above, in between or greater than current distance:

$$d = [d/2 + abs(l - n)]/2 \tag{2}$$

where d is distance, l is last sample, n is next sample; iv) storing the identified heel strikes in a buffer; v) calculation of the mean difference between heel strikes identified in the fourth step; vi) iteration through the heel strikes (indexes) to get the segments of the original discrete YY axis Euler signal between those heel strikes; vii) calculation of the heel strikes index for the original signal. After identifying the heel strikes, the strides can easily be obtained from the signal segments. Figure 4 shows the segments obtained from the acquired signal at each heel strike for the right and left legs. Each segment displayed in figure corresponds to a gait cycle.

A modified strides detection algorithm was developed and includes: i) calculation of the mean distance between the identified heel strikes of a YY Euler angle signal; ii) iteration through the heel strikes and extraction of the signals segments from the original signal contained in-between; iii) storage in a key-value list of the timestamp associated to the start of each segment and the segment itself; iv) elimination of all signal segments which length do not fit in the interval based on the mean segments distance: *[0.7*mean segments distance ;1.3*mean segments distance]*. By considering the segments of one of the two Euler angle signals, iteration through the segments and for each segment iteration through the key-value list of the other signal (other leg) that maps the start of an identified segment with a timestamp. If there is a segment timestamp that fits in the temporal window of the second signal extracted segment, then the current segment is accepted and the two segments from the different devices are considered as a gait stride, otherwise the segment is excluded.

Extremely Randomized Trees method was used for feature extraction. Following features were considered: *in time domain* - standard deviation, maximum, minimum, skew, kurtosis, displacement, interquantil range, mean absolute deviation and area; *in frequency domain* - Discrete Fast Fourier Transform (DFFT) (standard deviation of DFFT coefficients), Power Spectral Density – Welch (PSD) (standard deviation of PSD) and Discrete Wavelet Transform (Daubechies db4) (standard deviation of the approximation and detail coefficients); *in time-frequency domain:* Discrete Wavelet Transform (Daubechies db4) (standard deviation of the approximation and detail coefficients). A set of extracted features was used for training and other for testing the classification model.

4. Results and Discussions

The implemented gait assessment system was tested in laboratory with 4 volunteers [average and range age 38.8 years (33-45); average weight 78.8 Kg (67-87)] who performed normal and abnormal gait. A walking distance of nearly 40 meters for each gait was used to train the classifier, and of 10 meters to test the classification algorithm.

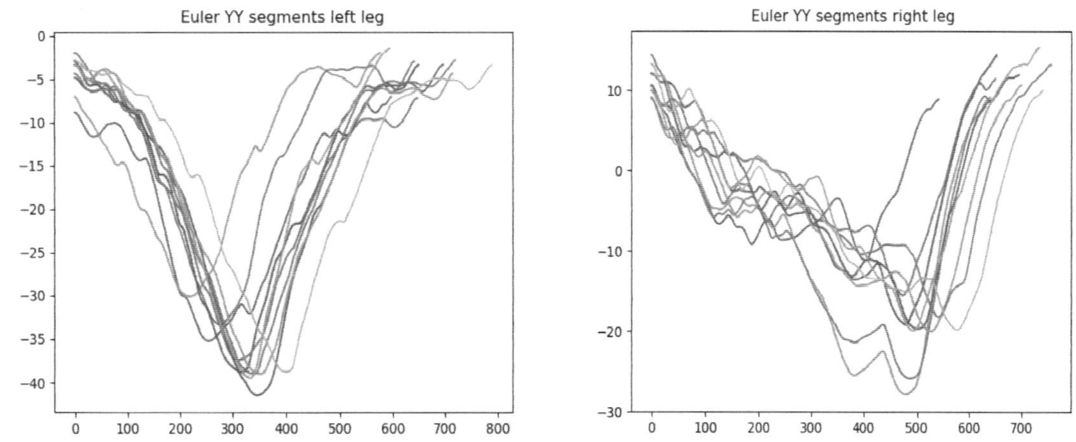

(a) (b)

Figure 4. Euler YY angle segments for left (a) and right legs (b).

Table 1. Classifier Scores.

Gait	Precision	Recall	F1
normal	1.000	0.951	0.974
v. slow	0.948	0.981	0.963
parkinsonian	1.000	1.000	1.000
hemipl. right	0.916	0.889	0.900
equinus	0.948	0.983	0.964
glut. max. right	0.972	0.935	0.952
glut. max. left	1.000	0.914	0.954
running	0.944	1.000	0.969
hemipl. left	0.972	1.000	0.985

The experiments were carried out by using 2 Shimmer3 devices (fs=51.2Hz) and an Android Smartphone Alcatel U5 (Quad-core 1.3 GHz, 8GB ROM + 1 GB RAM, Android 7.0 Nougat). The type of the tested gaits and the labels for training the classifier were: normal gait (label1), very slow gait (label 2), parkinsonian gait (label3), hemiplegic gait – right leg (label 4), equinus gait-jump knee (label 5), gluteus maximus - right leg (label 6), gluteus maximus - left leg (label 7), running gait (label 8), hemiplegic gait – left leg (label 9). The classifier was trained 6 times and its average scores Precision, Recall and F1 at stride level are illustrated in table 1. The data from 4 volunteers was used for training and testing the classifier. The data from walking tests for 10 meters was used to test the capacity of the system for correctly identify the gait features. For this test results the most frequent identified stride label was considered as the identified gait type. Volunteers 2 and 4 and had 9 gait types 100 % correctly identified. Volunteer 1 had one misidentified gait: gluteus maximus - right leg as gluteus maximus - left leg and Volunteer 3 had one misidentified gait: gluteus maximus - left leg as gluteus maximus - right leg. Although the precision of gait recognition was high, only the results from algorithm testing with a larger sample size would allow the comparison with existing results from other research studies.

5. Conclusions

In this work, a simple wearable, non-intrusive affordable mobile framework that allows remote patient monitoring during gait rehabilitation is described. The remote functionality provided by the app offers a tool for the health institutions for monitoring rehabilitation process. Heel strike and stride detection algorithms based on Euler angles and filtered signals disturbances including the alignment of IMUs signal data by means of a common temporal reference were developed and used for the classification and identification of several gait pathologies. More data from patients with pathologic and abnormal gait is intended to be included in the classifier in future work, to increase the sensitivity of the algorithm to predict the evolution of rehabilitation by considering the patient's gait session historic.

Acknowledgement

The work was supported by Fundação para a Ciência e Tecnologia, Project TailorPhy - PTDC/DTP-DES/6776/2014, Project UID/EEA/50008/2019 and Instituto de Telecomunicações.

References

[1] Lee C-Y and Lee J-J 2002 *Int. J. of Human-friendly Welfare Robotic Systems* http://koasas.kaist.ac.kr/handle/10203/8430
[2] Kong W *et al.* 2013 *Proc. IEEE Int. Conf. on Robotics and Biomimetics ROBIO (December 12-14, 2013, Shenzhen, China)*
[3] Muro-De-La-Herran A, Garcia-Zapirain B and Mendez-Zorrilla A 2014 *Sensors (Basel, Switzerland)* **14** 2 3362-94
[4] Nukala B T, Nakana T, Rodriguez A, Tsay J, Lopez J, Nguyen TQ, Zupancic S and Lie DYC 2016 *Biosensors (Basel, Switzerland)* **6** 4 58
[5] Alcaraz JC, Moghaddamnia S and Peissig J 2015 *Proc. 17th IEEE Int. Conf. on Networking, Application Services HealthCom (October 14 – 17, 2015, Boston, MA, USA)*
[6] Takeda R, Tadano S, Todoh M and Yoshinari S 2009 *Proc. 13th Int. Conf. on Biomedical Engineering CBME 2008 (December 3-6, 2008, Singapore)* (Berlin, Heidelberg: Springer) pp 1069-72
[7] Charry E, Lai D T, Begg R K and Palaniswami M 2009 *Proc. Conf. Proc. IEEE Engineering in Medicine and Biology Society (3-6 Sept. 2009, Minneapolis, USA)* 4824, 4826-4827
[8] Berke EC, Atasoy I, Acici K and Ogul H 2016 *Procedia Comput. Sci.* **98** 522-27
[9] Shimmer 2017 *9DoF Calibration User Manual* https://www.shimmersensing.com/images/uploads/docs/Shimmer_9DOF_Calibration_User_Manual_rev2.10a.pdf
[10] Shimmer 2016 *Shimmer Java/Android API User Manual* https://www.shimmersensing.com/images/uploads/docs/Shimmer_Java_Android_API_User_Manual_rev2.11.pdf
[11] Shimmer 2017 *MSS Android User Guide* https://www.shimmersensing.com/images/uploads/docs/MSS_Android_User_Manual_v2.8.pdf
[12] Knight C S 2008 *It's All in the Wrist: a Quantitative Characterization of Human Wrist Control* Massachusetts Institute of Technology Thesis http://hdl.handle.net/1721.1/45623
[13] Vidakovic J Z, Lazarevic M P, Kurgic V M, Dancuo Z Z and Ferenc G Z 2014 *FME Transactions* **42** 3 189-199

Modelling IoT technology for a cargo port

O Postolache[1], Y Yang[2], M Zhong[2], H Yao[2], F Yu[2], X Fu[2] and C Mi[2]

[1]Instituto de Telecomunicações and ISCTE-Instituto Universitário de Lisboa, Portugal
[2]Institute of Logistics Science and Engineering, Shanghai Maritime University, China

E-mail: opostolache@lx.it.pt

Abstract. A work on modelling Internet of Things architecture for cargo ports is described. The main things and their roles in a cargo port, different sensors and instrumentation that was used or might be used for sensing and control in a cargo port, and some examples of IoT technologies that might contribute for improving the safety of processes and the security of the port are presented. By increasing the accessibility and accuracy of relevant information, IoT technologies may increase efficiency and quality of services in cargo ports, the operational safety and security of the ports.

1. Introduction

Internet of Things (IoT) technologies by connecting sensors, devices, and intelligent operations (i.e., intelligent cargo movement) provide clear business benefit in terms of asset tracking and supply-chain management. This is achieved by means of continuous process synchronisation of supply-chain information, and seamless real-time tracking and tracing of objects [1]. Containers port become more and more important and have very fast yearly growth-rate. The goods management and security in the cargo ports are increasingly challenging. The expected increase of trade in the following years, the limited possibility to increase the space and traffic in ports and the need for improved security in the ports (that is high resources and time consuming) require solutions for better management of port asset and processes. Internet of Things technologies already is a solution adopted by some ports to increase the efficiency of management. Most of information systems for cargo ports were implemented for optimization and acceleration of yard stacking. In this paper is presented the work on modelling the IoT framework for cargo ports. What measures and what IoT architecture that might improve the asset and process management in a cargo ports were mainly investigated. In the following sections of the paper we present what are the main objects/things and their roles in a cargo port, what kind of sensors and instrumentation was used or might be used for sensing and control in a cargo port, and some examples of technologies that might contribute for improving the safety of processes and the security of the ports.

2. Port objects/things and their roles

Modelling of an IoT architecture for a domain require a thorough understanding of that domain and conceptualization/generalisation of the domain processes. A reference model can lead to a better understanding of targeted system and a tool for building architecture where design choices are based on best practices, in terms of functionality and information usage, to deal with conflicting requirements regarding functionality, performance, deployment and security. Following we describe a cargo port, listing several objects/things that generally can be found in a cargo port, and their main role.

2.1. Hinterland

Hinterland is defined as port's area that serve both for import and for export of goods, the inland region lying behind port or the area from which products are delivered to a port for shipping elsewhere. Hinterland includes the port authorities (e.g., custom authorities), logistic centre, the hinterland transportation infrastructure, storage yard. Different services are provided in the hinterland - custom clearance, administration of shipping documents, special treatments of goods, special inspection of containers, quarantine, shipping and receiving of containers, maintenance of the containers, loading and unloading of trucks, trains, feeders to/from the hinterland distribution, empty container handling, storage of goods.

2.2. Logistic centre

Logistic centre is a centre in a defined area within which all activities relating to transport, logistics and the distribution of goods (both for national and international transit), are carried out by various operators on a commercial basis. The Logistic Centre must be open to allow access to all companies involved in the activities of the port [2]. It includes offices of shipping companies, freight forwarders, stevedoring firm, waterway authorities, transportation and storage companies. It should comply with international and regional standards and quality performance to provide the framework for commercial and sustainable transport solutions. At the terminal controls tracking of things from ship unloading progress to positioning of containers for shipping, launching and scheduling for taking import containers should be integrated. At the terminal operators the cargo is managed to be sent to distribution centres.

2.3. Carrousel and Carriers

In a cargo port many types of carrousel and carriers can be found: automated guided vehicles (AVG); belt system; chain connected chassis system; overhead grid rail system; semi-trailer truck; trucks; trains; straddle carriers.

2.4. Cranes

Many types of cranes (see ELS classification of crane [3]) can be found in port. These may be classified taking into account the space where is placed (i.e., quay crane and storage yard crane) or the main functionality that a crane have (i.e., goliath gantry crane, transfer crane, block handling crane, floating dock crane, mobile harbour crane, lift trucks – reach stackers, fork lift trucks, container lift truck). In several ports the quay-crane is guided and controlled from central control office with cameras (nobody is on board of the crane, so the crane-crab can be accelerated faster, and the cycle time reduced). Many quay crane picks only one container in a stroke. In a few ports (e.g., Shanghai port) there are quay-cranes that can pick up 2 containers in a stroke and thereby double the capacity. The Chinese company ZPMC has a demonstration system in Shanghai with a fully automated quay-crane system and an automated container yard stacking-system in place.

2.5. Containers

There is a great variety of containers. This variety consists of the most common ones - the dry containers, and those being designed for the special kinds of the goods. The containers can be classified as: dry container; open type containers; open side platform containers; refrigerated containers, platform containers; live animal shipping containers; tank containers.

2.6. Buffer zones

In a cargo port can be found: loading buffer in front of the crane; buffer truck docking station; buffer storage.

2.7. Storage area
The containers are stored at the appropriate temperature required by different goods (i.e., at ambient temperature; fixed temperature; and cold storage/refrigerated). Facilities for different goods storage exists in many cargo ports (e.g., for liquid fuel, grains).

3. Measurements

3.1. The Sea Gauges and Weather Conditions
The weather conditions, particularly wind, ice and high sea waves have great importance during operations of ship berthing and loading. The instrument and equipment used for weather conditions can also be used to estimate the environmental impact of ports (e.g., pollution by noxious gases) and alert on natural disaster. The meteorological station, either on land or the sea, should have instruments for accurate weather data (i.e., barometer, anemometer, thermometer, rain gauge, SODAR - radar or sound waves that determine the wind speed and direction at various elevations above the ground; weather radar that locate precipitation, calculate its trajectory and estimate their types, direction and speed of wind).

In Las Palmas de Gran Canaria's Port, the main port in Spain, was designed and developed an information system that includes sensors for sea gauges and weather conditions [4]. It might support control tasks, through an alert management that monitors the value of many sensors.

3.2. Position and Displacement Detection
The measurement of position and displacement of physical objects is essential in many port scenarios (e.g., containers management, cranes controls, intrusion detection system, transportation traffic control). Position of the containers that could be managed using a RFID systems or radar devices might serve for intelligent decision for storage (in which multiple handling might be avoided), for monitoring and real time tracking of the containers, for detection of critical events, and also to stop goods' smuggling and theft. Inductive and magnetic sensors, acoustic or microwave-based sensors have applications for sea tide gauge measurement, liquid level in a tank, elevators displacement. Optical sensors that are insensitive to magnetic fields and electrostatic interferences have also many applications in a port. For example, a system based on optical character recognition for automatic number plate recognition (ANPR) and automatic freight-code recognition (AFCR) was proposed [5]. The system uses this information to tag the photographic evidence of the corresponding vehicles. The system also includes a colour-based indexing scheme, which automatically identify the representative colour of each vehicle captures and better vehicle boundary detection.

Detection of position and displacement of things and people in a port has high importance for safe operations in port and security of port. Issues related to security of a cargo port and safety of processes regard several aspects such as perimeter security, internal security and operative controls, maritime security, decision support systems security, prevention and emergency management, etc. For example, detection of illegal ships and oil slicks as well as protection of the ports from effects of explosives or other weapons of mass destruction are key issues for reducing its economic, environmental and social impact. Synthetic Aperture Radar (SAR) system was proposed for detection of ships and/or oil slicks for maritime surveillance [6]. The system might function well day and night, in stormy weather or rainy days. CetusII AUV and MIRIS high-resolution sonar (i.e., equipped with side-scanning sonar) was proposed for inspection of the ship hull and berthing [7]. A robot equipped with DIDSON for short-range imaging sonar and M900-2250 for long-range imaging sonar was developed and tested by [8]. The robot search and recognize objects by the dual imaging sonar. They suggest that effective underwater exploration might be carried out by using short-range imaging sonar with frequencies of 1MHz or higher (for good quality sonar image) together with long-range imaging sonar with frequencies bellow 1MHz. DoD's Architecture Framework that employs multisensory approach (i.e., ground surveillance radars and cameras) was proposed to detect, asses and track perimeter incursions [9]. Using these devices, the proposed system can perform the following processes: i) threat analysis - what threats are relevant to the facility, ordering these by their impact, criticality, etc.; ii) operational security -

providing real time situational awareness and incident resolution (i.e., intrusion detection and tracking; intrusion assessment; intrusion response).

Wireless sensors networks have been recognized as feasible means for object tracking. Wireless sensors network based on multi-sensor device Waspmote, developed by Liblium Company, was proposed for optimal management of containers [10]. The authors suggest that by using these devices various information on containers could be obtained: i) position and displacement of container; ii) container doors opening unexpectedly; iii) temperature; iv) air pressure; and v) illegal container content. Currently, different technology for wireless sensors network are commercialized (e.g., Tiny-mesh, Wirepas mesh). However, over the years the major concern related with wireless sensors networks remain the power consumption. Efforts have been made by various researchers to derive ways to extend the lifetime of power storage or to minimize the use of energy in the tracking process. A method for reduction the power consumption in wireless sensors network, that was inspired on Immune Network Theory was proposed by [11]. The algorithm determines the best sleeping policy in order to optimize the power consumption of the entire sensor network. By using artificial immune system when the number of sensors deployed in a sensor field and thus the corresponding density of sensor nodes increases, the average power consumption per sensor decreases.

3.3. Velocity and Acceleration Detection

Sensors for velocity and acceleration detection might have many applications in port (e.g., for traffic control in port). The advances in MEMS technology have produced an increase use of sensor for velocity and acceleration in multi-sensor devices. For example, the Bosh CISS (Connected Industrial Sensor solution) is a multi-sensor device based on MEMS technology that can detect acceleration, vibration, rotation, as well as environmental conditions. This device allows to activate or deactivate sensors and to configure the sampling rates according to specific needs (i.e., data streaming mode, event detection or time aggregation). Detection of the speed and location of illegal ships by using a wireless sensors networks equipped with Global Positioning Systems (GPS) was proposed by [12].

3.4. Chemical, Biological and Radiation Sensors

Multi-sensor technology might be used for real-time and accurate information on cargo (i.e., type of goods, good characteristics transported inside the container). This information might give information for cargo risk evaluation that might be used to plane the container inspection (i.e., chemical screening, biological screening or radiation screening). A system for detection hazardous chemical and biological contents of containers as well as contraband was proposed by [13]. Two screening methods, one external to the container while it is being moved from conveyance to ship or vice versa, and an alternative in-situ method to sample, detect, and report dangerous cargo from inside a sealed container while in transit was described.

3.5. Port Infrastructure and Machinery Health Monitoring

With increasing trade, the usage of port machinery is increasingly continuous, which might enhance the risk of equipment's damage. For safety management of the goods in port, the status of port machinery should be measured by placing on key location of port machinery, sensors for non-destructive detection of materials. Non-destructive evaluation techniques allow detection of the presence of cracks, discontinuities, mechanical fatigue, and other imperfections without material damage. For example, electrical conductivity can be used as an index of mechanical fatigue. The planar meander and mesh-type sensors respond very well to conducting and magnetic materials but no so well to dielectric materials. On the other hand, the interdigital sensor responds very well to dielectric materials [14]. A real-time strain monitoring system based on distributed Fibber Bragg Grating sensors was proposed by [15] for health monitoring of port machinery.

IoT technologies for health maintenance of port infrastructure and port machinery might: i) reduce equipment downtime (i.e., repair service can be performed before the equipment failure); ii) increase productive uptime (i.e., by predictively identifying problems, so repairs can be scheduled during production downtime, rather than peak periods); iii) reduce maintenance costs; iv) increase asset lifespan (i.e., by performing predictive maintenance before costly failures occur); v) improve worker safety.

4. IoT for cargo port

IoT community has developed many architectures, platforms, standards, frameworks, solutions for IoT based information systems.

IoT is a powerful tool for things management in a cargo port by having the capacity to gather insights from the field using sensors technologies as well as by supporting data analytics. In figure 1 a model of IoT technology for cargo port is presented. A function architecture that refer to ITU-T Y.2060 and several required measurements in the cargo port were considered.

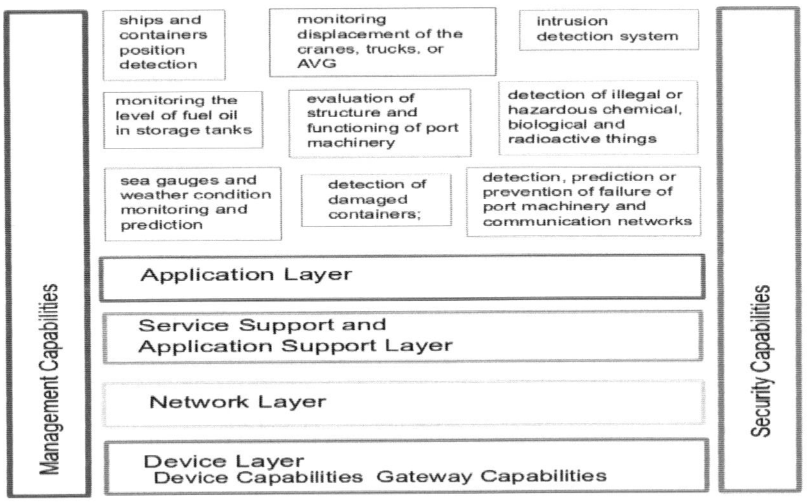

Figure 1. IoT model for cargo port.

The recent development of sensors having low power consumption, low cost, small size, limited radio range, and capacity for long-term working in harsh environments, increase the likelihood of IoT systems development for cargo ports, adapted to the various conditions and features of that port. Multi-model sensing that refers to sensing with different modalities and then exploiting the interdependence of these modalities to extract more useful data, should be considered in IoT system for cargo port. Information from sensors distributed in the port environment might be improved by data fusion and data meaning algorithms.

The communication of data in IoT framework might be supported by different types of wired or wireless technologies such as CAN bus, Zig Bee, Bluetooth, Wi-Fi. There are several hardware platforms available for sensor network deployment such as TelosB, Mica, IRIS, Waspmote, XBee ZB module, LoRa, BTLE, Tiny-mesh, Wirepasmesh, Xelba xOne Hub. Cellular connectivity like LTE is expected to increase in the future, especially with 5G and NB-IoT. The power of communications locations should consider buildings, height restrictions, and other topographic constraints.

Cloud computing, Hybrid Cloud or Microservices might be adopted for IoT architecture for cargo port for computing, storage, analytics and data visualization. Cloud Computing provides functional and also non-functional support (i.e., Service Level Agreements, low latency fault-tolerance, cost-optimization, security). These capabilities are provided as services (Platform-as-Service; Infrastructure as-a Service; Software-as-a-Service; etc.). Hybrid Cloud blends internal information system with cloud powered services. It might integrate private, public and managed cloud with the existing on-premises information system (e.g., Fujitsu MetaArc and Cloud Service K5). This may improve the agility, accessibility and security of the IoT systems for cargo ports. Both ETSI [16] team and Microsoft Azure [17] team recommend the use of Microservices in IoT architecture. Microservices are an architectural approach for developing applications as a set of small services, where each service is running as a separate process, communicating through simple mechanisms [16]. The main advantage of Microservice-based architecture is that one can develop, deploy, upgrade and scale every Microservice independently of the others. Moreover, developing Microservices separately enable the use of different technologies for each Microservice and to have distinctive Microservices for devices that communicate using different protocols. Also, this may support the safety and security of the system. For example, if a Microservice which communicates with a certain group of sensors crashes, the application provided by Microservices which communicate with other sensors will still be up and running.

The amount of data that can be acquired by IoT and the information provided by analytics tool in IoT frameworks are enormous. Nowadays, IoT data and information are susceptible to malicious usage by hackers but also by unauthorized competitors. Therefore, security of IoT system (i.e., protection of the confidentiality, integrity and availability of information asset) should be assured by design at every layer of the system architectures.

5. Conclusions

There are few studies on implemented information and communication technologies for cargo ports. IoT may contribute to increase efficiency and quality of services in cargo ports, the operational safety and security of the ports, by increasing the accessibility and accuracy of relevant information. More researches are needing to find what sensors and instrumentation and what communication and analytics technologies may contribute to design the optimal model and architecture of IoT platforms for better logistic (including security) of a cargo port. IoT for cargo ports should have an adaptive framework, which should allow multi-modalities sensing, modularity (i.e., flexibility to introduce new applications), should ensure adequate and safe connectivity to the existing information systems, and should process enormous quantity of data. As IoT data and information are nowadays susceptible to malicious usage by hackers but also by unauthorized competitors, design for cargo port information systems that support security of information at every layer of the information system architecture is required.

References

[1] IoT-A Internet of Things Architecture www.iot-a.eu/arm

[2] Feasibility Study on the network operation of Hinterland Hubs (Dry Port Concept) to improve and moderate ports' connections to the hinterland and to improve networking https://web.archive.org/web/20080413200609/http://inloc.info/internal/wp1/activity_15_feasibility_study_dry_ports.pdf

[3] ELS Engineered Lifting Systems & Equipment. https://www.engliftsystem.com

[4] Suárez JP, Trujillo A, Domínguez C, Santana J M and Fernandez P 2015 Managing and 3D visualization of real time big geo-referenced data from Las Palmas port through a flexible open source computer architecture *Proc. 1st Int. Conf. Geograph. Inform. Syst., Theory, Applications and Management (GISTAM)*

[5] Sadlier D A, Fergunson P, Conaire CO, O'Connor NE and Doyle K 2011 Image-based vehicle indexing for a seaport transportation surveillance system *Proc. 8th IEEE Int. Conf. Advanced Video and Signal-Based Surveillance* pp 367-72

[6] Jarabo-Amores P, Gonzalez-Bonilla MJ, Mata-Moya D, Martin-de-Nicolás J and Palma-Vazquez A 2012 Demonstrator of maritime SAR applications: Automatic ship detection results *Proc. IEEE Intern Geoscience and Remote Sensing Symposium*

[7] Trimble G M and Belcher E O 2002 Ship berthing and hull inspection using the Cetus II AUV and MIRIS high-resolution sonar *Proc. OCEANS'02 MTS/IEEE*

[8] Lee Y, Choi J, Jung J, Kim T and Choi H-T 2017 Underwater robot exploration and identification using dual imaging sonar: Basin test *Proc. IEEE Conf. Underwater Technologies*

[9] Hennin S, Germana G and Garcia L 2007 Integrated perimeter security system *Proc. IEEE Conf. Technologies for Homeland Security* pp 70-75

[10] Rezapour T Y, Atani R E and Abolghasemi M S 2014 Secure positioning for shipping containers in ports and terminals using WSN *Proc. 11th Int. ISC Conf. on Information Security and Cryptology* pp 10-14
[11] Lai W C and Lau H Y K 2010 A bio-inspired object tracking algorithm for minimizing power consumption *Proc. IEEE/ACM Int. Conf. Green Computing and Communication* pp 355-60
[12] Arifin A S and Firdaus T S 2017 Ship location detection using wireless sensor networks with cooperative nodes *Proc. 9th Int. Conf. on Ubiquitous and Future Networks (ICUFN)*
[13] Rudzinski C, Masters D, Buck A, Wall M, Tremblay D and Wack E 2010 Screening maritime shipping containers for weapons of mass destruction *Proc. IEEE Int. Conf. on Technologies for Homeland Security* pp 460-66
[14] Mukhopadhyay S C, Gooneratne C P, Gupta G S and Yamada S 2005 Characterization and comparative evaluation of novel planar electromagnetic sensors *IEEE Transactions on Magnetics* **41**(10) pp 3658-60
[15] Wu W and Liang L 2008 Long-term real-time monitoring system for port machinery *Proc. 1st Int. Conf. Intelligent Networks and Intelligent Systems*
[16] ETSI Technical Report TR103527. SmartM2M. Virtualized IoT Architectures with Cloud Back-ends. https://etsi.org/news-events
[17] Microsoft Azure IoT Reference Architecture. https://azure.microsoft.com/en-us/blog/azure-iot-reference-architecture-update

Problems and prospects of complex psychological phenomena measurement

S Kostromina[1] **and D Gnedykh**[1,2]

[1] Department of psychology, Saint-Petersburg University, 7/9 Universitetskaya nab., St. Petersburg, 199034, Russia
[2]Author to whom any correspondence should be addressed

E-mail: d.gnedyh@spbu.ru

Abstract. The article is dedicated to the key problems of complex psychological phenomena measurement: the difficulties of measurement systems correspondence in the evaluation of the psychological phenomenon by qualitative and quantitative measurement methods; the difficulties in the comparability of the measurement results in psychological phenomena studying. The problems that arise in the methodology of psychological science and in the organization of interdisciplinary research are discussed. In conclusion, the direction of solving the described problems aimed at joining efforts of psychologists and metrologists is proposed.

1. Introduction

One of the key methodological problems of psychology is the objectification of the studied psychological phenomena. The solution of this problem mainly determines the identity of psychology to Humanities or Natural sciences. The controversy of research methods used in psychology did not stop throughout the XX century which determined the formation of two main areas: psychometric and ideographic with their merits and limitations. The evidence of certain psychological patterns was based on the mathematical statistics methods (STATISTIKA, SPSS, R software package), computer processing of obtained data, complex multivariate methods of statistical analysis: regression, factor and discriminant analysis, structural equation modeling.

The development of psychophysiological and neurophysiological methods of nervous system and human brain investigation mainly contributed to the strengthening of biological approach and natural-scientific directions in psychological research. The question of linearity of the functional brain and mental activity remains open [1, 2]. The experience of the past decades accompanied by the study of complex psychological phenomenon using neurophysiological or genetic methods (e.g., the study of patriotism, or emotive qualities – kindness, aggression, etc.) showed the limitations of direct interpretation and the impossibility of transparent transfer of the received data to psychological phenomenon.

Thus, the main contradiction of the modern methodology of measurements in psychology is the impossibility of the integral assessment of psychological phenomena through the lens of linear functions and direct interpretations. Call for transdisciplinarity [3] and search for a new measurement paradigm different from traditional psychometric rise to a crescendo.

This call is based on the facts that, *firstly*, phenomena, which are comprehensive in its manifestations and cannot be 'measured' using a questionnaire or devices (for instance, consciousness, personality, psychological time, decision-making, psychology of choice, life scenarios, etc.), are increasingly in the focus of psychological research. Most of them can hardly be defined. *Secondly,* their evaluation requires not only quantitative but also qualitative methods. The need of development of multidimensional design of psychological research is determined by: the intersection of qualitative description of complex psychological phenomenon with those that are objectively and quantitatively recorded; the duality of psychological phenomena genesis, therefore it is difficult to identify the biological and social factors contribution. *Third,* most psychological research studies any mental phenomenon as a static condition, based on the idea that a personality or any other psychological phenomenon is a phenomenon with stable structure. Hence, in psychology there is the dominance of structuralism, traits theory and system approach, as well as the corresponding measurement technology: concept operationalization, indicators identification, selection of measurement tools and scale for each indicator; their calculation, comparison, and searching for the structure (correlation analysis), classes (clustering), types (discriminant analysis) or cause-and-effect relationships (regression analysis or structural equation modeling) between indicators.

Meanwhile, like any living system, the mental system is characterized by the processes of growth, relative stability, degradation and, sometimes, irreversibility. From the point of *view of dynamic approach, sequence and variability* are two dimensions of the mental processes nature. That means sequential movement in certain areas of mental development, but in different ways within and between areas.

The lack of methodology to conduct such studies leads to: 1- spontaneous combination of different measuring and evaluation instruments; 2 - discrepancy between studied psychological phenomenon content and measure procedure. As a result, there are unsustained interpretations and conclusions. An additional problem is the rapidly changing environment, in which people live, and mobility of mental structure. Therefore, researchers are faced with difficulties of replication of studied psychological phenomenon for new samples and temporal spans.

2. Measurement of complex psychological phenomena: problems and prospects

Based on the difficulties and contradictions mentioned above, we identified two main problems that arise when choosing of measuring methods and means in psychological research.

1) Difficulties of measuring systems correspondence in estimating of a phenomenon by means of qualitative and quantitative assessment methods.

2) Difficulties in comparability of the results of psychological phenomenon measurements in split-level interdisciplinary analysis.

The first problem is related to the fact that the complexity of a psychological phenomenon is the reason why researchers use several tools for its measurement and evaluation. This difficulty arises not only at the stage of their choice (the question of technique validity), but also at the stage of correspondence of the obtained data. Despite the use of standard methods of statistical analysis, the researchers need to develop a unified system of correspondence of disparate results and their interpretation. The difficulty is that in psychological science there are still no unified principles for the identification of such correspondence, or universal algorithm applicable to the interpretation of data based on the different measurement variables.

Exemplify the problem. The study by Zolotareva [4] aimed to establish similarities and differences in the understanding and description of shame and guilt among generations within a family. Shame and guilt are social emotions associated with the feeling of a mismatch between the act and the normative regulation (shame), or with a deep repentance activating the desire of a person to correct the situation and to change himself (guilt). The emotional context necessitates the use of such measuring means that allow to assess the degree of the emotional expression at the level of understanding (cognitive component), experience (sensory component) and action (behavioral component). In this regard, the most suitable methods are the psychodiagnostic test and the collection of information with a focus not on statistical measurements, but on understanding, explaining and interpreting information about the subject of the study. That means the combination of quantitative and qualitative methods.

However, if the quantitative measurement methodology is based on the answer to the question 'How much?', in qualitative methods not numbers are used, but words that allow to answer to the questions 'What?', 'How?', 'Why?' discovering the processes in the human mind. Such methods include interviews, narratives, focus-group method, discursive and ethnographic research strategies. As Melnikova and Khoroshilov [5] note, there are fundamental principles that allow to suggest the conceptual unity of qualitative research which are fundamentally different from quantitative ones: (1) contextual sensitivity; (2) understanding; (3) interpretative reconstruction; (4) reflexivity.

Contextual sensitivity implies the measurement of context specificity (historical, political, social, etc.), the separation of micro- and macro-social context, the identification of context and its ontological status [6]. The principle of understanding in qualitative methodology is implemented as a process of categorization and typology of the obtained data, and refers to the overcontextual generalization of repetitive semantic patterns/units (content analysis). Interpretative reconstruction of qualitative research involves the definition of meanings from the point of view of the subject (empathic-descriptive level of interpretation) and in depth understanding of the phenomenon based on existing theoretical traditions and prospects (critical-hermeneutic level of analysis). Reflexivity allows to evaluate critically the data analysis.

Thus, the methodology of qualitative research exists in the categories of meaning, word and context. It intersects with the quantitative thesaurus only in the frequency of use of certain words/meanings, and in the percentage expression of dominant judgments that determine different interpretations.

Existing integration models of different types of indicator variables do not always cope with the task of studying complex psychological phenomena. For example, the Rasch measurement model allows to analyze the qualitative data by quantitative methods [7]. However, the values of indicator variables of psychological phenomenon, which being subjected to linear transformation, are often behavioral characteristics. In this regard, the difficulty is that not all psychological phenomena (e.g., personality) have unambiguous "behavioral correlates", based on which it would be possible to predict their development and variability. Thus, this model can be more successfully applied in management and education (to assess the quality of education, achievement tests) [8-11] than in psychology.

The results of qualitative methods (interviews and descriptions of the situation) in the mentioned study was data on the frequency of answers matching in relation to the understanding of shame and guilt or the experience of these emotions in different situations (Likert scale from 0 to 10). When comparing the responses of representatives of different generations within one family, that kind of assessment allowed to conclude about the degree of proximity of their perception of the emotional context and the meaning of shame and guilt. The descriptions of situations, in which subjects experienced shame or guilt, allowed to define the types of situations and calculate the frequency of experience in each of them. Thus, there was the wide phenomenology of the obtained ontological data but their measurement model was limited.

In the described study there were following measuring steps. Data for each subject was represented in the set of numbers: the occurrence frequency of the answers to questions; the number of situations of shame or guilt in a particular area; the percentage of matches; the percentage of the differences in responses between the generations. Each set of numbers served as a model of the subject.

These models can be understood in different ways depending on the problem statements – as a linear space, a multidimensional (one–dimensional) random variable, etc. However, in psychology these models most often are limited to the calculation of frequencies, as well as the percentage of certain thematic blocks/answers. At the same time, even these manipulations do not guarantee the research validity. The attribution of numbers to subjects should make it possible to use these numbers in accordance with the requirements of mathematics. For a psychologist, this turns into a problem because not every numerical relation has the corresponding empiricism. And if some numerical relation does not correspond to the empirical object properties, its practical use becomes problematic. Especially in respect to scales of low types - nominal, ordinal, etc. [12].

In order to overcome the limitations of qualitative research, the results of interviews or narratives are supplemented by psychodiagnostic tests related to quantitative measurements. In the case of the shame and guilt study, two validated questionnaires 'Guilty Inventory' and 'Test of Self-Conscious Affect' were used. Both techniques use the classical psychometric approach with the identification of scales allowing to measure the distinct manifestation of shame and guilt key indicators. Thus, the research can be considered as the complementary study of a complex psychological phenomenon. However, as a result, these two kinds of measurement (qualitative and quantitative) exist as two parallels.

The qualitative strategy provided an adequate transition to the system of features describing the observed phenomena that lead to the opportunity for the identification of variables characterizing the processes of understanding and experiencing the shame and guilt. The quantitative parameters reflected the information about the degree of manifestation of shame and guilt presented in the form of the measuring scale. Is it possible to apply a new level of mathematical and statistical data processing? For instance, obtainment of the ontological facts of the second level – the calculation of correlations between parameters of the quantitative and qualitative parts of the research? Is it possible to combine them to calculate the regression model? A reasonable measurement methodology should be developed for this intersection, allowing to combine quantitative and qualitative estimating methods into a unique model [13, 14]. Otherwise, these methods will continue to exist as complementary but non-overlapping measurement strategies.

Another problem of the methodology of complex psychological phenomena measurement is interdisciplinary studies that require comparison of variables at different levels.

As an illustration of this problem, the study of neurophysiological features of cognitive control (CC) in the process of learning activity, for students with different levels of academic success is presented [15]. The main objective of neurophysiological studies of mental phenomena is the search and explanation of physiological mechanisms as the basis of human mental activity [1, 2, 16].

Cognitive control is the multi-level psychological phenomenon [17] with many components and functions [18], involved in regulation of behavior, thoughts and emotions [17]. Cognitive control provides the adjustment of cognitive processes (attention, memory, thinking) to the task solution depending on certain conditions.

The analysis of the methods, tools and instruments used to measure cognitive control allows to identify the main approaches:

- neurophysiological methods (e.g., electroencephalography through power (mV), waves frequency (Hz) and latency (ms) measurement);
- collecting behavioral data (recording of reaction time (ms); analysis of tasks results, e.g., number of correct / incorrect answers/decisions);
- psychodiagnostic empirical methods (psychodiagnostic tests and questionnaires) based on the participant's decisions about his own behavioristic characteristics (motivation, ability to planning, constancy to purpose, etc.).

Thus, for a comprehensive study, many indicators that allow to record cognitive control's occurrence and process, and its relationship with other phenomena should be taken into account.

In the mentioned study CC was studied at neurophysiological (by means of electroencephalograph), behavioral (recording of reaction time and number of incorrect answers, data on academic success) and personal (determination of personal characteristics using psychodiagnostic techniques) levels. Mathematical and statistical analysis was the following: nonparametric methods for comparison of independent samples, correlation analysis, MANOVA.

Despite of possibility to compare the data received on the basis of different scales and presented in different quantitative units of measurement, the full correspondence of the results, obtained at different levels of analysis of the same phenomenon, was not achieved. Registration of neurophysiological activity of the brain while subjects performed the task was aimed at identifying the characteristics of

CC through the measurement of physical quantities such as time (ms) and power (mV) of electric waves. Behavioral data related to the manifestation of cognitive control mechanisms reflected the time of decision–making (MS), and its correlation with the correct answer (using the nominal scale). Conclusions about students' personal characteristics were based on the analysis of scoring system data. The results of academic success (also presented in the scoring system) allowed to divide students into two groups (successful and not successful) and to compare them with each other.

By means of correlation analysis the relationships between certain indicators of cognitive control at the behavioral and personal levels were revealed. For instance, the more time a subject spends on experimental tasks, the more mistakes he makes; the more emotionally stable the student is, the better system of self-management is developed; the higher the student's need to communicate with people, the worse he passes the first exam period, etc.

However, it was difficult to correlate the results of neurophysiological part of the study with other data. In this case, it is not correct to establish the relationship between the power amplitude (mV) of evoked potentials (EP) (as the neurophysiological characteristics of cognitive control) and extroversion (as a personal characteristic associated with CC) not only from the point of view of comparability of these units of measurement, but also from the point of view of logic of interpretation and understanding of mental phenomena.

Regarding to behavioral data, such comparability seems more possible. For example, P300 component (EP at 300 ms) is recorded in the situation of identification, memorization, involvement and decision-making [19] and can by suggested as one of the neurophysiological indicators of cognitive control. Thus, the correlation can be as follow: the higher the P300 amplitude, the less mistakes in the control task are made by the participant of the study. In other words, the more a person is involved in the task, the more successfully he solves it.

Thus, the comparability of measurement results should be reasonable not only from the point of view of logic, but, first of all, from the point of view of Metrology [20]. Whereas these approaches to the measurement of cognitive control are based on different units of measurement, the results of behavioral measurements (at the level of personal characteristics) might not be comparable with neurophysiological ones. However, commonsense reasoning 'the more a person is involved in the task, the more successfully he solves it', which was made by the researcher, may seem quite logical. Nonetheless, there was no objective confirmation of this statement based on the assurance of uniformity of measurements of involvement and success (due to the impossibility to measure success in mV in comparison with involvement (which is associated with the component P300)).

Consequently, when measuring complex psychological phenomena, comparability of the results obtained at different levels of analysis becomes difficult. In this regard, there are also difficulties in correlating different results at the stage of their interpretation. Would be methodologically correct their generalization in the absence of mathematical and statistical basis? If the results, based on different methods of cognitive control measurements, are that students with high academic success have a high P300 amplitude (mV) and their self-management ability correlates with extroversion, can we conclude that the magnitude of P300 amplitude depends on students' desire to build social ties (or vice versa)? Obviously, until the mathematically and statistically reasonable confirmation will not obtained, such conclusion is not legitimate.

In our opinion, the key decision of these problems is the search for a new approach to the definition and development of 'universal' units of psychological phenomena measurement and measurement methods that can connect different levels of mathematical and psychological analysis of complex psychological phenomena.

3. Conclusions

Thus, despite achievements in the application of new statistical methods in psychology, 'zone of proximal development' of psychometrics is the development of an integrated methodology of measurement and evaluation of complex psychological phenomenon, establishing evidence-based principles for combining qualitative and quantitative research methods, psychological, sociological and neurophysiological techniques, as well as their interrelated mathematical and statistical processing. Psychological phenomena that exist at different levels, have different forms of manifestations, in certain contexts change their meanings, and characterized as self-developing systems are increasingly in the focus of psychologists. Nowadays, psychology is in need of a measurement methodology that allows to consider the processes of growth, irreversibility, variability, fluctuation and the potential of self-change of the mental system. That is, the methodology should base not on structuralism, but on procedurality (the idea of measuring complex multidimensional nonequilibrium systems).

Acknowledgments
Supported by RF Government grant contract No.14.W03.31.0010.

References
[1] Roik A O and Ivanickij G A 2011 Neurophysiological model of cognitive space *J. Higher Nerv. Act.* **61**(6) 688–696 (in Russian)
[2] Rose N 2016 Reading the human brain: how the mind became legible *Body Soc.* **22**(2) 140-177
[3] Guseltseva M S 2015 Psychology and new methodologies: epistemology of complex *Psych. Res.* **8**(42) 11 (in Russian)
[4] Zolotareva A S 2017 *Understanding of the Phenomena of Shame and Guilt by Representatives of Different Generations within One Family* (SPbU) p 109 (in Russian)
[5] Melnikova T A and Khoroshilov D A 2013 Methodological principles of qualitative research in psychology *Bulletin of Moscow University Ep. 14 Psychology* **3** 4–14 (in Russian)
[6] Hammersley M 2008 Context and Contextuality *The Sage Encyclopedia of Qualitative Res. Methods* Ed L M Given (Thousand Oaks, CA: Sage) vol 1 pp 122-123
[7] Mari L and Wilson M 2014 An introduction to the Rasch measurement approach for metrologists *Measurement* **51** 315-327
[8] Ingebo G S 1997 *Probability in the Measure of Achievement* (Chicago: MESAPress) p 148
[9] Maslak A, Moiseev S and Nasonova T 2018 Application of the Rasch method of evaluating latent variables in management and administration *Society. Integration. Education: Proc. Int. Scientific Conf.* **VI** 330-340
[10] Maslak A, Anisimova T, Rybkin A and Titenko E 2015 Measuring and monitoring of education infrastructure of southern federal district of the Russian Federation *Proc.-Social and Behavioral Sciences* **214** 1062-69
[11] Pozdnyakov S A 2007 *Study of Latent Variables Measurement Accuracy in Education* (Slavyansk-na-Kubani: Izd. centr SGPI) p 118 (in Russian)
[12] Tolstova YU N and Maslennikov E V 2000 Qualitative and quantitative strategies: empirical research as a dimension in the broad sense of the word *Sociol. Res.* **10** 101-108 (in Russian)
[13] Fisher W P Jr and Stenner A J 2011 Integrating qualitative and quantitative research approaches via the phenomenological method *Int. J. of Multiple Research Approaches* **5**(1) 89-103
[14] Laba L YA 2004 Approaches to integration of qualitative and quantitative methods *Sociological research* **2** 124-129 (in Russian)
[15] Kostromina S N, Mkrtychian N A, Kurmakaeva D M and Gnedykh D S 2017 The interrelationship between cognitive control and academic success of first-year students: An interdisciplinary study *Psych. in Russia: State of the Art* **10**(4) 60-75
[16] Ivanickij A M 2012 'Reading the brain': achievements, prospects and ethical problems *J. Higher Nerv. Act.* **62**(2) 133–142 (in Russian)
[17] Posner M I and Snyder C R R 1975 Attention and cognitive control *Information Processing and Cognition* Ed R L Solso (Hillsdale N Y: The Loyola Symposium. Hillsdale N Y) pp 55–85
[18] Miyake A, Friedman N P, Emerson M J, Witzki A H, Howerter A and Wager T D 2000 The unity and diversity of executive functions and their contributions to complex 'frontal lobe' tasks: a latent variable analysis *Cognitive Psychol.* **41** 49–100

[19] Gnezditsky V V and Shamshinova A M 2001 *Experience of Application of Evoked Potentials in Clinical Practice* (Moscow: AOZT 'Antidor') p 480 (in Russian)
[20] Biserova V A, Demidova N V and Jakoreva A S 2007 *Metrology, Standardization and Certification* (Moscow: EKSMO) p 10 (in Russian)

Measurement system based on virtual reference axis for evaluation form and location of surfaces and axes of rotary details

D Diakov, Hr Radev, V Vassilev and H Nikolova

Technical University of Sofia, Bulgaria

E-mail: hnikolova@tu-sofia.bg

Abstract: The system for measurement and control of the form and location of the surfaces and axes of large scaled rotational parts by means of a virtual reference base is used, by using a self-orientation multistage gauge allowing exclusion of the radial run-out of axis of rotation when access to the face surfaces of the workpiece is difficult or impossible.

1. Introduction

Deviation of the form and location of the surfaces and the axis of the rotational parts can be determined by measuring the radial run-out of a certain number of cross sections and subsequent processing of this primary measurement information by certain algorithms [1-4].

A basic requirement is the stability of the axis of rotation when measuring the radial run-out. Large scaled details due to their large mass and dimensions, the fulfillment of this requirement is very difficult, and in some cases practically impossible.

This problem is solved by bringing the run-out results to a virtual output base, a virtual base axis representing the straight line passing through the centroid of the points of the instantaneous axes of rotation, in the sections of two rings attached to the end faces of the workpiece approximately coaxially to its geometric axis [5]. The deviation from circularity of the two rings is previously known or negligible.

The processing of the results of the radial run-out measurement to the virtual output base is done by introducing corresponding corrections into these results, excluding the influence of the so-called run-out of axis of rotation (axis of rotation error motion) reflecting on the rotational axis inconsistency [6-8].

At measurement process the detail is based on prisms, its own bearing seats or centers, and the radial run-out is determined using universal length measurement tools.

Due to the specificity of the measured parts, usually and especially in repair and test operations, the measurements are made based on their own bearing seats, where the access to their end faces is difficult or impossible, respectively, the establishment of the reference rings is very difficult or impossible. This limits the application of the method described above [5].

The subject of this paper is a system for measuring variations in the form and location of the surfaces and axes of rotational large-scaled parts relative to a virtual output base with no reference rings attached to the end faces of the measured parts.

2. Measurement schemes

The basic measurement scheme is shown in figure 1. The part is based on prisms, centers, or its own bearing seats representing a base 5 on which radial run-out is measured.

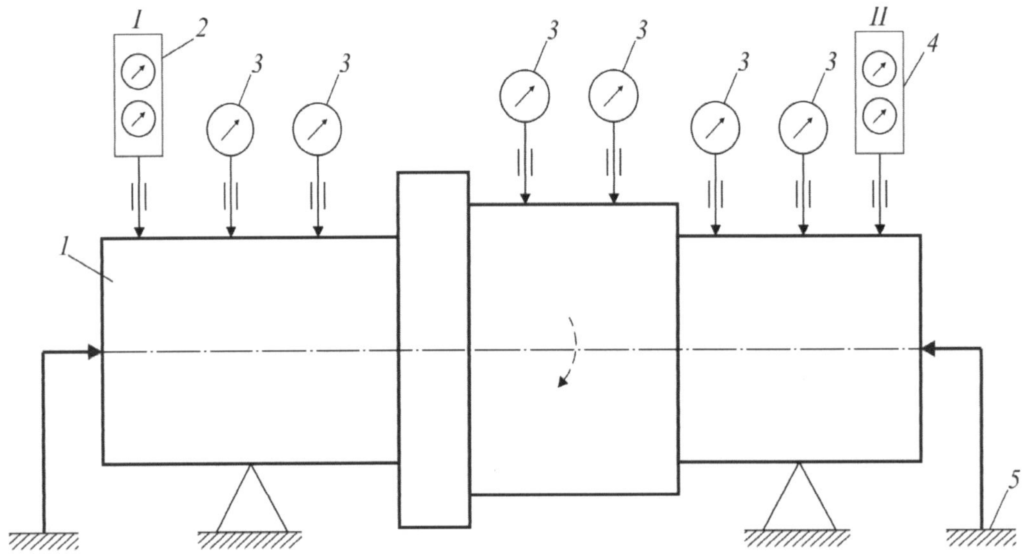

Figure 1. Measurement scheme.

With the measuring heads 3, the radial run-out in the respective cross sections is measured. This run-out includes within itself the deviation from the circularity of the measured profile, the eccentricity and the run-out of axis of rotation (axis of rotation error motion).

With the measuring devices 2 and 4 located in two pre-marked cross-sections I-I and II-II of the measured surface of the workpiece, the run-out of axis of rotation is determined. For the established run-out of axis of rotation, analogous to the method described in [5], corrections are made to the readings of all measuring heads, and so the corrected values of the radial run-out in the respective programs determine the deviations of form and location.

The main element of these measuring devices is a multistage self-orientating gauge (Figure 2) based on the measured surface in the respective cross-section. With a relative rotation of the gage relative to the shaft, each point in its body describes a circle equidistant to the median circle of the profile in the given cross section [9, 10]. This makes possible to define the run-out of axis of rotation as a deviation from circularity of the circular trajectory of a given point in the gauge body.

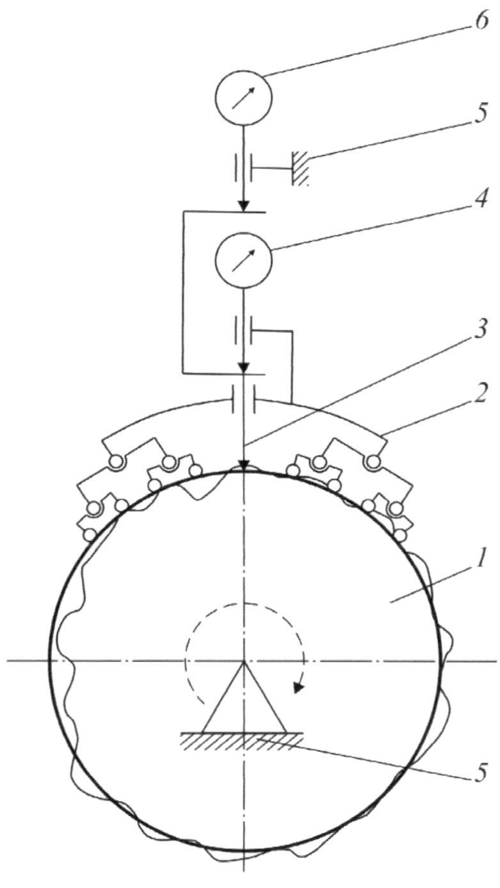

Figure 2. Measurement device based on multistage self-orientating gauge.

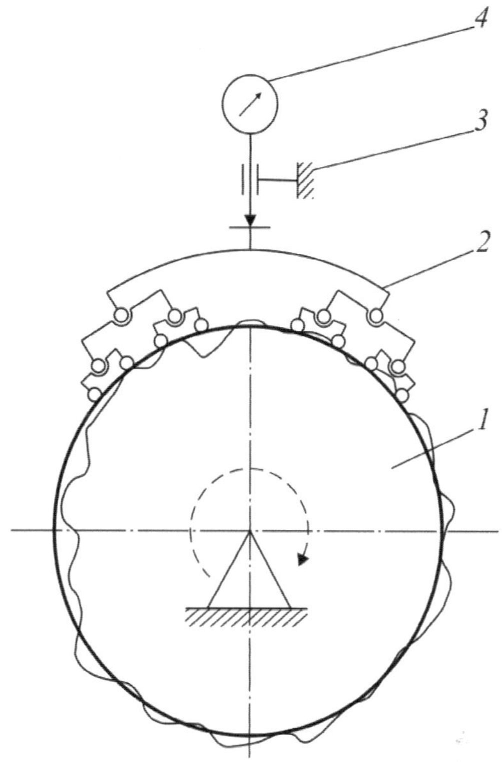

Figure 3. Measurement device based on multistage self-orientating gauge with one measuring head.

In the housing of the multistage self-orientating gauge 2 a measuring rod 3 is mounted whose measuring tip contacts the profile of the part 1 at the measured cross-section. Measuring heads 4 and 6, which are positioned along the measurement line simultaneously measure the run-out relative to the housing 2 of the gauge and the base 5. The readings A4 of the measurement head 4 represent the deviation from the circularity with respect to the median circle of the measured profile and the readings A6 of the measuring head 6 - deviations from circularity, eccentricity and run-out of axis of rotation. The difference between the two readings represent the eccentricity and the run-out of axis of rotation.

It is also possible to realize the measuring device (Figure 3), in which only the eccentricity and the run-out of axis of rotation of the average circle of the measured profile are determined by only one measuring head.

The choice of the particular realization of the measuring devices is determined by constructive, metrological and operational considerations, depending on the object and the measurement conditions.

3. System approbation

System approbation made in real environmental conditions at scheduled NPP Kozloduy turbine repairs confirmed its workability and good metrological characteristics.

4. Conclusion

The multistage self-orientating gauge measurement system allows the measurement of the deviations of the form and location of the surfaces and axes of large-scaled rotational parts at difficult or impossible access to their face surfaces.

References

[1] ISO 2017 *ISO 1101:2017 Geometrical Product Specifications (GPS) - Geometrical Tolerancing - Tolerances of Form, Orientation, Location and Run-out* (International Organisation for Standardization)

[2] Graham T Smith 2002 *Industrial Metrology: Surfaces and Roundness* ISBN-13: 978-1852335076, ISBN-10: 1852335076

[3] Whitehouse David J, *Handbook of Surface and Nanometrology* 2nd Edition, ISBN-13: 978-1420082012, ISBN-10: 9781420082012

[4] Zhelezarov I 2010 Analysis of measuring devices and measurement systems. *Mechanical Engineering in XXI Century* (Nis, Serbia) pp 199 –0. ISBN 978-86-6055-008-0

[5] Radev H 2008 Metod za izmervane na otklonenieto na formata i razpredelenie na ploshtta na rotatsionnite detaĭli na virtualna izkhodna baza, *Patent № 65461 V1 / 27.10.2008* (in Bulgarian).

[6] BDS 1976 *BDS 13872:1976 Mashini Metalorezheshti. Metodi za Proverka na Radialnoto Biene na Osta na Vŭrtene* (in Bulgarian)

[7] ISO 2012 *ISO 230-1:2012 Test Code for Machine Tools - Part 1: Geometric Accuracy of Machines Operating under No-load or Quasi-static Conditions* (International Organisation for Standardization)

[8] Dichev D, Koev H, Bakalova T and Louda P 2016 An algorithm for improving the accuracy of systems measuring parameters of moving objects *Metrol. Meas. Syst.* **23**(4) pp 555-65 DOI: 10.1515/mms-2016-0041

[9] Gebel I D 1973 *Meas.Tech.* 3 pp 347-53

[10] Dyakov D 2000 Optimizirane na parametrite na sistemata za izmervane na otklonenieto ot krŭga na bazata na mnogostepenna samozavisimost. *Sbornik dokladi ot 10-ti Natsionalen Nauchen Simpozium s mezhdunarodno uchastie "Metrologiya i metrologichno osiguryavane* 2000" pp 87-9, ISBN 954-438-229-3 (in Bulgarian)

Journal of Physics: Conference Series **1379** (2019) 012075 doi:10.1088/1742-6596/1379/1/012075

Measurement procedure: from detailed description to technology

G R Nezhikhovskiy

Head of sector, D.I. Mendeleyev Institute for Metrology, St. Petersburg, Russia

E-mail: grn30@mail.ru

Abstract. The tendency to the development of measurement procedures adapted by the user to the equipment, materials, software is considered. The necessity of estimating the measurement uncertainty with regard to the limits established in the procedure is shown.

Measurement procedures are developed as documents that allow operators to obtain results with uncertainty that does not exceed the target uncertainty. In accordance with the International Vocabulary of Metrology a description of measurement in the document should be detailed ("sufficiently detailed"). In practice, the degree of this detalization of measurement procedure is different. If we imagine a decreasing scale (see figure 1), then at its beginning there will be measurement procedure-prescription, implemented on specific instruments and punctually regulating the actions of operator. At the end of the scale there will be procedures that are easily adaptable to the equipment (reagents) and provide operators with a wide range of actions within the assigned limit values. (Such procedures are often considered as measuring technologies.) Measurement procedures-prescriptions border with measuring instruments, measurement procedures-technologies with measurement methods. In recent years, there has been a tendency for reducing the share of measurement procedures-prescriptions and increasing the share of measurement procedures with a lower degree of detalization.

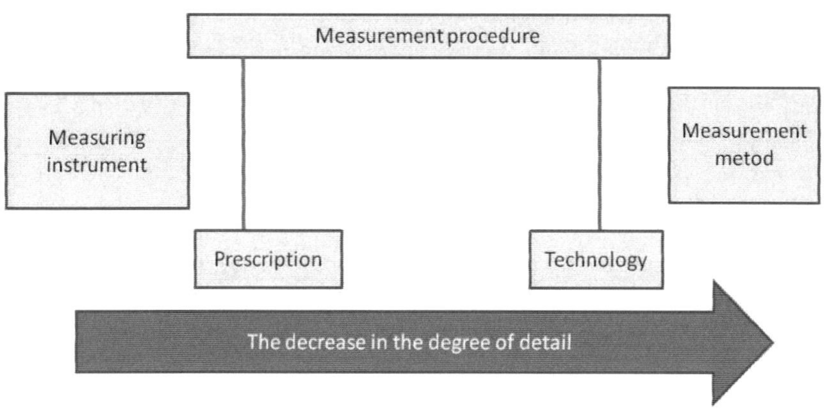

Figure 1. The degree of measurement procedure detalization.

There are several reasons for this trend. While in the past measurement procedures were usually developed for own use, now they are often developed by specialized organizations, including firms that supply universal measuring equipment. They are interested in the dissemination of typical solutions. The second reason is the availability for the user of a variety of auxiliary equipment, reagents and software, to which the measurement procedure can be adapted. The third reason is dissemination of a universal approach to expression and evaluation of measurement uncertainty. When a user purchases a measurement procedure with an uncertainty budget, he has the opportunity to assess the influence of certain changes on the combined standard uncertainty and expanded uncertainty of measurements himself. Another reason is the desire of many laboratories to apply a modern quality management system, one of the elements of which is the monitoring of the validity of the results. The availability of information on the stability of the results for each of the implemented measurement procedures allows to choose the optimal frequency of regulating and control operations.

Important stages in the development of measurement procedure-technologies are the formation of a complete mathematical model of measurement and agreement of the defined limit values (for example, limit values for the acceptance of calibration function) with the corresponding contributions to the uncertainty of measurements. It should be noted that the relationship of limit values with contributions to uncertainty has not been sufficiently addressed to date. The existing guidelines for measurement uncertainty assessment are more focused on the results of individual measurement experiments than on the results obtained by measurement procedures. At the same time, most laboratories are interested in assigning to all the results obtained by the measurement procedure the uncertainty value established in its development (validation, attestation). This interest is reflected in ISO/IEC 17025:2017. A note is included in the section «Evaluation of measurement uncertainty» for a particular procedure with established and confirmed uncertainty of measurement *"there is no need to estimate uncertainty of measurement for each test work once again, if the identified critical influencing factors are under control and remain unchanged"*.

Determination of measurement uncertainty as a metrological characteristic of the measurement procedures is one of the goals of attestation of measurement procedures within the framework of the system of ensuring the uniformity of measurements adopted in Russia. D.I. Mendeleyev Institute for Metrology attests more than 100 measurement procedures per year. Most of them are analytical measurement procedures.

For example, consider the procedure of measuring the mass fraction of an analyte in a sample of a solid object (W, %), which involves dissolving the sample of the object and subsequent measurement of the mass concentration of the analyte (C, mg/dm^3) using a universal atomic emission spectrometer with microwave plasma. Every day, before starting work, the user builds a calibration function as the dependence of the analytical signal (Y in conventional units) on C. In order to construct the calibration function he uses five ($m = 5$) calibration standards with different mass concentrations of the analyte (C_i, where i is the solution number). And the C_5/C_1 ratio does not exceed 50. To prepare calibration standards, a reagent (pure substance) is used, in which the mass fraction of the main component is not less than W^m (%). The procedure provides for the operations of checking the acceptability of the calibration function and periodic checks of its stability using the solution with a mass concentration of analyte C_4. To obtain the measurement result, all operations are carried out with two weighing, while the final result is calculated as the arithmetic average of two values of the mass fraction of the analyte, unless their difference does not exceed the established norm – the limit of repeatability of the results of single measurements.

The procedure developer has recorded the range of measurements of the mass fraction of the analyte (from 0.050 % to 2.5 %), the weight of the sample ($M \approx 200$ mg), established the requirements for weighing instruments, volumetric vessels and solvents, set the values of the mass concentration of the analyte in the calibration solutions (from 1,00 to 50 mg/dm^3) and confirmed the linearity of the calibration function $Y = a + bC$. At the same time, the user is given the opportunity to: a) select reagents available on the market with W^m from 97 % to 99.5 %; b) use spectrometers of various manufacturers;

Journal of Physics: Conference Series **1379** (2019) 012075 doi:10.1088/1742-6596/1379/1/012075

c) establish the time between stability checks of the calibration function. The mathematical model of measurement is represented by the expressions:

$$W = \frac{W^{*}+W^{**}}{2}, \quad (1)$$

$$W^{*} = \frac{C^{*}V^{*}}{M^{*}}100, \quad (2)$$

$$C^{*} = \frac{Y^{*}-a}{b}, \quad (3)$$

where superscripts * and ** correspond to the first and second single measurements. For the second single measurement, expressions (2) and (3) are valid with the superscript replaced by **.

Based on the approach outlined in [1], the relative total standard measurement uncertainty

$$u_{c}^{0} = \sqrt{(u_{r}^{0})^{2} + (u_{C}^{0})^{2} + (u_{V}^{0})^{2} + (u_{M}^{0})^{2}}, \quad (4)$$

where u_r^0 is the relative standard uncertainty due to the variation of measurement results under repeatability conditions, %;
u_C^0 – the relative standard uncertainty of measurements of the mass concentration of the analyte in solution, %;
$(u_V^0$ – the relative standard uncertainty associated with the volume of the solution, %;
u_M^0 – the relative standard uncertainty associated with the mass of the weight, % .

The spread of the measuring results was estimated in a series of experiments (12 pairs of measurements at five values of the mass fraction of the analyte). On the standard deviations found, the limits of repeatability of the results of single measurements of r (in %) were determined. The relative standard uncertainty was calculated using the expression:

$$u_r^0 = \frac{r}{2.8\cdot\sqrt{2}\cdot W}100, \quad (5)$$

For the lower limit of the measurement range (W = 0.050 %) r = 0.010 % is set. In this case, u_r^0 = 5 %.

u_C^0 was estimated on the assumption that the factor associated with the registration of analytical signals when measuring the mass concentration of an analyte is random and is covered by u_r^0. Thus, u_C^0 includes only contributions related to the calibration function

$$u_C^0 = \sqrt{(u_S^0)^2 + (u_g^0)^2 + (u_d^0)^2}, \quad (6)$$

where u_S^0 is the contribution due to the calibration solutions, %;
u_g^0 – the contribution associated with the establishment of the calibration function, %;
u_d^0 – the contribution related to possible instability of the calibration function, %.

u_S^0 is formed on the basis of a mathematical description of the calibration standard preparation process from contributions related to the purity of the initial reagent, metrological characteristics of the weighing instrument and volumetric vessels used. The maximum value of u_S^0 will be 2.5 % (with W^m = 97 %), the minimum – 1.8 % (with W^m = 99.5 %).

Preparation of the i-th calibration solution is described by the expression:

$$C_i = \frac{M_p\cdot W^m\cdot V_{ao}\cdot V_{az}}{100\cdot V_o\cdot V_z\cdot V_f}, \quad (7)$$

where M_p is the mass of the reagent weight (in mg) with the mass fraction of the main component W^m (in %);
V_o – volume of initial solution, dm^3;
V_{ao} – aliquot volume of initial solution, cm^3;

V_z – volume of intermediate solution, cm^3;
V_{az} – aliquot volume of intermediate solution, cm^3;
V_f – volume of calibration standard, cm^3.

The change in u_g^0 with the use of different spectrometers is most often due to the peculiarities of the algorithms for integrating analytical signals and determining regressions from experimental data. Spectrometers software rarely includes the calculation of u_g^0; usually, calibration results are accompanied by such parameters as correlation coefficient, residual standard deviation, absolute and relative deviations of the obtained analytical signals (or mass concentrations of the analyte) from the corresponding values on the calibration function. If we set the limit for the relative deviations of the analytical signals:

$$\frac{|Y_i - f(C_i)|}{f(C_i)} 100 \le g \, , \tag{8}$$

then for the linear dependence like $Y = a + bC$, the relationship between the norm and u_g^0 can be expressed by the formula (9)

$$u_S^0 \approx \frac{g}{2}\sqrt{\frac{1}{m} + \frac{(C_i - \bar{C})^2}{\sum_1^m (C_i - \bar{C})^2}} \, , \tag{9}$$

When m =5, (C_5/C_1) = 50 we have for i =1

$$u_S^0 \approx 0.4\, g \tag{10}$$

The calibration function is periodically checked for the stored solution with a mass concentration of analyte C_4. The calibration function is considered "stable", i.e. suitable for further use when the condition is met

$$\frac{|C_4^* - C_4|}{C_4} 100 \le d \, , \tag{11}$$

where C_4^* is the measured value of the mass concentration of the analyte in the fourth calibration standard, mg/dm^3.
The relationship between u_d^0 and the limit d is given by expression (12):

$$u_d^0 = \frac{d}{2\sqrt{3}} \tag{12}$$

In this example, limits g = 10% and d = 5% were set. After substitution in (6), we obtain:

$$u_f^0 = \sqrt{(2.5)^2 + (0.4 \cdot 10)^2 + \left(5/2\sqrt{3}\right)^2} = 4.9\,\% \, , \tag{13}$$

The estimation of u_V^0 was carried out guided by the specification of the applied flask with a capacity of 0.050 dm^3 (50 cm^3) with an admissible error Δ_V = 0.12 cm^3. Based on the assumption of a triangular distribution, the expression

$$u_V^0 = \frac{\Delta_V}{V \cdot \sqrt{6}} 100 \tag{14}$$

was used to receive u_V^0 = 0.1 %.

u_M^0 was estimated based on the limits of permissible error in operation Δ_M = 1.0 mg, specified in [2] for the selected weighing instrument. Based on the assumption of a rectangular distribution, the expression

$$u_M^0 = \frac{\Delta_M}{M \cdot \sqrt{3}} 100 \tag{15}$$

was used. For M =200 mg we receive u_M^0 = 0.3 %.

The uncertainty budget for the lower limit of the measurement range is shown in table 1.

Journal of Physics: Conference Series **1379** (2019) 012075 doi:10.1088/1742-6596/1379/1/012075

Table 1. The uncertainty budget of the analyte mass fraction measurement.

№	Uncertainty source		Contribution designation	Type of evaluation	Relative standard uncertainty (contribution to u_c^o), %	
1	Measurement variability		u_r^0	A	5	
2	Measuring of mass concentration of analyte in solution	Preparation of calibration standard	u_S^0	B	2.5	
3		Establishment of the calibration function	u_g^0	B	4.0	4.9
4		Possible instability of the calibration function	u_d^0	B	1.4	
5	Solution volume measuring		u_V^0	B	0.1	
6	Weigh mass measuring		u_M^0	B	0.3	
	Relative combined standard uncertainty, %		u_c^0		7	
	Expanded measurement uncertainty (with the coverage factor k = 2), %		U_c^0		14	

The relative expanded measurement uncertainty:

$$U_c^0 = k \cdot u_c^o \,, \qquad (16)$$

where k is the coverage factor (k = 2 at the level of confidence 0.95).

The calculated relative expanded uncertainty is 14 %, i.e. the target uncertainty was not exceeded.

In practice, the use of expressions (1) - (16) solves one of the two problems. The first is to establish in the measurement procedure limits r, g and d, which provide the requirement for the value u_c^0. The second is to check the limits established by the developer of the procedure [3]. In both cases, the ability to comply with the limits should be confirmed by experimental data available to the procedure developer. (Usually there are data of 4 or more experiments, including the construction and check of the stability of the calibration functions.) Of course, for individual users, the relative expanded uncertainty of measurements may be significantly lower. This circumstance does not reduce the effectiveness of the stated approach in the development of measurement procedures intended for broad application.

Conclusion

When determining the measurement uncertainty as a metrological characteristic of measurement procedures, in many cases it is necessary to ensure that the user can continuously monitor critical factors and at the same time give him some freedom to choose equipment, reagents, and frequency of checks. The solution to this problem is achieved by establishing norms and taking them into account when estimating uncertainty. Such a solution satisfies users of many analytical measurement procedures that have been attested by the D.I. Mendeleyev Institute for Metrology within the framework of the system of ensuring the uniformity of measurements adopted in Russia.

Journal of Physics: Conference Series **1379** (2019) 012075 doi:10.1088/1742-6596/1379/1/012075

References

[1] EURACHEM/CITAC 2012 *EURACHEM/CITAC Guide CG4: Quantifying Uncertainty in Analytical Measurement 3rd edition* ed S L R Ellison and A Williams (UK, EURACHEM/CITAC)

[2] OIML 2006 *OIML R 76-1:2006 Non-automatic Weighing Instruments Part 1: Metrological and Technical Requirements – Tests* (International Organization of Legal Metrology)

[3] Nezhikhovskiy G R 2012 Metrological research of quantitative chemical analysis procedures *Industrial laboratory. Diagnostics of materials* **78(2)** 69-75 (in Russian)

The status of and prospects for development of voltage quantum standards

Alexander Katkov

D.I. Mendeleyev Institute for Metrology, Moskovsky pr. 19, St. Petersburg, Russia

E-mail: a.s.katkov@vniim.ru

Abstract. The state and tendencies of the development of voltage standards based on the Josephson effect are considered. The methods of electric voltage reproduction are presented both in the field of dc and ac voltage of arbitrary shape based on the values of fundamental physical constants in the new SI, and directions of improvement and expansion of the field of application are outlined.

The discovery of the Josephson effect in 1962 allowed a revolutionary leap in the field of electrical measurements, which ultimately led to the transition of the international system of standards, in the form of physical artifacts kg, m, second, amperes, to standards based on quantum effects and on the values of fundamental physical constants.

By the development of the voltage standard one can judge about the cycle of transformations expected in this kind of measurement. From the beginning to the 70s of the 20th century the artifact reproducing the unit of electrical voltage had been a standard cell based on electrochemical processes producing a constant and stable electrical voltage. The voltage standard in many countries was represented by a group of standard cells shown in figure 1.

Figure 1. Group of standard cells.

The value of the unit stored by the group reference of the volt was determined by the Ohm's law in terms of the resistance unit reproduced by a mercury sample and the ampere reproduced by a silver voltameter (figure 2) based on the Faraday constant. Later on, the unit of resistance was determined be means of a calculable capacitor, and the current strength - by means of an ampere balance.

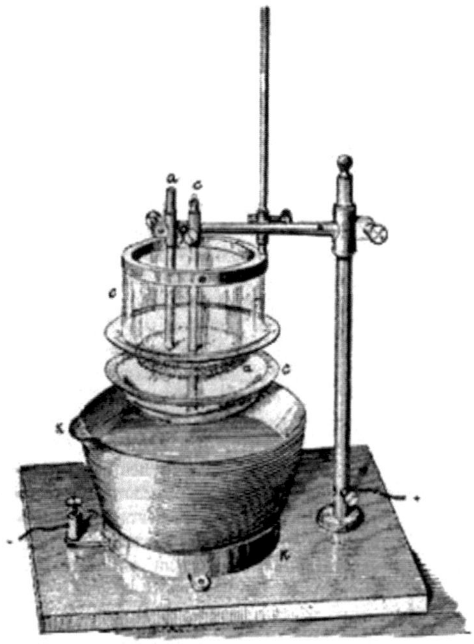

Figure 2. Silver voltameter.

The Josephson equation [1] obtained in 1962 can be explained in terms of known physical relations, in particular, in terms of a change of the magnetic flux per unit of time.

$U = d\Phi/dt$.

Replacing $d\Phi$ by the number of magnetic flux quanta $n\Phi_0$, and dt by its reciprocal value - frequency f - we obtain the well-known equation

$U = n\Phi_0 f = n(h/2e)f = nf/K_J$ (Figure 3),

where h is the Planck constant, $2e$ is the double electron charge, K_J is the Josephson constant.

The instrumental implementation of the Josephson quantum effect made it possible to use the obtained effect to create standards that reproduce voltage units with values attaining 20 V [2].

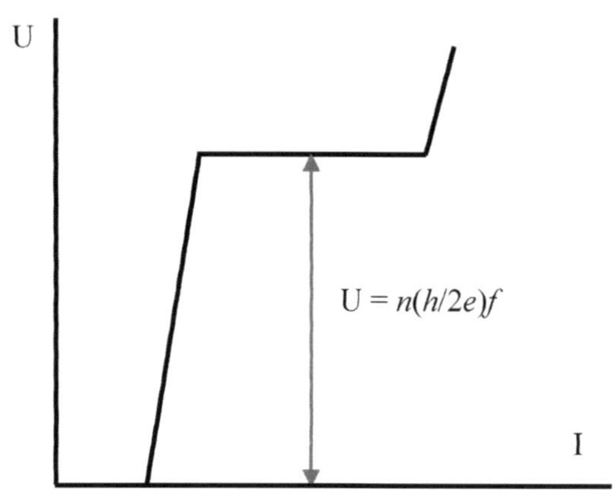

Figure 3. Volt-ampere characteristic of Josephson system.

The implementation of the Josephson effect in the metrological practice entailed issues with direct linkage to the SI due to a significant increase in the accuracy of relative voltage measurements. The transfer of the unit value from the ampere balance that have an accuracy at the level of several ppm became irrelevant and it was decided to refine the K_J value, which caused the introduction of the K_{J-90} with the uncertainty of 0.4 ppm in the SI (Recommendation 1 CI-1988).

Comparisons of standards based on the Josephson effect conducted by the BIPM showed discrepancies of 0.001 ppm or less. VNIIM developed transportable transfer standards of the voltage unit based on the Josephson effect (Figure 4) and made it available for international comparisons of national voltage standards [3-5]. Significant discrepancies between the uncertainty of relative measurements of voltage standards and the uncertainty of the Josephson constant in the SI was one of the reason for the revision of the existing SI.

Figure 4. Transportable Josephson voltage transfer standards.

Further development of Josephson voltage standards is associated with the idea of using exact values of voltage (quantum voltage steps) that depend only on the frequency of irradiation and the number of quantum steps. The transition to approximation of AC voltage with the help of quantum steps made it possible to create instruments [6] to measure alternating signals with an accuracy that allows us to measure the parameters of high-precision generators without using thermo-converters and comparing the AC voltage with the DC voltage (Figure 5). VNIIM participates in this research that by now has enabled us to measure voltages up to 10 V and frequencies up to 10 kHz [7]. Due to this method it is now possible to determine influencing factors when conducting high-precision measurements [8]. We have also investigated the possibility of measuring the parameters of high-precision analog-digital converters [9] (Figure 6). The issues related to transitions between quantum steps [10] that prevent us from reproducing variable signals with a high accuracy have been studied.

Figure 5. VNIIM JVS and PJVS voltage standards.

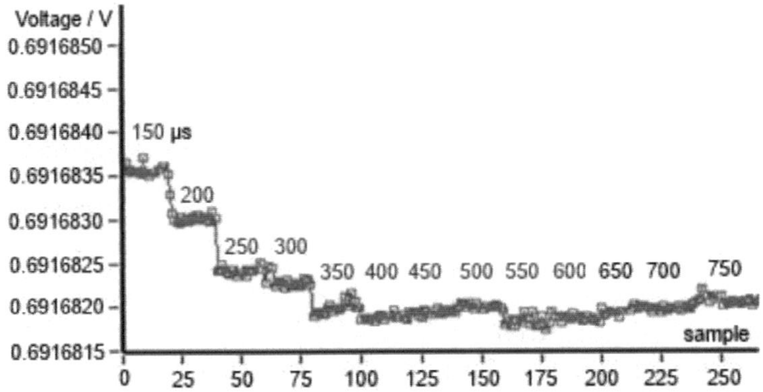

Figure 6. DVM voltage dependence from aperture time.

The achievements of the present period of development of the Josephson voltage standards are associated with the use of the pulsed method of irradiation of Josephson junctions [11], in contrast to the irradiation with sinusoidal signals in previous designs. The combination of pulsed-driven method with the delta-sigma modulation method allows us to reproduce the voltage with calculated parameters of the form - rms value, amplitude, amplitude of harmonics, etc., which creates a new level of metrological measurements that cannot be achieved by means of thermo-converters used in AC voltage standards.

The major problems related to this method of AC voltage unit realization that have not been resolved by now are increasing the reproducible voltage up to 10 V and increasing the frequency up to 1 GHz, the issues with hardware application, and the influence of connecting circuits.

The development and the use of installations that allow reproducing and measuring voltage with a high accuracy has led to the use of these devices in other areas of measurement. These include measurement of resistance ratios at DC and AC current, measurement of the ratio of inductances, the ratio of capacitances, measurement of power [12], and use of the noise temperature [13] in measurements.

The use of Josephson standards as reference sources allows us to measure AC voltages up to 100 V [14].

The development of Josephson systems technology provides the possibility for creation of practical devices based on high-temperature ceramics with a working temperature at the level of liquid nitrogen, which, unlike systems with a temperature at the level of liquid helium, allows these devices to be used in a wide metrological practice. The first samples of such voltage standards were fabricated in Russia [15].

The introduction of Josephson standards based on fundamental constants and therefore are not strictly connected with other standards, raises the question of conducting such metrological procedures as periodic comparisons and calibration. At present, the BIPM has come to the decision to carry out comparisons of Josephson DC voltage standards once every 20 years, which characterizes the approach to the gradual elimination of this type of metrological procedures that can lead to significant changes in metrological practice with the widespread introduction of lower precision meters methods. The development of quantum measurements and analysis of processes connecting micro- and macrocosm gives rise to ideas that try to identify the presence of quantum processes in living organisms [16], which can lead to revolutionary changes in many aspects of the development of society.

References

[1] Josephson B D 1962 Possible new effects in superconductive tunnelling *Phys. Lett.* **1**(7) pp 251-53

[2] Yamamori H, Yamada T, Sasaki H and Shoji A 2008 A 10 V programmable Josephson voltage standard circuit with a maximum output voltage of 20 V *Supercond. Sci. Tech.* **21**

DOI:10.1088/0953-2048/21/10/105007

[3] Behr R and Katkov 2005 A Comparison of Josephson array voltage standards by using a portable Josephson transfer standard *EUROMET.EM.BIPM-K10.a, Final Report*

[4] Katkov A, Lovtsus V and Behr R 2016 Portable Josephson voltage reference standard *Conf. digest CPEM 2016* 2 p. DOI: 10.1109/CPEM.2016.7540664

[5] Katkov A and Solve S 2012 Key comparison of the voltage standards of the VNIIM and the BIPM *Meas. Tech.* **54**(11) pp 1313-18

[6] Katkov A S., Lovtsyus V E, Bykov A I, Shevtsov V I and Novoderezhkin G V 2017 Reproduction of the volt based on SIS- and SNS-type Josephson junctions *Meas. Tech.* **60** p 589 https://doi.org/10.1007/s11018-017-1240-1

[7] Behr R, Katkov A S, Lee J, Bauer S, Kieler O F and Palafox L 2018 Frequency range extension of the AC quantum voltmeter *CPEM 2018 Conf. Digest* pp 984-85

[8] Katkov A, Gubler G, Lee J, Behr R and Nissilä J 2014 Influence of harmonics on AC measurements using a quantum voltmeter *Conf. Digest CPEM 2014* pp 526-27

[9] Gubler G and Katkov A 2012 Investigation of ADC-Aided AC measurement through the use of PJVS *Conf. Digest CPEM 2012* pp 64-65

[10] Katkov A, Behr R and Lee J 2008 Model for transient variation in stepwise synthesized Josephson sinewaves *Conf. Digest CPEM 2008* pp 380-381

[11] Benz S P, Hamilton C A, Burroughs C J and Harvey T E 1999 AC and dc bipolar voltage source using quantized pulses *IEEE Trans. Instrum. Meas.* **48** pp 266–69

[12] Behr R, Palafox L, Williams J M, Djordjevic S, Eklund G, Van Den Brom H E, Jeanneret B, Nissilä J, Katkov A and Benz S. P Binary Josephson array power standard 2010 *French College of Metrology* ISTE, London, UK. DOI: 10.1002/9780470611371.ch44

[13] Kamper R A and Zimmerman J. E 1971 Noise thermometry with the Josephson effect *J. Appl. Phys.* **42** p 132 DOI.org/10.1063/1.1659545

[14] Hagen T, Budovsky I, Benz S P and Burrough C J 2018 Calibration system for AC measurement standards using a pulse-driven Josephson voltage standard and an inductive voltage divider *CPEM 2018 Conf. Digest* DOI: 10.1109/CPEM.2012.6251108

[15] Khorshev S K *et al*. 2019 Voltage standard based on dry-cooled high-temperature superconductor Josephson junctions. *IEEE T. Instrum. Meas.* pp 1-8 DOI: 10.1109/TIM.2019.2896011

[16] Petrov M Quantum life https://chrdk.ru/sci/quantum_life

Journal of Physics: Conference Series **1379** (2019) 012077 doi:10.1088/1742-6596/1379/1/012077

Measuring systems designed for working with living organisms as biosensors. Features of their metrological maintenance

V A Lyubimtsev[1], S V Kholodkevich[1,2] and I I Druzhinin[3]

[1] Saint-Petersburg Scientific Research Centre for Ecological Safety Russian Academy of Sciences, 18 Corpusnaya str., St. Petersburg, 197110, Russia
[2] Saint-Petersburg State University, 7-9 Universitetskaya nab., St. Petersburg, 199034, Russia
[3] D.I. Mendeleev Institute for Metrology, 19 Moskovsky pr., St. Petersburg, 190005, Russia

E-mail: lyubimcev55@mail.ru

Abstract. The paper deals with the use of living organisms as biosensors for integrated assessing of the quality of their habitat. In particular, the paper considers the application of fish, shellfish, and crayfish for the water quality evaluation. By the example of a measuring system with crayfish, the features of metrological maintenance of systems with biosensors are described.

1. Introduction

Living organisms as biosensors in their natural habitat are widely used in solving environmental problems. First of all, this concerns a comprehensive evaluation of the rapid changes in the quality of this habitat which causes its biological danger. Integral behavioral and physiological characteristics, such as the mobility of organisms, frequency of respiratory movements and/or heart rate, are taken as parameters being measured directly.

In particular, the quality of the aquatic environment mainly takes into account the content of alkali and alkaline earth metal salts, humic and fulvic acids, dissolved oxygen, as well as pH value, the content of heavy metals that are foreign for human organisms (mercury, lead, etc.) and a number of highly toxic organic compounds. Among organic compounds, the most dangerous ones are pesticides, toxins of biological origin and chemical warfare agents. For implementing an instrumental quality control of the environment, the concentration of these compounds is determined individually or for a group of derivatives of the same compound with similar chemical properties. In doing so, the degree of biological hazard is determined by evaluating whether the concentration values exceed the values of the maximum permissible concentrations (MPC) set in regulations.

Regulations setting the procedure for determining the MPC values, on one hand, require lengthy (up to several months) experiments, and on the other hand, they do not allow the synergy and antagonism of influencing various toxicants to be taken into account. In Russia, the MPCs have been approved for at least 1000 toxicants. In water, to provide continuous or automated (with a period of no more than 10 - 30 minutes) instrumental control of them simultaneously, is unrealistic either for technical or economic reasons.

Journal of Physics: Conference Series **1379** (2019) 012077 doi:10.1088/1742-6596/1379/1/012077

2. Features of the use of hydrobionts as biosensors

Using hydrobionts (aquatic organisms), i.e. organisms living in water, (fish, bivalve mollusks, crayfish, crabs, etc.) as biosensors allows real-time monitoring of the entire spectrum of toxicants of different chemical kind and structure, taking into account synergism or antagonism of their effects. This is due to the fact that in such hydrobionts, all significant biochemical violations directly affect the work of their cardiovascular system, and movement characteristics. To monitor the toxicity of water, continuous recording and analysis of the mobility and frequency of respiratory movements in fish [1], the opening angle and frequency of periodic closing / opening valves in mollusk clam shells [2] and heart rate in crustaceans [3], are used. The use of adult hydrobiont specimens with a relatively long life cycle (several years) as biosensors allows minimizing the effect of age-related changes on the above characteristics.

The first stage of the reaction of aquatic organisms to abrupt variations in the quality of the environment is the process initiated by their sensory system. The further reaction depends on the behavior "program" inherent in a particular type of the biosensor when danger is detected. In many ways, it is identical to the responses to an abrupt change in illumination ("shadow") or to a short-time high-intensity noise.

Some examples: in bivalve mollusks, the response to exceeding the MPC of a toxicant consists in rapid closing the valves; in crayfish, this reaction lasts less than 1 minute, while the heart rate (*HR*) during this time interval compared to the *HR* value before the impact, can either decrease by 30 % or increase by 50 %. At the same time, both the qualitative and quantitative characteristics of such a reaction vary from one hydrobiont individual to another, but are well reproduced across time for each of them. In the case of a high concentration of the toxicant entering the water after a time interval inversely proportional to this concentration, a monotonic increase (or decrease) of the value of the integral characteristic under control, e.g., *HR*, starts.

The change direction depends on the biochemical mechanism of the toxicant influence. Thus, using the example of measuring the *HR* value in a wide range of crabs, shrimps and crayfish, it was shown that toxicants creating interference for transporting oxygen, cause the decrease of the *HR* value. On the contrary, while exposing to high concentrations of mercury ions which are the strongest nerve poison, the *HR* value during the first few hours increases significantly. However, in both cases, while exposing to high concentrations of toxicants, there is a trustworthy difference between the values of the *HR* measured during such an impact and before it.

If one accepts that at a crayfish rest state, under stable temperature (18 - 20 °C), the *HR* value fluctuates within the range of $\pm SD$ (*SD* is a standard deviation) around its average value HR_{av} and, e.g., for $SD/HR_{av} \approx 0.1$, then a 30 % increase or decrease of *HR* compared to HR_{av} can be considered as the sign of reliable changes of this parameter. Observations have shown that such an event under conditions of stable water quality can be recorded once per 3 - 4 hours. A similar situation takes place for both the value of the opening angle and frequency of the periodic closing / opening the valves of mollusks. Therefore, 30 % deviation of these characteristics from their average values is accepted as the criterion of a reliable deviation of a hydrobiont specimen behavior from the norm in a measuring system [2]. The coincidence of such deviations for both parameters in more than 50 % of the total number of mollusks in the system within 10 – 20 minutes is accepted as a criterion for changing the quality of water under control.

It should be noted that the above measuring systems for monitoring the quality of the aquatic environment are the "threshold" systems, since if the presence of significant concentrations of a toxicant in water causes an increase in the hydrobiont activity or circulation of its hemolymph, this fact further accelerates the toxicant inflow.

3. Measuring system designed to work with crayfish

The measuring system assigned for water quality control [3], installed at all water intakes of water supply stations of St. Petersburg, consists of 6 - 8 measuring channels (MC) of the same type, in which the crayfish are used as biosensors, *HR* being used as the measurand.

The principle of operation of a MC of a bio-electronic fiber-optic system is based on laser irradiating the outer surface of a crayfish in the zone of its heart localization followed by recording a radiation flux scattered in the opposite direction. To irradiate the outer surface and receive an optical signal carrying information about cyclic heart contractions of the crayfish, a thin fiber-optic sensor is used, which is non-invasively fixed on the crayfish shell. The flexibility and lightness of the fiber optic sensor provides the freedom of unhindered movements of the crayfish.

The change of no less than 30 % of the *HR* value compared with HR_{av}, at least, in 3 crayfish for 10 minutes, was taken as a criterion for changing the quality of water. Without a change in water quality, the probability of such an event is estimated by statistics no more than once per 1.5 months, and at the coincidence in 4 crayfish, the above takes place once per 2.5 years.

It was established experimentally that the *HR* change in response to exposure to toxicants with dangerous concentrations, varies between different biosensor specimens within the limits of ±10 %, and the values of SD/HR_{av} vary from 10 to 30 %. From this, it follows that to select specific specimens of biosensors, to use the criterion of the minimum values of both HR_{av} and SD/HR_{av} is necessary.

It was also found that the closer the *HR* value of a crayfish at a rest state is to the minimum value HR_{rest}, which is characteristic for a healthy crayfish at a comfortable water temperature *t*, the lower is the threshold of demonstrating a short-term sensory response, as well as the threshold of a response to "shadow" and noise. Thus, for the crayfish such as Astacys leptodactylus, in the temperature range from 16 to 26 °C, the value $HR_{rest} = (2.3t - 4)$ beats/min. This is due to the fact that in the same crayfish, the priorities of the biochemical resources distribution change depending on its functional state.

Thus, it was found that the closer the crayfish state in the daytime is to the rest state, the more pronounced is its sensory response and the smaller is its *SD* value. To achieve the state of the most complete rest of the crayfish, two conditions should be fulfilled: a non-invasive method of recording control characteristics (in this case, the optical method of receiving the cardioactivity signal is used) and the absence of interference for free movements (in this case, a flexible fiber-optic sensor is applied).

4. Metrological maintenance of measuring systems intended for work with crayfish

For successful functioning of such systems, not only properly working measuring equipment is necessary, but also a "workable" set of biosensors selected for this purpose according to certain criteria.

"Metrological health" of cancers in their application as biosensors, is provided by the following procedures carried out at the stage of preparing the MC for use and in the course of their operation:

1. Selection of crayfish based on their external (visual) parameters.

The need to control all 10 cancer limbs to be present is associated with experimentally established dependencies of the cancer's *HR* reaction changes in response to the addition of a number of model toxicants to water if such injuries are present. At the same time, there is a deterioration in the functional state of cancer, which is also manifested in an increase in HR_{av} in cancer in conditions of "operational rest" compared to the HR_{rest} values of (determined by the formula $HR_{rest} = 2.3t - 4$) and an increase in the concentration threshold of the sensory induced HR reaction in response to the model toxicants used (sodium, potassium and calcium chlorides, mineral and organic acids). The cancer antennal integrity monitoring is necessary concerned to normal operation of the primary chemo receptors. The selection of the approximately same weight (length) crayfish species Astacys leptodactylus: 30 - 40 g (100 - 120 mm), caught in the same water body, allows one to select the cancers to approximately the same age and biological properties and with the prospect of further growth within 1 - 2 years.

2. Selection of crayfish based on the level of signals resulting from cyclic contractions of the heart.

Comparison of the value of amplitudes A of pulsations caused by changes in the intensity of scattered light by the single-chamber heart tissues of cancer its compression/relaxation cycle for different instances of cancers show significant (up to 10 times) differences when using the same instance of a optical fiber and pulsation converter. In the instance of the same cancer at the same values of *HR* and water temperature, the A values vary in the range of ±20 %, only.

Figure 1a is a real example of a *HR* 20-second interval for the recorder's output ripple variation under the "test pulse" overlay and shows how the individual components of such ripple are determined in table 1.

Table 1. The example for a group of 12 Astacus leptodactylus cancers (n = 1 - 12) output ripple variation at (20 °C, stress, *HR* = 100 - 120) when using the same instance of an optical fiber and pulsation converter.

n	1	2	3	4	5	6	7	8	9	10	11	12
A, V	0.8	0.83	0.54	0.40	0.45	0.65	0.45	0.38	0.40	0.36	1.3	1.1
T, V	1.4	1.0	1.38	1.1	1.2	1.17	1.21	1.17	1.25	1.30	1.3	1.4
A/T	0.54	0.8	0.41	0.36	0.33	0.54	0.33	0.27	0.33	0.23	1.0	0.8
A/N	5.3	5.5	3.6	2.7	3.0	4.3	3.0	2.5	2.7	2.4	8.6	7.3

The following symbols are used in table 1: A - the amplitude of the cardio activity periodic signal pulsation; T - the amplitude of the "test pulse" caused by a change in (1.6±0.2) % of the laser radiation intensity (and the flow of light scattered by the outer wall of the cancer core and collected by the optical probe, respectively); A/T - the ratio of these amplitudes; A/N - the ratio of the amplitude of the signal of cardiac activity to the amplitude N of irregular "noise" caused by fluctuations in the light intensity, scattered by the cancer's core.

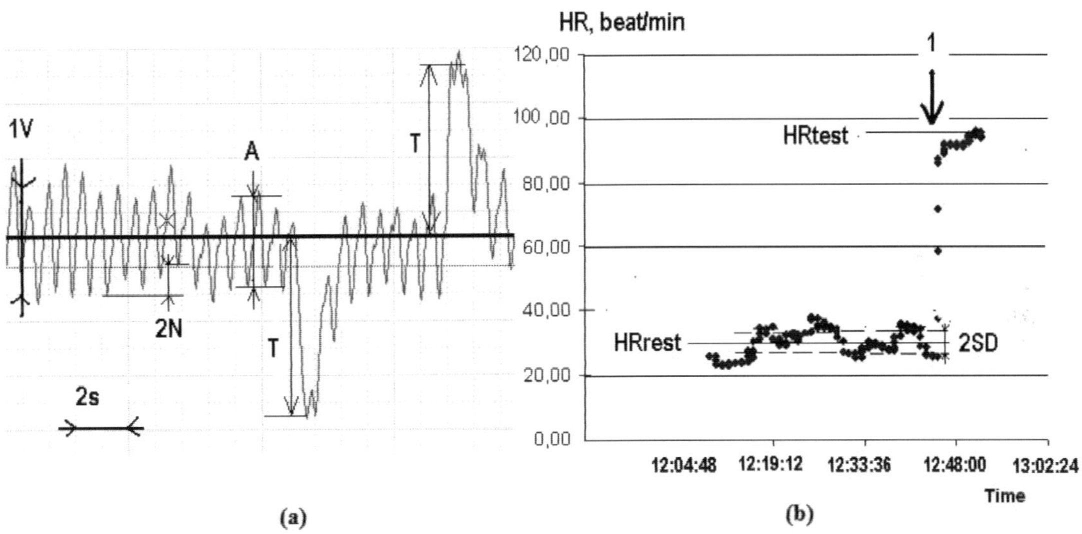

Figure 1. (a) - The example of the optical pulsation curve. The values of peak height T, amplitude A and double noise amplitude 2N shows the result of the real crayfish testing. The typical T, A and N values relations are presented in table 1. (b) - The example of the *HR* dynamics for the crayfish. The arrows 1 note the time of the 3 min mechanical stress action beginning. The HR_{rest} levels, HR_{test} levels and 2*SD* value, used in the table 2, are pointed out.

The table 1 data show that the variability ratio of the maximum value to the minimum value for the amplitude of the ripple A (1.3/0.36 = 3.6) practically determines the variability of A/T (1.0/0.23 = 4.3) and variability of A/N (8.6/2.4 = 3.6) - for the used samples of cancers. We have found previously [4], that the value of the measurement error *HR* is determined mainly by the value of the ratio A/N. Therefore, cancer selection to improve the accuracy of the measurement of the value of *HR* is to be made on the basis of the highest value A.

3. Use of individual settings of each MC in accordance with the physiological characteristics of the crayfish (*HR* and *SD* values at the rest state).

Almost all cancers are suitable as biological sensor to detect cases of accidental pollution of natural water by high concentrations of toxicants causing stress and 2 - 3 times increase in *HR* (see figure 1b), except those being in a state of severe stress, in which *HR* is already maximal HR_{test}. But when exposed to lower concentrations of toxicants there is a short-term (1 - 5 minutes) "sensory response", in which even the maximum value of *HR* is at the midpoint of the interval between HR_{rest} and HR_{test}. In this case, the statistical criterion: $(HR - HR_{rest}) > 3SD_{rest}$ may be used as the criterion of reliability of detection of such reaction against the background of fluctuation of *HR* value. A substantial individual variability of *HR* requires individual updates for each of the cancer the average value of HR_{rest} and value SD_{rest} (20 s) - for of *HR* values distribution, based on the number of heart beats over a interval of 20 second. Measurements for a large sample of cancers in the used temperature range 14 - 20 °C showed that the values of SD_{rest} (20 s) vary from day to day in the range (3 - 7) beats/min, almost the same for all cancers. In this case, a short-term (20 s) increase in the *HR* value relative to the average HR_{rest} value of more than 20 - 25 beats per minute characteristic of the sensory response can be reliably detected.

Average HR_{rest} values at the same temperature vary in the range of ±25 % for different cancers. A typical example of variation of the ratio of individual values to the average for the group value $HR_{rest}(n)/HR_{rest}(av)$ for different cancers № 1 - № 4 of the same size: № 1 - (1.20-1.25), № 2 - (1.02-1.07), № 3 - (0.75-0.80) and № 4 - (0.93-0.96).

"Metrological health" of cancers in the process of operation is ensured by periodic control testing of their reaction to the effects of characteristic toxicants, which enables making sure that a quantitative characteristic of the response to the selected test impact is maintained.

The rapid (30 - 40 seconds) increase in the concentration of sodium chloride in water to a final value of 30 g/l, but without changing the water circulation regime (see figure 1b) is used as crayfish periodic control test. At the test all cancers showed a rise in *HR* to values HR_{tox} equal to HR_{test} during the first 5 minutes - see table 2.

Table 2. An example of the maximum *HR* values achieved at a temperature of 20 °C for the same instances of Astacys leptodactylus cancers for one day under mechanical stress (HR_{test}) - after a short term "suspension" in water and under chemical stress (HR_{tox}) - with exposure to 30 g/l sodium chloride.

	1	2	3	4	$HR_{av} \pm SD$	SD / HR_{av}
HR_{test}, beat/min	97±1	84±1	96±1	87±1	91±6.5	0.060
HR_{tox}, beat/min	104±1	89±1	95±1	84±1	93±7.0	0.075
HR_{tox} / HR_{test}	1.07±0.02	1.06±0.02	0.99±0.02	0.96±0.02	1.020±0.045	0.045

Exposure to high (30 g/l) concentrations of sodium chloride creates a strong "osmotic shock" first of all - for the cancer's gills tissues. This causes cancer "avoidance", reaction with the maximal hormonal activation of motor activity possible for this temperature so leading to *HR* maximum values. Therefore - the average values of HR_{tox}, obtained under such influence [5], are to be the same, as the values of HR_{test} achieved during "handling" or short-term "suspension" above the bottom level in the aquarium water. This allowed to use more sophisticated periodic control of compliance the *HR*, maximum possible for one biosensor, with the temperature determined "normal" values HR_{test}. The fourfold repetition of the test procedure "suspension" within 2 - 3 weeks under the same conditions (17.5 °C), showed that 3 "healthy" cancers for the specified time period HR_{test} values (122±10), (113±6) and (102±12) beat/min vary from day to day (±*SD*) as significantly as individual HR_{test} values vary in this group with each of 4 tests: (117±17), (105±16), (116±6) and (111±9) beat/min. But weakened by cancer (deceased through 7 days after the test) have shown significantly lower HR_{test} values (87±8) beat/min, than the average group HR_{test} values of the "healthy" cancers - (112±10) beat/min.

Studies conducted in order to develop a biosensor simulator necessary for specifying and checking the metrological characteristics of the MCs of the systems intended for measuring the *HR* of crayfish, show that diffuse reflection from surface crayfish tissues is well reproduced in case of using a fluoroplastic plate of a certain thickness as a simulator of these tissues. When the optical sensor holder is mounted on such a plate instead of the crayfish shell, the signal entering the receiving fiber corresponds to the signal characteristic for a statistical average crayfish under the absence of contractions of its heart. In this case, the variable component of the signal is determined by the optical and electronic noise of all the elements included in the MC, which allows evaluating the level of this noise and controlling this level later in the process of operation.

To create a modulated optical signal that simulates the crayfish heartbeat, a fluoroplastic plate is used in the form of a sector, which is located at a certain distance from the plate simulating the superficial crayfish tissues. When this plate is rotated, a signal is generated that is similar to the signal arising from contractions of the crayfish heart. Frequency of rotation is set by a measurement standard. Comparison of this frequency and the frequency measured by the system equipment makes it possible to determine the error of the *HR* measurements. To check whether the optical properties of the simulator plate lie within the permissible limits, a coefficient of directed transmittance is measured with the help of a spectrophotometer.

In the course of operation of the system, in each MC, the automatic metrological self-check is carried out. It is based on monitoring the characteristic associated with the *SD* values of the crayfish at the rest state and at the state of deviation from the norm. If this characteristic exceeds the specified limits, a signal of the metrological malfunction of the corresponding MC is generated.

5. Conclusion

The above issues related to the metrological maintenance of the system intended for measuring the crayfish *HR* are also important of other measuring systems designed for working with biosensors. In the course of their operation, the necessity exists to monitor the biosensor performance as well as simulate the biosensor signals while determining the metrological characteristics of the systems.

References

[1] Sydor W J, Miller W R, Cairms J. and Gruber J D 1982 Use a box and line plots to assess fish vevtilatory behavior in biological monitoring systems *Can. J. Fish. Aquat. Sci.* **39(12)** 1719

[2] Zwart D D, Kramer K J M and Jenner H A 1995 Practical experiences with the biological early warning system mossel monitor *Environ. Toxic. Water* **10** 237

[3] Kholodkevich S V, Ivanov A V, Kurakin A S, Kornienko E L and Fedotov V P 2008 Real time biomonitoring of surface water toxicity level at water supply stations *J. Environmental Bioindicators* **3**(1) 23

[4] Lyubimtsev V, Kholodkevich S, Druzhinin I and Kurakin A 2017 Bioelectronic system with metrological self-check intended for diagnostics of acute water toxicity *Pribory* 10(208) pp 34-39 (In Russian)

[5] Luybimtsev V, Kholodkevich S, Taymanov R and Sapozhnikova K 2017 Bio-electronic System for Megapolis Water Supply Monitoring *13th Int. Symp. on Measurement Technology and Intelligent Instruments (ISMTII 2017)* (Xi'an China) pp 102-103

Journal of Physics: Conference Series **1379** (2019) 012078 doi:10.1088/1742-6596/1379/1/012078

Dynamic light scattering characterization of hard–soft gel phase transition in micellar crystal and comparison with rheological data

A F Kostko[1] and O A Pinchuk[2, 3]

[1] Faculty of Physics and Engineering, ITMO University, St. Petersburg, 197101, Russia
[2] Faculty of Cryogenic Engineering, ITMO University, St. Petersburg, 197101, Russia
[3] D.I. Mendeleyev Institute for Metrology VNIIM, St. Petersburg, 190005, Russia

E-mail: afkostko@yahoo.com; oapinchuk@yandex.ru

Abstract. We report the dynamic light scattering study of polymer dynamics in the broad vicinity of the hard-soft gel phase transition in micellar crystal. Inside the region of micellar crystal phase with increasing temperature the ratio of the magnitudes of two dynamic modes (slow and fast) monotonously decreases below 54 °C in the hard gel. Above 54 °C in the soft gel it is almost independent of temperature, remaining at a minimum. The method provides phenomenon details inaccessible with conventional rheological method, where simple detection of sharp phase transition is available. We show that the diffusion dynamic mode, which is common for a liquid phase, is also present in a micellar crystal despite the obvious difference in the rheological properties of micellar liquid and micellar crystal. This study demonstrates the advantages of the dynamic light scattering method at micro-scale level.

1. Introduction

Variety of measurement methods provides different aspects of phenomenon characterization. We discuss the application of dynamic light scattering method in the investigation of a micellar crystal that is already well studied with other conventional methods (neutron scattering, rheology, etc.) Block copolymers PEO-PPO-PEO (polyethylene oxide-polypropylene oxide-polyethylene oxide) are a family of non-ionic surfactants with variation of block length. The patented name of Poloxamers and several names associated with trademarks (Pluronics, Synperonics, Proxanols) are well-known. Solubility in water of PEO and PPO blocks is ensured by excess of oxygen atoms, while PPO blocks with additional methyl groups have lower than PEO blocks solubility in water. Aqueous Pluronic systems exhibit variety of phases, depending on parameters such as temperature, pressure and concentration. The presence of different phases in the micellar system is associated with their use in household chemicals, cosmetics, pharmaceuticals, medicine, biotechnology and in other fields. The gel is formed at sufficiently high concentrations, which is a micellar crystal with a cubic lattice well studied by neutron scattering methods (see, e.g., review [1]). While the dynamic light scattering research of liquid phases of Pluronic aqueous systems is reported in numerous papers (see references in [2]), the method was never before applied to Pluronic micellar crystal phase. This might be caused by expected difficulties with dynamic light scattering data interpretation associated with “frozen” fluctuations in gels [3].

New phase transition in the region of Pluronic F127 (EO_{100}-PO_{65}-EO_{100}) micellar crystal phase has been reported [4] in rheological research. It was found that the gel region in the phase diagram is divided into two regions of the hard gel and the soft gel with a phase transition between them. The rheological properties of the gel in these regions differ by an order of magnitude. We report the dynamic light scattering study of polymer dynamics in the broad vicinity of the hard-soft gel phase transition in aqueous system with Pluronic F108 (EO_{133}-PO_{55}-EO_{133}), which is rather similar to Pluronic F127. In contrast to rheological measurements with generated macroscopic shear stress, the dynamic light scattering method uses fluctuations relaxation at a microscopic scale of the order of wavelength.

2. Results and Discussion

Acquisition of intensity-intensity correlation functions was carried out with Photocor Complex instrument (laser wavelength – 654 nm). The scattering angle for all measurements was set to 90°. Accordingly, the probe distance of about 0.5 micrometer was used in DLS experiments.

The sample is Pluronic F108 (Sigma-Aldrich) aqueous system at concentration of 20 %. At this concentration the phase of a micellar crystal is detected between 28 °C and 86 °C. Lattice site is a micelle of about 11 nm diameter with less hydrophilic PPO blocks assembled in the core and the chains of PEO blocks flutter in the shell. About four tens of micelles in the micellar crystal lattice correspond to DLS probe distance.

Correlation functions were transformed using the method proposed in [3] for dynamic scattering of light in gels. As a result of this transformation, we obtain the correlation function without distortion caused by the interference contribution from static light scatterers, such as domains borders and defects in the crystal structure. The transformed correlation function is used to find the distribution of correlation times using the procedure CONTIN.

All obtained distributions have two dynamic modes, fast and slow. It is shown in [2] on the basis of data taken outside the phase region of the micellar crystal that the fast mode is diffusive and is associated with the relaxation of concentration fluctuations. Slow dynamic mode in a concentrated polymer solution is associated with polymer chains entanglements, both modes being bound [2, 5]. We found that at the micellar liquid – micellar crystal phase transition the shape of the correlation function and, accordingly, the distribution of correlation times does not exhibit significant transformation.

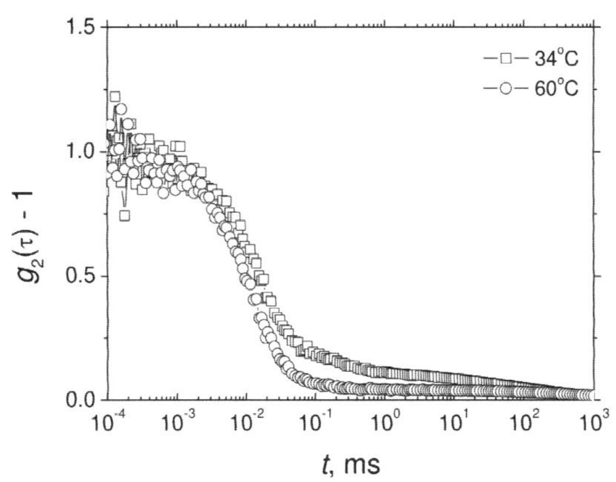

Figure 1. Correlation functions for micellar crystal at 34 °C and 60 °C. The angle of scattering is 90°.

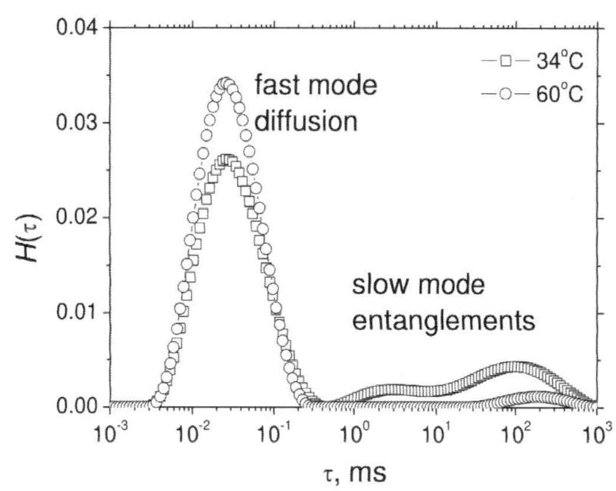

Figure 2. Distributions $H(\tau)$ obtained from correlation functions, shown in figure 1 for micellar crystal at 34 °C and 60 °C.

Two correlation functions obtained at temperatures below and above the hard-soft phase transition are shown in figure 1 where the correlation function at 34 °C exhibits the long-time tail that corresponds to slow dynamics of polymer chains. This long-time tail of the correlation function almost vanishes at 60 °C according to figure 1. We see also noticeable difference in the shapes of the long-time tail and the short-time step-like part of the correlation function obtained at 34 °C in the hard-gel phase. This shape of the tail is the evidence of lack of certain characteristic time in the dynamics slowdown by entanglements of the polymer chains. Accordingly, we observe rather broad entanglements mode in distribution, while short-time step-like part of the correlation function corresponds to a narrow diffusion mode. This is completely supported in figure 2.

The ratio of the magnitudes of the slow and the fast modes was calculated as the ratio of the areas under the curves that correspond to modes. Figure 2 shows that the ratio varies significantly with temperature. The dependence of this ratio on temperature is shown in figure 3 for the entire region of the micellar crystal from 28 °C up to 86 °C.

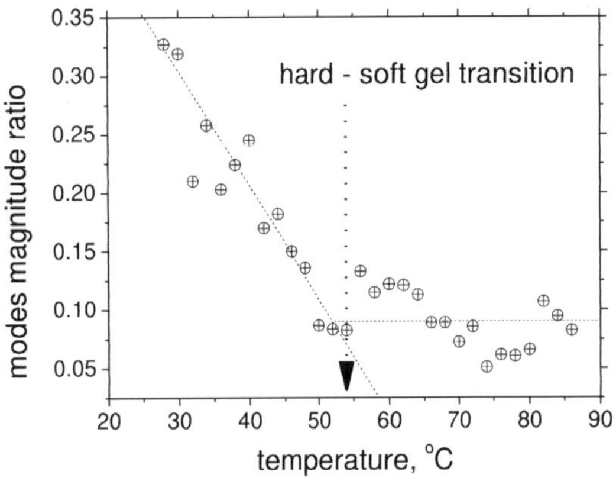

Figure 3. The ratio of the magnitudes of slow and fast dynamic modes for the phase region of the micellar crystal as a function of temperature.

Inside the region of micellar crystal phase two noticeable areas are observed. With increasing temperature the ratio of the magnitudes of the dynamic modes monotonously decreases below 54 °C, while above 54 °C it is almost independent of temperature, remaining at a minimum. This behavior can be explained by a monotonic micelle size decrease with heating associated with a gradual change in the interaction of the PEO end blocks with water. Consequently, the overlap area of entanglements of PEO blocks of neighbor micelles gradually reduces. At 54 °C micelle size corresponds to the touch concentration c^*, where the micelles do not overlap and the engagements of PEO block chains is minimal and is caused by thermal fluctuations. The change in polymer chains entanglements at a temperature of 54 °C corresponds to the phase transition between the hard gel, where neighbor micelles are essentially detained together by entanglements of PEO chains, and the soft gel, where entanglements of PEO chains are minimal.

3. Conclusion

Gradual decrease with heating of the micelles size built by PEO-PPO-PEO block copolymer is obtained with dynamic light scattering method. The method with a micrometer probe distance provides phenomenon details inaccessible with conventional rheological method with macro-scale, and where simple detection of sharp phase transition is available. We show that the diffusion dynamic mode, which is common for a liquid phase, is also present in a micellar crystal despite the obvious difference in the rheological properties of micellar liquid and micellar crystal. This study demonstrates the advantages of the dynamic light scattering method at micro-scale level.

Journal of Physics: Conference Series **1379** (2019) 012078 doi:10.1088/1742-6596/1379/1/012078

References

[1] Mortensen K 2001 PEO-related block copolymer surfactants *Colloid. Surface. A* **183–185** 277–92

[2] Kostko A F, Harden J L and McHugh M A 2009 Dynamic Light Scattering Study of Concentrated Triblock Copolymer Micellar Solutions under Pressure *Macromolecules* **42** 5328–38

[3] Rochas C and Geissler E 2014 Measurement of dynamic light scattering intensity in gels *Macromolecules* **47** 8012–17

[4] Liu S and Li L 2015 Multiple phase transition and scaling law for poly(ethylene oxide)–poly(propylene oxide)–poly(ethylene oxide) triblock copolymer in aqueous solution *ACS Appl. Mater. Interfaces* **7** 2688–97

[5] Kostko A F, Anisimov M A and Sengers J V 2007 Dynamics of critical fluctuations in polymer solutions *Phys. Rev.* E **76** 021804

AUTHOR INDEX